Operators Between Sequence Spaces and Applications

Bruno de Malafosse · Eberhard Malkowsky ·
Vladimir Rakočević

Operators Between Sequence Spaces and Applications

Springer

Bruno de Malafosse
University of Le Havre (LMAH)
Ore, France

Eberhard Malkowsky
Faculty of Management
Univerzitet Union Nikola Tesla
Beograd, Serbia

Vladimir Rakočević
Department of Mathematics
University of Niš
Niš, Serbia

ISBN 978-981-15-9744-2 ISBN 978-981-15-9742-8 (eBook)
https://doi.org/10.1007/978-981-15-9742-8

This Springer imprint is published by the registered company Springer Nature Singapore Pte Ltd.
The registered company address is: 152 Beach Road, #21-01/04 Gateway East, Singapore 189721, Singapore

Preface

The study of operators between sequence spaces is a wide field in modern summability. In general, summability theory deals with a generalization of the concept of convergence of sequences of complex numbers. One of the original ideas was to assign, in some way, a limit to divergent series, by considering a transform, in many cases defined by the use of matrices, rather than the original series. A central problem of interest is the characterization of classes of all operators between sequence spaces: The first result in this area was the famous Toeplitz theorem which established necessary and sufficient conditions on the entries of a matrix to transform to preserve convergence. The original proof used the analytical method of the gliding hump.

The introduction on a large scale of functional analytic methods to summability in the 1940s and the development of the *FK* and *BK* space theory made the study of matrix transformations and linear operators between sequence spaces a rapidly expanding field of interest in summability. More recently, in particular after 2000, the theory of measures on noncompactness was applied in the characterization of compact operators between *BK* spaces.

This book presents recent studies on bounded and compact operators, their underlying theories and applications to problems of the solution of infinite systems of linear equations in various sequence spaces. It consists of two parts.

In the first part, Chaps. 1–3, it presents the modern methods in functional analysis and operator theory and their applications in recent research in the representations and characterizations of bounded and compact linear operators between *BK* spaces and between matrix domains of triangles and row finite matrices in *BK* spaces. In the second part, Chaps. 4–7, we present applications and research results using the results of the first three chapters. This involves the study of whether infinite matrices $A = (a_{nk})_{n,k=1}^{\infty}$ are injective, surjective or bijective, when considered as operators between certain sequence spaces.

This book is unique, since it connects and presents the topics of the two parts in one volume for the first time, to the best of the authors' knowledge.

This book contains relevant parts of several of the authors' related lectures on graduate and postgraduate levels at universities in Australia, France, Germany, India, Jordan, Mexico, Serbia, Turkey, the USA and South Africa. A great number of illustrating examples and remarks were added concerning the presented topics.

The book could also be used as a textbook for graduate and postgraduate courses as a basis for and overview of research in the fields mentioned and is intended to address students, teachers and researchers alike.

Chapter 1 is mostly introductory and recalls the basic concepts from functional analysis needed in the book, such as linear metric and paranormed spaces, *FK*, *BK*, *AK* and *AD* spaces, multiplier spaces, matrix transformations, measures of noncompactness, in particular, the Hausdorff and Kuratowski measures of noncompactness, and measures of noncompactness of operators between Banach spaces. Although most of the material is standard, almost all proofs are presented of the vital results.

In Chap. 2, we study sequence spaces that have recently been introduced by the use of infinite matrices. They can be considered as the matrix domains in certain sequence spaces and can be used to define almost all the classical methods of summability as special cases. We apply the results and methods of Chap. 1 to determine their topological properties, bases and various duals; in particular, we establish a general result for the determination of the β-dual of arbitrary triangles in arbitrary *FK* spaces.

In Chap. 3, we characterize matrix transformations on the spaces of generalized weighted means and on matrix domains of triangles in *BK* spaces. We also establish estimates or identities for the Hausdorff measure of noncompactness of matrix transformations from arbitrary *BK* spaces with *AK* into c, c_0 and ℓ_1, and also from the matrix domains of an arbitrary triangle in ℓ_p, c and c_0 into c, c_0 and ℓ_1. Furthermore, we determine the classes of compact operators between the spaces just mentioned. Finally, we establish the representations of the general bounded linear operators from c into itself and from the space bv^+ of sequences of bounded variation into c, and the determination of the classes of compact operators between them.

In Chaps. 5 and 6, we obtain new results on sequence spaces inclusion and on sequence spaces equations using the results of the first part on matrix transformations. This study leads to a lot of results published until now, for instance, in the statistical convergence and in the spectral theory where the operators are defined by infinite matrices. We also deal with the solvability of infinite systems of linear equations in various sequence spaces. Here, we use the classical sequence spaces and the generalized Cesàro and difference operators to obtain many calculations and simplifications of complicated spaces involving these operators. We also consider the sum and the product of some linear spaces of sequences involving the sets of sequences that are "strongly bounded and summable to zero". Finally, we obtain new results on "statistical convergence".

In Chap. 7, we consider a Banach algebra in which we may obtain the inverse of an infinite matrix and obtain a new method to calculate the "Floquet exponent". Furthermore, we determine the solutions of the infinite linear system associated with the Hill equation with a second member and give a method to approximate them. Finally, we present a study of the Mathieu equation which can be written as an infinite tridiagonal linear system of equations.

Le Havre, France — Bruno de Malafosse
Niš, Serbia — Eberhard Malkowsky
Niš, Serbia — Vladimir Rakočević
August 2020

Contents

1 Matrix Transformations and Measures of Noncompactness 1
1.1 Linear Metric and Paranormed Spaces 2
1.2 FK and BK Spaces 10
1.3 Matrix Transformations into the Classical Sequence Spaces 15
1.4 Multipliers and Dual Spaces 19
1.5 Matrix Transformations Between the Classical Sequence Spaces 23
1.6 Crone's Theorem 27
1.7 Remarks on Measures of Noncompactness 34
1.8 The Axioms of Measures of Noncompactness 35
1.9 The Kuratowski and Hausdorff Measures of Noncompactness . . . 37
1.10 Measures of Noncompactness of Operators 42
References 44
Russian References. 45

2 Matrix Domains 47
2.1 General Results 47
2.2 Bases of Matrix Domains of Triangles 55
2.3 The Multiplier Space $M(X_\Sigma, Y)$ 66
2.4 The α-, β- and γ-duals of X_Σ 68
2.5 The α- and β-duals of $X_{\Delta^{(m)}}$ 78
2.6 The β-duals of Matrix Domains of Triangles in *FK* spaces 90
References 101

3 Operators Between Matrix Domains 105
3.1 Matrix Transformations on $W(u, v; X)$ 105
3.2 Matrix Transformations on X_T 112
3.3 Compact Matrix Operators 128
3.4 The Class $\mathcal{K}(c)$ 139

3.5 Compact Operators on the Space bv^+ . . . 145
References . . . 157

4 Computations in Sequence Spaces and Applications to Statistical Convergence . . . 159
4.1 On Strong τ-Summability . . . 160
4.2 Sum and Product of Spaces of the Form s_ξ, s_ξ^0, or $s_\xi^{(c)}$ . . . 162
4.3 Properties of the Sequence $C(\tau)\tau$ . . . 166
4.4 Some Properties of the Sets $s_\tau(\Delta)$, $s_\tau^0(\Delta)$ and $s_\tau^{(c)}(\Delta)$ . . . 172
4.5 The Spaces $w_\tau(\lambda)$, $\overset{\circ}{w}_\tau(\lambda)$ and $w_\tau^\bullet(\lambda)$ . . . 174
4.6 Matrix Transformations From $w_\tau(\lambda) + w_\nu(\mu)$ into s_γ . . . 179
4.7 On the Sets $c_\tau(\lambda, \mu)$, $\overset{\circ}{c}_\tau(\lambda, \mu)$ and $c_\tau^\bullet(\lambda, \mu)$ . . . 180
4.8 Sets of Sequences of the Form $[A_1, A_2]$ . . . 183
4.9 Extension of the Previous Results . . . 189
4.10 Sets of Sequences that are Strongly τ-Bounded With Index p . . . 192
4.11 Computations in W_τ and W_τ^0 and Applications to Statistical Convergence . . . 198
4.12 Calculations in New Sequence Spaces . . . 204
4.13 Application to $A-$Statistical Convergence . . . 207
4.14 Tauberian Theorems for Weighted Means Operators . . . 214
4.15 The Operator $C(\lambda)$ . . . 222
References . . . 226

5 Sequence Spaces Inclusion Equations . . . 229
5.1 Introduction . . . 230
5.2 The (SSIE) $F \subset E_a + F'_x$ with $e \in F$ and $F' \subset M(F, F')$ . . . 236
5.3 The (SSIE) $F \subset E_a + F'_x$ with $E, F, F' \in \{c_0, c, s_1, \ell_p, w_0, w_\infty\}$ . . . 238
5.4 Some (SSIE) and (SSE) with Operators . . . 245
5.5 The (SSIE) $F \subset E_a + F'_x$ for $e \notin F$ . . . 250
5.6 Some Applications . . . 253
References . . . 263

6 Sequence Space Equations . . . 265
6.1 Introduction . . . 266
6.2 The (SSE) $E_a + F_x = F_b$ with $e \in F$ . . . 271
6.3 Some Applications . . . 276
6.4 The (SSE) with Operators . . . 284
6.5 Some (SSE's) with the Operators Δ and Σ . . . 289
6.6 The Multiplier $M((E_a)_\Delta, F)$ and the (SSIE) $F_b \subset (E_a)_\Delta + F_x$ . . . 297
6.7 The (SSE) $(E_a)_\Delta + s_x^{(c)} = s_b^{(c)}$ . . . 300
6.8 More Applications . . . 307
References . . . 313

7 Solvability of Infinite Linear Systems 315
7.1 Banach Algebras of Infinite Matrices 315
7.2 Solvability of the Equation $Ax = b$ 320
7.3 Spectra of Operators Represented by Infinite Matrices 326
7.4 Matrix Transformations in $\chi(\Delta^m)$ 330
7.5 The Equation $Ax = b$, Where A Is a Tridiagonal Matrix 338
7.6 Infinite Linear Systems with Infinitely Many Solutions 343
7.7 The Hill and Mathieu Equations 349
References 356

Appendix: Inequalities 359

Index 363

About the Authors

Bruno de Malafosse was Full Professor at the Laboratoire de Mathématiques Appliquées Havrais, Université du Havre, France, until 2009. He has 35 years of teaching experience in most fields of analysis, probability theory, linear algebra and mathematics for informatics. He obtained the Doctorat of 3eme cycle at the Université du Havre Toulouse III, France, in 1980, with a thesis titled "Contribution à l'étude des systèmes infinis". He completed the Habilitation à Diriger des Recherches in 2004 at the Université du Havre, France, with a thesis titled "Sur la théorie des matrices infinies et applications". His research interests include the infinite matrix theory, summability, differential equations, theory of the sum of operators in the nondifferential case, numerical analysis, convergence of a numerical scheme optimization, quasi-Newton method, continuous fractions, spectral theory and sequence spaces. He is a member of the editorial boards of two reputed journals in Serbia and Jordan and a reviewer for the *Mathematical Reviews*. His list of publications includes 85 research papers published in international journals. He has been a supervisor/examiner for many doctoral theses in France, Algeria and India.

Eberhard Malkowsky is Full Professor at the Faculty of Management, University Union Nikola Tesla, Belgrade, Serbia. He completed his Ph.D. in Mathematics at Giessen University, Germany, in 1983 with the thesis titled "Toeplitz–Kriterien für Matrizenklassen bei Räumen absolut und stark limitierbarer Folgen". He also completed his Habilitation in Mathematics at Giessen University, Germany, in 1989, with the thesis titled "Matrix transformations in a new class of sequence spaces that includes spaces of absolutely and strongly summable sequences". He has 40 years of teaching experience in all fields of analysis, differential geometry and computer science at universities in Germany, South Africa, Serbia, Jordan and Turkey, and as a visiting professor in the USA, India, Hungary, France and Iran. He was also an invited lecturer at four summer schools of the German Academic Exchange Service (DAAD). He has supervised six Ph.D. theses and a number of B. Sc. and M.Sc. theses. His research interests include functional analysis, operator theory, summability, sequence spaces, matrix transformations and measures of noncompactness. He also has developed software for the visualization of topics in mathematics. His list of publications includes 164 research papers, 13 books and proceedings. He is a member of the editorial boards of 10 international journals and a reviewer for the *Mathematical Reviews* and *Zentralblatt der Mathematik*. Furthermore, he was the main organizer of 3 and participant of 8 research projects. He was the main organizer and a member of the organizing and scientific committees of 28 international conferences and has delivered more than 100 plenary keynote and invited talks. Moreover, he was an European Union expert for the evaluation of the Tempus Projects from 2004 to 2006.

Vladimir Rakočević is Full Professor at the Department of Mathematics, Faculty of Sciences and Mathematics, University of Niš, Serbia. He is also a corresponding member of the Serbian Academy of Sciences and Arts (SANU) in Belgrade, Serbia. He received his Ph.D. in Mathematics from the Faculty of Sciences, University of Belgrade, Serbia, in 1984; the title of his thesis was "Essential spectra and Banach algebras". His research interests include functional analysis, fixed point theory, operator theory, linear algebra and summability. He was a visiting professor at several universities and scientific institutions across the world. Furthermore, he participated as an invited/keynote speaker at numerous international scientific conferences and congresses. He is a member of the editorial boards of several international journals of repute. His list of publications includes 173 research papers in international journals. He was included in the Thomson Reuter's list of Highly Cited Authors in 2014. He is the co-author of 7 books and has supervised 7 Ph.D. and more than 50 B.Sc. and M.Sc. theses in mathematics.

Acronyms

ω	set of all complex sequences $x = (x_k)_{k=0}^{\infty}$
d_ω	metric on ω
c_0	set of all complex null sequences
c	set of all convergent complex sequences
ℓ_∞	set of all bounded complex sequences
$\|\cdot\|_\infty$	norm for c_0, c and ℓ_∞
ℓ_1	set of all absolutely convergent complex series
ℓ_p	$= \{x \in \omega : \sum_{k=0}^{\infty} \|x_k\|^p < \infty\}$ for $1 \leq p < \infty$
$\|\cdot\|_p$	norm for ℓ_p
$\ell(p)$	$= \{x \in \omega : \sum_{k=0}^{\infty} \|x_k\|^{p_k} < \infty\}$
$c_0(p)$	$= \{x \in \omega : \lim_{k\to\infty} \|x_k\|^{p_k} = 0\}$
$\ell_\infty(p)$	$= \{x \in \omega : \sup_k \|x\|^{p_k} < \infty\}$
g	paranorm on $\ell(p)$
g_0	paranorm on $c_0(p)$
$\mathcal{T}_X\|_Y$	relative topology of X on $Y \subset X$
$\mathrm{cl}_Y(E)$	closure of $E \subset Y$ in a topological space Y
X'	continuous dual of a Fréchet space space X
ϕ	set of finite complex sequences
A	$= (a_{nk})_{n,k=0}^{\infty}$ infinite matrix of complex entries
A_n	sequence in the nth row of the infinite matrix A
A^k	sequence in the kth column of the infinite matrix A
$A_n x$	$= \sum_{k=0}^{\infty} a_{nk} x_k$
Ax	$= (A_n x)_{n=0}^{\infty}$
(X, Y)	class of infinite matrices that map $X \subset \omega$ into $Y \subset \omega$
$e = (e_k)_{k=0}^{\infty}$	sequence with $e_k = 1$ for all k
$e^{(n)} = (e_k^{(n)})_{k=0}^{\infty}$	sequence with $e_n^{(n)} = 1$ and $e_k^{(n)} = 0$ for $k \neq n$
$x^{[m]}$	$= \sum_{n=0}^{m} x_n e^{(n)}$, m-section of the sequence $x = (x_k)_{k=0}^{\infty}$

$\bar{B}_\delta(x), \bar{B}_{X,\delta}(x)$	closed ball of radius δ and centre in x_0 in a metric space (X, d)
$\| a \|_\delta^* = \| a \|_{X,\delta}^*$	$= \sup_{x \in \bar{B}_\delta[0]} \left\|\sum_{k=0}^\infty a_k x_k\right\|$
$\mathcal{B}(X, Y)$	space of bounded linear operators between the Banach spaces X and Y
$\mathcal{K}(X, Y)$	set of compact operators in $\mathcal{B}(X, Y)$
X^*	continuous dual of the normed space X
cs	set of all convergent complex series
bs	set of all bounded complex series
$\| \cdot \|_{bs}$	norm for bs and cs
$M(X, Y)$	multiplier of the set $X \subset \omega$ in the set $Y \subset \omega$
X^α	α-dual of the set $X \subset \omega$
X^β	β-dual of the set $X \subset \omega$
X^γ	γ-dual of the set $X \subset \omega$
X^f	functional dual of the set $X \subset \omega$
$\mathcal{M}_X$	class of bounded sets in the metric space X
$\mathcal{M}_X^c$	subclass of closed sets in $\mathcal{M}_X$
diam (S)	diameter of the set S in a metric space
$\alpha(Q)$	Kuratowski measure of noncompactness of the set $Q \in \mathcal{M}_X$
$\chi(Q)$	Hausdorff measure of noncompactness of the set $Q \in \mathcal{M}_X$
$\mathrm{co}(S)$	convex hull of the subset S of a linear space
$\bar{B}_X$	closed unit ball in the normed space X
B_X	open unit ball in the normed space X
S_X	unit sphere in the normed space X
$\| L \|_\chi$	Hausdorff measure on noncompactness of the operator L
X_A	$= \{x \in \omega : Ax \in X\}$, matrix domain of A in X
$(X, (p_n))$	vector space X with its metrizable topology given by the sequence (p_n) of seminorms in the sense of Theorem 2.1
$z^{-1} * Y$	$= \{a \in \omega : a \cdot z = (a_k z_k)_{k=0}^\infty \in Y\}$
z^α	$= z^{-1} * \ell_1$
z^β	$= z^{-1} * cs$
z^γ	$= z^{-1} * bs$
$\Sigma = (\sigma_{nk})_{n,k=0}^\infty$	triangle of the partial sums with $\sigma_{nk} = 1$ for $(0 \le k \le n)$ and $\sigma_{nk} = 0$ for $k > n$ $(n = 0, 1, \ldots)$
$\Sigma^{(m)}$	triangle of the mth iterated partial sums
$\Delta = (\Delta_{nk})_{n,k=0}^\infty$	triangle of the backward differences with $\Delta_{n,n} = 1$, $\Delta_{n-1,n} = -1$ and $\Delta_{nk} = 0$ otherwise
$\Delta^{(m)}$	triangle of the mth iterated backward differences
$c_0\left((p), \Delta^{(m)}\right)$	$= (c_0(p))_{\Delta^{(m)}}$
$bv(p)$	$= (\ell(p))_\Delta$

$c_0\left(\Delta^{(m)}\right)$	$=(c_0)_{\Delta^{(m)}}$
$c\left(\Delta^{(m)}\right)$	$=c_{\Delta^{(m)}}$
$\ell_\infty\left(\Delta^{(m)}\right)$	$=(\ell_\infty)_{\Delta^{(m)}}$
bv_p	$=(\ell_p)_\Delta$
bv	$=(\ell_1)_\Delta=\{x\in\omega:\sum_{k=0}^\infty\lvert x_k-x_{k-1}\rvert<\infty\}$, space of sequences of bounded variation
$\mathcal{U}$	$=\{u\in\omega:u_k\neq 0\,for\,all\,k=0,1,\ldots\}$
$1/u$	$=(1/u_k)_{k=0}^\infty$ for $u\in\mathcal{U}$
C_1	triangle of the arithmetic means or the Cesàro means of order 1
$\bar{N}_q$	triangle of the weighted or Riesz means
$(\bar{N},q)_0$	$=(c_0)_{\bar{N}_q}$
$(\bar{N},q)$	$=c_{\bar{N}_q}$
$(\bar{N},q)_\infty$	$=(\ell_\infty)_{\bar{N}_q}$
$c_0(u\Delta)$	$=(u^{-1}*c_0)_\Delta$ for $u\in\mathcal{U}$
$c(u\Delta)$	$=(u^{-1}*c)_\Delta$ for $u\in\mathcal{U}$
$\ell_\infty(u\Delta)$	$=(u^{-1}*\ell_\infty)_\Delta$ for $u\in\mathcal{U}$
$E^r=(e^r_{nk})_{n,k=0}^\infty$	Euler matrix $0<r<1$, the triangle with $e^r_{nk}=\binom{n}{k}(1-r)^{n-k}r^k$ for $0\leq k\leq n$
e^r_p	$=(\ell_p)_{E^r}$ for $1\leq p<\infty$
e^r_0	$=(c_0)_{E^r}$
e^r_c	$=c_{E^r}$
e^r_∞	$=(\ell_\infty)_{E^r}$
$T^-=(t^-_{nk})_{n,k=0}^\infty$	matrix with $t^-_{n,n-1}=1$ and $t^-_{nk}=0$ $(k\neq 0)$ for $n=0,1,\ldots$
$\Delta^+=(\Delta^+_{nk})_{n,k=0}^\infty$	matrix of the forward differences with $\Delta^+_{nn}=1$, $\Delta^+_{n,n+1}=-1$ and $\Delta^+_{nk}=0$ otherwise
$W(u,v;X)$	$=v^{-1}*(u^{-1}*X)_\Sigma=\{x\in\omega:u\cdot\Sigma(v\cdot x)\in X\}$ for $u,v\in\mathcal{U}$, the set of generalized weighted means
$\mathbf{n}+1$	$=(n+1)_{n=0}^\infty$
bv^+	$=(\ell_1)_{\Delta^+}$
bv^+_0	$=bv^+\cap c_0$
$\mathcal{U}^+$	class of sequences with positive real terms
D_u	diagonal matrix with the sequence u on its diagonal
$\frac{1}{u}$	$=(1/u_k)_{k=1}^\infty$ for $u\in\mathcal{U}$
$(\frac{1}{u})^{-1}*E$	$=D_uE=\{y=(y_n)_{n=1}^\infty\in\omega:\frac{y}{u}=(\frac{y_n}{u_n})_{n=1}^\infty\in E\}$ for $u\in\mathcal{U}$ and $E\subset\omega$

E_τ	$= D_\tau E$ for $\tau \in \mathcal{U}^+$ and $E \subset \omega$
s_τ	$= (\frac{1}{\tau})^{-1} * \ell_\infty$ for $\tau \in \mathcal{U}^+$
s_τ^0	$= (\frac{1}{\tau})^{-1} * c_0$ for $\tau \in \mathcal{U}^+$
$s_\tau^{(c)}$	$= (\frac{1}{\tau})^{-1} * c$ for $\tau \in \mathcal{U}^+$
$\Vert x \Vert_{s_\tau}$	$= \sup_n \vert x_n/\tau_n \vert$ for $x \in s_\tau, s_\tau^0, s_\tau^{(c)}$ and $\tau \in \mathcal{U}^+$
$S_{\tau,\nu}$	$= (s_\tau, s_\nu)$ for $\tau, \nu \in \mathcal{U}^+$
$\Vert A \Vert_{S_{\tau,\nu}}$	$= \sup_n (1/\nu_n) \sum_{k=1}^\infty \vert a_{nk} \vert \tau_k$ for $\tau, \nu \in \mathcal{U}^+$
S_τ	$= S_{\tau,\tau}$ for $\tau \in \mathcal{U}^+$
Γ_τ	$= \{A \in S_\tau : \Vert I - A \Vert_{S_\tau} < 1\}$
s_r	$= s_\tau$ for $\tau = (r^n)_{n=1}^\infty$ and $r > 0$
s_r^0	$= s_\tau^0$ for $\tau = (r^n)_{n=1}^\infty$ and $r > 0$
$s_r^{(c)}$	$= s_\tau^{(c)}$ for $\tau = (r^n)_{n=1}^\infty$ and $r > 0$
S_r	$= S_\tau$ for $\tau = (r^n)_{n=1}^\infty$ and $r > 0$
Γ_r	$= \Gamma_\tau$ for $\tau = (r^n)_{n=1}^\infty$ and $r > 0$
$\mathcal{U}_1^+$	$= \{x \in \mathcal{U}^+$ with $x_k \leq 1$ for all $k\}$
$c(1)$	$= \{x \in \omega : \lim_{k\to\infty} x_k = 1\}$
$E * F$	$= \{xy = (x_n y_n)_{n=1}^\infty \in \omega : x \in E$ and $y \in F\}$
$C(\lambda) = ((C(\lambda))_{nk})_{n,k=1}^\infty$	for $\lambda \in \mathcal{U}$, where $(C(\lambda))_{nk} = 1/\lambda_n$ for $1 \leq k \leq n$ and $(C(\lambda))_{nk} = 0$ for $k > n$
$\Delta(\lambda) = ((\Delta(\lambda))_{nk})_{n,k=1}^\infty$	for $\lambda \in \mathcal{U}$, where $(\Delta(\lambda))_{nn} = \lambda_n$ $(\Delta(\lambda))_{n,n-1} = -\lambda_{n-1}$ and $(\Delta(\lambda))_{n,k} = 0$ for $k \neq n, n-1$
$\tau^\bullet$	$= (\tau_{n-1}/\tau_n)_{n=1}^\infty$ for $\tau \in \mathcal{U}^+$
$\widehat{C_1}$	$= \{\tau \in \mathcal{U}^+ : ((1/\tau_n)(\sum_{k=1}^n \tau_k))_{n=1}^\infty \in \ell_\infty\}$
$\hat{C}$	$= \{\tau \in \mathcal{U}^+ : ((1/\tau_n)(\sum_{k=1}^n \tau_k))_{n=1}^\infty \in c\}$
$\widehat{C_1^+}$	$= \left\{\tau \in \mathcal{U}^+ \bigcap cs : (1/\tau_n)(\sum_{k=n}^\infty \tau_k) = O(1)\ (n \to \infty)\right\}$
Γ	$= \{\tau \in \mathcal{U}^+ : \limsup_{n\to\infty} \tau_n^\bullet < 1\}$
Γ^+	$= \{\tau \in \mathcal{U}^+ : \limsup_{n\to\infty} (\tau_{n+1}/\tau_n) < 1\}$
$\hat{\Gamma}$	$= \{\tau \in \mathcal{U}^+ : \lim_{n\to\infty} \tau_n^\bullet < 1\}$
$c^\bullet$	$= \{\tau \in \mathcal{U}^+ : \tau^\bullet \in c\}$
$\vert x \vert$	$= (\vert x_n \vert)_{n=1}^\infty$ for $x = (x_n)_{n=1}^\infty \in \omega$
$w_\tau(\lambda)$	$= \{x \in \omega : C(\lambda)(\vert x \vert) \in s_\tau\}$
$w_\tau^\circ(\lambda)$	$= \{x \in \omega : C(\lambda)(\vert x \vert) \in s_\tau^0\}$
$w_\tau^\bullet(\lambda)$	$= \{x \in \omega : x - le \in w_\tau^\circ(\lambda)$ for some $l \in \mathbb{C}\}$
$c_\tau(\lambda, \mu)$	$= \{x \in \omega : C(\lambda)(\vert \Delta(\mu)x \vert) \in s_\tau\}$ for $\tau\lambda, \mu \in \mathcal{U}^+$ and $\mu \in \omega$
$c_\tau^\circ(\lambda, \mu)$	$= \{x \in \omega : C(\lambda)(\vert \Delta(\mu)x \vert) \in s_\tau^0\}$ for $\tau \in \mathcal{U}^+$ and $\mu \in \omega$
$c_\tau^\bullet(\lambda, \mu)$	$= \{x \in \omega : x - le \in c_\tau^\circ(\lambda\mu)$ for some $l \in \mathbb{C}\}$ for $\tau \in \mathcal{U}^+$ and $\mu \in \omega$
$[C, C]$	$= \{x \in \omega : (1/\lambda_n)(\sum_{m=1}^n \vert (1/\mu_m)(\sum_{k=1}^m x_k) \vert) = \tau_n O(1)(n \to \infty)\}$
$[C, \Delta]$	$= \{x \in \omega : (1/\lambda_n)(\sum_{k=1}^n \vert \mu_k x_k - \mu_{k-1} x_{k-1} \vert) = \tau_n O(1)(n \to \infty)\}$
$[\Delta, C]$	$= \{x \in \omega : -\lambda_{n-1} \vert (1/\mu_{n-1})(\sum_{k=1}^{n-1} x_k) \vert + \lambda_n \vert (1/\mu_n)(\sum_{k=1}^n x_k) \vert = \tau_n O(1)\ (n \to \infty)\}$
$[\Delta, \Delta]$	$= \{x \in \omega : -\lambda_{n-1} \vert \mu_{n-1} x_{n-1} - \mu_{n-2} x_{n-2} \vert + \lambda_n \vert \mu_n x_n - \mu_{n-1} x_{n-1} \vert = \tau_n O(1)(n \to \infty)\}$

$[\Delta, \Delta^+]$	$= \{x : \lambda_n \lvert \mu_n(x_n - x_{n+1}) \rvert - \lambda_{n-1} \lvert \mu_{n-1}(x_{n-1} - x_n) \rvert = \tau_n O(1)(n \to \infty)\}$
$[\Delta, C^+]$	$= \{x : \lambda_n \lvert \sum_{i=n}^{\infty}(x_i/\mu_i) \rvert - \lambda_{n-1} \lvert \sum_{i=n-1}^{\infty}(x_i/\mu_i) \rvert = \tau_n O(1)(n \to \infty)\}$
$[\Delta^+, \Delta]$	$= \{x : \lambda_n \lvert \mu_n x_n - \mu_{n-1} x_{n-1} \rvert - \lambda_n \lvert \mu_{n+1} x_{n+1} - \mu_n x_n \rvert = \tau_n O(1)(n \to \infty)\}$
$[\Delta^+, C]$	$= \left\{x : \lambda_n \left((1/\mu_n) \lvert \sum_{i=1}^{n} x_i \rvert - (1/\mu_{n+1}) \lvert \sum_{i=1}^{n+1} x_i \rvert \right) = \tau_n O(1)(n \to \infty)\right\}$
$[\Delta^+, \Delta^+]$	$= \{x : \lambda_n \mu_n \lvert x_n - x_{n+1} \rvert - \lambda_n \mu_{n+1} \lvert x_{n+1} - x_{n+2} \rvert = \tau_n O(1)(n \to \infty)\}$
$[C^+, C]$	$= \left\{x : \sum_{k=n}^{\infty} \left((1/\lambda_k) \left\lvert (1/\mu_k) \sum_{i=1}^{k} x_i \right\rvert \right) = \tau_n O(1)(n \to \infty)\right\}$
$[C^+, C^+]$	$= \{x : \sum_{k=n}^{\infty} ((1/\lambda_k) \lvert \sum_{i=k}^{\infty}(x_i/\mu_i) \rvert) = \tau_n O(1)(n \to \infty)\}$
$w_\tau^p(\lambda)$	$= \{x \in \omega : C(\lambda)(\lvert x \rvert^p) \in s_\tau\}$ for $0 < p < \infty$
$w_\tau^{\circ p}(\lambda)$	$= \{x \in \omega : C(\lambda)(\lvert x \rvert^p) \in s_\tau^0\}$ for $0 < p < \infty$
$w_\tau^{+p}(\lambda)$	$= \{x \in \omega : C^+(\lambda)(\lvert x \rvert^p) \in s_\tau\}$ for $0 < p < \infty$
$\overset{\circ}{w}{}_\tau^{+p}(\lambda)$	$= \{x \in \omega : C^+(\lambda)(\lvert x \rvert^p) \in s_\tau^0\}$ for $0 < p < \infty$
$c_\tau^p(\lambda, \mu)$	$= (w_\tau^p(\lambda))_{\Delta(\mu)} = \{x \in \omega : C(\lambda)(\lvert \Delta(\mu) x \rvert^p) \in s_\tau\}$ for $0 < p < \infty$
$\dot{c}_\tau^{+p}(\lambda, \mu)$	$= (w_\tau^p(\lambda))_{\Delta^+(\mu)} = \{x \in \omega : C(\lambda)(\lvert \Delta^+ \rvert(\mu) x \rvert^p) \in s_\tau\}$ for $0 < p < \infty$
$c_\tau^{+\cdot p}(\lambda, \mu)$	$= (w_\tau^{+p}(\lambda))_{\Delta(\mu)} = \{x : C^+(\lambda)(\lvert \Delta(\mu) x \rvert^p) \in s_\tau\}$ for $0 < p < \infty$
$c_\tau^{+p}(\lambda, \mu)$	$= (w_\tau^{+p}(\lambda))_{\Delta^+(\mu)} = \{x : C^+(\lambda)(\lvert \Delta^+(\mu) x \rvert^p) \in s_\tau\}$ for $0 < p < \infty$
$c_\tau^p(\lambda, \mu)$	$= \{x = (x_n)_{n=1}^{\infty} : \sup_n [(1/\lvert \lambda_n \rvert \tau_n)(\sum_{k=1}^{n} \lvert \mu_k x_k - \mu_{k-1} x_{k-1} \rvert^p)] < \infty\}$ for $0 < p < \infty$
$\dot{c}_\tau^{+p}(\lambda, \mu)$	$= \{x = (x_n)_{n=1}^{\infty} : \sup_n [(1/\lvert \lambda_n \rvert \tau_n)(\sum_{k=1}^{n} \lvert \mu_k (x_k - x_{k+1}) \rvert^p)] < \infty\}$ for $0 < p < \infty$
$c_\tau^{+\cdot p}(\lambda, \mu)$	$= \{x = (x_n)_{n=1}^{\infty} : \sup_n [(1/\tau_n) \sum_{k=n}^{\infty} ((1/\lvert \lambda_k \rvert) \lvert \mu_k x_k - \mu_{k-1} x_{k-1} \rvert^p)] < \infty\}$ for $0 < p < \infty$
$\widetilde{c_\tau^{+p}}(\lambda, \mu)$	$= \{x = (x_n)_{n=1}^{\infty} : \lim_{n \to \infty} [(1/\tau_n) \sum_{k=n}^{\infty} ((1/\lvert \lambda_k \rvert) \lvert \mu_k (x_k - x_{k+1}) \rvert^p)] = 0\}$ for $0 < p < \infty$
$w_\infty(\lambda)$	$= \{x = (x_n)_{n=1}^{\infty} \in \omega : \sup_n (1/\lambda_n) \sum_{k=1}^{n} \lvert x_k \rvert < \infty\}$
$w^0(\lambda)$	$= \{x = (x_n)_{n=1}^{\infty} \in \omega : \lim_{n \to \infty} (1/\lambda_n) \sum_{k=1}^{n} \lvert x_k \rvert = 0\}$
$\lVert x \rVert_\lambda$	$= \lVert x \rVert_{w_\infty(\lambda)} = \sup_n \left((1/\lambda_n) \sum_{k=1}^{n} \lvert x_k \rvert \right)$ for $x \in w_\infty(\lambda), w^0(\lambda)$ and $\lambda \in \mathcal{U}^+$
$\lVert A \rVert_{(w_\infty(\lambda), w_\infty(\lambda))}$	$= \sup_{x \neq 0} (\lVert Ax \rvert_\lambda / \lVert x \rVert_\lambda)$
W_τ	$= \{x \in \omega : \lVert x \rVert_{W_\tau} = \sup_n \left((1/n) \sum_{k=1}^{n} \lvert x_k \rvert / \tau_k \right) < \infty\}$ for $\tau \in \mathcal{U}^+$
W_τ^0	$= \left\{x \in \omega : \lim_{n \to \infty} \left((1/n) \sum_{k=1}^{n} \lvert x_k \rvert / \tau_k \right) = 0\right\}$ for $\tau \in \mathcal{U}^+$
$W_\tau(\Delta(\lambda))$	$= \{x \in \omega : \sup_n \left((1/n) \sum_{k=1}^{n} (1/\tau_k) \lvert \lambda_k x_k - \lambda_{k-1} x_{k-1} \rvert \right) < \infty\}$ for $\lambda, \tau \in \mathcal{U}^+$
$W_\tau(C(\lambda))$	$= \{x \in \omega : \sup_n (1/n) \sum_{m=1}^{n} \left((1/\lambda_m \tau_m) \lvert \sum_{k=1}^{m} x_k \rvert \right) < \infty\}$ for $\lambda, \tau \in \mathcal{U}^+$

$W_\tau(C^+(\lambda))$	$=\{x\in\omega:\sup_n(1/n)\sum_{m=1}^n(1/\tau_m)\left(\left\lvert\sum_{k=1}^m x_k\right\rvert/\lambda_k\right)<\infty\}$ for $\lambda,\tau\in\mathcal{U}^+$
$[C,\Delta]_{W_\tau}$	$=\{x\in\omega:\sup_n[(1/n)\sum_{m=1}^n((1/\lambda_m\tau_m)\sum_{k=1}^m\lvert\mu_kx_k-\mu_{k-1}x_{k-1}\rvert)]<\infty\}$ for $\lambda,\mu,\tau\in\mathcal{U}^+$
$[C,C]_{W_\tau}$	$=\{x\in\omega:\sup_n\left[(1/n)\sum_{m=1}^n\left((1/\lambda_m\tau_m)\sum_{k=1}^m(1/\mu_k)\left\lvert\sum_{i=1}^k x_i\right\rvert\right)\right]<\infty\}$ for $\lambda,\mu,\tau\in\mathcal{U}^+$
$[C^+,\Delta]_{W_\tau}$	$=\{x\in\omega:\sup_n[(1/n)\sum_{m=1}^n((1/\tau_m)\sum_{k=m}^\infty(1/\lambda_k)\lvert\mu_kx_k-\mu_{k-1}x_{k-1}\rvert)]<\infty\}$ for $\lambda,\mu,\tau\in\mathcal{U}^+$
$[C^+,C]_{W_\tau}$	$=\{x\in\omega:\sup_n\left[(1/n)\sum_{m=1}^n\left((1/\tau_m)\sum_{k=m}^\infty(1/\lambda_k)(1/\mu_k)\left\lvert\sum_{i=1}^k x_i\right\rvert\right)\right]<\infty\}$ for $\lambda,\mu,\tau\in\mathcal{U}^+$
$[C^+,C^+]_{W_\tau}$	$=\{x\in\omega:\sup_n[(1/n)\sum_{m=1}^n((1/\tau_m)\sum_{k=m}^\infty(1/\lambda_k)\lvert\sum_{i=k}^\infty(x_i/\mu_i)\rvert)]<\infty\}$ for $\lambda,\mu,\tau\in\mathcal{U}^+$
$\lVert(a_n)_{n=1}^\infty\rVert_{\mathcal{M}}$	$=\sum_{v=1}^\infty 2^v\max_{2^v\le k\le 2^{v+1}-1}\lvert a_k\rvert$
$R_{\mathcal{E}}$	equivalence relation defined for any $\mathcal{E}\subset\omega$ by $xR_{\mathcal{E}}y$ if and only if $\mathcal{E}_x=\mathcal{E}_y$ for $x,y\in\mathcal{U}^+$
$cl^{\mathcal{E}}(b)$	equivalence class of $b\in\mathcal{U}^+$ and the equivalence relation $R_{\mathcal{E}}$
$\mathcal{I}_a(E,F,F')$	$=\{x\in\mathcal{U}^+:F\subset E_a+F'_x\}$, where $E,F,F'\subset\omega$ are linear spaces and $a\in\mathcal{U}^+$
$\bar{\chi}$	$=\{x\in\mathcal{U}^+:1/x\in\chi\}$ for any $\chi\subset\omega$
$B(\tilde{r},\tilde{s})$	the bidiagonal matrix with $[B(\tilde{r},\tilde{s})]_{nn}=r_n$ for all n and $[B(\tilde{r},\tilde{s})]_{n,n-1}=s_{n-1}$ for all $n\ge 2$
G^+	$=G\cap\mathcal{U}^+$ for any $G\subset\omega$
$\mathcal{U}_1^+$	$=\{x\in\mathcal{U}^+:x_n\le 1\text{ for all }n\}$
$\mathbf{\Lambda}$	$=\{x=(x_n)_{n=1}^\infty\in\omega:\sup_n(\lvert x_n\rvert^{1/n})<\infty\}$
$\mathbf{\Gamma}$	$=\{x=(x_n)_{n=1}^\infty\in\omega:\lim_{n\to\infty}(\lvert x_n\rvert^{1/n})=0\}$
$\hat{C}_\Lambda$	$=\{a\in\mathcal{U}^+:[C(a)a]_n\le k^n\text{ for all }n\text{ and some }k>0\}$
cs_0	$=(c_0)_\Sigma=\{x\in\omega:(\sum_{k=1}^n x_k)_{n=1}^\infty\in c_0\}$
bv_p	$=(\ell_p)_\Delta=\{x\in\omega:\sum_{k=1}^\infty\lvert y_k-y_{k-1}\rvert^p<\infty\}$ for $0<p<\infty$
$\mathbb{S}_{R,\bar{R}}$	$=\{x\in\mathcal{U}^+:(s_R^0)_\Delta+s_x^{(c)}=s_{\bar{R}}^{(c)}\}$
$\mathbb{B}(s_\tau)$	$=\mathcal{B}(s_\tau)\bigcap(s_\tau,s_\tau)$
$\mathcal{L}$	the set of lower triangular infinite matrices, that is, $A\in\mathcal{L}$ if $a_{nk}=0$ for $k>n$ and all n
$N_{p,\tau}(A)$	$=[\sum_{n=1}^\infty(\sum_{k=1}^\infty(\lvert a_{nk}\rvert(\tau_k/\tau_n))^q)^{p-1}]^{1/p}$
$\widehat{\mathcal{B}}_p(\tau)$	$=\{A=(a_{nk})_{n,k=1}^\infty:N_{p,\tau}(A)<\infty\}$
$\Gamma'_{p,\tau}$	$=\{A=(a_{nk})_{n,k=1}^\infty\in((\ell_p)_\tau,(\ell_p)_\tau):N_{p,\tau}(I-A)<1\}$
$\sigma(A,X)$	spectrum of an operator $A\in\mathcal{B}(X,X)$
$\rho(A,X)$	$=\mathbb{C}\backslash\sigma(A,X)$ resolvent set of an operator $A\in\mathcal{B}(X,X)$

Chapter 1
Matrix Transformations and Measures of Noncompactness

The major part of this chapter is introductory and included as a reference for the reader's convenience; it recalls the concepts and results from the theories of sequence spaces, matrix transformations in Sects. 1.1–1.3 and 1.5 and measures of noncompactness in Sects. 1.7–1.10 that are absolutely essential for the book. Although the results of this chapter may be considered as standard in the modern theories of matrix transformations, we included their proofs with the exception of those in Sect. 1.4 on the relations between various kind of duals; these results are mainly included for the sake of completeness and not directly used in the remainder of the book.

We refer the reader interested in matrix transformations to [1, 5, 14, 19, 23, 25, 27, 33–37] and the survey paper [29], and, in measures of noncompactness, to [2–4, 10, 12, 21–23, 30]. Although the concepts and results of this chapter are standard, we decided to cite from [23] in almost all cases, if necessary.

Sections 1.3 and 1.5 contain results that are less standard. They concern the characterizations of matrix transformations from arbitrary FK spaces into the spaces c_0, c, ℓ_∞ and ℓ_1 of all null, convergent and bounded sequences and of all absolutely convergent series, and all known characterizations of classes matrix transformations between the classical sequence spaces c_0, c, ℓ_∞ and $\ell_p = \{x = (x_k) : \sum_k |x_k|^p < \infty\}$ for $1 \leq p < \infty$. For completeness sake, we also prove Crone's theorem which characterizes the class of all matrix transformations of ℓ_2 into itself in Sect. 1.6.

The underlying fundamental concepts for the theories of sequence spaces and matrix transformations are those of linear metric and paranormed spaces of FK, BK and AK spaces, and of various kinds of dual spaces. The general results related to these concepts are used in the characterizations of matrix transformations between the *classical sequence spaces*. By this, we mean to give necessary and sufficient conditions on the entries of an infinite matrix to map one classical sequence space into another.

We also present an axiomatic introduction of measures of noncompactness on bounded sets of complete metric spaces, recall the definitions and essential proper-

B. de Malafosse et al., *Operators Between Sequence Spaces and Applications*,
https://doi.org/10.1007/978-981-15-9742-8_1

ties of the *Kuratowski* and *Hausdorff measures of noncompactness*, which are the most prominent measures on noncompactness, and the famous theorem by *Goldenštein, Go'hberg and Markus*, which gives an estimate for the Hausdorff measure of noncompactness of bounded sets in Banach spaces with a Schauder basis. Finally, we recall the definition of the measure of noncompactness of operators and list some important related properties.

1.1 Linear Metric and Paranormed Spaces

Here, we recall the concepts of *linear metric* and *paranormed spaces*, which are fundamental in the theory of sequence spaces and matrix transformations.

The concept of a *linear* or *vector space* involves an algebraic structure given by the definition of two operations, namely, the sum of any two of its elements, also called *vectors*, and the product of any *scalar* with any vector. It is clear that the set ω of all complex sequences $x = (x_k)_{k=0}^{\infty}$ is a linear space with respect to the addition and scalar multiplication defined termwise, that is,

$$x + y = (x_k + y_k)_{k=0}^{\infty} \text{ and } \lambda x = (\lambda x_k)_{k=0}^{\infty}$$
$$\text{for all } x = (x_k)_{k=0}^{\infty}, y = (y_k)_{k=0}^{\infty} \in \omega \text{ and all } \lambda \in \mathbb{C}.$$

On the other hand, a topological structure of a set may be given by a metric. For instance, ω is a metric space with its metric d defined by

$$d(x, y) = \sum_{k=0}^{\infty} \frac{1}{2^k} \frac{|x_k - y_k|}{1 + |x_k - y_k|} \text{ for all } x, y \in \omega. \tag{1.1}$$

If a set is both a linear and metric space, then it is natural to require the algebraic operations to be continuous with respect to the metric. The continuity of the algebraic operations in a linear metric space (X, d) means the following: If (x_n) and (y_n) are sequences in X and (λ_n) is a sequence of scalars with $x_n \to x$, $y_n \to y$ and $\lambda_n \to \lambda$ $(n \to \infty)$, then it follows that $x_n + y_n \to x + y$ and $\lambda_n x_n \to \lambda x$ $(n \to \infty)$.

A complete linear metric space is called a *Fréchet space*. Unfortunately, this terminology is not universally agreed on. Some authors call a complete linear metric space an F space, and a locally convex F space a Fréchet space (e.g. [28, p. 8] or [15, p. 208]) which Wilansky calls an F space. We follow Wilansky's terminology of a Fréchet space being a complete linear metric space [31–33]. We will see later that ω is a Fréchet space with the metric defined in (1.1).

The concept of a *paranorm* is closely related to linear metric spaces. It is a generalization of that of absolute value. The paranorm of a vector may be thought of as its distance from the origin. We recall the definition of a paranormed space for the reader's convenience.

Definition 1.1 Let X be a linear space.
(a) A function $p : X \to \mathbb{R}$ is called a *paranorm*, if

$$p(0) = 0, \tag{P.1}$$

$$p(x) \geq 0 \text{ for all } x \in X, \tag{P.2}$$

$$p(-x) = p(x) \text{ for all } x \in X, \tag{P.3}$$

$$p(x + y) \leq p(x) + p(y) \text{ for all } x, y \in X \ (triangle\, inequality) \tag{P.4}$$

if (λ_n) is a sequence of scalars with $\lambda_n \to \lambda \ (n \to \infty)$ and (x_n) is a sequence of vectors with $p(x_n - x) \to 0 \ (n \to \infty)$ then it follows that $p(\lambda_n x_n - \lambda x) \to 0 \ (n \to \infty)$ $(continuity\, of\, multiplication\, by\, scalars)$. (P.5)

If p is a paranorm on X, then (X, p), or X for short, is called a *paranormed space.* A paranorm p for which $p(x) = 0$ implies $x = 0$ is called *total.*
(b) For any two paranorms p and q, p is called *stronger* than q if, whenever (x_n) is a sequence with $p(x_n) \to 0 \ (n \to \infty)$, then also $q(x_n) \to 0 \ (n \to \infty)$. If p is stronger than q, then q is said to be *weaker* than p. If p is stronger than q and q is stronger than p, then p and q are called *equivalent.* If p is stronger than q, but p and q are not equivalent, then p is called *strictly stronger* than q, and q is called *strictly weaker* than p.

If p is a total paranorm for a linear space X, then it is easy to see that $d(x, y) = p(x - y) \ (x, y \in X)$ defines a metric on X, thus *every totally paranormed space is a linear metric space.* The converse is also true. *The metric of any linear metric space is given by some total paranorm* [31, Theorem 10.4.2, p. 183].

The next well-known result shows how a sequence of paranorms may be used to define a paranorm.

Theorem 1.1 *Let $(p_k)_{k=0}^{\infty}$ be a sequence of paranorms on a linear space X. We define the so-called* Fréchet combination *of $(p_k)_{k=0}^{\infty}$ by*

$$p(x) = \sum_{k=0}^{\infty} \frac{1}{2^k} \frac{p_k(x)}{1 + p_k(x)} \text{ for all } x \in X. \tag{1.2}$$

Then, we have

(a) p is a paranorm on X and satisfies

$$\begin{gathered} p(x_n) \to 0 \ (n \to \infty) \\ \text{if and only if} \\ p_k(x_n) \to 0 \ (n \to \infty) \text{ for each } k. \end{gathered} \tag{1.3}$$

(b) p is the weakest paranorm stronger than every p_k.
(c) p is total if and only if the set $\{p_k : k = 0, 1, \dots\}$ is total.

(We recall that a set Φ if functions from a linear space X to a linear space is said to be *total* if given $x \in X \setminus \{0\}$, there exists $f \in \Phi$ with $f(x) \neq 0$.)

We obtain as an immediate consequence of Theorem 1.1.

Corollary 1.1 *The set ω is a complete, totally paranormed space with its paranorm p defined by*

$$p(x) = \sum_{k=0}^{\infty} \frac{1}{2^k} \frac{|x_k|}{1+|x_k|} \text{ for all } x \in \omega. \tag{1.4}$$

Thus, ω is a Fréchet space with its natural *metric d_ω given by*

$$d_\omega(x,y) = \sum_{k=0}^{\infty} \frac{1}{2^k} \frac{|x_k - y_k|}{1+|x_k + y_k|} \text{ for all } x, y \in \omega. \tag{1.5}$$

Furthermore, convergence in (ω, d_ω) and coordinatewise convergence are equivalent, that is,

$$d_\omega(x^{(n)}, x) \to 0 \ (n \to \infty) \text{ if and only if } \lim_{n\to\infty} x_k^{(n)} = x_k \text{ for each } k.$$

We close this section with an example we prove in detail for the reader's convenience, since part of it will be used in Part (b) of Examples 1.2, 2.3, and in Sect. 6,

Example 1.1 Let $p = (p_k)_{k=0}^{\infty}$ be a sequence of positive reals and

$$\ell(p) = \left\{ x \in \omega : \sum_{k=0}^{\infty} |x_k|^{p_k} < \infty \right\}, \ c_0(p) = \left\{ x \in \omega : \lim_{k\to\infty} |x_k|^{p_k} = 0 \right\}$$

and

$$\ell_\infty(p) = \left\{ x \in \omega : \sup_k |x|^{p_k} < \infty \right\}.$$

(a) Then $\ell(p)$, $c_0(p)$ and $\ell_\infty(p)$ are linear spaces if and only if the sequence p is bounded.

(b) Let the sequence p be bounded and $M = \max\{1, \sup_k p_k\}$. Then $\ell(p)$ and $c_0(p)$ are complete, totally paranormed spaces with their natural paranorms g and g_0 given by

$$g(x) = \left(\sum_{k=0}^{\infty} |x_k|^{p_k} \right)^{1/M} \quad \text{for all } x \in \ell(p)$$

and

$$g_0(x) = \sup_k |x_k|^{p_k/M} \text{ for all } x \in c_0(p),$$

and g and g_0 are strictly stronger than the natural paranorm of ω on $\ell(p)$ and $c_0(p)$, respectively.

But g_0 is a paranorm for $\ell_\infty(p)$ if and only if

$$m = \inf_k p_k > 0, \tag{1.6}$$

in which case $\ell_\infty(p)$ reduces to the classical space ℓ_∞ of bounded complex sequences.

Proof (a) First we assume that the sequence p is bounded. We write $X(p)$ for any of the sets $\ell(p)$, $c_0(p)$ and $\ell_\infty(p)$. Let $x, y \in X(p)$ and $\lambda \in \mathbb{C}$ be given.

(a.i) First we show $x + y \in X(p)$.
Putting $\alpha_k = p_k/M \leq 1$ for all k, we obtain $|x_k + y_k|^{\alpha_k} \leq |x_k|^{\alpha_k} + |y_k|^{\alpha_k}$ by inequality (A.1).
If $X(p) = c_0(p)$ or $X(p) = \ell_\infty(p)$, then

$$\sup_k |x_k + y_k|^{\alpha_k} \leq \sup_k |x_k|^{\alpha_k} + \sup_k |y_k|^{\alpha_k}, \tag{1.7}$$

which implies $x + y \in X(p)$.
If $X(p) = \ell(p)$, we get applying Minkowski's inequality (A.4)

$$\begin{aligned}\left(\sum_{k=0}^{\infty} |x_k + y_k|^{p_k}\right)^{1/M} &= \left(\sum_{k=0}^{\infty} |x_k + y_k|^{\alpha_k M}\right)^{1/M} \\ &\leq \left(\sum_{k=0}^{\infty} (|x_k|^{\alpha_k} + |y_k|^{\alpha_k})^M\right)^{1/M} \\ &\leq \left(\sum_{k=0}^{\infty} |x_k|^{\alpha_k M}\right)^{1/M} + \left(\sum_{k=0}^{\infty} |y_k|^{\alpha_k M}\right)^{1/M} \\ &= \left(\sum_{k=0}^{\infty} |x_k|^{p_k}\right)^{1/M} + \left(\sum_{k=0}^{\infty} |y_k|^{p_k}\right)^{1/M}.\end{aligned}$$

Thus, we have shown

$$\left(\sum_{k=0}^{\infty} |x_k + y_k|^{p_k}\right)^{1/M} \leq \left(\sum_{k=0}^{\infty} |x_k|^{p_k}\right)^{1/M} + \left(\sum_{k=0}^{\infty} |y_k|^{p_k}\right)^{1/M}, \tag{1.8}$$

and so $x + y \in \ell(p)$. This completes Part (a.i) of the proof.

(a.ii) Now we show $\lambda x \in X(p)$.
We put $\Lambda = \max\{1, |\lambda|^M\} < \infty$. Then $|\lambda_k|^{p_k} \leq \Lambda$ for all k and so

$$\sum_{k=0}^{\infty} |\lambda x_k|^{p_k} \le \Lambda \sum_{k=0}^{\infty} |x_k|^{p_k} < \infty$$

and

$$\sup_k |\lambda x_k|^{p_k} \le \Lambda \sup_k |x_k|^{p_k} < \infty,$$

hence $\lambda x \in X(p)$. This completes Part (a.ii) of the proof.

Thus, we have shown that if the sequence p is bounded then $\ell(p)$ is a linear space.
To show the converse part, we assume that the sequence p is not bounded. Then there exists a subsequence $(p_{k(j)})_{j=0}^{\infty}$ of the sequence p such that $p_{k(j)} > j+1$ for all j. We define the sequence $x = (x_k)_{k=0}^{\infty}$ by

$$x_k = \begin{cases} \dfrac{1}{(j+1)^{2/p_{k(j)}}} & (k = k(j)) \\ 0 & (k \neq k(j)) \end{cases} \quad (j = 0, 1, \ldots).$$

Then, we have

$$\sum_{k=0}^{\infty} |x_k|^{p_k} = \sum_{j=0}^{\infty} \frac{1}{(j+1)^2} < \infty,$$

hence $x \in \ell(p)$, but

$$\sup_k |2x_k|^{p_k} > \frac{2^{j+1}}{(j+1)^2} \text{ for } j = 0, 1, \ldots,$$

that is, $2x \notin \ell_\infty(p)$. This shows that if the sequence p is not bounded, then the spaces $X(p)$ are not linear spaces. This completes the proof of Part (a).

(b) Now let the sequence p be bounded.

(b.1) First we show that g is a total paranorm for $X(p) = \ell(p)$ and g_0 is a total paranorm for $X(p) = c_0(p)$.
We write $g_{X(p)} = g$ for $X(p) = \ell(p)$ and $g_{X(p)} = g_0$ for $X(p) = c_0(p)$. Then, we obviously have $g_{X(p)} : X(p) \to \mathbb{R}$, $g_{X(p)}(0) = 0$, $g_{X(p)}(x) \ge 0$, $g_{X(p)}(x) > 0$ implies $x \neq 0$ and $g_{X(p)}(-x) = g_{X(p)}(x)$, and (1.8) and (1.7) are the triangle inequalities for $\ell(p)$ and $c_0(p)$, respectively.
To show the condition in (P.5) of Definition 1.1, let $(\lambda_n)_{n=0}^{\infty}$ be a sequence of scalars with $\lambda_n \to \lambda$ $(n \to \infty)$ and $(x^{(n)})_{n=0}^{\infty}$ be a sequence of elements $x^{(n)} \in X(p)$ with $g_{X(p)}(x^{(n)} - x) \to 0$ $(n \to \infty)$. We observe that by (1.7) or (1.8)

$$g_{X(p)}\left(\lambda_n x^{(n)} - \lambda x\right) \le g_{X(p)}\left((\lambda_n - \lambda)(x^{(n)} - x)\right) + g_{X(p)}\left(\lambda(x^{(n)} - x)\right) + g_{X(p)}\left((\lambda_n - \lambda)x\right). \quad (1.9)$$

It follows from $\lambda_n \to \lambda$ $(n \to \infty)$ that $|\lambda_n - \lambda| < 1$ for all sufficiently large n, hence

$$g_{X(p)}\left((\lambda_n - \lambda)(x^{(n)} - x)\right) \leq g_{X(p)}(x^{(n)} - x) \to 0 \ (n \to \infty).$$

Furthermore, we have

$$g_{X(p)}\left(\lambda(x^{(n)} - x)\right) \leq \Lambda g_{X(p)}(x^{(n)} - x) \to 0 \ (n \to \infty).$$

Therefore, the first two terms on the right in (1.9) tend to zero as n tends to infinity.
To show that the third term on the right also tend to zero as n tend to infinity, let $\varepsilon > 0$ be given with $\varepsilon < 1$.

(α) First, we consider the case of $X(p) = \ell(p)$.
Since $x \in \ell(p)$, we can choose a non-negative integer k_0 such that

$$\left(\sum_{k=k_0+1}^{\infty} |x_k|^{p_k}\right)^{1/M} < \varepsilon/2.$$

Now we choose a non-negative integer n_0 such that

$$|\lambda_n - \lambda| \leq 1 \text{ and } \max_{0 \leq k \leq k_0} |\lambda_n - \lambda|^{p_k} < \left(\frac{\varepsilon}{2(g(x)+1)}\right)^M$$

for all $n \geq n_0$. Then we obtain for all $n \geq n_0$

$$g\left((\lambda_n - \lambda)x\right) = \left(\sum_{k=0}^{\infty} |\lambda_n - \lambda|^{p_k} |x_k|^{p_k}\right)^{1/M} \leq$$

$$\left(\sum_{k=0}^{k_0} |\lambda_n - \lambda|^{p_k} |x_k|^{p_k}\right)^{1/M} + \left(\sum_{k=k_0+1}^{\infty} |\lambda_n - \lambda|^{p_k} |x_k|^{p_k}\right)^{1/M} <$$

$$\frac{\varepsilon}{2(g(x)+1)} \left(\sum_{k=0}^{\infty} |x_k|^{p_k}\right)^{1/M} + \left(\sum_{k=k_0+1}^{\infty} |x_k|^{p_k}\right)^{1/M} \leq \varepsilon/2 + \varepsilon/2 = \varepsilon.$$

Hence, the third term in (1.9) also tends to zero as n tends to infinity.

(β) Now we consider the case of $X(p) = c_0(p)$.
Since $x \in c_0(p)$, we can choose a non-negative integer k_0 such that

$$|x_k|^{\alpha_k} \leq \varepsilon/2 \text{ for all } k \geq k_0.$$

Now we choose a non-negative integer n_0 such that

$$|\lambda_n - \lambda| \leq 1 \text{ and } \max_{0 \leq k \leq k_0} |\lambda_n - \lambda|^{p_k} < \left(\frac{\varepsilon}{2(g_0(x)+1)}\right)^M.$$

Then we obtain for all $n \geq n_0$

$$(|\lambda_n - \lambda| \cdot |x_k|)^{\alpha_k} \leq |x_k|^{\alpha_k} \text{ for all } k \geq k_0 + 1$$

and

$$(|\lambda_n - \lambda| \cdot |x_k|)^{\alpha_k} \leq \frac{\varepsilon}{2(g_0(x)+1)} \cdot |x_0|^{\alpha_k} \text{ for } 0 \leq k \leq k_0.$$

Hence, the third term in (1.9) also tends to zero as n tends to infinity. Thus, we have shown that $g_{X(p)}$ is a total paranorm on $X(p)$. This completes Part (b.1) of the proof.

(b.2) Now we show that convergence in $X(p)$ is strictly stronger than coordinatewise convergence, when $X(p) = \ell(p)$ or $X(p) = c_0(p)$.

(b.2.i) First, we show that $g_{X(p)}(x^n - x) \to 0$ $(n \to \infty)$ implies $x_k^{(n)} \to x_k$ $(n \to \infty)$ for each k.
We fix k. Then, we have

$$|x^{(n)} - x_k| \leq \left(g_{X(p)}(x^{(n)} - x)\right)^{M/p_k} \text{ for all } n = 0, 1, \ldots$$

and so $g_{X(p)}(x^n - x) \to 0$ $(n \to \infty)$ implies $x_k^{(n)} \to x_k$ $(n \to \infty)$, hence convergence in the paranorm of $X(p)$ implies coordinatewise convergence. This completes Part (b.2.i) of the proof.

(b.2.ii) Now we show that the converse is not true.
We define the sequence $(x^{(n)})_{n=0}^{\infty}$ by $x^{(n)} = e^{(n)}$ $(n = 0, 1, \ldots)$. Then obviously $|x_k^{(n)}| = 0$ for all $n > k$, hence $\lim_{n\to\infty} x_k^{(n)} = 0$ for each k, but for all distinct values of n and m, we have $g(x^{(n)} - x^{(m)}) = 2^{1/M}$ and $g_0(x^{(n)} - x^{(m)}) \geq 1$, hence $(x^{(n)})_{n=0}^{\infty}$ is not a Cauchy sequence, and so not convergent. This shows that coordinatewise convergence does not imply convergence in the paranorm of $X(p)$, in general.
Applying Corollary 1.1, we conclude that the paranorm on $X(p)$ is strictly stronger than that of ω on $X(p)$. This completes Part (b.2.ii) of the proof.

This completes Part (b.2) of the proof.

(b.3) Now we show that $X(p)$ is complete, when $X(p) = \ell(p)$ or $X(p) = c_0(p)$.
Let $(x^{(n)})_{n=0}^{\infty}$ be a Cauchy sequence in $X(p)$. Then $(x_k^{(n)})_{n=0}^{\infty}$ is a Cauchy sequence in $\mathbb{C}$ for each k, by Part (b.2.i) of the proof, hence convergent,

$$x = \lim_{n\to\infty} x_k^{(n)}, \text{ say.}$$

(b.3.i) First we consider the case of $X(p) = \ell(p)$.
Let $\varepsilon > 0$ be given and K be an arbitrary non-negative integer. Then there is a non-negative integer N such that

$$\left(\sum_{k=0}^{K} |x_k^{(n)} - x_k^{(m)}|^{p_k}\right)^{1/M} \leq g(x^{(n)} - x^{(m)}) < \varepsilon \text{ for all } m, n \geq N$$

which implies

$$\left(\sum_{k=0}^{K} |x_k^{(n)} - x_k|^{p_k}\right)^{1/M} = \lim_{m\to\infty} \left(\sum_{k=0}^{K} |x_k^{(n)} - x_k^{(m)}|^{p_k}\right)^{1/M} \leq \varepsilon$$

for all $n \geq N$.

Since K was arbitrary, it follows that

$$g(x^{(n)} - x) \leq \varepsilon \text{ for all } n \geq N, \tag{1.10}$$

in particular, $x^{(n)} - x \in \ell(p)$. Since $\ell(p)$ is a linear space by Part (a), we have $x \in \ell(p)$, and we conclude from (1.10) that $x^{(n)} \to x$ $(n \to \infty)$ in $\ell(p)$. Thus, we have shown that $\ell(p)$ is complete.

(b.3.ii) Now we consider the case of $X(p) = c_0(p)$.
Let $\varepsilon > 0$ be given. Then there exists a non-negative integer N such that

$$|x_k^{(n)} - x_k^{(m)}|^{\alpha_k} \leq g_0(x^{(n)} - x^{(m)}) < \varepsilon \text{ for all } n, m \geq N \text{ and for all } k,$$

and so

$$|x_k^{(n)} - x_k|^{\alpha_k} = \lim_{m\to\infty} |x_k^{(n)} - x_k^{(m)}|^{\alpha_k} \leq \varepsilon \text{ for all } n \geq N \text{ and all } k,$$

that is,

$$g_0(x^{(n)} - x) \leq \varepsilon \text{ for all } n \geq N. \tag{1.11}$$

Since $x^{(N)} \in c_0(p)$, there exists a non-negative integer k_0 such that $|x_k^{(N)}|^{p_k} < \varepsilon$ for all $k \geq k_0$, hence

$$|x_k|^{\alpha_k} \leq \left|x_k^{(N)} - x_k\right|^{\alpha_k} + \left|x_k^{(N)}\right|^{\alpha_k} \leq \varepsilon^{1/M} + \varepsilon^{1/M} \leq 2\varepsilon^{1/M}$$

and so $|x_k|^{p_k} \leq 2^{1/M}\varepsilon \leq 2\varepsilon$ for all $k \geq k_0$. This and (1.11) together imply $x^{(n)} \to x$ $(n \to \infty)$ in $c_0(p)$. Thus, we have shown that $c_0(p)$ is complete.

(b.4) Finally, we show that g_0 is a paranorm for $\ell_\infty(p)$ if and only if the condition (1.6) is satisfied.

If $0 < m \le p_k \le M$ for all k, then obviously $\ell_\infty(p) = \ell_\infty$, and ℓ_∞ is a Banach space.
Conversely, if $m = 0$, then there is a subsequence $(p_{k(j)})_{j=0}^\infty$ of the sequence $(p_k)_{k=0}^\infty$ such that $p_{k(j)} < 1/(j+1)$ for $j = 0, 1, \dots$. We define the sequence $x = (x_k)_{k=0}^\infty$ by

$$x_k = \begin{cases} 1 & (k = k(j)) \\ 0 & (k \neq k(j)) \end{cases} \quad (j = 0, 1, \dots).$$

Then obviously $x \in \ell_\infty(p)$. Let $\lambda_n = 1/n$ for all n. Then $\lim_{n\to\infty} \lambda_n = 0$, but

$$g_0(\lambda_n x) = \sup_k |\lambda_n x_k|^{p_k} \ge \lim_{j\to\infty} \left(\frac{1}{n}\right)^{1/(j+1)} = 1 \text{ for all } n \in \mathbb{N},$$

hence $g_0(\lambda_n x) \not\to 0$ $(n \to \infty)$. Thus, g_0 cannot be a paranorm for $\ell_\infty(p)$.

This completes the proof. □

1.2 FK and BK Spaces

Here we give a short overview concerning the theory of FK and BK spaces which is the most powerful tool in the characterization of matrix transformations between sequence spaces. The fundamental result of this section is Theorem 1.4 which states that matrix maps between FK spaces are continuous. Most of the results in this section can be found in [21, 23, 31, 33]. We also refer the reader to [14, 34].

We provide the proofs of Theorems 1.2–1.7, and Corollary 1.2 for the reader who may not be too familiar with the theory of BK spaces.

We start with a slightly more general definition.

Definition 1.2 Let H be a linear space and a Hausdorff space. An *FH space* is a Fréchet space X such that X is a subspace of H and the topology of X is stronger than the restriction of the topology of H on X.

If $H = \omega$ with its topology given by the metric d_ω of (1.5) in Corollary 1.1, then an FH space is called an *FK space*.

A *BH space* or a *BK space* is an FH or FK space which is a Banach space.

Remark 1.1 (a) If X is an FH space, then the inclusion map $\iota : X \to H$ with $\iota(x) = x$ for all $x \in X$ is continuous. Therefore, X is continuously embedded in H.

(b) Since convergence in (ω, d_ω) and coordinatewise convergence are equivalent by Corollary 1.1, convergence in an FK space implies coordinatewise convergence.

(c) The letters *F*, *H*, *K* and *B* stand for *F*réchet, *H*ausdorff, *K*oordinate, the German word for coordinate, and *B*anach.

(d) Some authors include local convexity in the definition of an FH space. Since most of the theory presented here can be developed without local convexity, we follow Wilansky [33] and do not include it in our definition. If local convexity, however, is needed then it will explicitly be mentioned.

Example 1.2 (a) Trivially ω is an FK space with its natural metric of (1.5) in Corollary 1.1.

(b) If $p = (p_k)_{k=0}^{\infty}$ is a bounded sequence of positive reals, then $\ell(p)$ and $c_0(p)$ are FK spaces with their metrics d given by the their paranorms, since they are Fréchet subspaces of ω with their metrics stronger than the natural metric of ω on them by Example 1.1.

(c) We consider the spaces ℓ_∞, c and c_0 of all bounded, convergent and null sequences, and

$$\ell_p = \left\{ x \in \omega : \sum_{k=0}^{\infty} |x_k|^p < \infty \right\} \text{ for } 1 \leq p < \infty.$$

It is well known that ℓ_∞, c and c_0 are BK spaces with

$$\|x\|_\infty = \sup_k |x_k|;$$

since $|x_k| \leq \|x\|_\infty$ for all x and all k, those spaces are BK spaces.
Since ℓ_p obviously is the special case of $\ell(p)$ with the sequence $p = p \cdot e$, and

$$\|x\|_p = \left(\sum_{k=0}^{\infty} |x_k|^p \right)^{1/p}$$

is a norm on ℓ_p for $1 \leq p < \infty$, ℓ_p is a BK space by Example 1.1.
We refer to spaces in this part as the *classical sequence spaces*.

Remark 1.2 We have seen in Example 1.1 (b) that g_0 is only a paranorm for $\ell_\infty(p)$ when the condition in (1.6) holds, in which case $\ell_\infty(p) = \ell_\infty$. Grosse–Erdmann [9] determined a linear topology for $\ell_\infty(p)$. He showed in [9, Theorem 2 (ii)] that $\ell_\infty(p)$ is an IBK space, that is, it can be written as the union of an increasing sequence of BK spaces, and is endowed with the inductive limit topology.

The following results are fundamental. Their proofs use well-known theorems from functional analysis, which are included in Appendix A.2 for the reader's convenience.

Theorem 1.2 *Let X be a Fréchet space, Y be an FH space, $\mathcal{T}_Y$, $\mathcal{T}_H|_Y$ denote FH topologies on Y and of H on Y, and $f : X \to Y$ be linear. Then $f : X \to (Y, \mathcal{T}_H|_Y)$ is continuous, if and only if $f : X \to (Y, \mathcal{T}_Y)$ is continuous.*

Proof (i) First, we assume that $f : X \to (Y, \mathcal{T}_Y)$ is continuous. Since Y is an FH space, we have $\mathcal{T}_H|_Y \subset \mathcal{T}_Y$, and so $f : X \to (Y, \mathcal{T}_H|_Y)$ is continuous.

(ii) Conversely, we assume that $f : X \to (Y, \mathcal{T}_H|_Y)$ is continuous. Then it has closed graph by Theorem A.1 in Appendix A.2. Since Y is an FH space, we again have $\mathcal{T}_H|_Y \subset \mathcal{T}_Y$, and so $f : X \to (Y, \mathcal{T}_Y)$ has closed graph. Consequently, $f : X \to (Y, \mathcal{T}_Y)$ is continuous by the closed graph theorem, Theorem A.2 in Appendix A.2. □

We obtain as an immediate consequence of Theorem 1.2.

Corollary 1.2 *Let X be a Fréchet space, Y be an FK space, $f : X \to Y$ be linear, and the* coordinates *$P_n : X \to \mathbb{C}$ for $n = 0, 1, \ldots$ be defined by $P_n(x) = x_n$ for all $x \in X$. If $P_n \circ f : X \to \mathbb{C}$ is continuous for every n, then $f : X \to Y$ is continuous.*

Proof Since convergence and coordinatewise convergence are equivalent in ω by Corollary 1.1, the continuity of $P_n \circ f : X \to \mathbb{C}$ for all n implies the continuity of $f : X \to \omega$, and so $f : X \to Y$ is continuous by Theorem 1.2. □

By ϕ, we denote the set of all finite sequences. Thus, $x = (x_k)_{k=0}^\infty \in \phi$ if and only if there is an integer k such that $x_j = 0$ for all $j > k$.

Let X be a Fréchet spaces. Then we denote by X' the set of all continuous linear functionals on X; X' is called the *continuous dual of X*.

Theorem 1.3 *Let $X \supset \phi$ be an FK space. If the series $\sum_{k=0}^\infty a_k x_k$ converge for all $x \in X$, then the linear functional f_a defined by $f_a(x) = \sum_{k=0}^\infty a_k x_k$ for all $x \in X$ is continuous.*

Proof We define the functionals $f_a^{[n]}$ for all $n \in \mathbb{N}_0$ by $f_a^{[n]}(x) = \sum_{k=0}^n a_k x_k$ for all $x \in X$. Since X is an FK space and $f_a^{[n]} = \sum_{k=0}^n a_k P_k$ is a finite linear combination of the continuous coordinates P_k ($k = 0, 1, \ldots$), we have $f_a^{[n]} \in X'$ for all n. By hypothesis, the limits $f_a(x) = \lim_{n\to\infty} f_a^{[n]}(x)$ exist for all $x \in X$, hence $f_a \in X'$ by the Banach–Steinhaus theorem, Theorem A.3. □

The next result is one of the most important ones in the theory of matrix transformations between sequence spaces. We need the following notations. Let X and Y be subsets of ω, $A = (a_{nk})_{n,k=0}^\infty$ be an infinite matrix of complex numbers and $x \in \omega$. Then, we write

$$A_n x = \sum_{k=0}^\infty a_{nk} x_k \text{ for } n = 0, 1, \ldots \text{ and } Ax = (A_n x)_{n=0}^\infty$$

provided all the series converge; Ax is called the *A transform of the sequence x*. We write (X, Y) for the class of all infinite matrices that map X into Y, that is, for which the series $A_n x$ converge for all n and all $x \in X$, and $Ax \in Y$ for all $x \in X$.

Theorem 1.4 *Any matrix map between FK spaces is continuous.*

Proof Let X and Y be FK spaces, $A \in (X, Y)$ and $L_A : X \to Y$ be defined by $L_A(x) = Ax$ for all $x \in X$. Since the maps $P_n \circ L_A : X \to \mathbb{C}$ are continuous for all n by Theorem 1.3, $L_A : X \to Y$ is continuous by Corollary 1.2. □

It turns out as a consequence of Theorem 1.2 that the FH topology of an FH space is unique, more precisely, we have the following.

Theorem 1.5 *Let $(X, \mathcal{T}_X)$ and $(Y, \mathcal{T}_Y)$ be FH spaces with $X \subset Y$, and $\mathcal{T}_Y|_X$ denote the topology of Y on X. Then*

$$\mathcal{T}_X \supset \mathcal{T}_Y|_X\,; \tag{1.12}$$

$$\mathcal{T}_X = \mathcal{T}_Y|_X \text{ if and only if } X \text{ is a closed subspace of } Y. \tag{1.13}$$

In particular, the topology of an FH space is unique.

Proof (i) First we show the inclusion in (1.12).
Since X is an FH space, the inclusion map $\iota : (X, \mathcal{T}_X) \to (H, T_H)$ is continuous by Remark 1.1 (a). Therefore, $\iota : (X, \mathcal{T}_X) \to (Y, T_Y)$ is continuous by Theorem 1.2. Thus, the inclusion in (1.12) holds.

(ii) Now we show the identity in (1.13).
Let $\mathcal{T}$ and $\mathcal{T}'$ be FH topologies for an FH space. Then it follows by what we have just shown in Part (i) that $\mathcal{T} \subset \mathcal{T}' \subset \mathcal{T}$.

(α) If X is closed in Y, then X becomes an FH space with $\mathcal{T}_Y|_X$. It follows from the uniqueness that $\mathcal{T}_X = \mathcal{T}_Y|_X$.

(β) Conversely, if $\mathcal{T}_X = \mathcal{T}_Y|_X$, then X is a complete subspace of Y, and so closed in Y. □

The next results are also useful.

Theorem 1.6 *Let X, Y and Z be FH spaces with $X \subset Y \subset Z$. If X is closed in Z, then X is closed in Y.*

Proof Since X is closed in $(Y, \mathcal{T}_Z|_Y)$, it is closed in $(Y, \mathcal{T}_Y)$ by Theorem 1.5. □

Let Y be a topological space, and $E \subset Y$. Then we write $\mathrm{cl}_Y(E)$ for the *closure of E in Y.*

Theorem 1.7 *Let X and Y be FH spaces with $X \subset Y$, and E be a subset of X. Then, we have*

$$cl_Y(E) = cl_Y(\,cl_X(E)),\ in\ particular\ cl_X(E) \subset cl_Y(E).$$

Proof Since X is closed in $(Y, \mathcal{T}_Z|_Y)$, it is closed in $(Y, \mathcal{T}_Y)$ by Theorem 1.5. □

Example 1.3 (a) Since c_0 and c are closed in ℓ_∞, their BK topologies are the same; since ℓ_1 is not closed in ℓ_∞, its BK topology is strictly stronger than that of ℓ_∞ on ℓ_1 (Theorem 1.5).

(b) If c is not closed in an FK space X, then X must contain unbounded sequences (Theorem 1.6).

Now we recall the definition of the AD and AK properties.

Definition 1.3 Let $X \supset \phi$ be an FK space. Then X is said to have
(a) AD if $\mathrm{cl}_X(\phi) = X$;
(b) AK if every sequence $x = (x_k)_{k=0}^{\infty} \in X$ has a unique representation

$$x = \sum_{k=0}^{\infty} x_k e^{(k)},$$

that is, if every sequence x is the limit of its *m-sections*

$$x^{[m]} = \sum_{k=0}^{m} x_k e^{(k)}.$$

The letters A, D and K stand for *a*bschnitts*d*icht, the German word for sectionally dense, and Abschnitts*k*onvergenz, the German word for sectional convergence.

Example 1.4 (a) Every FK space with AK obviously has AD.
(b) An Example of an FK space with AD which does not have AK can be found in [33, Example 5.2.14].
(c) The spaces ω, $c_0(p)$, $\ell(p)$ for $p = (p_k)_{k=0}^{\infty} \in \ell_\infty$, in particular, c_0 and ℓ_p ($1 \leq p < \infty$) have AK.
(d) The space c does not have AK, since $e \in c$.

Now we recall the concept of a *Schauder basis*. We refer the reader to [18, 24] for further studies.

Definition 1.4 A *Schauder basis* of a linear metric space X is a sequence (b_n) of vectors such that for every vector $x \in X$ there is a unique sequence (λ_n) of scalars with

$$\sum_{n=0}^{\infty} \lambda_n b_n = x, \text{ that is, } \lim_{m \to \infty} \sum_{n=0}^{m} \lambda_n b_n = x.$$

For finite-dimensional spaces, the concepts of Schauder and algebraic bases coincide. In most cases of interest, however, the concepts differ. Every linear space has an algebraic basis, but there are linear spaces without a Schauder basis which we will soon see.

We recall that a metric space (X, d) is said to be *separable* if it has a *countable dense subset*; this means there is a countable set $A \subset X$ such that for all $x \in X$ and all $\varepsilon > 0$ there is an element $a \in A$ with $d(x, a) < \varepsilon$.

The next result is well known from elementary functional analysis.

Theorem 1.8 *Every complex linear metric space with a Schauder basis is separable.*

We close this section with two well-known important examples.

Example 1.5 The space ℓ_∞ has no Schauder basis, since it is not separable.

Example 1.6 We put $b^{(-1)} = e$ and $b^{(k)} = e^{(k)}$ for $k = 0, 1, \ldots$. Then the sequence $(b^{(n)})_{n=-1}^{\infty}$ is a Schauder basis of c; more precisely every sequence $x = (x_k)_{k=0}^{\infty} \in c$ has a unique representation

$$x = \xi\, e + \sum_{k=0}^{\infty} (x_k - \xi) e^{(k)} \text{ where } \xi = \xi(x) = \lim_{k \to \infty} x_k. \tag{1.14}$$

1.3 Matrix Transformations into the Classical Sequence Spaces

Now we apply the results of the previous section to characterize the classes (X, Y) where X is an arbitrary FK space and Y is any of the spaces ℓ_∞, c, c_0 and ℓ_1.

Let (X, d) be a metric space, $\delta > 0$ and $x_0 \in X$. Then, we write

$$\overline{B}_\delta(x_0) = \overline{B}_{X,\delta}(x_0) = \{x \in X : d(x, x_0) \leq \delta\}$$

for the *closed ball of radius* δ *with its centre in* x_0.

If $X \subset \omega$ is a linear metric space and $a \in \omega$, then we write

$$\|a\|_\delta^* = \|a\|_{X,\delta}^* = \sup_{x \in \overline{B}_\delta[0]} \left| \sum_{k=0}^{\infty} a_k x_k \right|, \tag{1.15}$$

provided the expression on the right hand exists and is finite which is the case whenever the series $\sum_{k=0}^{\infty} a_k x_k$ converge for all $x \in X$ (Theorem 1.3).

Let us recall that a subset S of a linear space X is said to be *absorbing* if, for every $x \in X$, there is $\varepsilon > 0$ such that $\lambda x \in S$ for all scalars λ with $|\lambda| \leq \varepsilon$.

The next statement is well known.

Remark 1.3 ([23, Remark 3.3.4 (j)]) Let (X, p) be a paranormed space. Then the open and closed neighbourhoods $N_r(0) = \{x \in X : p(x) < r\}$ and $\overline{N}_r(0) = \{x \in X : p(x) \leq r\}$ of 0 are absorbing for all $r > 0$.

The first result characterizes the class (X, ℓ_∞) for arbitrary FK spaces X.

Theorem 1.9 *Let X be an FK space. Then we have $A \in (X, \ell_\infty)$ if and only if*

$$\|A\|_\delta^* = \sup_n \|A_n\|_\delta^* < \infty \textit{ for some } \delta > 0. \tag{1.16}$$

Proof First, we assume that (1.16) is satisfied. Then the series $A_n x$ converge for all $x \in \overline{B}_\delta(0)$ and for all n, and $Ax \in \ell_\infty$ for all $x \in \overline{B}_\delta(0)$. Since the set $\overline{B}_\delta(0)$ is absorbing by Remark 1.3, we conclude that the series $A_n x$ converge for all n and all $x \in X$, and $Ax \in \ell_\infty$ for all $x \in X$.

Conversely, we assume $A \in (X, \ell_\infty)$. Then the map $L_A : X \to \ell_\infty$ defined by

$$L_A(x) = Ax \text{ for all } x \in X \tag{1.17}$$

is continuous by Theorem 1.4. Hence, there exist a neighbourhood N of 0 in X and a real $\delta > 0$ such that $\overline{B}_\delta(0) \subset N$ and $\|L_A(x)\|_\infty < 1$ for all $x \in N$. This implies (1.16). □

Let X and Y be Banach spaces. Then we use the standard notation $\mathcal{B}(X, Y)$ for the set of all bounded linear operators from X to Y. It is well known that $\mathcal{B}(X, Y)$ is a Banach space with the *operator norm* defined by

$$\|L\| = \sup\{\|L(x)\| : \|x\| = 1\} \text{ for all } L \in \mathcal{B}(X, Y);$$

we write X^* for X' with the norm defined by

$$\|f\| = \sup\{|f(x)| : \|x\| = 1\} \text{ for all } f \in X'.$$

Theorem 1.10 *Let X and Y be BK spaces.*

(a) Then $(X, Y) \subset \mathcal{B}(X, Y)$, that is, every $A \in (X, Y)$ defines an operator $L_A \in \mathcal{B}(X, Y)$ by (1.17).
(b) If X has AK then $\mathcal{B}(X, Y) \subset (X, Y)$, that is, for each $L \in \mathcal{B}(X, Y)$ there exists $A \in (X, Y)$ such that (1.17) *holds.*
(c) We have $A \in (X, \ell_\infty)$ if and only if

$$\|A\|_{(X,\ell_\infty)} = \sup_n \|A_n\|_X^* = \sup_n \left(\sup\{|A_n x| : \|x\| = 1\}\right) < \infty; \tag{1.18}$$

if $A \in (X, \ell_\infty)$, then

$$\|L_A\| = \|A\|_{(X,\ell_\infty)}. \tag{1.19}$$

Proof (a) This is Theorem 1.4.
(b) Let $L \in \mathcal{B}(X, Y)$ be given. We write $L_n = P_n \circ L$ for all n, and put $a_{nk} = L_n(e^{(k)})$ for all n and k. Let $x = (x_k)_{k=0}^\infty \in X$ be given. Since X has AK, we have $x = \sum_{k=0}^\infty x_k e^{(k)}$, and since Y is a BK space, it follows that $L_n \in X^*$ for all n. Hence, we obtain $L_n(x) = \sum_{k=0}^\infty x_k L_n(e^{(k)}) = \sum_{k=0}^\infty a_{nk} x_k = A_n x$ for all n, and so $L(x) = Ax$.
(c) This follows from Theorem 1.9 and the definition of $\|A\|_{(X,\ell_\infty)}$. □

In many cases, the characterizations of the classes (X, c) and (X, c_0) can easily be obtained from the characterization of the class (X, ℓ_∞).

Theorem 1.11 *Let X be an FK space with AD, and Y and Y_1 be FK spaces with Y_1 a closed subspace of Y. Then $A \in (X, Y_1)$ if and only if $A \in (X, Y)$ and $Ae^{(k)} \in Y_1$ for all k.*

Proof First, we assume $A \in (X, Y_1)$. Then $Y_1 \subset Y$ implies $A \in (X, Y)$, and $e^{(k)} \in X$ for all k implies $Ae^{(k)} \in Y_1$ for all k.

Conversely, we assume $A \in (X, Y)$ and $Ae^{(k)} \in Y_1$ for all k. We define the map $L_A : X \to Y$ by (1.17). Then $Ae^{(k)} \in Y_1$ implies $L_A(\phi) \subset Y_1$. By Theorem 1.4, L_A is continuous, hence $L_A(\text{cl}_X(\phi)) \subset \text{cl}_Y(L_A(\phi))$. Since Y_1 is closed in Y, and ϕ is dense in the AD space X, we have $L_A(X) = L_A(\text{cl}_X(\phi)) \subset \text{cl}_Y(L_A(\phi)) \subset \text{cl}_Y(Y_1) = \text{cl}_{Y_1}(Y_1) = Y_1$ by Theorem 1.5. □

The following result is sometimes referred to as *improvement of mapping*.

Theorem 1.12 (Improvement of mapping) *Let X be an FK space, $X_1 = X \oplus e = \{x_1 = x + \lambda e : x \in X,\ \lambda \in \mathbb{C}\}$, and Y be a linear subspace of ω. Then $A \in (X_1, Y)$ if and only if $A \in (X, Y)$ and $Ae \in Y$.*

Proof First, we assume $A \in (X_1, Y)$. Then $X \subset X_1$ implies $A \in (X, Y)$, and $e \in X_1$ implies $Ae \in Y$.

Conversely, we assume $A \in (X, Y)$ and $Ae \in Y$. Let $x_1 \in X_1$ be given. Then there are $x \in X$ and $\lambda \in \mathbb{C}$ such that $x_1 = x + \lambda e$, and it follows that $Ax_1 = A(x + \lambda e) = Ax + \lambda Ae \in Y$. □

We need Lemma [26] for the characterization of the class (X, ℓ_1). Although it is elementary, we provide its proof here.

Lemma 1.1 ([26]) *Let $a_0, a_1, \dots, a_n \in \mathbb{C}$. Then the following inequality holds:*

$$\sum_{k=0}^{n} |a_k| \le 4 \cdot \max_{N \subset \{0,\dots,n\}} \left| \sum_{k \in N} a_k \right|. \tag{1.20}$$

Proof First we consider the case when $a_0, a_1, \dots, a_n \in \mathbb{R}$.

We put $N^+ = \{k \in \{0, \dots, n\} : a_k \ge 0\}$ and $N^- = \{k \in \{0, \dots, n\} : a_k < 0\}$, and obtain

$$\sum_{k=0}^{n} |a_k| = \left| \sum_{k \in N^+} a_k \right| + \left| \sum_{k \in N^-} a_k \right| \le 2 \cdot \max_{N \subset \{0,\dots,n\}} \left| \sum_{k \in N} a_k \right|.$$

Now we assume $a_0, a_1, \dots, a_n \in \mathbb{C}$.

We write $a_k = \alpha_k + i\beta_k$ for $k = 0, 1, \dots, n$ and, for any subset N of $\{0, \dots, n\}$, we put

$$x_N = \sum_{k \in N} \alpha_k,\ y_N = \sum_{k \in N} \beta_k \text{ and } z_N = x_N + iy_N = \sum_{k \in N} a_k.$$

Now we choose subsets N_r, N_i and N_* of $\{0, \dots, n\}$ such that

$$\left|x_{N_r}\right| = \max_{N\subset\{0,\dots,n\}} |x_N|, \ \left|y_{N_i}\right| = \max_{N\subset\{0,\dots,n\}} |y_N| \text{ and } \left|z_{N_*}\right| = \max_{N\subset\{0,\dots,n\}} |z_N|.$$

Then, we have for all subsets N of $\{0, \dots, n\}$

$$|x_N|, |y_N| \le \left|z_{N_*}\right| \text{ and } \left|x_{N_r}\right| + \left|y_{N_i}\right| \le 2 \cdot \left|z_{N_*}\right|.$$

Finally, it follows by the first part of the proof that

$$\begin{aligned}\sum_{k=0}^{n} |a_k| &\le \sum_{k=0}^{n} |\alpha_k| + \sum_{k=0}^{n} |\beta_k| \le \left(\left|x_{N_r}\right| + \left|x_{N_i}\right|\right)\\ &\le 4 \cdot \left|z_{N_*}\right| = 4 \cdot \max_{N\subset\{0,\dots,n\}} \left|\sum_{k\in N} a_k\right|. \qquad \square\end{aligned}$$

Now we give the characterization of the class (X, ℓ_1) for arbitrary FK spaces. The proof is similar to that of Theorem 1.9.

Theorem 1.13 ([20, Satz 1]) *Let X be an FK space. Then $A \in (X, \ell_1)$ if and only if*

$$\|A\|_{(X,\ell_1)} = \sup_{\substack{N \subset \mathbb{N}_0 \\ N \text{ finite}}} \left\| \left(\sum_{n\in N} a_{nk}\right)_{k=0}^{\infty} \right\|_{X,\delta}^{*} < \infty \text{ for some } \delta > 0. \tag{1.21}$$

Proof (i) First we show the sufficiency of the condition in (1.21).
We assume that (1.21) is satisfied. Then the series $A_n x$ converge for all $x \in B_\delta[0]$ and for all n. Let $m \in \mathbb{N}_0$ be given. Then we have by Lemma 1.1

$$\begin{aligned}\sum_{n=0}^{m} |A_n x| &\le 4 \cdot \max_{N_m\subset\{0,\dots,m\}} \left|\sum_{n\in N_m}\sum_{k=0}^{\infty} a_{nk}x_k\right|\\ &= 4 \cdot \max_{N_m\subset\{0,\dots,m\}} \left|\sum_{k=0}^{\infty}\left(\sum_{n\in N_m} a_{nk}\right) x_k\right|\\ &\le 4 \cdot \max_{N_m\subset\{0,\dots,m\}} \left\|\left(\sum_{n\in N_m} a_{nk}\right)_{k=0}^{\infty}\right\|_{X,\delta}^{*}\\ &\le 4 \cdot \|A\|_{(X,\ell_1)} < \infty,\end{aligned}$$

for all $x \in \overline{B}_\delta(0)$. Since $m \in \mathbb{N}_0$ was arbitrary, $Ax \in \ell_1$ for all $x \in \overline{B}_\delta(0)$. Since the set $\overline{B}_\delta(0)$ is absorbing by Remark 1.3, the series $A_n x$ converge for all n and all $x \in X$, and $Ax \in \ell_1$ for all $x \in X$. Therefore, we have $A \in (X, \ell_1)$. Thus, we have shown the sufficiency of the condition in (1.21)

(ii) Now we show the necessity of the condition in (1.21).
We assume $A \in (X, \ell_1)$. Then the series $A_n x$ converge for all n and all $x \in X$, and L_A is continuous by Theorem 1.4. Hence there exist a neighbourhood U of 0 in X and a real $\delta > 0$ such that $\overline{B}_\delta(0) \subset U$ and $\|L_A(x)\|_1 < 1$ for all $x \in U$. Since

$$\left|\sum_{k=0}^{\infty}\left(\sum_{n\in N} a_{nk}\right)\right| = \left|\sum_{n\in N} A_n x\right| \leq \|A_n x\|_1 < \infty$$

for all finite subsets N of $\mathbb{N}_0$, we conclude that the condition in (1.21) holds. □

Remark 1.4 It follows from the proof of Theorem 1.13 that if $A \in (X, \ell_1)$, where X is an FK space, then

$$\|A\|_{(X,\ell_1)} \leq \|L_A\| \leq 4 \cdot \|A\|_{(X,\ell_1)}. \tag{1.22}$$

1.4 Multipliers and Dual Spaces

Here we consider the so-called *multipliers* and *β-duals*, the latter of which are of greater interest than the continuous duals in the theory of matrix transformations. They naturally arise in the characterizations of matrix transformations in connection with the convergence of the series $A_n x$.

We also consider α-, γ-, *functional* and *continuous duals* of sequence spaces and some relations between them.

Finally, we recall an important result, Theorem 1.22, that connects the properties of a matrix A with those of its transpose A^T. This theorem has several applications in the characterizations of matrix transformations considered in Sect. 1.5.

Let cs and bs denote the sets of all convergent and bounded series.

The β-duals of sequence spaces are special cases of *multiplier spaces*.

Definition 1.5 Let X and Y be subsets of ω. The set

$$M(X, Y) = \left\{a \in \omega : a \cdot x = (a_k x_k)_{k=0}^{\infty} \in Y \text{ for all } \in X\right\}$$

is called the *multiplier space of* X *in* Y.

Special cases are $X^\alpha = M(X, \ell_1)$, $X^\beta = M(X, cs)$ and $X^\gamma = M(X, bs)$, the α-, β- and γ-duals of X.

The following simple results are fundamental, and will frequently be used in the sequel.

Proposition 1.1 *Let X, $\tilde{X}$, Y and $\tilde{Y}$ be subsets of ω, and $\{X_\delta : \delta \in I\}$, where I is an indexing set, be a collection of subsets of ω. Then, we have*

(i) $Y \subset \tilde{Y}$ *implies* $M(X,Y) \subset M(X,\tilde{Y})$;
(ii) $X \subset \tilde{X}$ *implies* $M(\tilde{X},Y) \subset M(X,Y)$;
(iii) $X \subset M(M(X,Y),\ Y)$;
(iv) $M(X,Y) = M\left(M(M(X,Y),\ Y),\ Y\right)$;
(v) $M\left(\bigcup_{\delta\in I} X_\delta, Y\right) = \bigcap_{\delta\in I} M(X_\delta, Y)$.

The following result is an immediate consequence of Proposition 1.1.

Corollary 1.3 ([23, Corollary 9.4.3]) *Let* X *and* $\tilde{X}$ *be subsets of* ω, $\{X_\delta : \delta \in I\}$, *where* I *is an indexing set, be a collection of subsets of* ω, *and* $\dagger$ *denote any of the symbols* α, β *or* γ. *Then, we have*

(i) $X^\alpha \subset X^\beta \subset X^\gamma$; *(ii)* $X \subset \tilde{X}$ *implies* $\tilde{X}^\dagger \subset X^\dagger$;
(iii) $X \subset X^{\dagger\dagger} = (X^\dagger)^\dagger$; *(iv)* $X^\dagger = X^{\dagger\dagger\dagger}$;

$$(v)\ \left(\bigcup_{\delta\in I} X_\delta\right)^\dagger = \bigcap_{\delta\in I} X_\delta^\dagger.$$

Example 1.7 ([13, Lemma 1]) We have

$$(i) M(c_0, c) = \ell_\infty; \quad (ii) M(c, c) = c; \quad (iii) M(\ell_\infty, c) = c_0.$$

Proof (i) If $a \in \ell_\infty$, then $a \cdot x \in c$ for all $x \in c_0$, and so $\ell_\infty \subset M(c_0, c)$.
Conversely, we assume $a \notin \ell_\infty$. Then there is a subsequence $(a_{k(j)})_{j=0}^\infty$ of the sequence a such that $|a_{k(j)}| > j+1$ for all $j = 0, 1, \ldots$. We define the sequence x by

$$x_k = \begin{cases} \dfrac{(-1)^j}{a_{k(j)}} & (k = k(j)) \\ 0 & (k \neq k(j)) \end{cases} \quad (j = 0, 1, \ldots). \tag{1.23}$$

Then, we have $x \in c_0$ and $a_{k(j)} x_{k(j)} = (-1)^j$ for all $j = 0, 1, \ldots$, hence $ax \notin c$.
This shows $M(c_0, c) \subset \ell_\infty$.
(ii) If $a \in c$, then $a \cdot x \in c$ for all $x \in c$, and so $c \subset M(c, c)$.
Conversely, if $a \in M(c, c)$, then $a \cdot x \in c$ for all $x \in c$, in particular, for $x = e \in c$, $a \cdot e = a \in c$, and so $M(c, c) \subset c$.
(iii) If $a \in c_0$, then $a \cdot x \in c$ for all $x \in \ell_\infty$, and so $c_0 \subset M(\ell_\infty, c_0)$.
Conversely, we assume $a \notin c_0$. Then there are a real $b > 0$ and a subsequence $(a_{k(j)})_{j=0}^\infty$ of the sequence a such that $|a_{k(j)}| > b$ for all $j = 0, 1, \ldots$. We define the sequence x as in (1.23). Then, we have $x \in \ell_\infty$ and $a_{k(j)} x_{k(j)} = (-1)^j$ for all $j = 0, 1, \ldots$, hence $a \notin M(\ell_\infty, c)$. This shows $M(\ell_\infty, c) = c_0$. □

Example 1.8 Let $\dagger$ denote any of the symbols α, β or γ. Then we have $\omega^\dagger = \phi$, $\phi^\dagger = \omega$, $c_0^\dagger = c^\dagger = \ell_\infty^\dagger = \ell_1$, $\ell_1^\dagger = \ell_\infty$ and $\ell_p^\dagger = \ell_q$ $(1 < p < \infty;\ q = p/(p-1))$.

Another dual space frequently arises in the theory of sequence spaces.

Definition 1.6 Let $X \supset \phi$ be an FK space. Then the set

$$X^f = \{(f(e^{(n)}))_{n=0}^{\infty} : f \in X'\}$$

is called the *functional dual of* X.

Many of the results in the remainder of this section and their proofs can also be found in [33, Sects. 7.2 and 8.3]. The following results, Theorems 1.14, 1.15, 1.19 and 1.20 and Examples 1.9, 1.10 and 1.13, concern the relations between the various duals; they are not absolutely essential, but interesting in themselves and included for completeness sake. Along with Theorem 1.21 the theorems in the previous sentence are needed to prove Theorem 1.22.

We start with a relation between the β- and continuous duals of FK spaces. The first part of the next result is Theorem 1.3.

Theorem 1.14 ([23, Theorem 9.5.2]) *Let $X \supset \phi$ be an FK space. Then $X^\beta \subset X'$; this means that there is a linear one-to-one map $T : X^\beta \to X'$. If X has AK then T is onto.*

Theorem 1.15 ([23, Theorem 9.5.5]) *(a) Let $X \supset \phi$ be an FK space. Then we have $X^f = (cl_X(\phi))^f$.*
(b) Let $X, Y \supset \phi$ be FK spaces. If $X \subset Y$, then $X^f \supset Y^f$. If X is closed in Y, then $X^f = Y^f$.

It might be expected from $X \subset X^{\dagger\dagger}$ in Part (iii) of Corollary 1.3 that X is contained in X^{ff}, but Example 1.9 will show that this is not the case, in general. We will, however, see in Theorem 1.20 that $X \subset X^{ff}$ for BK spaces with AD.

If X is a linear space, S is a subset of X and $y \in X \setminus S$, then we write

$$S \oplus y = \{x : x = s + \lambda y \text{ for some } s \in S \text{ and some scalar } \lambda\}.$$

Example 1.9 ([23, Example 9.5.6]) Let $X = c_0 \oplus z$ with z unbounded. Then X is a BK space, $X^f = \ell_1$ and $X^{ff} = \ell_\infty$, so $X \not\subset X^{ff}$.

Theorem 1.16 ([23, Theorem 9.5.7]) *Let $X \supset \phi$ be an FK space.*

(a) We have $X^\beta \subset X^\gamma \subset X^f$.
(b) If X has AK, then $X^\beta = X^f$.
(c) If X has AD, then $X^\beta = X^\gamma$.

A relation between the functional and continuous duals of an FK space is given by the next result.

Theorem 1.17 ([23, Theorem 9.5.8]) *Let $X \supset \phi$ be an FK space.*
(a) Then the map $q : X' \to X^f$ given by $q(f) = (f(e^{(k)}))_{k=0}^{\infty}$ is onto. Moreover, if $T : X^\beta \to X'$ denotes the map of Theorem *1.14, then $q(Ta) = a$ for all $a \in X^\beta$.*
(b) Then $X^f = X'$, that is, the map q of Part (a) is one-to-one, if and only if X has AD.

Example 1.10 ([23, Example 9.5.9]) We have $c^\beta = c^f = \ell_1$. The map T of Theorem 1.14 is not onto. We consider $\lim \in X'$. If there were $a \in X^f$ with $\lim a = \sum_{k=0}^\infty a_k x_k$, then it would follow that $a_k = \lim e^{(k)} = 0$, hence $\lim x = 0$ for all $x \in c$, contradicting $\lim e = 1$. Also then map q of Theorem 1.17 is not onto, since $q^\circ \lim = 0$.

It will turn out in Theorems 1.18 and 1.19 that the multiplier spaces and the functional duals of BK spaces are again BK spaces. These results *do not extend to FK spaces*, in general, as we will see in Example 1.12.

Theorem 1.18 ([23, Theorem 9.4.6]) *Let $X \supset \phi$ and $Y \supset \phi$ be BK spaces. Then $Z = M(X, Y)$ is a BK space with*

$$\|z\| = \sup_{x \in S_X} \|x \cdot z\| \text{ for all } z \in Z.$$

We obtain as an immediate consequence of Theorem 1.18.

Corollary 1.4 ([23, Corollary 9.4.7]) *The α-, β- and γ-duals of a BK space X are BK spaces with*

$$\|a\|_\alpha = \sup_{x \in S_X} \|a \cdot x\|_1 = \sup_{x \in S_X} \left(\sum_{k=0}^\infty |a_k x_k| \right) \text{ for all } a \in X^\alpha,$$

and

$$\|a\|_\beta = \sup_{x \in S_X} \|a\|_{bs} = \sup_{x \in S_X} \left(\sup_n \left| \sum_{k=0}^n a_k x_k \right| \right) \text{ for all } a \in X^\beta, X^\gamma.$$

Furthermore, X^β is a closed subspace of X^γ.

Example 1.11 ([23, Example 9.5.3 and (4.50) and (4.56) in Theorem 4.7.2 for c^*]) The continuous duals c_0^*, ℓ_1^* and ℓ_p^* for $1 < p < \infty$ are norm isomorphic with ℓ_1, ℓ_∞ and ℓ_q where $q = p/(p-1)$, respectively.

We also have $f \in c^*$ if and only if $f(x) = \chi(f) \cdot \xi + \sum_{k=0}^\infty a_k x_k$ where $\chi : c^* \to \mathbb{C}$ is defined by $\chi(f) = f(e) - \sum_{k=0}^\infty f(e^{(k)})$, $\xi = \lim_{k\to\infty} x_k$ and $a = (f(e^{(k)}))_{k=0}^\infty \in \ell_1$, and we have

$$\|f\| = |\chi(f)| + \|a\|_1. \tag{1.24}$$

Finally, we have $\|a\|_{\ell_\infty}^* = \|a\|_1$ for all $a \in \ell_\infty^\beta$.

Theorem 1.18 fails to hold for FK spaces, in general.

Example 1.12 ([23, Example 9.4.8]) The space ω is an FK space, and $\omega^\alpha = \omega^\beta = \omega^\gamma = \phi$, but ϕ has no Fréchet metric.

Theorem 1.19 ([23, Theorem 9.5.10]) *Let $X \supset \phi$ be a BK space. Then X^f is a BK space.*

The next theorem gives a sufficient condition for $X \subset X^{ff}$ to hold.

Theorem 1.20 ([23, Theorem 9.5.11]) *Let $X \supset \phi$ be a BK space. Then $X^{ff} \supset cl_X(\phi)$. Hence, if X has AD, then $X \subset X^{ff}$.*

The condition that X has AD is not necessary for $X \subset X^{ff}$, in general.

Example 1.13 ([23, Example 9.5.12]) Let $X = c_0 \oplus z$ with $z \in \ell_\infty$. Then, we have $X^{ff} = \ell_1^f = \ell_\infty \supset X$, but X does not have AD.

The next results connect the properties of A with those of its transpose A^T. This is related to but different from the adjoint map. They are useful in the characterizations of some classes of matrix transformations between classical sequence spaces.

Theorem 1.21 ([23, Theorem 9.6.1]) *Let $X \supset \phi$ be an FK space and Y be any set of sequences. If $A \in (X, Y)$ then $A^T \in (Y^\beta, X^f)$. If X and Y are BK spaces and Y^β has AD then we have $A^T \in (Y^\beta, cl_{X^f}(X^\beta))$.*

Theorem 1.22 ([23, Theorem 9.6.3]) *Let X and Z be BK spaces with AK and $Y = Z^\beta$. Then we have $(X, Y) = (X^{\beta\beta}, Y)$; furthermore, $A \in (X, Y)$ if and only if $A^T \in (Z, X^\beta)$.*

1.5 Matrix Transformations Between the Classical Sequence Spaces

We apply the results of the previous sections to give necessary and sufficient conditions on the entries of a matrix A to be in any of the classes (X, Y) where X and Y are any of the classical sequence spaces ℓ_p $(1 \le p \le \infty)$, c_0 and c with the exceptions of (ℓ_p, ℓ_r) where both $p, r \neq 1, \infty$ (the characterizations are not known), and of (ℓ_∞, c) (Schur's theorem) and (ℓ_∞, c_0) [29, **21 (21.1)**] (no functional analytic proof seems to be known). The proof is purely analytical and uses the method of the *gliding hump*. A proof of Schur's theorem can be found, for instance, in [19] or [33].

The class (ℓ_2, ℓ_2) was characterized by Crone ([6] or [27, pp. 111–115]). We characterize the class (ℓ_2, ℓ_2) in Sect. 1.6.

Theorem 1.23 *Let $1 < p, r < \infty$, $q = p/(p-1)$ and $s = r/(r-1)$. Then the necessary and sufficient conditions for $A \in (X, Y)$ can be read from the following table:*

From / To	ℓ_∞	c_0	c	ℓ_1	ℓ_p
ℓ_∞	**1.**	**2.**	**3.**	**4.**	**5.**
c_0	**6.**	**7.**	**8.**	**9.**	**10.**
c	**11.**	**12.**	**13.**	**14.**	**15.**
ℓ_1	**16.**	**17.**	**18.**	**19.**	**20.**
ℓ_r	**21.**	**22.**	**23.**	**24.**	**unknown**

1., 2., 3. *(1.1) where*

(1.1) $\sup_n \sum_{k=0}^{\infty} |a_{nk}| < \infty$

4. *(4.1) where*

(4.1) $\sup_{n,k} |a_{nk}| < \infty$

5. *(5.1) where*

(5.1) $\sup_n \sum\limits_{k=0}^{\infty} |a_{nk}|^q < \infty$

6. *(6.1) where*

(6.1) $\lim_{n\to\infty} \sum_{k=0}^{\infty} |a_{nk}| = 0$

7. *(1.1), (7.1) where*

(7.1) $\lim_{n\to\infty} a_{nk} = 0$ *for every* k

8. *(1.1), (7.1), (8.1) where*

(8.1) $\lim_{n\to\infty} \sum_{k=0}^{\infty} a_{nk} = 0$

9. *(4.1), (7.1)*

10. *(5.1), (7.1)*

11. *(11.1), (11.2) where*

(11.1) $\sum_{k=0}^{\infty} |a_{nk}|$ *converges uniformly in* n

(11.2) $\lim_{n\to\infty} a_{nk} = \alpha_k$ *exists for every* k

12. *(1.1), (11.2)*

13. *(1.1), (11.2), (13.1) where*

(13.1) $\lim_{n\to\infty} \sum_{k=0}^{\infty} a_{nk} = \alpha$ *exists*

14. *(4.1), (11.2)*

15. *(5.1), (11.2)*

16., 17., 18. *(16.1) where*

(16.1) $\sup\limits_{\substack{N \subset \mathbb{N}_0 \\ N \text{ finite}}} \left(\sum_{k=0}^{\infty} \left|\sum_{n\in N} a_{nk}\right|\right) < \infty$

19. *(19.1) where*

(19.1) $\sup_k \sum_{n=0}^{\infty} |a_{nk}| < \infty$

20. *(20.1) where*

(20.1) $\sup\limits_{\substack{N \subset \mathbb{N}_0 \\ N \text{ finite}}} \left(\sum_{k=0}^{\infty} \left|\sum_{n\in N} a_{nk}\right|^q\right) < \infty$

21., 22., 23. *(21.1) where*

(21.1) $\sup\limits_{\substack{K \subset \mathbb{N}_0 \\ K \text{ finite}}} \sum_{n=0}^{\infty} \left|\sum_{k\in K} a_{nk}\right|^r < \infty$

24. *(24.1) where*

(24.1) $\sup_k \sum_{n=0}^{\infty} |a_{nk}|^r < \infty.$

Proof The condition (1.1) in **1.** and **2.**, and the conditions (4.1) and (5.1) in **4.** and **5.** follow from Part (c) in Theorem 1.10 and Example 1.11. Since $c_0 \subset c \subset \ell_\infty$, we also obtain the condition in (1.1) of **3.**.

Theorem 1.11 and the conditions in **2.**, **4.** and **5.** yield those in **7.** and **12.**, **9.** and **14.**, and in **10.** and **15.**. Now the conditions in **8.** and **13.** follow from those in **7.** and **8.** by Theorem 1.12.

The conditions in **6.** and **11.** are in [29, **21 (21.1)**] and [33, Theorem 1.18 (i)]]; no functional analytic proofs seem to be known in these cases.

The conditions (16.1) in **16.** and **17.**, and (20.1) in **20.** follow from Theorem 1.13 and Example 1.11. Since $c_0 \subset c \subset \ell_\infty$, we also obtain the condition (16.1) in **18.**. Furthermore, the condition (19.1) in **19.** follows from Theorem 1.22 with $Z = c_0$, hence $Y = c_0^\beta = \ell_1$, and **1.**.

The conditions (21.1) in **21.** and **22.**, and (24.1) in **24.** follow from Theorem 1.22 and those in **20.** and **1.**. Since $c_0 \subset c \subset \ell_\infty$, we also obtain the condition in (21.1) in **23.**. □

Remark 1.5 We would obtain from Theorem 1.13 and Example 1.11 that $A \in (\ell_1, \ell_1)$ if and only if

$$\sup_{\substack{N \subset \mathbb{N}_0 \\ N \text{ finite}}} \left(\sup_k \left| \sum_{n \in N} a_{nk} \right| \right) < \infty. \tag{1.25}$$

It is easy to see that conditions (16.1) of Theorem 1.23 and in (1.25) are equivalent.

Remark 1.6 The results of Theorem 1.23 can be found in [29, 33]; many references to the original proofs can be found in [29]. The characterizations of Parts **1.**, **2.** and **3.** of Theorem 1.23 are given in [33, Example 8.4.5A, p. 129] or [29, (1.1) in 1.], of **4.** in [33, Example 8.4.1A, p. 126] or [29, (6.1) in 6.], of **5.** [33, Example 8.4.5D, p. 129] or [29, (5.1) in 5.], of **6.** in [33, Theorem 1.7.19, p. 17] or [29, (21.1) in 21.], of **7.** in [33, Example 8.4.5A, p. 129] or [29, (1.1), (11.2) in 23.], of **8.** in [33, Example 8.4.5A, p. 129] or [29, (1.1), (11.2), (22.1) in 22.], of **9.** in [33, Example 8.4.1A, p. 126] or [29, (6.2), (11.2) in 28.], of **10.** in [33, Example 8.4.6D, p. 129] or [29, (5.1), (11.2) in 27.] of **11.** in [33, Theorem 1.17.8, p.15] or [29, (10.1), (10.4) in 10.], of **12.** in [33, Example 8.4.5A, p. 129] or [29, (1.1), (10.1) in 12], of **13.** in [33, Example 8.4.5A, p. 129 or Theorem 1.3.6, p. 6] or [29, (1.1), (10.1), (11.1) in 11.], of **14.** in [33, Example 8.4.1A, p. 126] or [29, (6.1), (10.1) in 17.], of **15.** in [33, Example 8.4.5D, p. 129] or [29, (5.1), (10.1) in 16.], of **16.**, **17.** and **18.** in [33, Example 8.4.9A, p. 130] or [29, (72.2) in 72.], of **19.** in [33, Example 8.4.1D, p. 126] or [29, (77.1) in 77.], of **20.** in [33, Example 8.4.1D, p. 126] or [29, (76.1) in 76.], of **21.**, **22.** and **23.** in [33, Example 8.4.8A, p. 131] or [29, (63.1) in 63.] and of **24.** in [33, Example 8.4.1D, p. 126] or [29, (68.1) in 68.].

The conditions for the class (ℓ_∞, c_0) in [33, Theorem 1.7.19, p. 17] are

$$\sum_{k=0}^{\infty} |a_{nk}| \text{ converges uniformly in } n \tag{i}$$

and (7.1) in **7.** of Theorem 1.23. Two more sets of equivalent conditions for the class (ℓ_∞, c) are given in [33, Theorem 1.17.8, p.15], they are

$$\begin{cases} \mathbf{11.}\,(11.2)\ \sum_{k=0}^{\infty}|a_{nk}|<\infty \text{ for all } n, \\ \sum_{k=0}^{\infty}|\alpha_k|<\infty \text{ with } \alpha_k \text{ from } \mathbf{11.}\,(11.2)\ \textit{and} \\ \lim_{n\to\infty}\sum_{k=0}^{\infty}|a_{nk}-\alpha_k|=0 \end{cases} \quad \text{(ii)}$$

or

$$\begin{cases} \mathbf{11.}\,(11.2)\ \textit{and} \\ \lim_{n\to\infty}\sum_{k=0}^{\infty}|a_{nk}|=\sum_{k=0}^{\infty}|\alpha_k|, \text{ both series being convergent.} \end{cases} \quad \text{(iii)}$$

Alternative equivalent conditions for the classes (ℓ_∞,ℓ_1), (c_0,ℓ_1) and (c,ℓ_1) are given in [33, Example 8.4.9A, p. 130], namely,

$$\sup_{\substack{K\subset\mathbb{N}_0\\ K \text{ finite}}}\sum_{n\to\infty}\left|\sum_{k\in K}a_{nk}\right|<\infty \quad \text{(iv)}$$

and in [29, 72.], namely,

$$\sup_{\substack{N,K\subset\mathbb{N}_0\\ N,K \text{ finite}}}\left|\sum_{n\in N}\sum_{k\in K}a_{nk}\right|<\infty, \quad (72.1)$$

or (72.3), which is (iv), or

$$\sum_{n=0}^{\infty}\left|\sum_{k\in K_*}a_{nk}\right| \text{ converges for all } K_*\subset\mathbb{N}_0. \quad (72.4)$$

If we apply Theorem 1.22 with $X=Z=c_0$ and $Y=c_0^\beta=\ell_1$ then we have $A\in(c_0,\ell_1)$ if and only if $A^T\in(c_0,\ell_1)$, that is, the conditions are symmetric in n and k, and (iv) follows in the same way as **16.** (16.1). It can easily be shown that the other conditions given here are equivalent.

An alternative condition for the class (ℓ_1,ℓ_p) is given in [33, Example 8.4.1D, p. 126], namely,

$$\sup_k\sum_{n=0}^{\infty}|a_{nk}|^p<\infty, \quad \text{(v)}$$

which can be obtained by applying Theorem 1.22 with $X=\ell_1$, $Z=\ell_q$ and $Y=\ell_p$ and then using **5** (5.1) with A and p replaced by A^T and q.

Finally, an alternative condition for the classes (c_0, ℓ_r), (c, ℓ_r) and (ℓ_∞, ℓ_r) is given in [29, 63.], namely,

$$\sum_{n=0}^{\infty} \left| \sum_{k \in K_*} a_{nk} \right| \text{ converges for all } K_* \subset \mathbb{N}_0. \tag{63.1}$$

1.6 Crone's Theorem

In this section, we prove *Crone's theorem* which characterizes the class (ℓ_2, ℓ_2). The conditions are substantially different from those for the characterizations of the classes in the previous section.

We will need some concepts and results from the theory of Hilbert spaces.

We recall that if $A = (a_{nk})$ is a (finite or infinite) matrix of complex entries, then $A^* = (a^*_{nk})$ denotes its conjugate complex transpose, that is,

$$a^*_{nk} = \bar{a}_{kn} \text{ for all } n \text{ and } k.$$

An *eigenvalue* of an $n \times n$ matrix A is a number λ for which

$$Ax = \lambda x \text{ for some vector } x \neq 0;$$

the vector x is called *eigenvector of* A. It is known that every $n \times n$ matrix A has at least one eigenvalue.

Although the next result is well known, we include its proof for the reader's convenience.

Lemma 1.2 *The norm of an $n \times n$ matrix A as a map from n-dimensional Euclidean space into itself is equal to the square root of the largest eigenvalue of A^*A.*

Proof (i) First we show that each eigenvalue of A^*A is non-negative.

Let λ be an eigenvalue of A^*A. Then there exists a vector $x \neq 0$ such that $(A^*A)x = \lambda x$. Since $x \neq 0$ it follows that

$$\lambda = \frac{1}{\|x\|_2^2} \cdot x^* \lambda x = \frac{1}{\|x\|_2^2} \left(x^* A^* A x\right) = \frac{\|Ax\|^2}{\|x\|^2} \geq 0.$$

(ii) Now we prove the conclusion of the lemma.

It is a well-known result from linear algebra that since A^*A is Hermitian, there is a matrix C such that $C^{-1} = C^*$ and

$$C^{-1} A^* A C = D,$$

where D is a diagonal matrix with the eigenvalues λ_k of A^*A on its diagonal. It follows that

$$\|A\|^2 = \sup_{x \neq 0} \frac{x^*A^*Ax}{\|x\|^2} = \sup_{x \neq 0} \frac{x^*CC^{-1}A^*ACC^{-1}x}{\|x\|^2} = \sup_{x \neq 0} \frac{(C^*x)^*D(C^*x)}{\|x\|^2}$$

$$= \sup_{y \neq 0} \frac{y^*Dy}{\|y\|^2} = \sup_{y \neq 0} \left(\frac{\sum_{k=1}^{n} \lambda_k |y_k|^2}{\sum_{k=1}^{n} |y_k|^2} \right) \leq \lambda_{max}, \tag{1.26}$$

where we put $y = C^*x$ and λ_{max} is the largest eigenvalue of A^*A.
Let x_{max} be an eigenvector of λ_{max}. Then, we have

$$x_{max}^* A^* A x_{max} = \lambda_{max} \|x_{max}\|^2,$$

hence

$$\lambda_{max} \leq \|A\|^2. \tag{1.27}$$

It follows from (1.26) and (1.27) that $\|A\|^2 = \lambda_{max}$.
Finally, since $\lambda_{max} \geq 0$ by Part (i) of the proof, we conclude

$$\|A\| = \sqrt{\lambda_{max}}.$$

□

In the sequel, the norm of a matrix will always mean its norm as a map in Euclidean or Hilbert space.

If $A = (a_{nk})_{n,k=0}^{\infty}$ is an infinite matrix and $m \in \mathbb{N}_0$, then $A^{[m]} = (a_{nk}^{[m]})_{n,k=0}^{\infty}$ denotes the matrix with

$$a_{nk}^{[m]} = \begin{cases} a_{nk} & (0 \leq n, k \leq m) \\ 0 & (n, k \geq m+1). \end{cases}$$

Remark 1.7 One may regard $A^{[m]}$ either as a map from ℓ_2 into ℓ_2, or from $m+1$-dimensional space into itself. The norm in either case is the same.

Proof Let $x = (x_k)_{k=0}^{\infty} \in \ell_2$ be given, and $y = (x_0, x_1, \ldots, x_m)$. We obtain for the m-section $x^{[m]} = \sum_{k=0}^{n} x_k e^{(k)}$ of the sequence x

$$\left(x^{[m]}\right)^* \left(A^{[m]}\right)^* A^{[m]} x^{[m]} = \sum_{n=0}^{\infty} \left(\left(A^{[m]}x^{[m]}\right)^*\right)_n \left(A^{[m]}x^{[m]}\right)_n$$

$$= \sum_{n=0}^{m} \left[\left(\sum_{j=0}^{m} \bar{x}_k \left(a_{jn}^{[m]}\right)^* \right) \left(\sum_{k=0}^{m} a_{nk}^{[m]} \bar{x}_k \right) \right]$$

$$= \sum_{n=0}^{m} \left[\left(\sum_{j=0}^{m} \bar{x}_j \left(a_{jn}^{[m]}\right)^* \right) \left(\sum_{k=0}^{m} a_{nk}^{[m]} x_k \right) \right]$$

$$= \sum_{n=0}^{m} \left(A^{[m]}x\right)_n^* \left(A^{[m]}x\right)_n$$

$$= x^* \left(A^{[m]}\right)^* A^{[m]}x.$$

We also obtain for the $(m+1) \times (m+1)$ matrix $B = (b_{nk})_{n,k=0}^{m}$ with $b_{nk} = a_{nk}$ $(n, k = 0, 1, \ldots, m)$

$$y^* B^* B y = x^* \left(A^{[m]}\right)^* A^{[m]}x, \text{ hence } \|B\| = \|A^{[m]}\|. \qquad \square$$

We need some lemmas.

We recall that, since ℓ_2 is a BK space with AK, by Theorem 1.10 (a) and (b), $\mathcal{B}(\ell_2, \ell_2)$ can be identified with the class (ℓ_2, ℓ_2), in particular, if $A \in (\ell_2, \ell_2)$, then we write $L_A \in \mathcal{B}(\ell_2, \ell_2)$ for the operator with $L_A(x) = Ax$ for all $x \in \ell_2$, and $\|A\| = \|L_A\|$.

Lemma 1.3 *We have $A \in (\ell_2, \ell_2)$ if and only if $\sup_m \|L_{A^{[m]}}\| < \infty$; in this case,*

$$\|L_A\| = \sup_m \|L_{A^{[m]}}\| \text{ and } \lim_{m\to\infty} A^{[m]}x = Ax.$$

Proof (i) First, we show that $A \in (\ell_2, \ell_2)$ implies

$$\sup_m \left\|A^{[m]}\right\| < \infty. \tag{1.28}$$

Let $A \in (\ell_2, \ell_2)$. Then there exists a constant M such that

$$\|Ax\|_2 = \|L_A(x)\|_2 \le M\|x\|_2 \text{ for all } x \in \ell_2.$$

Let $x \in \bar{B}_{\ell_2} = \{x \in \ell_2 : \|x\|_2 \le 1\}$ and $m \in \mathbb{N}_0$ be given. Then we have $x^{[m]} \in \bar{B}_{\ell_2}$ and

$$\left\|A^{[m]}x\right\|_2^2 = \sum_{n=0}^{\infty} |A_n^{[m]}x|^2 \le \sum_{n=0}^{\infty} |A_n x^{[m]}|^2 \le \|A\|^2 \cdot \|x^{[m]}\|_2^2 \le M^2,$$

hence

$$\left\|A^{[m]}x\right\|_2 \le M.$$

This implies

$$\left\|A^{[m]}\right\| = \sup_{x\in\bar{B}_{\ell_2}} \left\|A^{[m]}x\right\|_2 \le M \text{ for all } m,$$

and (1.28) is an immediate consequence.

(ii) Now we show that the condition in (1.28) implies $A \in (\ell_2, \ell_2)$.
We assume that the condition in (1.28) is satisfied, that is, $\sup_m \|A^{[m]}\| = M < \infty$. Let $\varepsilon > 0$ and $x \in \ell_2$ be given, and $x^{[j]}$ for $j \in \mathbb{N}_0$ be the j section of the sequence x. We choose an integer k_0 such that

$$\left\|x^{[k_2]} - x^{[k_1]}\right\|_2 < \frac{\varepsilon}{M+1} \text{ for all } k_2 > k_1 > j. \tag{1.29}$$

Let n be an arbitrary integer. We choose $k_0 \geq n$. Then it follows from (1.29) that

$$\begin{aligned}\left|\sum_{k=k_1+1}^{k_2} a_{nk}x_k\right| &= \left|A_n^{[k_2]}x - A_n^{[k_1]}x\right| = \left|A_n^{[k_2]}\left(x^{[k_2]} - x^{[k_1]}\right)\right| \\ &\leq \left\|A^{[k_2]}\right\| \cdot \left\|x^{[k_2]} - x^{[k_1]}\right\|_2 < M \cdot \frac{\varepsilon}{M+1} < \varepsilon\end{aligned}$$

for all $k_2 > k_1 > k_0$.
Thus, $A_n x$ exists for all n so that $A_n = (a_{nk})_{k=0}^{\infty} \in \ell_2^{\beta} = \ell_2$ for all n.
Since $e^{(k)} \in \bar{B}_{\ell_2}$ for $k = 0, 1, \ldots$, it follows that

$$\left(\sum_{n=0}^{m} |a_{nk}|^2\right)^{1/2} = \left(\sum_{n=0}^{\infty} \left|A_n^{[m]}e^{(k)}\right|^2\right)^{1/2} = \left\|A^{[m]}e^{(k)}\right\|_2 \leq \left\|A^{[m]}\right\| \leq M$$

for all $m \geq k$, hence

$$\left(\sum_{n=0}^{\infty} |a_{nk}|^2\right)^{1/2} \leq M,$$

and so also $A^k = (a_{nk})_{n=0}^{\infty} \in \ell_2$ for the columns of the matrix A. This implies

$$\left\|A^{[m]}e^{(k)} - Ae^{(k)}\right\|_2 = \left(\sum_{n=m+1}^{\infty} |a_{nk}|^2\right)^{1/2} \to 0 \ (m \to \infty),$$

that is,

$$\lim_{m\to\infty} A^{[m]}e^{(k)} = Ae^{(k)} \text{ for } k = 0, 1, \ldots. \tag{1.30}$$

Let j and m be arbitrary integers. Then we have for $j > m$

$$\left(\sum_{n=0}^{m}\left|\sum_{k=0}^{j} a_{nk}x_k\right|^2\right)^{1/2} \leq \left\|A^{[j]}\right\| \cdot \|x\|_2 \leq M \cdot \|x\|_2.$$

It follows from $A_n \in \ell_2^{\beta}$ for $n = 0, 1, \ldots$ that, for all m,

$$\left(\sum_{n=0}^{m}|A_n x|^2\right)^{1/2} = \lim_{\ell\to\infty}\left(\sum_{n=0}^{m}\left|\sum_{k=0}^{j}a_{nk}x_k\right|^2\right)^{1/2} \leq M\cdot\|x\|_2,$$

and so

$$\|Ax\|_2 = \left(\sum_{n=0}^{\infty}|A_n x|^2\right)^{1/2} \leq M\cdot\|x\|_2, \tag{1.31}$$

thus $Ax \in \ell_2$.

This completes Part (ii) of the proof.

(iii) Now we show that $\sup_m \|A^{[m]}\| < \infty$ implies

$$\lim_{m\to\infty} A^{[m]}x = Ax \text{ for all } x \in \ell_2. \tag{1.32}$$

We assume that $\sup_m \|A^{[m]}\| < \infty$. Then $A \in (\ell_2, \ell_2)$ by Part (ii) of the proof, and so $L_A \in \mathcal{B}(\ell_2, \ell_2)$. Let $x \in \phi$, that is, $x = \sum_{k=0}^{j} x_k e^{(k)}$ for some $j \in \mathbb{N}_0$. It follows from (1.30) that

$$\begin{aligned}\lim_{m\to\infty} A^{[m]}x &= \lim_{m\to\infty}\left(\sum_{k=0}^{j} x_k A^{[m]}e^{(k)}\right) = \sum_{k=0}^{j} x_k \lim_{m\to\infty} A^{[m]}e^{(k)}\\ &= \sum_{k=0}^{j} x_k A e^{(k)} = Ax.\end{aligned}$$

Since ℓ_2 has AK, ϕ is dense in ℓ_2, and since L_A is continuous on ℓ_2, we obtain (1.32).

(iv) Finally, we show that $\sup_m \|A^{[m]}\| < \infty$ implies

$$\|A\| = \sup_m \left\|A^{[m]}\right\|. \tag{1.33}$$

We assume $\sup_m \|A^{[m]}\| < \infty$. First, it follows from Part (ii) of the proof that $A \in (\ell_2, \ell_2)$, and so we have for all $x \in \ell_2$ and all $m \in \mathbb{N}_0$

$$\left\|A^{[m]}x\right\|_2 = \left\|A^{[m]}x^{[m]}\right\|_2 \leq \left\|Ax^{[m]}\right\|_2 \leq \|A\|\cdot\|x\|_2,$$

hence

$$\sup_m \left\|A^{[m]}\right\| \leq \|A\|.$$

By (1.31), we also have

$$\|A\| \leq \sup_m \left\|A^{[m]}\right\| \leq \|A\|,$$

and so (1.33) follows.

This concludes the proof of the lemma. □

Lemma 1.4 *We have $A \in (\ell_2, \ell_2)$ if and only if A^*A exists and*

$$\sup_m \left\| \left(A^*A\right)^{[m]} \right\| < \infty. \tag{1.34}$$

Proof (i) First, we assume $A \in (\ell_2, \ell_2)$. Then we $A^* \in (\ell_2, \ell_2)$ for the adjoint matrix, so A^*A exists and $A^*A \in (\ell_2, \ell_2)$. Now (1.34) follows by Lemma 1.3.

(ii) Conversely, we assume that (1.34) is satisfied.

Let $x \in \phi$ with $\|x\|_2 \leq 1$, that is, $x = \sum_{k=0}^{m} x_k e^{(k)}$ for some $m \in \mathbb{N}_0$. Then, we have

$$\begin{aligned}
\|Ax\|_2^2 &= \left(x^{[m]}\right)^* A^*Ax^{[m]} = \sum_{j=0}^{\infty} \left(\left(x^{[m]}\right)^* A^*\right)_j A_j x^{[m]} \\
&= \sum_{j=0}^{\infty} \left(\sum_{k=0}^{m} \bar{x}_k a_{kj}^*\right) \left(\sum_{n=0}^{m} a_{jn} x_n\right) = \sum_{k,n=0}^{m} \left(\sum_{j=0}^{\infty} a_{kj}^* a_{jn}\right) \bar{x}_k x_n \\
&= \sum_{k,n=0}^{m} \bar{x}_k \left(\left(A^*A\right)^{[m]}\right)_{kn} x_n = x^* \left(A^*A\right)^{[m]} x \\
&\leq \|x^*\|_2 \cdot \left\| \left(A^*A\right)^{[m]} x \right\|_2 \leq \left\| \left(A^*A\right)^{[m]} \right\| \cdot \|x\|_2 \\
&\leq \left\| \left(A^*A\right)^{[m]} \right\| \leq \sup_m \left\| \left(A^*A\right)^{[m]} \right\| < \infty.
\end{aligned}$$

Thus, A defines a bounded linear operator $L : (\phi, \|\cdot\|_2) \to \ell_2$, where $L(x) = Ax$ for all $x \in \phi$. Since ϕ is dense in ℓ_2, it is easy to see that L can be extended to an operator form all of ℓ_2 into ℓ_2. □

Now we establish the characterization of the class (ℓ_2, ℓ_2).

Theorem 1.24 (Crone) ([6]) *We have $A \in (\ell_2, \ell_2)$ if and only if*

$$\left(A^*A\right)^m \text{ exists for all } m = 0, 1, \ldots \tag{1.35}$$

and

$$C = \sup_m \left[\sup_n \left(\left(\left(A^*A\right)^m \right)_{nn} \right)^{1/2m} \right] < \infty; \tag{1.36}$$

in this case, $\|A\| = C$.

Proof (i) First, we show the necessity of the conditions in (1.35) and (1.36).

Let $A \in (\ell_2, \ell_2)$. This implies $A^* \in (\ell_2, \ell_2)$ and so $(A^*A)^m$ exists for each m and maps ℓ_2 into itself. We obtain for all $m, k = 0, 1, \ldots$

$$\left((A^*A)^m\right)_{kk} = \left(e^{(k)}\right)^* \left(A^*A\right)^m e^{(k)},$$

for if we put $B = (A^*A)^m$, then we have

$$\left(e^{(k)}\right)^* Be^{(k)} = \sum_{j=0}^{\infty} \left(e^{(k)}\right)_j^* B_j e^{(k)} = B_k e^{(k)} = b_{kk}.$$

It follows that

$$\begin{aligned}\left|\left((A^*A)^m\right)_{kk}\right| &= \left|\left(e^{(k)}\right)^* \left(A^*A\right)^m e^{(k)*}\right| \le \left\|\left(e^{(k)}\right)^*\right\|_2 \cdot \left\|\left(A^*A\right)^m\right\| \cdot \left\|e^{(k)}\right\|_2 \\ &\le \left\|A^*A\right\|^m \le \left\|A^*\right\|^m \cdot \|A\|^m = \|A\|^{2m}.\end{aligned}$$

Since

$$\left(A^*A\right)_{kk} = \sum_{j=0}^{\infty} a^*{}_{kj^*} a_{jk} = \sum_{j=0}^{\infty} \bar{a}_{jk} a_{jk} = \sum_{j=0}^{\infty} |a_{jk}|^2 \ge 0 \text{ for } k = 0, 1, \dots,$$

it follows that

$$\sup_k \left[\left((A^*A)^m\right)_{kk}\right]^{1/2m} = \sup_k \left|\left((A^*A)^m\right)_{kk}\right|^{1/2m} \le \|A\| \text{ for } m = 0, 1, \dots$$

and

$$C = \sup_m \left[\sup_k \left[\left((A^*A)^m\right)_{kk}\right]^{1/2m}\right] \le \|A\|.$$

(ii) Now we show the sufficiency of the conditions in (1.35) and (1.36).
It is easy to see that

$$\left(\left(A^*A\right)^2\right)^{[m]} - \left(\left(A^*A\right)^{[m]}\right)^2 \tag{1.37}$$

is a Hermitian matrix with non-negative entries on its diagonal and it is known from linear algebra that the matrix in (1.37) is positive and semi-definite. Thus, we have

$$x^* \left[\left(\left(A^*A\right)^2\right)^{[m]} - \left(\left(A^*A\right)^{[m]}\right)^2\right] \ge 0 \text{ for all } x \in \ell_2$$

so that for $x \in \bar{B}_{\ell_2}$

$$x^* \left(\left(A^*A\right)^{[m]}\right)^2 x \le x^* \left(\left(A^*A\right)^2\right)^{[m]} x \le \left\|\left(\left(A^*A\right)^2\right)^{[m]}\right\|$$

and

$$\left\|\left(\left(A^*A\right)^{[m]}\right)^2\right\| \le \left\|\left(\left(A^*A\right)^2\right)^{[m]}\right\|.$$

Furthermore

$$x^* \left((A^*A)^{[m]} \right)^2 x = x^* \left((A^*A)^{[m]} \right)^* (A^*A)^{[m]} x$$

implies

$$\left\| \left((A^*A)^{[m]} \right)^2 \right\| = \left\| (A^*A)^{[m]} \right\|^2 \leq \left\| \left((A^*A)^2 \right)^{[m]} \right\|.$$

It follows by mathematical induction that

$$\left\| (A^*A)^{[m]} \right\|^{2^j} \leq \left\| \left((A^*A)^{2^j} \right)^{[m]} \right\| \text{ for } j = 1, 2, \ldots. \tag{1.38}$$

The sum of the diagonal entries of an $m \times m$ Hermitian matrix is equal to the sum of the eigenvalues of the matrix. Thus, for an $m \times m$ Hermitian matrix B (for which the entries on the diagonal are non-negative), we have

$$\|B\| = \max\{\lambda : \lambda \text{ is an eigenvalue of } B\} \leq m \cdot \max_k b_{kk}.$$

We conclude from this and (1.38)

$$\left\| (A^*A)^{[m]} \right\| \leq \left\| \left((A^*A)^{2^j} \right)^{[m]} \right\|^{2^{-j}} \leq (m+1)^{2^{-j}} \cdot \left[\max_k \left(\left((A^*A)^{2^j} \right)^{[m]} \right)_{kk} \right]^{2^{-j}}$$
$$(m+1)^{2^{-j}} \cdot C^2 \text{ for } j = 0, 1, \ldots.$$

Since this holds for all j, it follows that

$$\left\| (A^*A)^{[m]} \right\| \leq C^2 \text{ for } m = 0, 1, \ldots$$

and so $A \in (\ell_2, \ell_2)$ by Lemma 1.4. □

1.7 Remarks on Measures of Noncompactness

The known characterizations of matrix transformations between the classical sequence spaces were given in Theorem 1.23. This was mainly achieved by applying the theory of FK and BK spaces; the fundamental result was Theorem 1.4 which states that matrix transformations between FK spaces are continuous.

The characterization of compact matrix operators can be achieved by applying the theory of *measures of noncompactness*, in particular, the *Hausdorff measure of noncompactness* and its properties. In the remainder of this section, we list the essential concepts and and results in this field.

First, we recall that a linear operator L from Banach space X to a Banach space Y is said to be *compact*, or *completely continuous*, if $D_L = X$ for the domain D_L of L and, for every bounded sequence $(x_n) \in X$, the sequence $(L(x_n))$ has a convergent subsequence in Y. We write $\mathcal{K}(X, Y)$ for the set of all compact operators from X to Y, and $\mathcal{K}(X) = \mathcal{K}(X, X)$, for short. It is clear that a compact operator is bounded, that is, $\mathcal{K}(X, Y) \subset \mathcal{B}(X, Y)$.

Measures of noncompactness are also very useful tools in functional analysis, for instance, in metric fixed point theory and the theory of operator equations in Banach spaces. They are also used in the studies of functional equations, ordinary and partial differential equations, fractional partial differential equations, integral and integro-differential equations, optimal control theory and in the characterizations of compact operators between Banach spaces.

In the next section, we will give an axiomatic introduction to measures of noncompactness of bounded sets in complete metric spaces, and establish some of their most import properties. In particular, we study the *Kuratowski*, *Hausdorff* and *separation measures of noncompactness*. Furthermore, we prove the famous *Goldenštein–Goh'berg–Markus* theorem which gives an estimate for the Hausdorff measure of noncompactness of bounded sets in Banach spaces with a Schauder basis.

The first measure of noncompactness, denoted by α, was defined and studied by Kuratowski [16] in 1930. In 1955, Darbo [7] used the function α to prove his fixed point theorem, which is a very important generalization of Schauder's fixed point theorem, and includes the existence part of Banach's fixed point theorem.

Other important measures of noncompactness are the *ball* or *Hausdorff measure of noncompactness* χ introduced by Goldenštein, Goh'berg and Markus [39] and later studied by Goldenštein and Markus [40] in 1968, and the *separation measure of noncompactness* β introduced by Istrăţescu [11] in 1972.

These measures of noncompactness are studied in detail and their use is discussed, for instance, in the monographs [2–4, 12, 17, 21–23, 30, 38].

1.8 The Axioms of Measures of Noncompactness

In this section, we introduce the *axioms of a measure of noncompactness* on the class $\mathcal{M}_X$ of *bounded subsets of a complete metric space* (X, d). It seems that the axiomatic approach is the best way of dealing with measures of noncompactness. It is possible to use several systems of axioms which are not necessarily equivalent.

However, the notion of a measure of noncompactness was originally introduced in metric spaces. So we are going to give our axiomatic definition in this class of spaces. This approach is convenient for our purposes, and can be found in [21, 23, 30].

In the books [2, 3], for instance, two different patterns are provided for the axiomatic introduction of measures of noncompactness in Banach spaces. We are going to work mainly with the Hausdorff measure of noncompactness in Banach spaces which satisfies the axioms of either axiomatic approach.

We recall that a set S in a topological space is said to be *relatively compact* or *pre-compact*, if its closure $\bar{S}$ is compact.

Definition 1.7 ([23, Definition 7.5.1]) Let (X, d) be a complete metric space. A set function $\phi : \mathcal{M}_X \to [0, \infty)$ is called a *measure of noncompactness on* $\mathcal{M}_X$, if it satisfies the following conditions for all $Q, Q_1, Q_2 \in \mathcal{M}_X$:

(MNC.1) $\phi(Q) = 0$ if and only if Q is relatively compact (regularity);

(MNC.2) $\phi(Q) = \phi(\overline{Q})$ (invariance under closure);

(MNC.3) $\phi(Q_1 \cup Q_2) = \max\{\phi(Q_1), \phi(Q_2)\}$ (semi-additivity).

The number $\phi(Q)$ is called the *measure of noncompactness of the set* Q.

The following properties can easily be deduced from the axioms in Definition 1.7. By $\mathcal{M}_X^c$, we denote the subclass of all closed sets in $\mathcal{M}_X$.

Proposition 1.2 ([23, Proposition 7.5.3]) *Let ϕ be a measure of noncompactness on a complete metric space (X, d). Then ϕ has the following properties :*

$$Q_1 \subset Q_2 \text{ implies } \phi(Q_1) \le \phi(Q_2) \text{ (monotonicity).} \tag{1.39}$$

$$\phi(Q_1 \cap Q_2) \le \min\{\phi(Q_1), \phi(Q_2)\} \text{ for all } Q_1, Q_2 \in \mathcal{M}_X. \tag{1.40}$$

$$\text{If } Q \text{ is finite then } \phi(Q) = 0 \text{ (non-singularity).} \tag{1.41}$$

$$\left\{\begin{array}{l} \text{Generalized Cantor's intersection property} \\ \text{If } (Q_n) \text{ is a decreasing sequence of nonempty sets in } \mathcal{M}_X^c \text{ and} \\ \lim_{n\to\infty} \phi(Q_n) = 0, \text{ then the intersection} \\ \qquad\qquad Q_\infty = \bigcap Q_n \neq \emptyset \\ \text{is compact.} \end{array}\right. \tag{1.42}$$

Remark 1.8 If X is a Banach space then a measure of noncompactness may have some additional properties related to the linear structure of a normed space such as *homogeneity*, *subadditivity*, *translation invariance* and *invariance under the passage to the convex hull* (cf. (8)–(11) in Part of Theorem 1.25).

1.9 The Kuratowski and Hausdorff Measures of Noncompactness

In this section, we reall the definitions and some properties of the *Kuratowski* and Hausdorff measures of noncompactness.

As usual, $\text{diam}(S) = \sup_{x,y \in S} d(x, y)$ is the diameter of a set S in a metric space (X, d).

Definition 1.8 Let (X, d) be a complete metric space.

(a) The function $\alpha : \mathcal{M}_X \to [0, \infty)$ with

$$\alpha(Q) = \inf \left\{ \varepsilon > 0 : Q \subset \bigcup_{k=1}^{n} S_k, \ S_k \subset X, \ \text{diam}(S_k) < \varepsilon \ (k = 1, 2, \ldots, n \in \mathbb{N}) \right\}$$

is called the *Kuratowski measure of noncompactness (KMNC)*, and the real number $\alpha(Q)$ is called the *Kuratowski measure of noncompactness of Q*.

(b) The function $\chi : \mathcal{M}_X \to [0, \infty)$ with

$$\chi(Q) = \inf \left\{ \varepsilon > 0 : Q \subset \bigcup_{k=1}^{n} B_{r_k}(x_k), \ x_k \in X, \ r_k < \varepsilon \ (k = 1, 2, \ldots, n \in \mathbb{N} \right\}$$

is called the *Hausdorff* or *ball measure of noncompactness*, and the real number $\chi(Q)$ is called the *Hausdorff* or *ball measure of noncompactness of Q*.

Remark 1.9 Therefore, $\alpha(Q)$ is the infimum of all positive real numbers ε such that Q can be covered by a finite number of sets of diameters less than ε, and $\chi(Q)$ is the infimum of all positive real numbers ε such that Q can be covered by a finite number of open balls of radius less than ε.

Similarly, $\chi(Q)$ is the infimum of all positive real numbers ε such that Q has a finite ε-net.

It is not supposed in the definition of $\chi(Q)$ that the centres of the balls which cover Q belong to Q.

If S is a subset of a linear space then

$$\text{co}(S) = \bigcap \{C : C \supset S \text{ convex } \}$$

denotes the convex hull of S.

Theorem 1.25 ([30, Proposition II.2.3 and Theorem II.2.4] or [21, Lemma 2.11 and Theorem 2.12]) *(a) Let (X, d) be a complete metric space and $Q, Q_1, Q_2 \in \mathcal{M}_X$ and $\psi = \alpha$ or $\psi = \chi$. Then, we have*

$$\psi(Q) = 0 \text{ if and only if } Q \text{ is relatively compact} \tag{1}$$

$$\psi(Q) = \psi(\overline{Q}) \tag{2}$$

$$\psi(Q_1 \cup Q_2) = \max\{\psi(Q_1), \psi(Q_2)\} \tag{3}$$

$$Q_1 \subset Q_2 \text{ implies } \psi(Q_1) \le \psi(Q_2) \tag{4}$$

$$\psi(Q_1 \cap Q_2) \le \min\{\psi(Q_1), \psi(Q_2)\} \tag{5}$$

$$\text{If } Q \text{ is finite then } \psi(Q) = 0 \tag{6}$$

$$\begin{cases} \text{Generalized Cantor's intersection property} \\ \text{If } (Q_n) \text{ is a decreasing sequence of nonempty sets in } \mathcal{M}_X^c \\ \text{and } \lim_{n\to\infty} \psi(Q_n) = 0\text{, then the intersection} \\ \quad Q_\infty = \bigcap_n Q_n \neq \emptyset \\ \text{is compact.} \end{cases} \tag{7}$$

(b) If X is a Banach space, then we have for all $Q, Q_1, Q_2 \in \mathcal{M}_X$

$$\psi(Q_1 + Q_2) \le \psi(Q_1) + \psi(Q_2) \tag{8}$$

$$\psi(x + Q) = \psi(Q) \text{ for all } x \in X \tag{9}$$

$$\psi(\lambda Q) = |\lambda|\psi(Q) \text{ for each scalar } \lambda \tag{10}$$

$$\begin{cases} \text{invariance under the passage to the convex hull} \\ \qquad \psi(co(Q)) = \psi(Q). \end{cases} \tag{11}$$

Remark 1.10 The conditions in (1), (2) and (3) show that the Kuratowski and Hausdorff measures of noncompactness are a measure of noncompactness in the sense of Definition 1.7, the conditions in (4), (5), (6) and (7) are analogous to those of Proposition 1.2, and the conditions in (8), (9), (10) and (11) are analogous to those mentioned in Remark 1.8.

The following result is well known.

Theorem 1.26 *Let X be an infinite-dimensional Banach space and $\overline{B_X}$ denote the closed unit ball in X. Then, we have*

$$\alpha(\overline{B_X}) = 2 \text{ and } \chi(\overline{B_X}) = 1.$$

Proof We write $\bar{B} = \overline{B_X}$, for short.

(i) First, we show $\alpha(\bar{B}) = 2$.
We clearly have $\alpha(\bar{B}) \le 2$.
If $\alpha(\bar{B}) < 2$, then, by Definition 1.8 (b), there exist bounded, closed subsets Q_k of X with $\operatorname{diam}(Q_k) < 2$ for $k = 1, 2, \dots, n$ such that $\bar{B} \subset \bigcup_{k=1}^n Q_k$. Let $X_n = \{x_1, x_2, \dots, x_n\}$ be a linearly independent subset of X, and E_n be the set of all real linear combinations of elements of X_n. Clearly E_n is a real

n-dimensional normed space, where the norm on E_n is, of course, the restriction of the norm of X on E_n. Let $S_n = \{x \in E_n : \|x\| = 1\}$ denote the unit sphere on E_n. We note that $S_n \subset \bigcup_{k=1}^{n}(S_n \cap Q_k)$, $\operatorname{diam}(S_n \cap Q_k) < 2$ and $S_n \cap Q_k$ is a closed subset of E_n for each $k = 1, 2, \ldots, n$. This is a contradiction to the well-known *Ljusternik–Šnirleman–Borsuk theorem* [8, pp. 303–307], which states that if S_n is the unit sphere of an n-dimensional real normed space E_n, F_k is a closed subset of E_n for each $k = 1, 2, \ldots, n$ and $S_n \subset \bigcup_{k=1}^{n} F_k$, then there exists a $k_0 \in \{1, 2, \ldots, n\}$ such that the set $S_n \cap F_{k_0}$ contains a pair of antipodal points, that is, there exists $x_0 \in S_n \cap F_{k_0}$ such that $\{x_0, -x_0\} \subset S_n \cap F_{k_0}$.

(ii) Now we show $\chi(\bar{B}) = 1$.
We clearly have $\chi(\bar{B}) \leq 1$.
If $\chi(\bar{B}) = q < 1$, then we choose $\varepsilon > 0$ such that $q + \varepsilon < 1$. By Definition 1.8 (b), there exists a $(q + \varepsilon)$-net of $\bar{B}$, say $\{x_1, x_2, \ldots, x_n\}$. Hence

$$\bar{B} \subset \bigcup_{k=1}^{n}(x_k + (q + \varepsilon)\bar{B}). \tag{1.43}$$

Now it follows from Theorem 1.25, (4), (3) and (9) that

$$q = \chi(\bar{B}) \leq \max_{1 \leq k \leq n} \chi\left((x_k + (q + \varepsilon)\bar{B})\right) = (q + \varepsilon)q.$$

Since $q + \varepsilon < 1$, we have $q = 0$ by (1.43), that is, $\bar{B}$ is a totally bounded set. But this is impossible since X is is an infinite-dimensional space.
Thus, we have $\chi(\bar{B}) = 1$. □

Now we are going to prove the famous *Goldenštein–Goh'berg–Markus theorem* which gives an estimate for the Hausdorff measure of noncompactness of bounded sets in Banach spaces with Schauder bases.

Theorem 1.27 (Goldenštein, Goh'berg, Markus) ([39]) *Let X be a Banach space with a Schauder basis (b_n), $\mathcal{P}_n : X \to X$ for $n \in \mathbb{N}$ be the projector onto the linear span of $\{b_1, b_2, \ldots, b_n\}$, that is,*

$$\mathcal{P}_n(x) = \sum_{k=1}^{n} \lambda_k b_k \text{ for all } x = \sum_{k=1}^{\infty} \lambda_k b_k \in X,$$

and $\mathcal{R}_n = I - \mathcal{P}_n$. Furthermore, let the function $\mu : \mathcal{M}_X \to [0, \infty)$ be defined by

$$\mu(Q) = \limsup_{n \to \infty} \left(\sup_{x \in Q} \|\mathcal{R}_n(x)\| \right). \tag{1.44}$$

Then the following inequalities hold for all $Q \in \mathcal{M}_X$:

$$\frac{1}{a} \cdot \mu(Q) \leq \chi(Q) \leq \inf_n \left(\sup_{x \in Q} \|\mathcal{R}_n(x)\| \right) \leq \mu(Q), \tag{1.45}$$

where $a = \limsup_{n\to\infty} \|\mathcal{R}_n\|$ denotes the basis constant of the Schauder basis.

Proof We obviously have for any $n \in \mathbb{N}$

$$Q \subset \mathcal{P}_n(Q) + \mathcal{R}_n(Q) \text{ for all } Q \in \mathcal{M}_X. \tag{1.46}$$

It follows from Theorem 1.25 (4) and (8)

$$\chi(Q) \leq \chi\left(\mathcal{P}_n(Q)\right) + \chi\left(\mathcal{R}_n(Q)\right) \leq \sup_{x \in Q} \|\mathcal{R}_n(x)\|.$$

Now we obtain

$$\chi(Q) \leq \inf_n \left(\sup_{x \in Q} \|\mathcal{R}_n(x)\| \right) \leq \mu(Q).$$

Hence, it suffices to show the first inequality in (1.45).

Let $\varepsilon > 0$ be given and $\{y_1, y_2, \ldots, y_n\}$ be a $(\chi(Q) + \varepsilon)$-net of Q. Then it follows that

$$Q \subset \{y_1, y_2, \ldots, y_n\} + (\chi(Q) + \varepsilon)\,\bar{B},$$

where $\bar{B}$ denotes the closed unit ball in X. This implies that, for any $x \in Q$, there exist $y \in \{y_1, y_2, \ldots, y_n\}$ and $z \in \bar{B}$ such that $x = y + (\chi(Q) + \varepsilon)z$, and so

$$\sup_{x \in Q} \|\mathcal{R}_n(x)\| \leq \sup_{1 \leq k \leq n} \|\mathcal{R}_n(y_k)\| + (\chi(Q) + \varepsilon)\,\|\mathcal{R}_n\|,$$

hence

$$\limsup_{n\to\infty} \left(\sup_{x \in Q} \|\mathcal{R}_n(x)\| \right) \leq (\chi(Q) + \varepsilon) \limsup_{n\to\infty} \|\mathcal{R}_n\|.$$

Since $\varepsilon > 0$ was arbitrary, this yields the first inequality in (1.45). □

Remark 1.11 ([23, Remark 7.9.4]) If we write in Theorem 1.27

$$\tilde{\mu}(Q) = \inf_n \left(\sup_{x \in Q} \|\mathcal{R}_n(x)\| \right),$$

then we obtain from the inequalities in (1.45)

$$\frac{1}{a}\tilde{\mu}(Q) \leq \chi(Q) \leq \tilde{\mu}(Q) \text{ for all } Q \in \mathcal{M}_X. \tag{1.47}$$

Remark 1.12 ([23, Remark 7.9.5]) The measure of noncompactness μ in Theorem 1.27 is equivalent to the Hausdorff measure of noncompactness by (1.45).

If $a = 1$ then we obtain from (1.45)

$$\tilde{\mu}(Q) = \mu(Q) = \chi(Q) \text{ for all } Q \in \mathcal{M}_X.$$

If $X = \ell_p$ $(1 \leq p < \infty)$ or $X = c, c_0$, then the limit in (1.44) exists.

Example 1.14 (a) If $X = c_0$ or $X = \ell_p$ for $1 \leq p < \infty$ with the standard Schauder basis $(e^{(n)})_{n=1}^{\infty}$ and the natural norms, then $a = 1$ and we have

$$\chi(Q) = \begin{cases} \lim_{n\to\infty}\left(\sup_{x\in Q}\left(\sup_{k\geq n+1}|x_k|\right)\right) & (Q \in \mathcal{M}_{c_0}) \\ \lim_{n\to\infty}\left(\sup_{x\in Q}\left(\sum_{k=n+1}^{\infty}|x_k|^p\right)^{1/p}\right) & (Q \in \mathcal{M}_{\ell_p}). \end{cases}$$

(b) If $X = c$, then $a = 2$ and we have

$$\frac{1}{2}\cdot\mu(Q) = \frac{1}{2}\cdot\lim_{n\to\infty}\left(\sup_{x\in Q}\|\mathcal{R}_n(x)\|_\infty\right) \leq \chi(Q) \leq \mu(Q) \text{ for every } Q \in \mathcal{M}_X, \tag{1.48}$$

where for each $x \in c$ with $\xi_x = \lim_{k\to\infty} x_k$

$$\|\mathcal{R}_n(x)\|_\infty = \sup_{k\geq n+1}|x_k - \xi_x|.$$

We close this section with defining one more measure of noncompactness which is also equivalent to the Hausdorff measure of noncompactness.

Theorem 1.28 ([23, Theorem 7.9.8]) *Let X be a Banach space with a Schauder basis. Then the function $\nu : \mathcal{M}_X \to [0, \infty)$ with*

$$\nu(Q) = \liminf_{n\to\infty}\left(\sup_{x\in Q}\|\mathcal{R}_n(x)\|\right) \text{ for all } Q \in \mathcal{M}_X$$

is a measure of noncompactness on X which is invariant under the passage to the convex hull. Moreover, the following inequalities hold:

$$\frac{1}{a}\cdot\nu(Q) \leq \chi(Q) \leq \nu(Q) \text{ for all } Q \in \mathcal{M}_X.$$

The properties of ν now follow from those of μ, or can be proved similarly. □

Remark 1.13 ([23, Remark 7.9.9]) It is obvious from Theorem 1.28 that $\nu = \mu$ if

$$\lim_{n\to\infty}\left(\sup_{x\in Q}\|\mathcal{R}_n(x)\|\right)$$

exists for all $Q \in \mathcal{M}_X$, for instance, for $X = c_0$ or $X = \ell_p$ $(1 \le p < \infty)$ with the standard basis by Part (a) of Example 1.14.

When $X = c$ with the supremum norm and the standard basis, then we have $\nu \neq \mu$ by Part (b) of Example 1.14.

1.10 Measures of Noncompactness of Operators

In this section, we recall the definition of *measures of noncompactness of operators* between Banach spaces and list some important properties of the Hausdorff measure of noncompactness of such operators.

Definition 1.9 Let ϕ and ψ be measures of noncompactness on the Banach spaces X and Y, respectively.
(a) An operator $L : X \to Y$ is said to be (ϕ, ψ)-*bounded*, if

$$L(Q) \in \mathcal{M}_Y \text{ for all } Q \in \mathcal{M}_X, \tag{1.49}$$

and if there is a non-negative real number c such that

$$\psi(L(Q)) \le c \cdot \phi(Q) \text{ for all } Q \in \mathcal{M}_X. \tag{1.50}$$

(b) If an operator L is (ϕ, ψ)-bounded, then the number

$$\|L\|_{(\phi,\psi)} = \inf\{c \ge 0 : (1.50) \text{ holds}\} \tag{1.51}$$

is called the (ϕ, ψ)-*operator norm of* L, or (ϕ, ψ)-*measure of noncompactness of* L.

If $\psi = \phi$, we write $\|L\|_\phi = \|L\|_{(\phi,\phi)}$, for short.

First, we recall the formula for the Hausdorff measure of noncompactness of a bounded linear operator between infinite-dimensional Banach spaces.

Theorem 1.29 *Let X and Y be infinite-dimensional Banach spaces and $L : X \to Y$ be a bounded linear operator. Then, we have*

$$\|L\|_\chi = \chi(L(S_X)) = \chi(L(\overline{B}_X)), \chi(L(B_X)) \tag{1.52}$$

where B_X and $\overline{B}_X$ and S_X denote the open and closed unit balls, and the unit sphere in X.

Proof We write $\bar{B} = \overline{B}_X$ and $S = S_X$ for the closed unit ball and the unit sphere in X.

Since co(S) $= \bar{B}$ and $L(\text{co}(S)) = \text{co}(L(S)))$, it follows from Theorem 1.25 (2) and (11) that

$$\chi(L(\bar{B})) = \chi(L(\text{co}(S))) = \chi(\text{co}(L(S))) = \chi(L(S)).$$

Hence, we have by (1.51) and Theorem 1.26

$$\chi(L(\bar{B})) \leq \|L\|_\chi \cdot \chi(\bar{B}) = \|L\|_\chi \cdot 1 = \|L\|_\chi.$$

Now, we show

$$\|L\|_\chi \leq \chi(L(\bar{B})). \tag{1.53}$$

Let $Q \in \mathcal{M}_X$ and $\{x_1, x_2, \ldots, x_n\}$ be a finite r-net of Q. Then $Q \subset \bigcup_{k=1}^n B_r(x_k)$ and obviously

$$L(Q) \subset \bigcup_{k=1}^n L\left(B_r(x_k)\right).$$

It follows from this Theorem 1.25 (4), (3), the additivity of L, (8), (6), (10)

$$\begin{aligned}
\chi(L(Q)) &\leq \chi\left(\bigcup_{k=1}^n L(B_r(x_k))\right) \leq \max_{1\leq k\leq n} \chi\left(L(B_r(x_k))\right) \\
&= \max_{1\leq k\leq n} \chi\left(L(\{x_k\} + B_r(0))\right) = \max_{1\leq k\leq n} \chi\left(L(\{x_k\}) + L\left(B_r(0)\right)\right) \\
&\leq \max_{1\leq k\leq n} \chi\left(L(\{x_k\})\right) + \left(L\left(B_r(0)\right)\right) = \chi\left(L\left(B_r(0)\right)\right) \\
&= \chi\left(L(rB_1(0))\right) = \chi(rL(B)) \\
&= r\chi(L(B)).
\end{aligned}$$

This implies

$$\chi(L(Q)) \leq \chi(Q) \cdot \chi(L(Q)),$$

and so (1.53). □

Finally, we need the following useful known results.

Theorem 1.30 *Let X, Y and Z be Banach spaces, $L \in \mathcal{B}(X, Y)$ and $\tilde{L} \in \mathcal{B}(Y, Z)$. Then $\|L\|_\chi$ is a seminorm on $\mathcal{B}(X, Y)$ and*

$$\|L\|_\chi = 0 \text{ if and only if } L \in \mathcal{K}(X, Y); \tag{1.54}$$

$$\|L\|_\chi \leq \|L\|; \tag{1.55}$$

$$\|\tilde{L} \circ L\|_\chi \leq \|\tilde{L}\|_\chi \cdot \|L\|_\chi. \tag{1.56}$$

Proof (i) The proof that $\|\cdot\|_\chi$ is a seminorm is elementary.

(ii) The statement in (1.54) follows from the observation that an operator $L : X \to Y$ is compact if and only if it is continuous and maps bounded sets into relatively compact sets.

(iii) Now we show (1.55).
We have by (1.52) in Theorem 1.29

$$\|L\|_\chi = \chi(L(S_X)) \leq \sup_{x \in S_X} \|L(x)\| = \|L\|.$$

(iv) Finally, we show the inequality in (1.56).
We have for all $Q \in \mathcal{M}_X$ by Definition 1.9

$$\chi\left((\tilde{L} \circ L)(Q)\right) = \chi\left(\tilde{L}(L(Q))\right) \leq \|\tilde{L}\|_\chi \cdot \chi((L(Q))) \leq \|\tilde{L}\|_\chi \cdot \|L\|_\chi \cdot \chi(Q),$$

and the inequality in (1.56) is an immediate consequence. □

References

1. Aasma, A., Dutta, H., Natarajan, P.N.: An Introductory Course in Summability Theory. Wiley, New York (2017)
2. Akhmerov, R.R., Kamenskii, M.I., Potapov, A.S., Rodkina, A.E., Sadovskii, B.N.: Measures of Noncompactness and Condensing Operators. Operator Theory. Advances and Applications, vol. 55. Springer, Basel (1992)
3. Banaś, J., Goebel, K.: Measures of Noncompactness in Banach Spaces. Lecture Notes in Pure and Applied Mathematics, vol. 60. Marcel Dekker Inc., New York (1980)
4. Banaś, J., Mursaleen, M.: Sequence Spaces and Measures of Noncompactness with Applications to Differential and Integral Equations. Springer, New Delhi (2014)
5. Boos, J.: Classical and Modern Methods in Summability. Oxford University Press, Oxford (2000)
6. Crone, L.: A characterization of matrix mappings on ℓ^2. Math. Z. **123**, 315–317 (1971)
7. Darbo, G.: Punti uniti in transformazioni a condominio non compatto. Rend. Sem. Math. Univ. Padova **24**, 84–92 (1955)
8. Furi, M., Vignoli, A.: On a property of the unit sphere in a linear normed space. Bull. Acad. Pol. Sci. Sér. Sci. Math. Astron. Phys. **18**(6), 333–334 (1970)
9. Grosse-Erdmann, K.-G.: The structure of sequence spaces of Maddox. Can. J. Math. **44**(2), 298–307 (1992)
10. Hadžić, O.: Fixed point theory in topological vector spaces. University of Novi Sad, Institute of Mathematics, Novi Sad, Serbia (1984)
11. Istrăţescu, V.: On a measure of noncompactness. Bull. Math. Soc. Sci. Math. R. S. Roumanie (N.S.) **16**, 195–197 (1972)
12. Istrăţescu, V.: Fixed Point Theory, an Introduction. Reidel Publishing Company, Dordrecht (1981)
13. Jarrah, A.M., Malkowsky, E.: BK spaces, bases and linear operators. Rend Circ. Mat. Palermo, Ser. II, Suppl. **52**, 177–191 (1998)
14. Kamthan, P.K., Gupta, M.: Sequence Spaces and Series. Marcel Dekker, New York (1981)
15. Köthe, G.: Topologische lineare Räume I. Springer, Heidelberg (1966)
16. Kuratowski, K.: Sur les espaces complets. Fund. Math. **15**, 301–309 (1930)

17. Kuratowski, K.: Topologie. Warsaw (1958)
18. Lindenstrauss, J., Tzafriri, L.: Classical Banach Spaces I, Sequence Spaces. Springer, Heidelberg (1977)
19. Maddox, I.J.: Elements of Functional Analysis. Cambridge University Press, Cambridge (1971)
20. Malkowsky, E.: Klassen von Matrixabbildungen in paranormierten FK- Räumen. Analysis **7**, 275–292 (1987)
21. Malkowsky, E., Rakočević, V.: An Introduction into the Theory of Sequence Spaces and Measures of Noncompactness. Zbornik radova, Matematčki institut SANU, vol. 9(17), pp. 143–234. Mathematical Institute of SANU, Belgrade (2000)
22. Malkowsky, E., Rakočević, V.: On some results using measures of noncompactness. Advances in Nonlinear Analysis via the Concept of Measure of Noncompactness, pp. 127–180. Springer, Berlin (2017)
23. Malkowsky, E., Rakočević, V.: Advanced Functional Analysis. CRC Press, Taylor & Francis Group, Boca Raton (2019)
24. Marti, J.M.: Introduction to the Theory of Bases. Springer, Heidelberg (1967)
25. Mursaleen, M.: Applied Summability Methods. Springer, Cham (2014)
26. Peyerimhoff, A.: Über ein Lemma von Herrn Chow. J. Lond. Math. Soc. **32**, 33–36 (1957)
27. Ruckle, W.H.: Sequence Spaces. Research Notes in Mathematics, vol. 49. Pitman, Boston (1981)
28. Rudin, W.: Functional Analysis. McGraw Hill, New York (1973)
29. Stieglitz, M., Tietz, H.: Matrixtransformationen in Folgenräumen. Eine Ergebnisübersicht. Math. Z. **154**, 1–16 (1977)
30. Ayerbe Toledano, J.M., Dominguez Benavides, T., Lopez Acedo, G.: Measures of Noncompactness in Metric Fixed Point Theory. Operator Theory Advances and Applications, vol. 99. Birkhäuser Verlag, Basel (1997)
31. Wilansky, A.: Functional Analysis. Blaisdell Publishing Company, New York (1964)
32. Wilansky, A.: Modern Methods in Topological Vector Spaces. McGraw Hill, New York (1978)
33. Wilansky, A.: Summability Through Functional Analysis. Mathematical Studies, vol. 85. North-Holland, Amsterdam (1984)
34. Zeller, K.: Abschnittskonvergenz in FK-Räumen. Math. Z. **55**, 55–70 (1951)
35. Zeller, K.: Allgemeine Eigenschaften von Limitierungsverfahren. Math. Z. **53**, 463–487 (1951)
36. Zeller, K.: Matrixtransformationen von Folgenräumen. Univ. Rend. Mat. **12**, 340–346 (1954)
37. Zeller, K., Beekmann, W.: Theorie der Limitierungsverfahren. Springer, Heidelberg (1968)

Russian References

38. Р. Р. Ахмеров, М. И. Каменский, А. С. Потапов и др., Меры некомпактности и уплотняющие операторы, *Новосибирск, Наука*, 1986
39. Л. С. Гольденштейн, И. Ц. Гохберг и А. С. Маркус, Исследование некоторых свойств линейных ограниченных операторов в связи с их q-нормой, *Уч. зап. Кишиневского гос. ун-та*, **29**, (1957), 29–36
40. Л. С. Гольденштейн, А. С. Маркус, О мере некомпактности ограниченных множеств и линейных операторов, *В кн.: Исследование по алгебре и математическому анализу, Кишинев: Картя Молдавеняске* (1965) 45–54

Chapter 2
Matrix Domains

In Chap. 2, we study sequence spaces that have recently been introduced by the use of infinite matrices. They can be considered as the *matrix domains* of particular triangles in certain sequence spaces. This seems to be natural in view of the fact that most classical methods of summability are given by triangles.

There are a large number of publications in this area. Here we list some some of them related to the topics of this section and Sect. 3 [3, 4, 6, 7, 9, 14, 14, 15, 20, 22–24, 28–33, 35, 46–48, 50, 51, 58–64].

We also mention more of the following recent publications related to the topics covered in this chapter, for instance, [1, 8, 17, 21, 26, 34, 35, 45, 49, 52–56].

The results in these publications are proved for each sequence space separately. We use the relevant theory presented in Chap. 1 to provide a general, unified approach for matrix domains of arbitrary triangles in arbitrary FK spaces and some special cases.

In particular, we obtain some fundamental results concerning topological properties, Schauder bases, the β-duals of matrix domains of triangles in the classical sequence spaces and the characterizations of matrix transformations between these spaces. We also consider some important special cases.

Our general results reduce the determination of the β-duals of the matrix domains in the classical sequence space and the characterizations of matrix transformations between them to the β-duals of the spaces and the matrix transformations between the spaces themselves.

For further related studies, we refer the interested reader to the text book [25].

2.1 General Results

In this section, we study the most important fundamental general results for matrix domains related to their topological properties. Many of the results of this section can be found in [65, Sects. 4.2 and 4.3].

B. de Malafosse et al., *Operators Between Sequence Spaces and Applications*,
https://doi.org/10.1007/978-981-15-9742-8_2

It turns out that the theory of FK spaces of the previous section applies to most matrix domains.

Throughout, we assume that all the FK spaces are locally convex.

Definition 2.1 Let X be a subset of ω and A be an infinite matrix. Then the set

$$X_A = \{x \in \omega : Ax \in X\}$$

is called the *matrix domain of A in X*, in particular, the set c_A is called the *convergence domain of A*.

We will frequently use the following well-known result from functional analysis.

Theorem 2.1 ([44, Corollary 3.10.9] or [65, 4.0.1]) *Every locally convex metrizable space X has its topology defined by a finite or infinite sequence (p_n) of seminorms p_n; this means that if $(x^{(k)})$ is a sequence in X, then $x^{(k)} \to 0$ $(k \to \infty)$ if and only if $p_n(x^{(k)}) \to 0$ $(k \to \infty)$ for each n.*

We use the notation $(X, (p_n))$ for a vector space X with its metrizable topology given by the sequence (p_n) of semi-norms p_n in the sense of Theorem 2.1.

Example 2.1 (a) The space $(\omega, (|P_n|))$ is an FK space where (P_n) is the sequence of coordinates, and $x^{(m)} \to x$ $(m \to \infty)$ in ω if and only if $x_n^{(m)} \to x_n$ $(m \to \infty)$ for each n (Corollary 1.1).

(b) The space c is a BK space with $p(x) = \|x\|_\infty$; there is only one semi-norm, a norm in this case, and $x^{(m)} \to 0$ $(m \to \infty)$ in c if and only if $\|x_m\|_\infty \to 0$ $(m \to \infty)$ (Example 1.2).

The next result is fundamental.

Theorem 2.2 *Let $(X, (p_n))$ and $(Y, (q_n))$ be FK spaces, A be a matrix defined on X, that is, $X \subset \omega_A$, and*

$$Z = X \cap Y_A = \{x \in \omega : Ax \in Y\}.$$

Then Z is an FK space with $(p_n) \cup (q_n \circ A)$; this means that the topology of Z is given by the sequences (p_n) and $(q_n \circ A)$ of all the seminorms p_n and $q_n \circ A$.

Proof The countable set $\{p_n, q_n \circ A : n \in \mathbb{N}\}$ of seminorms yields a metrizable topology larger than that of X, hence of ω, since $(X, (p_n))$ is an FK space.
We have to show that Z is complete.
Let $(x^{(m)})_{m=0}^\infty$ be a Cauchy sequence in Z. Then it clearly is a Cauchy sequence in X which is convergent by the completeness of X,

$$\lim_{m\to\infty} x^{(m)} = t \text{ in } X, \text{ say,}$$

that is,

$$\lim_{m\to\infty} p_n\left(x^{(m)} - t\right) = 0 \text{ for each } n.$$

Since $x^{(m)} \in Y_A$, it follows that $Ax^{(m)} \in Y$, and so $(Ax^{(m)})$ is a Cauchy sequence in Y, since $q_n(Ax^{(m)}) = (q \circ A)(x^{(m)})$. So the sequence $(Ax^{(m)})$ is convergent by the completeness of Y,

$$\lim_{m\to\infty} Ax^{(m)} = b \text{ in } Y, \text{ say,}$$

that is,

$$\lim_{m\to\infty} q_n \left(Ax^{(m)} - b\right) = 0 \text{ for each } n.$$

Since $X \subset \omega_A$ and the matrix map A from X to ω is continuous by Theorem 1.4, it follows that

$$\lim_{m\to\infty} Ax^{(m)} = At \text{ in } \omega.$$

We also have

$$\lim_{m\to\infty} Ax^{(m)} = b \text{ in } \omega,$$

since the topology of the FK space is stronger than that of ω on Y. This yields $b = At$, since ω is a metric space, and so

$$t \in Z, \ \lim_{m\to\infty} p_n \left(x^{(m)} - t\right) = 0$$

and

$$\lim_{m\to\infty} (q \circ A) \left(x^{(m)} - t\right) = \lim_{m\to\infty} q \left(Ax^{(m)} - b\right) = 0 \text{ for each } n.$$

□

Let $A = (a_{nk})_{n,k=0}^{\infty}$ be an infinite matrix and $A_n = (a_{nk})_{k=0}^{\infty}$ and $A^k = (a_{nk})_{n=0}^{\infty}$ denote the sequences in the n-th row and the k-th column of A. Then A is said to be *row finite* if $A_n \in \phi$ for all n; it is said to be *column finite* if $A^k \in \phi$ for all k. An infinite matrix $T = (t_{nk})_{n,k=0}^{\infty}$ is called a *triangle* if $t_{nn} \neq 0$ and $t_{nk} = 0$ $(k > n)$ for all $n = 0, 1, \ldots$. It is known that any triangle T has a unique inverse S which also is a triangle and $x = S(Tx) = T(Sx)$ ([65, 1.4.8, p. 9], [13, Remark 22 (a), p. 22]).

The next result is a consequence of Theorem 2.2.

Theorem 2.3 *Let A be a row-finite matrix. Then $(c_A, (p_n))$ is an FK space where*

$$p_{-1} = \|\cdot\|_A, \textit{ that is, } p_{-1}(x) = \|Ax\|_\infty, \textit{ and } p_n(x) = |x_n| \textit{ for } n = 0, 1, 2, \ldots.$$

If A is a triangle, then (c_A, p_{-1}) is a BK space.

Proof We apply Theorem 2.2 with $X = \omega$ and $Y = c$ so that $Z = c_A$. The seminorms on $X = \omega$ are $|P_n|$ with $P_n(x) = x_n$ for $n = 0, 1, \ldots$ by Part (a) of Example 2.1. Also $Y = c$ is a BK space with $q = \|x\|_\infty$ by Part (b) of Example 2.1. We put $p_{-1} = q \circ A$. Then the statement is clear from Theorem 2.2.
If A is a triangle, then the matrix map $A : c_A \to x$ is one to one, linear and onto. So c_A becomes a Banach space equivalent to c with the norm $\|\cdot\|_{c_A}$ defined by

$$\|x\|_{c_A} = \|Ax\|_\infty \text{ for all } x \in c_A.$$

To see that the coordinates are continuous let $B = A^{-1}$ be the inverse of A, also a triangle. It follows that

$$|x_n| = |B_n(Ax)| = \left|\sum_{k=0}^{n} b_{nk}A_k x\right| \leq \sum_{k=0}^{n} |b_{nk}| \cdot \|Ax\|_\infty = M_n \cdot \|x\|_{c_A} \text{ for all } n,$$

where $M_n = \sum_{k=0}^{n} |b_{nk}|$. This shows that the coordinates are continuous. □

Let z be any sequence and Y be a subset of ω. Then we write

$$z^{-1} * Y = \left\{a \in \omega : a \cdot z = (a_k z_k)_{k=0}^{\infty} \in Y\right\}.$$

Part (a) of the next result is a special case of Theorem 2.2 when A is a diagonal matrix with the sequence z on the diagonal.

Theorem 2.4 *Let (Y, q) be an FK space and z be a sequence.*

(a) Then $z^{-1} * Y$ is an FK space with $(p_n) \cup (h_n)$ where

$$p_n = |x_n| \text{ and } h_n = q_n(z \cdot x) \text{ for all } n.$$

(b) If Y has AK then $z^{-1} * Y$ has AK also.

Proof (a) We define the diagonal matrix $D = (d_{nk})_{n,k=0}^{\infty}$ by $d_{nn} = z_n$ and $d_{nk} = 0$ $(k \neq n)$ for $n = 0, 1, \ldots,$ and apply Theorem 2.2 with $X = \omega$.

(b) We fix n. Then we have $p_n(x - x^{[m]}) = 0$ for all $m > n$, that is,

$$\lim_{m \to \infty} p_n(x - x^{[m]}) = 0.$$

Since Y has AK, we also obtain

$$h_n(x - x^{[m]}) = q_n(z \cdot (x - x^{[m]})) = q_n(z \cdot x - z \cdot x^{[m]}) \to 0 \ (m \to \infty).$$

Therefore, it follows that $x^{[m]} \to x$ in $z^{-1} * Y$, and so $z^{-1} * X$ has AK.

□

Example 2.2 Let $\alpha = (\alpha_k)_{k=0}^{\infty}$ be a sequence of positive reals, and $z = 1/\alpha = (1/\alpha_k)_{k=0}^{\infty}$. The spaces

$$\ell_\alpha^p = z^{-1} * \ell_p = \left\{ x \in \omega : \sum_{k=0}^{\infty} \left(\frac{|x_k|}{\alpha_k} \right)^p < \infty \right\} \text{ for } 1 \le p < \infty,$$

$$s_\alpha^0 = z^{-1} * c_0 = \left\{ x \in \omega : \lim_{k \to \infty} \frac{x_k}{\alpha_k} = 0 \right\},$$

$$s_\alpha^{(c)} = z^{-1} * c = \left\{ x \in \omega : \lim_{k \to \infty} \frac{x_k}{\alpha_k} = \xi \text{ exists} \right\}$$

and

$$s_\alpha = z^{-1} * \ell_\infty = \left\{ x \in \omega : \sup_k \frac{|x_k|}{\alpha_k} < \infty \right\}$$

were studied in [16]. They are BK spaces with

$$\|x\| = \left(\sum_{k=0}^{\infty} \left(\frac{|x_k|}{\alpha_k} \right)^p \right)^{1/p} \text{ for } \ell_\alpha^p \quad \text{and } \|x\| = \sup_k \left(\frac{|x_k|}{\alpha_k} \right) \text{ in the other cases.}$$

Furthermore, s_α^0 has AK and every sequence $x = (x_k)_{k=0}^\infty \in s_\alpha^c$ has a unique representation

$$x = \alpha \cdot \xi + \sum_{k=0}^{\infty} (x_k - \alpha_k \xi) e^{(k)}, \text{ where } \xi = \lim_{k \to \infty} \frac{x_k}{\alpha_k}.$$

We write

$$z^\alpha = \{x \in \omega : x \cdot z \in \ell_1\} = z^{-1} * \ell_1, \; z^\beta = z^{-1} * cs \text{ and } z^\gamma = z^{-1} * bs.$$

Theorem 2.5 *Let z be a given sequence. Then $(z^\beta, (p_n))$ is an AK space with*

$$p_{-1}(x) = \|z \cdot x\|_{bs} = \sup_n \left| \sum_{k=0}^{n} z_k x_k \right| \text{ and } p_n(x) = |x_n| \text{ for all } n \ge 0. \tag{2.1}$$

For any k for which $z_k \neq 0$, the seminorm p_k may be omitted.
If $z \in \phi$ then the seminorm p_{-1} may be omitted.

Proof We write $\Sigma = (\sigma_{nk})_{n,k=0}^\infty$ for the triangle with $\sigma_{nk} = 1$ for $(0 \le k \le n)$ and $\sigma_{nk} = 0$ for $k > n$ $(n = 0, 1, \dots)$. Then $cs = c_\Sigma$ is a BK space with respect to the norm $\|\cdot\|_{bs}$ defined by

$$\|x\|_{bs} = \|\Sigma x\|_\infty = \sup_n \left| \sum_{k=0}^{n} x_k \right| \text{ for all } x \in cs$$

by Theorem 2.3. Also it is easy to see that cs has AK. Therefore, the space $(z^\beta, (p_n))$ is an AK space with the seminorms given in (2.1) by Theorem 2.4 and the definition

of z^β.
If $z_k \neq 0$ and the matrix $A = (a_{nk})$ is defined by $a_{nk} = z_k$ for $0 \leq k \leq n$ and $a_{nk} = 0$ for $k > n$ $(n = 0, 1, \dots)$, then we obtain

$$|x_k| = \frac{|A_k x - A_{k-1} x|}{|z_k|} \leq 2\frac{p_{-1}(x)}{|z_k|},$$

and so p_k is redundant.
If $z \in \phi$ then $z^\beta = \omega$. □

We need the following result from functional analysis:

(I) ([65, 4.0.9]) Let a vector space X have a collection C of locally convex topologies. For each $\mathcal{T} \in C$ let $P(\mathcal{T})$ be the set of seminorms that generates $\mathcal{T}$ as in Theorem 2.1. Let $\mathcal{P} = \bigcup\{P(\mathcal{T}) : \mathcal{T} \in C\}$. Then $\mathcal{P}$ generates the so-called $sup - topology$ $\sup C$ with the property that

$$x \to 0 \text{ if and only if } p(x) \to 0 \text{ for all } p \in P(\mathcal{T}) \text{ and all } \mathcal{T} \in C.$$

Thus, $x \to 0$ if and only if $x \to 0$ in $(X, \mathcal{T})$ for each $\mathcal{T} \in C$.
If each $\mathcal{T} \in C$ is a norm topology and C has finitely many members, $\sup C$ is given by a norm, namely, the sum of these norms.

Theorem 2.6 *The intersection of countably many FH spaces is an FH space. If each X_n is an FK space with AK, then $X = \bigcap_{n=1}^\infty X_n$ has AK.*

Proof Let $\mathcal{T}_n$ be the topology on X_n. We place on X the topology $\mathcal{T} = \sup\{\mathcal{T}_n : X_n\}$. If each X_n is an FK space, then the coordinates are continuous in each $\mathcal{T}_n$, hence in the larger topology $\mathcal{T}$.
If (x_m) is a $\mathcal{T}$-Cauchy sequence in X, then it is a $\mathcal{T}_n$-Cauchy sequence for each n, and hence convergent by the completeness of X_n,

$$x^{(m)} \to t_n \in X_n, \text{ say } (m \to \infty) \text{ for each } n.$$

Then, we have

$$x^{(m)} \to t_n \in H \ (m \to \infty) \text{ for each } n,$$

since the FH topology of X_n is stronger than that of H on X_n. Thus, all the t_n are the same, since H is a Hausdorff space, $t_n = t$, say, for all n. Thus, we clearly have $t \in X$ and $x^{(m)} \to t$ $(m \to \infty)$ in $\mathcal{T}_n$ for each n, so $x \to t$ in $(X, \mathcal{T})$ by (I). □

Now we are able to prove the next result.

Theorem 2.7 *Let A be a matrix. Then $(\omega_A, (p_n) \cup (h_n))$ is an AK space with*

$$p_n(x) = |x_n| \text{ and } h_n(x) = \sup_m \left| \sum_{k=0}^m a_{nk} x_k \right| \text{ for all } n.$$

For any k such that A^k has at least one nonzero term, p_k may be omitted.
For any k such that $A_n \in \phi$, h_n may be omitted.

Proof We observe that $\omega_A = \bigcap_{n=0}^{\infty} A_n^{\beta}$, and each space A_n^{β} is an AK space with $p_n(x) = |x_n|$ and $h_n(x) = \sup_m |\sum_{k=0}^{m} a_{nk}x_k|$ by Theorem 2.5. Also, the intersection of countably many AK spaces is an AK space by Theorem 2.6.
If $a_{nk} \neq 0$ then we have $|x_k| = 2 \cdot h_n(x)/|a_{nk}|$ as in the proof of Theorem 2.5, so p_k is redundant.
If $A_n \in \phi$ then h_n can be omitted by the last part of Theorem 2.5. □

Theorem 2.8 *Let $(Y, (q_n))$ be an FK space with FK and A be a matrix. Then Y_A is an FK space with $(p_n) \cup (h_n) \cup (q_n \circ A)$ where the sequences (p_n) and (h_n) are as in* Theorem 2.7.
For any k such that A^k has at least one nonzero term, p_k may be omitted.
For any n such that $A_n \in \phi$, h_n may be omitted.
If A is a triangle, only $(q_n \circ A)$ is needed.

Proof We apply Theorem 2.2 with $X = \omega_A$, which is an FK space by Theorem 2.7. Then $Z = Y_A$ and the seminorms are obtained from Theorems 2.2 and 2.7.
The remaining parts follow from Theorem 2.7 and the fact that if A is a triangle then the map $A : Y_A \to Y$ is an equivalence. □

Example 2.3 We write $\Sigma^{(1)} = \Sigma$, where Σ is the triangle of the proof of Theorem 2.5, and $\Delta = \Delta^{(1)} = (\Delta_{nk})_{n,k=0}^{\infty}$ for the matrix with $\Delta_{nn} = 1$, $\Delta_{n-1,n} = -1$ and $\Delta_{nk} = 0$ otherwise. Let $m \in \mathbb{N} \setminus \{1\}$. Then we write $\Sigma^{(m)} = \Sigma^{(m-1)} \cdot \Sigma$ and $\Delta^{(m)} = \Delta^{(m-1)} \cdot \Delta$. Since the matrices Δ and Σ are obviously inverse to one another and matrix multiplication is associative for triangles ([65, Corollary 1.4.5]), the matrices $\Delta^{(m)}$ and $\Sigma^{(m)}$ are also inverse to one another. It is well known that

$$\Delta_n^{(m)} x = \sum_{k=0}^{m} (-1)^k \binom{m}{k} x_{n-k} = \sum_{k=\max\{0,n-m\}}^{n} (-1)^{n-k} \binom{m}{n-k} x_k \tag{2.2}$$

and

$$\Sigma_n^{(m)} x = \sum_{k=0}^{n} \binom{m+n-k-1}{n-k} x_k \text{ for } n = 0, 1, \ldots. \tag{2.3}$$

(a) Let $p = (p_k)_{k=0}^{\infty}$ be a bounded sequence of positive reals, $M = \max\{1, \sup_k p_k\}$ and $m \in \mathbb{N}$. By Theorem 2.8, Part (b) of Example 1.2, (2.3) and (2.2), the spaces $(\ell(p))_{\Sigma^{(m)}}$, $(\ell(p))_{\Delta^{(m)}}$ and $c_0((p), \Delta^{(m)}) = (c_0(p))_{\Delta^{(m)}}$ are FK spaces with the total paranorms

$$g_{(\ell(p))_{\Sigma^{(m)}}}(x) = \left(\sum_{k=0}^{\infty}\left|\sum_{j=0}^{k}\binom{m+k-j-1}{k-j}x_j\right|^{p_k}\right)^{1/M},$$

$$g_{(\ell(p))_{\Delta^{(m)}}}(x) = \left(\sum_{k=0}^{\infty}\left|\sum_{j=\max\{0,k-m\}}^{k}(-1)^{k-j}\binom{m}{k-j}x_j\right|^{p_k}\right)^{1/M}$$

and

$$g_{c_0((p),\Delta^{(m)})}(x) = \left(\sup_k\left|\sum_{j=\max\{0,k-m\}}^{k}(-1)^{k-j}\binom{m}{k-j}x_j\right|^{p_k}\right)^{1/M}.$$

If $m = 1$, then we write $bv(p) = (\ell(p))_{\Delta}$, and the spaces $(\ell(p))_{\Sigma}$, $bv(p)$ and $c_0((p), \Delta)$ are FK spaces with the total paranorms

$$g_{(\ell(p))_{\Sigma}}(x) = \left(\sum_{k=0}^{\infty}\left|\sum_{j=0}^{k}x_j\right|^{p_k}\right)^{1/M} \text{ and } g_{bv(p)}(x) = \left(\sum_{k=0}^{\infty}|x_k - x_{k-1}|^{p_k}\right)^{1/M}$$

and

$$g_{c_0((p),\Delta)}(x) = \left(\sup_k|x_k - x_{k-1}|^{p_k}\right)^{1/M}.$$

(b) Since c_0, c and ℓ_∞ are BK spaces with $\|\cdot\|_\infty$ by Part (c) of Example 1.2, the sets $c_0(\Delta^{(m)}) = (c_0)_{\Delta^{(m)}}$, $c(\Delta^{(m)}) = c_{\Delta^{(m)}}$ and $\ell_\infty(\Delta^{(m)}) = (\ell_\infty)_{\Delta^{(m)}}$ are BK spaces with

$$\|x\|_{(\ell_\infty)_{\Delta^{(m)}}} = \sup_k\left|\sum_{j=\max\{0,k-m\}}^{k}(-1)^{k-j}\binom{m}{k-j}x_j\right| \tag{2.4}$$

and the sets $c_{\Sigma^{(m)}}$ and $(\ell_\infty)_{\Sigma^{(m)}}$ are BK spaces with

$$\|x\|_{(\ell_\infty)_{\Sigma^{(m)}}} = \sup_k\left|\sum_{j=0}^{k}\binom{m+k-j-1}{k-j}x_j\right|. \tag{2.5}$$

Also, the sets $(\ell_p)_{\Delta^{(m)}}$ and $(\ell_p)_{\Sigma^{(m)}}$ are BK spaces with

$$\|x\|_{(\ell_p)_{\Delta^{(m)}}} = \left(\sum_{k=0}^{\infty}\left|\sum_{j=\max\{0,k-m\}}^{k}(-1)^{k-j}\binom{m}{k-j}x_j\right|^{p}\right)^{1/p} \tag{2.6}$$

and

$$\|x\|_{(\ell_p)_{\Sigma^{(m)}}} = \left(\sum_{k=0}^{\infty}\left|\sum_{j=0}^{k}\binom{m+k-j-1}{k-j}x_j\right|^p\right)^{1/p}. \tag{2.7}$$

For $m = 1$, the norms in (2.4)–(2.7) reduce to

$$\|x\|_{(\ell_\infty)_\Delta} = \sup_k |x_k - x_{k-1}|, \quad \|x\|_{(\ell_\infty)_\Sigma} = \sup_k \left|\sum_{j=0}^{k} x_j\right| = \|x\|_{bs}$$

$$\|x\|_{bv_p} = \left(\sum_{k=0}^{\infty}|x_k - x_{k-1}|^p\right)^{1/p} \text{ and } \|x\|_{(\ell_p)_\Sigma} = \left(\sum_{k=0}^{\infty}\left|\sum_{j=0}^{k} x_j\right|^p\right)^{1/p}.$$

If $p = 1$, we write $bv = bv_1$ for the set of *sequences of bounded variation.*

The next result is an immediate consequence of Theorem 2.8; it is a generalization of Theorem 2.3.

Corollary 2.1 *Let A be a matrix. Then c_A is an FK space with $(p_n) \cup (h_n)$ where*

$p_{-1}(x) = \|Ax\|_\infty$ *and p_n and $h_n (n = 0, 1, \dots)$ are defined as in* Theorem 2.7.

For any k such that A^k has at least one nonzero term, p_k may be omitted.
For any n such that $A_n \in \phi$, h_n may be omitted (Theorem 2.3).

Theorem 2.9 *Let X and Y be FK spaces, A be a matrix, and X be a closed subspace of Y. Then X_A is a closed subspace of Y_A.*

Proof Since Y is an FK space, Y_A is an FK space by Theorem 2.8. So the map $L : Y_A \to Y$ with $L(x) = Ax$ for all $x \in Y_A$ is continuous by Theorem 1.4, and so $X_A = L^{-1}(X)$ is closed. □

Example 2.4 We apply Theorem 2.9 and obtain that $(c_0)_{\Sigma^{(m)}}$ is a closed subspace of $c_{\Sigma^{(m)}}$ and $c_{\Sigma^{(m)}}$ is a closed subspace of $(\ell_\infty)_{\Sigma^{(m)}}$; also $(c_0)_{\Delta^{(m)}}$ is a closed subspace of $c_{\Delta^{(m)}}$ and $c_{\Delta^{(m)}}$ is a closed subspace of $(\ell_\infty)_{\Delta^{(m)}}$.

2.2 Bases of Matrix Domains of Triangles

In this section, we study the bases of matrix domains of triangles.

The main results are Theorem 2.12 and Corollary 2.5. Theorem 2.12 expresses the Schauder bases of matrix domains X_T of triangles T in linear metric sequence spaces X with a Schauder basis in terms of the Schauder basis of X. Corollary 2.5 is the special case of Theorem 2.12, where X is an FK space with AK and gives explicit representations of the sequences in X_T, $X_T \oplus e$ and $(X \oplus e)_T$ with respect to the sequences $e^{(n)}$ $(n = 0, 1, \dots)$ and e. These results would yield the Schauder bases for the matrix domains in the classical sequence spaces (with the exception of ℓ_∞ (Corollary 2.4 and Example 1.5)) and in $c_0(p)$ and $\ell(p)$ (Example 2.7) in a great number of recent publications.

We start with some special results of interest.

Let $\mathcal{U} = \{u \in \omega : u_k \neq 0 \text{ for all } k = 0, 1, \dots\}$. If $u \in \mathcal{U}$ then we write $1/u = (1/u_k)_{k=0}^\infty$.

Theorem 2.10 ([27, Theorem 2]) *Let X be a BK space and $u \in \mathcal{U}$.*

(a) If $(b^{(k)})_{k=0}^\infty$ is a basis of X and $c^{(k)} = (1/u) \cdot b^{(k)}$ for $k = 0, 1, \dots,$ then $(c^{(k)})_{k=0}^\infty$ is a basis of $Y = u^{-1} * X$.

(b) Let

$$|u_0| \leq |u_1| \leq \dots \text{ and } |u_n| \to \infty \ (n \to \infty),$$

and $T = (t_{nk})_{n,k=0}^\infty$ be the triangle with

$$t_{nk} = \begin{cases} \dfrac{1}{u_n} & (0 \leq k \leq n) \\ 0 & (k > n) \end{cases} \quad (n = 0, 1, \dots).$$

Then $(c_0)_T$ has AK.

Proof (a) Let $\|\cdot\|$ be the BK norm of X. Then Y is a BK space with respect to $\|\cdot\|_u$ defined by $\|y\|_u = \|u \cdot y\|$ $(y \in Y)$ by Theorem 2.8 and Part (a) of Theorem 2.4. Also $u \cdot c^{(k)} = b^{(k)} \in X$ implies $c^{(k)} \in Y$ for all k. Finally, let $y \in Y$ be given. Then $u \cdot y = x \in X$ and there exists a unique sequence $(\lambda_k)_{k=0}^\infty$ of scalars such that

$$x^{<m>} = \sum_{k=0}^{m} \lambda_k b^{(k)} \to x \ (m \to \infty) \text{ in } X.$$

We put

$$y^{<m>} = (1/u) \cdot x^{<m>} = \sum_{k=0}^{m} \lambda_k c^{(k)} \text{ for } m = 0, 1, \dots.$$

Then we have $u \cdot y^{<m>} \to x = u \cdot y$ $(m \to \infty)$ in X, hence $y^{<m>} \to y$ $(m \to \infty)$ in Y, that is, $y = \sum_{k=0}^\infty \lambda_k c^{(k)}$. Obviouly this representation is unique.

(b) By Theorem 2.1, $(c_0)_T$ is a BK space with respect to the norm $\|\cdot\|_{(c_0)_T}$ defined by

$$\|x\|_{(c_0)_T} = \sup_n |T_n x| = \sup_n \left| \frac{1}{u_n} \sum_{k=0}^{n} x_k \right| \text{ for all } x \in (c_0)_T.$$

Also $|u_n| \to \infty$ $(n \to \infty)$ implies $\phi \subset (c_0)_T$. Let $\varepsilon > 0$ and $x \in (c_0)_T$ be given. Then there is a non-negative integer n_0 such that $|T_m x| < \varepsilon/2$ for all $m > n_0$. Then it follows that

$$\|x - x^{[m]}\|_{(c_0)_T} = \sup_{n \geq m+1} \left| \frac{1}{u_n} \sum_{k=m+1}^{n} x_k \right| \leq \sup_{n \geq m+1} (|T_n x| + |T_m x|) < \varepsilon,$$

hence

$$x = \sum_{k=0}^{\infty} x_k e^{(k)}. \tag{2.8}$$

Obviously, the representation in (2.8) is unique.

□

We apply Theorem 2.10 to determine the expansions of sequences in the matrix domians of the arithmetic means in c_0 and c.

Example 2.5 Let $C_1 = (a_{nk})_{n,k=0}^{\infty}$ be the *matrix of the arithmetic means* or *Cesàro matrix of order* 1, that is,

$$a_{nk} = \begin{cases} \dfrac{1}{n+1} & (0 \leq k \leq n) \\ 0 & (k > n) \end{cases} \quad (n = 0, 1, \dots).$$

We write $C_0^{(1)} = (c_0)_{C_1}$, $C^{(1)} = c_{C_1}$ and $C_\infty^{(1)} = (\ell_\infty)_{C_{(1)}}$ for the sets of sequences that are summable to 0, summable and bounded by the C_1 method. Then the space $C_0^{(1)}$ has AK and every sequence $x = (x_k)_{k=0}^{\infty}$ has a unique representation

$$x = \xi \cdot e + \sum_{k=0}^{\infty} (x_k - \xi) e^{(k)} \text{ where } \xi = \lim_{n \to \infty} \frac{1}{n+1} \sum_{k=0}^{n} x_k. \tag{2.9}$$

Proof We apply Part (b) of Theorem 2.10 with $u_n = (n + 1)$ $(n = 0, 1, \dots)$ and obtain that ${C_0}^{(1)}$ has AK.
If $x = (x_k)_{k=0}^{\infty} \in C^{(1)}$, then there exists a unique complex number such that

$$\frac{1}{n+1} \sum_{k=0}^{n} x_k - \xi = \frac{1}{n+1} \sum_{k=0}^{n} (x_k - \xi) \to 0 \ (n \to \infty),$$

hence $x^{(0)} = x - \xi \cdot e = (x_k - \xi)_{k=0}^{\infty} \in C_0^{(1)}$, and so $x^{(0)} = \sum_{k=0}^{\infty} (x_k - \xi) e^{(k)}$, since $C_0^{(1)}$ has AK. Therefore, it follows that

$$x = x^{(0)} + \xi \cdot e = \xi \cdot e + \sum_{k=0}^{\infty} (x_k - \xi) e^{(k)},$$

which is (2.9). □

Now we consider some spaces related to the so-called *weighted* or *Riesz means*.

Example 2.5 and the results presented following Definition 2.2 were applied in [41] to simplify the results in [11, 12, 18, 19, 43].

Definition 2.2 Let $q = (q_k)_{k=0}^{\infty}$ be a sequence of non-negative real numbers with $q_0 > 0$ and $Q_n = \sum_{k=0}^{n} q_k$ for $n = 0, 1, \ldots$. Then the matrix $\bar{N}_q = ((\bar{N}_q)_{nk})_{n,k=0}^{\infty}$ of the *weighted means* or *Riesz means* is defined by

$$\left(\bar{N}_q\right)_{nk} = \begin{cases} \dfrac{q_k}{Q_n} & (0 \leq k \leq n) \\ 0 & (k > n) \end{cases} \quad (n = 0, 1, \ldots).$$

We write

$$\left(\bar{N}, q\right)_0 = (c_0)_{\bar{N}_q}, \ \left(\bar{N}, q\right) = c_{\bar{N}_q} \text{ and } \left(\bar{N}, q\right)_{\infty} = (\ell_{\infty})_{\bar{N}_q}$$

for the sets of all sequences that are summable to 0, summable and bounded by the method $\bar{N}_q$. If $p = e$, then these sets reduce to the sets $C_0^{(1)}$, $C^{(1)}$ and $C_{\infty}^{(1)}$ of Example 2.5.

Corollary 2.2 ([27, Corollary 1]) *Each of the sets* $(\bar{N}, q)_0$, $(\bar{N}, q)$ *and* $(\bar{N}, q)_{\infty}$ *is a* BK *space with respect to the norm* $\| \cdot \|_{\bar{N},q}$ *defined by*

$$\|x\|_{\bar{N}_q} = \sup_n \left| \frac{1}{Q_n} \sum_{k=0}^{n} q_k x_k \right|,$$

$(\bar{N}, q)_0$ *is a closed subspace of* $(\bar{N}, q)$ *and* $(\bar{N}, q)$ *is a closed subspace of* $(\bar{N}, q)_{\infty}$. *Furthermore, if* $Q_n \to \infty$ $(n \to \infty)$, *then* $(\bar{N}, q)_0$ *has* AK, *and every sequence* $x = (x_k)_{k=0}^{\infty} \in (\bar{N}, q)$ *has a unique representation*

$$x = \xi \cdot e + \sum_{k=0}^{\infty} (x_k - \xi) e^{(k)}, \quad \text{where } \xi = \lim_{n \to \infty} \left(\frac{1}{Q_n} \sum_{k=0}^{n} q_k x_k \right). \tag{2.10}$$

Proof The first part follows from Theorem 2.9 and the facts that c_0 is closed in c and c is closed in ℓ_{∞}.

We define the matrix $T = (t_{nk})_{n,k=0}^{\infty}$ by

$$t_{nk} = \begin{cases} \dfrac{1}{Q_n} & (0 \leq k \leq n) \\ O & (k > n) \end{cases} \quad (n = 0, 1, \ldots).$$

Then we have $(\bar{N}, q)_0 = q^{-1} * (c_0)_T$ and $(\bar{N}, q)_0$ has AK by Part (b) of Theorem 2.10 and Theorem 2.4.

Finally, let $x = (x_k)_{k=0}^{\infty} \in (\bar{N}, q)$. Then there exists a unique complex number ξ such that

$$\lim_{n\to\infty}\left(\frac{1}{Q_n}\sum_{k=0}^{n} q_k x_k - \xi\right) = \lim_{n\to\infty}\left(\frac{1}{Q_n}\sum_{k=0}^{n} q_k (x_k - \xi)\right) = 0,$$

that is, $x^{(0)} = x - \xi \cdot e \in (\bar{N}, q)_0$. Hence $x^{(0)} = \sum_{k=0}^{\infty}(x_k - \xi)e^{(k)}$, since $(\bar{N}, q)_0$ has AK. Therefore, it follows that

$$x = x^{(0)} + \xi \cdot e = \xi \cdot e + \sum_{k=0}^{\infty}(x_k - \xi)e^{(k)},$$

which is (2.10). □

Now we consider some *generalized difference sequence spaces.*

Definition 2.3 For any $u \in \mathcal{U}$, we define the *generalized difference sequence spaces*

$$c_0(u\Delta) = \left(u^{-1} * c_0\right)_\Delta, \; c(u\Delta) = \left(u^{-1} * c\right)_\Delta \text{ and } \ell_\infty(u\Delta) = \left(u^{-1} * \ell_\infty\right)_\Delta;$$

if $u = e$ then we obtain the following spaces introduced and studied in [36]

$$c_0(\Delta) = (c_0)_\Delta, \; c(\Delta) = c_\Delta \text{ and } \ell_\infty(\Delta) = (\ell_\infty)_\Delta.$$

Corollary 2.3 ([27, Corollary 2]) *Let $u \in \mathcal{U}$. Then each of the spaces $c_0(u\Delta)$, $c(u\Delta)$ and $\ell_\infty(u\Delta)$ is a BK space with respect to the norm $\|\cdot\|_{\ell_\infty(u\Delta)}$ defined by*

$$\|x\|_{\ell_\infty(u\Delta)} = \sup_k |u_k(x_k - x_{k-1})| \text{ where } x_{-1} = 0;$$

$c_0(u\Delta)$ is a closed subspace of $c(u\Delta)$ and $c(u\Delta)$ is a closed subspace of $\ell_\infty(u\Delta)$.

Proof This follows from Theorems 2.8 and 2.9. □

Now we determine a basis for each of the spaces $c_0(u\Delta)$ and $c(u\Delta)$.

Theorem 2.11 ([27, Theorem 3]) *Let $u \in \mathcal{U}$,*

$$b^{(-1)} = \Sigma(1/u) = \left(\sum_{k=0}^{n}\frac{1}{u_k}\right)_{n=0}^{\infty} \text{ and } b^{(k)} = e - \sum_{j=0}^{k-1} e^{(j)} \text{ for } k \geq 0.$$

Then

(a) $(b^{(k)})_{k=0}^{\infty}$ is a basis of $c_0(u\Delta)$ and every sequence $x = (x_k)_{k=0}^{\infty} \in c_0(u\Delta)$ has a unique representation

$$x = \sum_{k=0}^{\infty}(\Delta_k x)\, b^{(k)} = \sum_{k=0}^{\infty}(x_k - x_{k-1})b^{(k)}; \tag{2.11}$$

(b) $(b^{(k)})_{k=-1}^{\infty}$ *is a basis of* $c(u\Delta)$ *and every sequence* $x = (x_k)_{k=0}^{\infty} \in c(u\Delta)$ *has a unique representation*

$$x = \xi \cdot b^{(-1)} + \sum_{k=0}^{\infty} \left(x_k - x_{k-1} - \frac{1}{u_k} \right) b^{(k)}, \text{ where } \xi = \lim_{k\to\infty} u_k(x_k - x_{k-1}). \tag{2.12}$$

Proof (a) We obviously have $b^{(k)} \in c_0(u\Delta)$ for $k = 0, 1, \dots$. Let $x = (x_k)_{k=0}^{\infty} \in c_0(u\Delta)$ and $\varepsilon > 0$ be given. Then there is a non-negative integer m_0 such that

$$|u_m(x_m - x_{m-1})| < \varepsilon \text{ for all } m \geq m_0.$$

Let $m \geq m_0$ and

$$x^{<m>} = \sum_{k=0}^{m} (x_k - x_{k-1}) b^{(k)}.$$

Since

$$x_n^{<m>} = \sum_{k=0}^{m} (x_k - x_{k-1}) \left(e_n - \sum_{j=0}^{k-1} e_n^{(j)} \right) = x_m - \sum_{j=0}^{m-1} e_n^{(j)} \sum_{k=j+1}^{m} (x_k - x_{k-1})$$
$$= \begin{cases} x_m & (n \geq m) \\ x_n & (n \leq m-1), \end{cases}$$

we have

$$x_n^{<m>} - x_{n-1}^{<m>} = \begin{cases} 0 & (n \geq m+1) \\ x_n - x_{n-1} & (n \leq m), \end{cases}$$

and so $\|x - x^{<m}\|_{\ell_\infty(u\Delta)} \leq \varepsilon$ for all $m \geq 0$. Thus, (2.11) is satisfied, and obviously this representation is unique.
(b) Obviously $b^{(k)} \in c(u\Delta)$ for all $k = -1, 0, 1, \dots$. Let $x = (x_k)_{k=0}^{\infty} \in c(u\Delta)$. We put $y = x - \xi \cdot b^{(-1)}$ where $\xi = \lim_{k\to\infty} u_k(x_k - x_{k-1})$. Then $y_k - y_{k-1} = x_k - x_{k-1} - \xi/u_k$ for $k = 0, 1, \dots$, hence $y \in c_0(u\Delta)$, and (2.12) follows from (2.11).

□

We have seen in Theorem 2.4 that if Y is an FK space with AK then $z^{-1} * Y$ is an FK space with AK; $z^{-1} * Y$ is the matrix domain of the diagonal matrix $D = (d_{nk})_{n,k=0}^{\infty}$ in Y where $d_{nn} = z_n$ for $n = 0, 1, \dots$. This result does not extend to matrix domains of triangles, in general. If Y has AK then Y_T need not have AK, for instance, c_0 has AK, but $c_0(\Delta)$ has not. Also, the matrix domain of a triangle in a space which is not an AK space may have AK, for instance, c does not have AK but $cs = c_\Sigma$ has AK.

We will, however, see in Corollary 2.4 that the matrix domain in a sequence space has a Schauder basis if and only if the space itself has a Schauder basis.

The first result gives a relation between the basis of a linear metric sequence space X and the matrix domain of a triangle in X.

Now let T be an arbitrary triangle and S be its inverse.

Theorem 2.12 ([43, Proposition 2.1]) *If $(b^{(n)})_{n=0}^{\infty}$ is a basis of a linear metric sequence space (X, d) then $(Sb^{(n)})_{n=0}^{\infty}$ is a basis for $Z = X_T$ with the metric d_T defined by*

$$d_T(z, \tilde{z}) = d(Tz, T\tilde{z}) \text{ for all } z, \tilde{z} \in Z.$$

Proof We write $c^{(n)} = Sb^{(n)}$ for $n = 0, 1, \ldots$. Then, we have $c^{(n)} \in Z$ for all n since

$$Tc^{(n)} = T(Sb^{(n)}) = (TS)b^{(n)} = b^{(n)} \in X.$$

Let $z \in Z$ be given. Then $x = Tz \in X$ and there is a unique sequence $(\lambda_n)_{n=0}^{\infty}$ of scalars such that $x^{<m>} = \sum_{n=0}^{m} \lambda_n b^{(n)} \to x$ $(m \to \infty)$. We put

$$z^{<m>} = \sum_{n=0}^{m} \lambda_n c^{(n)} \text{ for } m = 0, 1, \ldots.$$

Then it follows that

$$Tz^{<m>} = \sum_{n=0}^{m} \lambda_n Tc^{(n)} = \sum_{n=0}^{m} \lambda_n b^{(n)} = x^{<m>} \text{ for } m = 0, 1, \ldots,$$

hence

$$d_T(z^{<m>}, z) = d(Tz^{<m>}, Tz) = d(x^{<m>}, x) \to 0 \ (m \to \infty).$$

□

Since matrix multiplication of triangles is associative ([65, Corollary 1.4.5]), it follows that $X = (X_T)_S$, and an application of Theorem 2.12 yields the following.

Corollary 2.4 *The matrix domain X_T of a linear metric sequence space has a basis if and only if X has a basis.*

Example 2.6 Since ℓ_∞ has no basis by Example 1.5, the spaces $bs = (\ell_\infty)_\Sigma$ and $\ell_\infty(\Delta) = (\ell_\infty)_\Delta$ have no bases by Corollary 2.4.

Now we consider a few special cases of Theorem 2.12, in particular, when X has AK.

Corollary 2.5 ([43, Corollary 2.3]) *Let X be an FK space with AK and the sequences $c^{(n)}$ $(n = 0, 1, \ldots)$ and $c^{(-1)}$ be defined by*

$$c_k^{(n)} = \begin{cases} 0 & (0 \le k \le n-1) \\ s_{kn} & (k \ge n) \end{cases} \quad \textit{and} \quad c_k^{(-1)} = \sum_{j=0}^{k} s_{kj} \ (k = 0, 1, \ldots). \tag{2.13}$$

(a) *Then every sequence* $z = (z_n)_{n=0}^{\infty} \in Z = X_T$ *has a unique representation*

$$z = \sum_{n=0}^{\infty} (T_n z) c^{(n)}. \tag{2.14}$$

(b) *Then every sequence* $v = (v_n)_{n=0}^{\infty} \in V = X_T \oplus e$ *has a unique representation*

$$v = \xi \cdot e + \sum_{n=0}^{\infty} T_n(v - \xi e) c^{(n)}, \tag{2.15}$$

where ξ *is the uniquely determined complex number such that* $v = z + \xi e$ *for* $z \in Z = X_T$.

(c) *Then every sequence* $w = (w_n)_{n=0}^{\infty} \in W = (X \oplus e)_T$ *has a unique representation*

$$w = \xi \cdot c^{(-1)} + \sum_{n=0}^{\infty} (T_n w - \xi) c^{(n)}, \tag{2.16}$$

where ξ *is the uniquely determined complex number such that* $Tw - \xi e \in X$.

Proof First, we note that $c^{(n)} = Se^{(n)}$ for $n = 0, 1, \ldots,$ and $c^{(-1)} = Se$. Hence the sequences $(c^{(n)})_{n=0}^{\infty}$ and $(c^{(n)})_{n=-1}^{\infty}$ are bases of Z and W, respectively, by Theorem 2.12.

(a) Let $z = (z_n)_{n=0}^{\infty} \in Z$ be given. Then $x = Tz \in X$ and (2.14) follows if we take $\lambda_n = T_n z$ $(n = 0, 1, \ldots)$ in the proof of Theorem 2.12.
(b) Let $v = (v_n)_{n=0}^{\infty} \in V = X_T \oplus e$ be given. Then there are uniquely determined $z \in Z$ and $\xi \in \mathbb{C}$ such that $v = z + \xi e$, and we have $z = \sum_{n=0}^{\infty} (T_n z) c^{(n)}$ by Part (a). It follows that $v = \xi e + z = \xi e + \sum_{n=0}^{\infty} T_n(v - \xi e) c^{(n)}$.
(c) Let $w = (w_n)_{n=0}^{\infty} \in W$. Then $u = Tw \in U = X \oplus e$, and there are uniquely determined $x \in X$ and $\xi \in \mathbb{C}$ such that $u = x + \xi e$. We put $z = w - \xi c^{(-1)}$. Then $z \in Z$, since $Tz = T(w - \xi c^{(-1)}) = Tw - \xi T c^{(-1)} = u - \xi e = x \in X$. So we have $z = \sum_{n=0}^{\infty} (T_n z) c^{(n)} = \sum_{n=0}^{\infty} (T_n w - \xi) c^{(n)}$ by Part (a). Now (2.16) is an immediate consequence, since $w = z + \xi c^{(-1)}$.

□

Now we give some applications of Corollary 2.5.

Example 2.7 (a) We consider the spaces $(\ell(p))_{\Delta^{(m)}}$ and $(c_0((p), \Delta^{(m)}))$ for bounded sequences $p = (p_k)_{k=0}^{\infty}$ of positive reals and $m \in \mathbb{N}$. We put $T = \Delta^{(m)}$. Then $S = \Sigma^{(m)}$. Since the spaces $\ell(p)$ and $c_0(p)$ are FK spaces by Part (b) of Example 1.2 and it is easy to see that they have AK, the sequences $c^{(n)}$ of the

Schauder bases $(c^{(n)})_{n=0}^{\infty}$ of $(\ell(p))_{\Delta^{(m)}}$ and $c_0((p), \Delta^{(m)})$ are given by (2.3) in Example 2.3 and (2.13) in Corollary 2.5 by

$$c_k^{(n)} = \begin{cases} 0 & (0 \le k \le n-1) \\ \binom{m+k-n-1}{k-n} & (k \ge n), \end{cases}$$

and so every sequence $x \in (\ell(p))_{\Delta^{(m)}}$ or $x \in c_0((p), \Delta^{(m)})$ has a unique representation, by (2.14) in Part (a) of Corollary 2.5, $x = \sum_{n=0}^{\infty}(\Delta_n^{(m)}x) \cdot c^{(n)}$ ([42, Theorem 1]) for $(c_0((p), \Delta^{(m)})))$.
If $m = 1$ then we obtain $c^{(n)} = e - e^{[n-1]}$ for all $n \in \mathbb{N}_0$, where $e^{[-1]} = (0)_{k=0}^{\infty}$, and every sequence $x \in (\ell(p))_{\Delta}$ or $x \in c_0((p), \Delta)$ has a unique representation

$$x = \sum_{n=0}^{\infty}(x_n - x_{n-1}) \cdot (e - e^{[n-1]}) \text{ where } x_{-1} = 0. \tag{2.17}$$

(b) We consider the space $(\ell(p))_{\Sigma^{(m)}}$ for bounded sequences $p = (p_k)_{k=0}^{\infty}$ of positive reals and $m \in \mathbb{N}$. We put $T = \Sigma^{(m)}$. Then $S = \Delta^{(m)}$. Since the space $\ell(p)$ is an FK space with AK, the sequences $c^{(n)}$ of the Schauder basis $(c^{(n)})_{n=0}^{\infty}$ of $(\ell(p))_{\Sigma^{(m)}}$ are given by (2.2) in Example 2.3 and (2.13) in Corollary 2.5 by

$$c_k^{(n)} = \begin{cases} (-1)^{k-n}\binom{m}{k-n} & (k \ge n) \\ 0 & (0 \le k \le n-1) \end{cases} \quad \text{if } n < m,$$

and

$$c_k^{(n)} = \begin{cases} (-1)^{k-n}\binom{m}{k-n} & (n+m \ge k \ge n) \\ 0 & (0 \le k \le n-1 \text{ or } k > n+m) \end{cases} \quad \text{if } n \ge m. \tag{2.18}$$

But since $\binom{m}{k-n} = 0$ for $k \ge m+n+1$, the sequences $c^{(n)}$ are given by (2.18) for all n.
If $m = 1$ then we obtain $c^{(n)} = e^{(n)} - e^{(n+1)}$ for all $n = 0, 1, \ldots$ and every sequence $x = (x_n)_{n=0}^{\infty} \in (\ell(p))_{\Sigma}$ has a unique representation

$$x = \sum_{n=0}^{\infty}\left(\sum_{k=0}^{n} x_k\right)(e^{(n)} - e^{(n+1)})$$

by (2.14) in Part (a) of Corollary 2.5.

(c) We consider the BK space $c(\Delta) = (c_0 \oplus e)_{\Delta}$ of Part (b) of Example 2.3. Then the sequence $c^{(-1)}$ in (2.13) of Corollary 2.5 is obviously given by

$$c_k^{(-1)} = \sum_{j=0}^{k} s_{kj} = k+1 \text{ for } k = 0, 1, \ldots.$$

If we write $(\mathbf{k}+\mathbf{1})$ for the sequence $(k+1)_{k=0}^{\infty}$, then every sequence $w \in c(\Delta)$ has a unique representation

$$w = \lim_{n\to\infty} \Delta w_n \cdot (\mathbf{k}+\mathbf{1}) + \sum_{n=0}^{\infty} (\Delta w_n - \lim_{n\to\infty} \Delta w_n)(e - e^{[n-1]})$$

by (2.16) in part c of Corollary 2.5.

(d) Finally, we consider the space $cs = (c_0 \oplus e)_{\Sigma}$. Then we obviously have $c^{(-1)} = e^{(0)}$ for the sequence in (2.13) of Corollary 2.5. Now every sequence $w \in cs$ has a unique representation by (2.16) in Part (c) of Corollary 2.5

$$\begin{aligned} w &= \sum_{n=0}^{\infty} w_n \cdot e^{(0)} + \sum_{n=0}^{\infty} \left(\sum_{k=0}^{n} w_k - \sum_{k=0}^{\infty} w_k \right) (e^{(n)} - e^{(n+1)}) \\ &= \sum_{n=0}^{\infty} w_n \cdot e^{(0)} + \sum_{n=0}^{\infty} \left(\sum_{k=n+1}^{\infty} w_k \right) \cdot (e^{(n+1)} - e^{(n)}). \end{aligned} \tag{2.19}$$

We write $y = Tw - \xi \cdot e$ where $\xi = \lim_{n\to\infty} T_n w = \lim_{n\to\infty} \sum_{k=0}^{n} w_k$. Then $y \in c_0$ and so the series $\sum_{n=0}^{\infty} y_n e^{(n)}$ and $\sum_{n=0}^{\infty} y_n e^{(n+1)}$ converge (in the ℓ_∞-norm), and it follows from (2.19)

$$\begin{aligned} w &= \xi \cdot e^{(0)} + \sum_{n=0}^{\infty} y_n (e^{(n)} - e^{(n+1)}) = \xi \cdot e^{(0)} + \sum_{n=0}^{\infty} y_n e^{(n)} - \sum_{n=0}^{\infty} y_n e^{(n+1)} \\ &= \xi \cdot e^{(0)} + y_0 \cdot e^{(0)} + \sum_{n=1}^{\infty} (y_n - y_{n-1}) e^{(n)} = w_0 \cdot e^{(0)} + \sum_{n=1}^{\infty} w_n \cdot e^{(n)} \\ &= \sum_{n=0}^{\infty} w_n \cdot e^{(n)}, \end{aligned}$$

that is, cs has AK.

We close this section with one more example to obtain the results given for the Schauder bases of the spaces in [2, 5, 12].

Example 2.8 (a) The *Euler sequence spaces* were studied in [2, 5]; they are defined as follows: Let $0 < r < 1$ and $E^r = (e_{nk}^r)_{n,k=0}^{\infty}$ be the *Euler matrix of order* r with

$$e_{nk}^r = \begin{cases} \binom{n}{k}(1-r)^{n-k} r^k & (0 \le k \le n) \\ 0 & (k > n) \end{cases} \quad (n = 0, 1, \dots).$$

Then the Euler sequence spaces are defined by $e_p^r = (\ell_p)_{E^r}$ for $1 \le p < \infty$, $e_0^r = (c_0)_{E^r}$, $e_c^r = c_{E^r}$ and $e_\infty^r = (\ell_\infty)_{E^r}$. Writing $T = E^r$, for short, we observe that the inverse $S = (s_{nk})_{n,k=0}^{\infty}$ of the triangle T is given by

$$s_{nk} = \begin{cases} \binom{n}{k}(r-1)^{n-k}r^{-n} & (0 \le k \le n) \\ 0 & (k > n) \end{cases} \quad (n = 0, 1, \dots). \tag{2.20}$$

Now [2, Theorem (i), (ii)] is an immediate consequence of (2.14) and (2.15) in Parts (a) and (b) of Corollary 2.5; Part (a) also yields Schauder bases for the spaces e_p^r $(1 \le p < \infty)$.

(b) The spaces $a_0^r(\Delta)$ and $a_c^r(\Delta)$ were introduced and studied in [12] as follows. If $T = (t_{nk})_{n,k=0}^{\infty}$ is the triangle defined by

$$t_{nk} = \begin{cases} \dfrac{1}{n+1}(r^k - r^{k+1}) & (0 \le k \le n-1) \\ \dfrac{r^{n+1}}{n+1} & (k = n) \\ 0 & (k > n) \end{cases} \quad (n = 0, 1, \dots),$$

then $a_0^r(\Delta) = (c_0)_T$ and $a_c^r = c_T$. Since the inverse matrix $S = (s_{nk})_{n,k=0}^{\infty}$ of T is given by

$$s_{nk} = \begin{cases} (k+1)\left(\dfrac{1}{1+r^k} - \dfrac{1}{1+r^{k+1}}\right) & (0 \le k \le n-1) \\ \dfrac{n+1}{1+r^n} & (k = n) \\ 0 & (k > n) \end{cases} \quad (n = 0, 1, \dots), \tag{2.21}$$

[12, Theorem 3 (a), (b)] is an immediate consequence of (2.8) and Parts (a) and (b) of Corollary 2.5.

Remark 2.1 We write $\mathbf{1/(n+1)} = (1/(n+1))_{n=0}^{\infty}$. It was shown in [41, Corollary 2.5] that

$$a_0^r(\Delta) = (\mathbf{1/(n+1)})^{-1} * c_0, \; a_c^r(\Delta) = (\mathbf{1/(n+1)})^{-1} * c$$

and

$$a_\infty^r(\Delta) = (\mathbf{1/(n+1)})^{-1} * \ell_\infty.$$

By [41, Remark 2.6] $a_0^r(\Delta)$ has AK, and every sequence $x = (x_k)_{k=0}^{\infty} \in a_c^r(\Delta)$ has a unique representation

$$x = (\mathbf{n+1}) \cdot \xi - \sum_{n=0}^{\infty} (x_n - (n+1)\xi)e^{(n)}, \text{ where } \xi = \lim_{n \to \infty} \frac{x_n}{n+1}.$$

We note that the last two statements are special cases of $\alpha_n = n + 1$ $(n = 0, 1, \dots)$ of the corresponding statements in Example 2.2.

2.3 The Multiplier Space $M(X_\Sigma, Y)$

In this and the following sections, we determine some dual spaces of matrix domains of triangles. Among other things, we reduce the determination of $(X_T)^\beta$ for certain FK spaces to that of the determination of X^β and the characterization of the class (X, c_0).

We start with an almost trivial, but useful result that gives the multiplier of the matrix domain of a diagonal matrix and a sequence space.

Proposition 2.1 *Let $u, v \in \mathcal{U}$. Then we have for arbitrary subsets X and Y of ω*

$$M(u^{-1} * X, Y) = (1/u)^{-1} * M(X, Y),$$

in particular, if $\dagger$ denotes any of the symbols α, β or γ, then

$$(u^{-1} * X)^\dagger = (1/u)^{-1} * X^\dagger.$$

We also have

$$A \in (u^{-1} * X, v^{-1} * Y) \text{ if and only if } B \in (X, Y),$$

where

$$b_{nk} = \frac{v_n a_{nk}}{u_k} \text{ for } n, k = 0, 1, \ldots.$$

Proof Since $x \in u^{-1} * X$ if and only if $y = u \cdot x \in X$ and $a \cdot x = b \cdot y$ where $b = a/u = a \cdot (1/u)$, it follows that $a \in M(u^{-1} * X, Y)$ if and only if $b \in M(X, Y)$, that is, $a \in (1/u)^{-1} * M(X, Y)$. □

First we determine the multiplier space of X_Σ in Y. From this, we will easily obtain the α-, β- and γ-duals of X_T in the special case $T = \Sigma$.

A subset X of ω is said to be *normal*, if $x \in X$ and $|y_k| \le |x_k|$ for all k together imply $y \in X$.

Let $T^- = (t^-_{nk})_{n,k=0}^\infty$ be the matrix with $t^-_{n,n-1} = 1$ and $t^-_{nk} = 0$ $(k \neq 0)$ for $n = 0, 1, \ldots$, that is, T^- is the matrix of the *right shift operator*. We also write $\Delta^+ = (\Delta^+_k)_{n,k=0}^\infty$ for the matrix of the *forward differences* with

$$\Delta^+_{nk} = \begin{cases} 1 & (k = n) \\ -1 & (k = n+1) \\ 0 & (k \neq n, n+1) \end{cases} \qquad (n = 0, 1, \ldots).$$

We obviously have $\Delta = -\Delta^+ \cdot T^- = -T^- \cdot \Delta^+$.

Theorem 2.13 ([39, Theorem 2.3]) *Let X be any set of sequences, Y be a linear subspace of ω and $Y \subset Y_{T^-}$. We put*

$$Z_1 = (M(X,Y))_{\Delta^+}\,,\quad Z_2 = M(X, Y_\Delta) \text{ and } Z_3 = M(X,Y).$$

Then, we have

$$Z_1 \cap Z_2 \subset M\left(X_\Sigma, Y\right). \tag{2.22}$$

If, in addition, X and Y are normal and $Y_{T^-} \subset Y$ then

$$M\left(X_\Sigma, Y\right) = Z_1 \cap Z_3. \tag{2.23}$$

Proof We write $Z = X_\Sigma$ and observe that $z \in Z$ if and only if $x = \Sigma z \in X$; furthermore, we have $z = \Delta x$. We obtain

$$\begin{aligned} T^-(x \cdot \Delta^+ a) + \Delta(a \cdot x) &= (T^- x) \cdot (T^-(\Delta^+ a)) + a \cdot x - T^-(a \cdot x) \\ &= -(T^- x) \cdot (\Delta a) + a \cdot x - (T^- a) \cdot (T^- x) \\ &= -(T^- x) \cdot (\Delta a + T^- a) + a \cdot x \\ &= a \cdot (x - T^- x) = a \cdot \Delta x = a \cdot z, \end{aligned}$$

that is,

$$a \cdot z = T^-(x \cdot \Delta^+ a) + \Delta(a \cdot x). \tag{2.24}$$

(i) First we show the inclusion in (2.22)
We assume $a \in Z_1 \cap Z_2$. Let $z \in Z$ be given. Then $x \in X$, and $a \in Z_1$ implies $\Delta^+ \in M(X,Y)$, hence $x \cdot \Delta^+ a \in Y \subset Y_{T^-}$, that is, $T^-(x \cdot \Delta^+ a) \in Y$. Also, $a \in Z_2$ implies $a \cdot x \in Y_\Delta$, hence $\Delta(a \cdot x) \in Y$. Since Y is a linear space, (2.22) follows from (2.24).

(ii) Now we show that $Y \subset Y_{T^-}$ implies

$$Z_3 \subset Z_2. \tag{2.25}$$

Let $a \in Z_3$ and $x \in X$ be given. Then we have $a \cdot x \in Y \subset Y_{T^-}$, hence $T^-(a \cdot x) \in Y$, and so

$$\Delta(a \cdot x) = a \cdot x - T^-(a \cdot x) \in Y, \text{ since } Y \text{ is a linear space.}$$

Thus, we have $a \in M(X, Y_\Delta) = Z_2$ and we have shown (2.25).

(iii) Now we show that if X and Y are normal and $Y_{T^-} \subset Y$, then

$$M(Z,Y) \subset Z_1 \cap Z_3. \tag{2.26}$$

We assume $a \in M(Z,Y)$. Let $x \in X$ be given. Then we have $a \cdot \Delta x = a \cdot z \in Y$. We define the sequence $\tilde{x}$ by

$$\tilde{x}_n = (-1)^n |x_n| \text{ for } n = 0, 1, \ldots.$$

Since X is normal, it follows that $\tilde{x} \in X$, hence $\tilde{z} = \Delta\tilde{x} \in Z$, and consequently

$$a \cdot \tilde{z} = \left((-1)^n a_n \left(|x_n| + |x_{n-1}|\right)\right)_{n=0}^{\infty} \in Y.$$

Furthermore, $|a_n x_n| \leq |a_n \tilde{z}_n|$ for all n implies $a \cdot x \in Y$, since Y is normal. This shows $a \in Z_3 = M(X, Y)$. By (2.25), this implies $a \in Z_2$, that is, $\Delta(a \cdot x) \in Y$. Therefore, we obtain from (2.24)

$$T^-(x \cdot \Delta^+ a) \in Y, \text{ since } Y \text{ is a linear space,}$$

and thus $x \cdot (\Delta a) \in Y_{T^-} \subset Y$, that is, $a \in Z_1$. Thus, we have shown (2.26).

Finally, we obtain (2.23) from (2.26), (2.25) and (2.22). □

2.4 The α-, β- and γ-duals of X_Σ

Now we apply Theorem 2.13 to determine the α-, β- and γ-duals of X_Σ.

Since ℓ_1 is a normal linear space with $(\ell_1)_{T^-} = \ell_1$ we immediately obtain from Theorem 2.13.

Corollary 2.6 ([39, Corollary 2.1]) *For any subset X of ω, we have*

$$(X^\alpha)_{\Delta^+} \cap X^\alpha \subset (X_\Sigma)^\alpha. \tag{2.27}$$

If X is normal, then we have

$$(X_\Sigma)^\alpha = (X^\alpha)_{\Delta^+} \cap X^\alpha. \tag{2.28}$$

Remark 2.2 ([39, Remark 2.1]) If X is normal with $X_{T^-} = X$, then we have $(X_\Sigma)^\alpha = X^\alpha$.

Proof We show that $X \subset X_{T^-}$ implies $X^\alpha \subset (X^\alpha)_{\Delta^+}$. Then statement follows from (2.28).

Let $a \in X^\alpha$ and $x \in X$ be given. Then, we have $a \cdot x \in \ell_1$, and $a \cdot T^- x \in \ell_1$, since $X \subset X_{T^-}$. It also follows from

$$T_n^-(x \cdot \Delta^+ a) = x_{n-1}\Delta_{n-1}^+ = x_{n-1}a_{n-1} - x_{n-1}a_n = T_n^-(x \cdot a) - a_n T_n^- x$$

that

$$\left|T_n^-(x \cdot \Delta^+ a)\right| \leq \left|T_n^-(x \cdot a)\right| + \left|a_n T_n^- x\right| \text{ for } n = 0, 1, \ldots.$$

Therefore, we obtain $T^-(x \cdot \Delta^+ a) \in \ell_1$, hence $x \cdot \Delta^+ a \in \ell_1$ and thus $a \in (X^\alpha)_{\Delta^+}$. □

Example 2.9 We obtain from Remark 2.2

$$\left((\ell_p)_\Sigma\right)^\alpha = \begin{cases} \ell_\infty & (p = 1) \\ \ell_q & (1 < p < \infty; q = p/(p-1)), \end{cases}$$
$$((\ell_\infty)_\Sigma)^\alpha = bs^\alpha = \ell_\infty^\alpha = \ell_1 = c_0^\alpha = ((c_0)_\Sigma)^\alpha$$

and $\ell_1 = bs^\alpha \subset cs^\alpha \subset ((c_0)_\Sigma)^\alpha = \ell_1$ implies $cs^\alpha = \ell_1$.

Now we apply Theorem 2.13 to determine the β- and γ-duals of X_Σ.

Corollary 2.7 ([39, Corollary 2.2]) *Let X be any subset of ω. We write*

$$Z_1^\dagger = \left(X^\dagger\right)_{\Delta^+} \textit{ for } \dagger = \beta, \gamma, \ Z_2^\beta = M(X, c), \ Z_2^\gamma = M(X, \ell_\infty) \textit{ and } Z_3 = M(X, c_0).$$

Then we have

$$Z_1^\dagger \cap Z_2^\dagger \subset (X_\Sigma)^\dagger \textit{ for } \dagger = \beta, \gamma. \tag{2.29}$$

If, in addition, X is normal, then we have

$$(X_\Sigma)^\beta = Z_1^\beta \cap Z_3 \textit{ and } (X_\Sigma)^\gamma = Z_1^\gamma \cap Z_2^\gamma. \tag{2.30}$$

If $a \in (X_\Sigma)^\beta$, then we have

$$\sum_{k=0}^\infty a_k z_k = \sum_{k=0}^\infty \left(\Delta_k^+ a\right)(\Sigma_k z) \textit{ for all } z \in X_\Sigma. \tag{2.31}$$

Proof We put $Z = X_\Sigma$, $Y(1; \beta) = cs$, $Y(1; \gamma) = bs$ and $Y(2; \dagger) = (Y(1; \dagger))_\Delta$. Since cs and bs are linear spaces with $cs \subset cs_{T^-}$ and $bs \subset bs_{T^-}$, and since $cs_\Delta = c$ and $bs_\Delta = \ell_\infty$, (2.22) in Theorem 2.13 implies for $\dagger = \beta, \gamma$

$$\left(X^\dagger\right)_{\Delta^+} \cap M\left(X, (Y(1; \dagger))_\Delta\right) = \left(X^\dagger\right)_{\Delta^+} \cap M\left(X, Y(2; \dagger)\right) = Z_1^\dagger \cap Z_2^\dagger \subset Z^\dagger.$$

Now let X be normal and $a \in Z^\dagger$. We write $\tilde{Y}(2; \beta) = c_0$ and $\tilde{Y}(2; \gamma) = \ell_\infty$. First $a \in Z^\dagger$ implies $a \cdot z \in \tilde{Y}(2; \dagger)$ for all $z \in Z$. Since $\tilde{Y}(2; \dagger)$ is normal, we conclude $a \in M(X, \tilde{Y}(2; \dagger))$ by (2.23) in Theorem 2.13. We obtain from (2.24) with $x = \Sigma z$

$$\Sigma(a \cdot z) = T^-\left(x \cdot \Delta^+ a\right) + a \cdot x. \tag{2.32}$$

Now $\Sigma(a \cdot z) \in Y(2; \dagger)$ and $a \cdot x \in Y(2; \dagger)$ together imply $(\Delta^+ a) \cdot x \in Y(1; \dagger)$ for all $x \in X$, that is,

$$Z^\beta \subset Z_1^\beta \cap Z_3 \text{ and } Z^\gamma \subset Z_1^\gamma \cap Z_2^\gamma. \tag{2.33}$$

Since $Z_3 = M(X, c_0) \subset M(x, c) = Z_2^{\beta}$, the identities in (2.30) follow from (2.33) and (2.29).
Finally, (2.32) and (2.30) together imply (2.31). □

As a first application of Corollary 2.7, we determine the α-, β- and γ-duals of the matrix domain of Σ in the classical sequence spaces ℓ_p $(1 \le p \le \infty)$, c_0 and c.

We need the following result.

Proposition 2.2 ([39, Lemma 3.1]) *We have*

(a) $M(c_0, c_0) = \ell_\infty$,
(b) $M(c, c) = c$,
(c) $M(\ell_\infty, c_0) = c_0$,
(d) $M(\ell_p, c_0) = \ell_\infty$ $(1 \le p < \infty)$,
(e) $M(\ell_p, \ell_\infty) = \ell_\infty$ $(1 \le p \le \infty)$,
(f) $M(c_0, \ell_\infty) = \ell_\infty$.

Proof (b) This is Part (ii) in Example 1.7.
(a), (c) We have by Part (i) in Proposition 1.1, Parts (i) and (iii) in Example 1.7

$$M(c_0, c_0) \subset M(c_0, c) = \ell_\infty \text{ and } M(\ell_\infty, c_0) \subset M(\ell_\infty, c) = c_0.$$

Conversely, if $a \in \ell_\infty$, then we have $a \cdot x \in c_0$ for all $x \in c_0$, that is, $a \in M(c_0, c_0)$; and if $a \in c_0$ then $a \cdot x \in c_0$ for all $x \in \ell_\infty$, that is, $a \in M(\ell_\infty, c_0)$. Thus, we have shown Parts (a) and (c).
(d) Let $1 \le p < \infty$. Since $\ell_p \subset c_0$, it follows from Part (ii) of Proposition 1.1 and Part (a) that

$$M(\ell_p, c_0) \supset M(c_0, c_0) = \ell_\infty. \tag{2.34}$$

Conversely, if $a \notin \ell_\infty$, then there exists a subsequence $(a_{k(j)})_{j=0}^{\infty}$ of the sequence a such that $|a_{k(j)}| > (j+1)^2$ for $j = 0, 1, \ldots$. We put

$$x_k = \begin{cases} \dfrac{1}{a_{k(j)}} & (k = k(j)) \\ 0 & (k \ne k(j)) \end{cases} \quad (j = 0, 1, \ldots).$$

Then we have

$$\sum_{k=0}^{\infty} |x_k|^p = \sum_{j=0}^{\infty} \left(\frac{1}{|a_{k(j)}|} \right)^p < \sum_{j=0}^{\infty} \frac{1}{(j+1)^{2p}} < \infty,$$

that is, $x \in \ell_p$. But $a_{k(j)} x_{k(j)} = 1$ for $j = 0, 1, \ldots$, hence $a \cdot x \notin c_0$, that is, $a \notin M(\ell_p, c_0)$. This shows

$$M(\ell_p, c_0) \subset \ell_\infty. \tag{2.35}$$

Now Part (d) follows from (2.34) and (2.35).
(e) Since trivially $\ell_\infty \subset M(\ell_\infty, \ell_\infty)$ it follows from Part (ii) in Proposition 1.1 that

$$\ell_\infty \subset M(\ell_\infty, \ell_\infty) \subset M(\ell_p, \ell_\infty) \text{ for } 1 \le p \le \infty. \tag{2.36}$$

Conversely, we assume $a \notin \ell_\infty$. Then there is a subsequence $(a_{k(j)})_{j=0}^\infty$ of the sequence a such that $|a_{k(j)}| > 2^j$ for $j = 0, 1, \dots$. We put

$$x_k = \begin{cases} \dfrac{1}{\sqrt{|a_{k(j)}|}} & (k = k(j)) \\ 0 & (k \neq k(j)) \end{cases} \quad (k = 0, 1, \dots)$$

Then we have

$$\sum_{k=0}^\infty |x_k| = \sum_{j=0}^\infty \frac{1}{\sqrt{|a_{k(j)}|}} < \sum_{j=0}^\infty \frac{1}{\left(\sqrt{2}\right)^j} < \infty,$$

that is, $x \in \ell_1 \subset \ell_p$ $(1 \le p \le \infty)$. But

$$a_{k(j)}x_{k(j)} = \sqrt{|a_{k(j)}|} > \left(\sqrt{2}\right)^j \text{ for } j = 0, 1, \dots,$$

hence $a \cdot x \notin \ell_\infty$, that is, $a \notin M(\ell_p, c_0)$. Thus, we have shown

$$M(\ell_p, \ell_\infty) \subset \ell_\infty \text{ for } 1 \le p \le \infty. \tag{2.37}$$

Now Part (e) follows from (2.36) and (2.37).
(f) We obtain from Part (ii) in Proposition 1.1 and Part (e)

$$\ell_\infty \subset M(\ell_\infty, \ell_\infty) \subset M(c_0, \ell_\infty) \subset M(\ell_1, \ell_\infty) = \ell_\infty.$$

□

Now we can determine the duals $(X_\Sigma)^\dagger$ for $\dagger = \beta, \gamma$ and $X = \ell_p, c_0, c$ $(1 \le p \le \infty)$.

Corollary 2.8 *We have*

(a)

$$((\ell_p)_\Sigma)^\beta = \begin{cases} \ell_\infty & (p = 1) \\ (\ell_q)_{\Delta^+} \cap \ell_\infty & (1 < p < \infty; q = p/(p-1)), \end{cases}$$

$$((c_0)_\Sigma)^\beta = (c_\Sigma)^\beta = cs^\beta = bv \text{ ([65, Theorem, 7.3.5 (v))]},$$

$$((\ell_\infty)_\Sigma)^\beta = bs^\beta = bv \cap c_0 \text{ ([65, Theorem 7.3.5 (vi)]);}$$

(b)

$$((\ell_p)_\Sigma)^\gamma = ((\ell_p)_\Sigma)^\beta \text{ for } 1 \le p < \infty$$

and

$$((c_0)_\Sigma)^\gamma = cs^\gamma = bs^\gamma = bv \text{ ([65, Theorem 7.3.5 (v), (vii)]).}$$

Proof Since ℓ_p $(1 \le p \le \infty)$ and c_0 are normal, we can apply (2.30) in Corollary 2.7 to obtain

$$((\ell_p)_\Sigma)^\beta = \left(\ell_p^\beta\right)_{\Delta^+} \cap M(\ell_p, c_0) \text{ and } ((c_0)_\Sigma)^\beta = \left(c_0^\beta\right)_{\Delta^+} \cap M(c_0, c_0) \quad (2.38)$$

and

$$((\ell_p)_\Sigma)^\gamma = \left(\ell_p^\gamma\right)_{\Delta^+} \cap M(\ell_p, \ell_\infty) \text{ and } ((c_0)_\Sigma)^\gamma = \left(c_0^\gamma\right)_{\Delta^+} \cap M(c_0, \ell_\infty). \quad (2.39)$$

We observe that $\ell_p^\beta = \ell_p^\gamma = \ell_q$ for $1 \le p \le \infty$ where $q = \infty$ for $p = 1$ and $q = 1$ for $p = \infty$, $c_0^\beta = c_0^\gamma = \ell_1$, $M(\ell_p, c_0) = M(\ell_p, \ell_\infty) = \ell_\infty$ for $1 \le p < \infty$ by Parts (d) (e) in Proposition 2.2, and $M(c_0, c_0) = M(c_0, \ell_\infty) = \ell_\infty$ by Parts (a) and (f) of Proposition 2.2. Hence, it follows from (2.38) and (2.39) that

$$((\ell_p)_\Sigma)^\beta = ((\ell_p)_\Sigma)^\gamma = \left(\ell_q\right)_{\Delta^+} \cap \ell_\infty \text{ for } 1 \le p < \infty$$

and

$$((c_0)_\Sigma)^\beta = ((c_0)_\Sigma)^\gamma = (\ell_1)_{\Delta^+} \cap \ell_\infty = bv \cap \ell_\infty.$$

Since obviously $\ell_\infty \subset (\ell_\infty)_{\Delta^+}$ and $bv \subset \ell_\infty$, we have

$$((\ell_1)_\Sigma)^\beta = \ell_\infty \text{ and } ((c_0)_\Sigma)^\beta = bv.$$

Now if $p = \infty$ then we have by (2.38) and Part (c) of Proposition 2.2

$$((\ell_\infty)_\Sigma)^\beta = (\ell_1)_{\Delta^+} \cap M(\ell_\infty, c_0) = bv \cap c_0$$

and by (2.39) and Part (e) of Proposition 2.2

$$((\ell_\infty)_\Sigma)^\gamma = (\ell_1)_{\Delta^+} \cap M(\ell_\infty, \ell_\infty) = bv \cap \ell_\infty = bv.$$

It remains to show the statements for cs.
We obtain from (2.29) in Theorem 2.7 and Part (b) of Proposition 2.2

$$cs^\beta = (c_\Sigma)^\beta \supset (c^\beta)_{\Delta^+} \cap M(c, c) = bv \cap c,$$

but $bv \subset c$, so $cs^\beta \supset bv$. Conversely, we have $cs^\beta = (c_\Sigma)^\beta \subset ((c_0)_\Sigma)^\beta = bv$. Finally,

$$bv = bs^\gamma \supset cs^\gamma \supset ((c_0)_\Sigma)^\gamma = bv$$

yields $cs^\gamma = bv$. □

Unlike as in the case of $p = 1$, the β- and γ-duals of $(\ell_p)_\Sigma$ cannot be reduced for $1 < p < \infty$.

Remark 2.3 We have

$$(\ell_p)_{\Delta^+} \not\subset \ell_\infty \text{ and } \ell_\infty \not\subset (\ell_p)_{\Delta^+} \text{ for } 1 < p < \infty.$$

Proof We define the sequence x by $x_k = k^{1/2q}$ $(k = 0, 1, \dots)$, where $q = p/(p-1)$. Then, by the inequality in (A.5), there exists a constant M such that

$$\left|\Delta_k^+ x\right| = |x_k - x_{k+1}| \le M \cdot (k+1)^{1/2q-1} = M \cdot (k+1)^{1/q-1-1/2q} \text{ for } k = 1, 2, \dots,$$

hence

$$\left|\Delta_k^+ x\right|^p \le M^p \cdot (k+1)^{p(-(1-1/q))-p/2q} = M^p \cdot (k+1)^{-1-p/2q} \text{ for } k = 1, 2, \dots.$$

Since $p/2q > 0$, it follows that

$$\sum_{k=1}^{\infty} \left|\Delta_k^+ x\right|^p < \infty,$$

hence $x \in (\ell_p)_{\Delta^+}$, but clearly $x \notin \ell_\infty$. This shows $(\ell_p)_{\Delta^+} \not\subset \ell_\infty$.
Now we define the sequence x by $x_k = (-1)^k$ $(k = 0, 1, \dots)$. Then clearly $x \in \ell_\infty$, but $|\Delta_k^+ x| = |x_k - x_{k+1}| = 2$ for all k, hence $x \notin (\ell_p)_{\Delta^+}$. This shows $\ell_\infty \not\subset (\ell_p)_{\Delta^+}$. □

Now we consider the so-called spaces of *generalized weighted means*.

Definition 2.4 Let $u, v \in \mathcal{U}$ and $X \in \omega$. Then the sets $W(u, v; X)$ of *generalized weighted means* are defined by

$$W(u, v; X) = v^{-1} * (u^{-1} * X)_\Sigma = \{x \in \omega : u * \Sigma(v \cdot x) \in X\}.$$

There are the following special cases.

Example 2.10 (a) We have $W(e, e; c) = cs$ and $W(e, e; \ell_\infty) = bs$.
(b) If $v = q$ is a non-negative sequence with $q_0 > 0$ and $u = 1/Q$ with $Q_n = \sum_{k=0}^n q_k$ for $n = 0, 1, \dots$, then we obtain the sets of weighted means of Definition 2.2

$$W(1/Q, q; c_0) = (\bar{N}, q)_0, \quad W(1/Q, q; c) = (\bar{N}, q) \text{ and } W(1/Q, q; \ell_\infty) = (\bar{N}, q)_\infty.$$

(c) If $v = (1 + r^k)_{k=0}^{\infty}$ for some fixed r with $0 < r < 1$ and $u = (1/(n+1))_{n=0}^{\infty}$, then we obtain the sets

$$a_p^r = W(u, v; \ell_p) \ (1 \leq p < \infty) \ ([10]),$$

and

$$a_0^r = W(u, v; c_0), \ a^r = W(u, v; c) \text{ and } a_\infty^r = W(u, v; \ell_\infty) \ ([11]).$$

(d) If $v = e$ and $u = (1/(n+1))_{n=0}^{\infty}$, then we obtain the sets of Example 2.5

$$W(u, v; c_0) = C_0^{(1)}, \ W(u, v; c) = C^{(1)} \text{ and } W(u, v; \ell_\infty) = C_\infty^{(1)},$$

and the *Cesàro sequence spaces of non-absolute type* $X_p = W(u, v; \ell_p)$ for $1 \leq p \leq \infty$ ([57]).

Now we give the α-, β- and γ-duals of the generalized weighted means.

Corollary 2.9 ([39, Theorem 2.4]) *Let* $u, v \in \mathcal{U}$, $b = (1/u) \cdot \Delta^+(a/v)$ *and* $d = a/(u \cdot v)$. *Then, we have*

(a)
$$\begin{aligned}
(W(u, v; X))^\alpha &\supset \{a \in \omega : b \in X^\alpha \text{ and } d \in X^\alpha\},\\
(W(u, v; X))^\beta &\supset \{a \in \omega : b \in X^\beta \text{ and } d \in M(X, c)\},\\
(W(u, v; X))^\gamma &\supset \{a \in \omega : b \in X^\gamma \text{ and } d \in M(X, \ell_\infty)\}.
\end{aligned}$$

(b) If, in addition, X is normal, then

$$\begin{aligned}
(W(u, v; X))^\alpha &= \{a \in \omega : b \in X^\alpha \text{ and } d \in X^\alpha\},\\
(W(u, v; X))^\beta &= \{a \in \omega : b \in X^\beta \text{ and } d \in M(X, c_0)\},\\
(W(u, v; X))^\gamma &= \{a \in \omega : b \in X^\gamma \text{ and } d \in M(X, \ell_\infty)\}.
\end{aligned}$$

If $a \in (W(u, v; X))^\beta$, *then we have*

$$\sum_{k=0}^{\infty} a_k z_k = \sum_{k=0}^{\infty} \Delta_k^+(a/v) \Sigma_k(v \cdot z) \text{ for all } z \in W(u, v; X). \tag{2.40}$$

Proof Writing $W = W(u, v; X)$ we have by Proposition 2.1

$$W^\dagger = (1/v)^{-1} * ((u^{-1} * X)_\Sigma)^\dagger \text{ for } \dagger = \alpha, \beta, \gamma.$$

(i) We put $Y = u^{-1} * X$ and obtain by the inclusion in (2.27) of Corollary 2.6 and Proposition 2.1

$$(Y_\Sigma)^\alpha \supset (Y^\alpha)_{\Delta^+} \cap Y^\alpha = \left((1/u)^{-1} * X^\alpha\right)_{\Delta^+} \cap (1/u)^{-1} * X^\alpha. \tag{2.41}$$

Hence if

$$a \in (1/v)^{-1} * \left[\left((1/u)^{-1} * X^\alpha\right)_{\Delta^+} \cap ((1/u)^{-1} * X^\alpha)\right],$$

that is, $(1/u) \cdot \Delta^+(a/v) = b \in X^\alpha$ and $(1/v) \cdot (1/u) = 1/(u \cdot v) = d \in X^\alpha$, then $a \in W^\alpha$.
This shows

$$W^\alpha \supset \{a \in \omega : b \in X^\alpha \text{ and } d \in X^\alpha\}.$$

Also, if X is normal then there is equality in (2.41) by (2.28) in Corollary 2.6, and so $(1/v) \cdot (1/u) = 1/(u \cdot v) = d \in X^\alpha$.
Thus, we have shown the first statements in Parts (a) and (b).

(ii) Now we obtain by the inclusion in (2.29) of Corollary 2.7 and Proposition 2.1

$$(Y_\Sigma)^\dagger \supset (Y^\dagger)_{\Delta^+} \cap M(Y, c) = \left((1/u)^{-1} * X^\dagger\right)_{\Delta^+} \cap \left((1/u)^{-1} * M(X, c)\right).$$

Arguing as at the end of the proof of Part (i), we obtain the second and third inclusions in Part (a).
Furthermore, if X is normal then we obtain by the first identity in (2.30) of Corollary 2.7 and Proposition 2.1

$$(Y_\Sigma)^\beta = (Y^\beta)_{\Delta^+} \cap M(Y, c_0) = \left((1/u)^{-1} * X^\beta\right) \cap \left((1/u)^{-1} * M(X, c_0)\right),$$

and by the second identity in (2.30) of Corollary 2.7 and Proposition 2.1

$$(Y_\Sigma)^\gamma = (Y^\gamma)_{\Delta^+} \cap M(Y, \ell_\infty) = \left((1/u)^{-1} * X^\gamma\right) \cap \left((1/u)^{-1} * M(X, \ell_\infty)\right).$$

Now the second and third identities in Part (b) follow as before.

(iii) Finally, we show (2.40).
Let $a \in W^\beta$ and $z \in W$ be given. Then we have $x = u \cdot \Sigma(v \cdot z) \in X$, $z = (1/v) \cdot \Delta(x/u)$ and

$$\begin{aligned}\sum_{k=0}^{n} a_k z_k &= \sum_{k=0}^{n} \frac{a_k}{v_k} \Delta_k(x/u) = \sum_{k=0}^{n-1} \Delta_k^+(a/v) \cdot \frac{x_k}{u_k} + \frac{a_n x_n}{u_n v_n} \\ &= \sum_{k=0}^{n-1} \Delta_k^+(a/v) \cdot \Sigma_k(v \cdot z) + d_n x_n \text{ for } n = 0, 1, \ldots.\end{aligned} \tag{2.42}$$

It follows from $d \in M(X, c_0)$ that $d \cdot x \in c_0$, and so (2.40) follows. □

Now we consider the special case of Corollary 2.9 when X is any of the classical sequence spaces ℓ_p $(1 \le p \le \infty)$, c_0 and c.

Corollary 2.10 ([39, Theorem 3.1]) *We write $q = \infty$ for $p = 1$, $q = p/(p-1)$ for $1 < p < \infty$ and $q = 1$ for $p = \infty$. Using the notations of* Corollary 2.9, *we obtain*

$$\begin{aligned}
\left(W(u,v;\ell_p)\right)^\alpha &= \left\{a \in \omega : b \in \ell_q \text{ and } d \in \ell_q\right\} \text{ for } 1 \le p \le \infty,\\
(W(u,v;c))^\alpha &= (W(u,v;c_0))^\alpha = \{a \in \omega : b \in \ell_1 \text{ and } d \in \ell_1\},\\
\left(W(u,v;\ell_p)\right)^\beta &= \left\{a \in \omega : b \in \ell_q \text{ and } d \in \ell_\infty\right\} \text{ for } 1 \le p < \infty,\\
(W(u,v;\ell_\infty))^\beta &= \{a \in \omega : b \in \ell_1 \text{ and } d \in c_0\},\\
(W(u,v;c_0))^\beta &= \{a \in \omega : b \in \ell_1 \text{ and } d \in \ell_\infty\},\\
(W(u,v;c))^\beta &= \{a \in \omega : b \in \ell_1 \text{ and } d \in c\},\\
\left(W(u,v;\ell_p)\right)^\gamma &= \left\{a \in \omega : b \in \ell_q \text{ and } d \in \ell_\infty\right\} \text{ for } 1 \le p \le \infty
\end{aligned}$$

and

$$(W(u,v;c))^\gamma = (W(u,v;c_0))^\gamma = \{a \in \omega : b \in \ell_1 \text{ and } d \in \ell_\infty\}.$$

Proof All the assertions except for the duals of $W(u,v;c)$ are immediate consequences of Corollary 2.9 and Proposition 2.2.
Furthermore,

$$W(u,v;c_0) \subset W(u,v;c) \subset W(u,v;\ell_\infty) \text{ and } (W(u,v;c_0))^\alpha = (W(u,v;\ell_\infty))^\alpha$$

imply $(W(u,v;c_0))^\alpha = (W(u,v;c))^\alpha$.
We also have by Part (ii) of Proposition 1.1 and the second identity of Part (b) of Corollary 2.9

$$\begin{aligned}
&(W(u,v;c))^\beta \subset (W(u,v,c_0))^\beta = \left\{a \in \omega : (1/u)\cdot\Delta^+(a/v) \in \ell_1 \text{ and } d \in \ell_\infty\right\}\\
&\subset \left\{a \in \omega : (1/u)\cdot\Delta^+(a/v) \in c_0\right\}.
\end{aligned}$$

So $a \in (W(u,v;c))^\beta$ implies $d \in c$ by (2.42), hence

$$(W(u,v;c))^\beta \subset (W(u,v,c_0))^\beta = \left\{a \in \omega : (1/u)\cdot\Delta^+(a/v) \in \ell_1 \text{ and } d \in c\right\}.$$

Conversely, we have by (2.29) in Corollary 2.7

$$(W(u,v;c))^\beta \supset \left\{a \in \omega : (1/u)\cdot\Delta^+(a/v) \in \ell_1 \text{ and } d \in c\right\}.$$

The part concerning the γ-duals is proved similarly. □

Corollary 2.10 yields many special cases.

Example 2.11 ([39, Example 3.1])

(a) ([27, Theorem 6] for the β-duals)
Let $(q_k)_{k=0}^{\infty}$ be a non-negative sequence with $q_0 > 0$ and $Q_n = \sum_{k=0}^{n} q_k$ for $n = 0, 1, \ldots$. We put

$$b_k = Q_k \left(\frac{a_k}{q_k} - \frac{a_{k+1}}{q_{k+1}} \right) \text{ and } d_k = \frac{Q_k a_k}{q_k} \text{ for } k = 0, 1, \ldots .$$

Then it follows from Corollary 2.10 that

$$\begin{aligned}
\left((\bar{N}, q)_0\right)^{\alpha} = (\bar{N}, q)^{\alpha} = \left((\bar{N}, q)_\infty\right)^{\alpha} &= \{a \in \omega : b \in \ell_1 \text{ and } d \in \ell_1\}, \\
\left((\bar{N}, q)_0\right)^{\beta} &= \{a \in \omega : b \in \ell_1 \text{ and } d \in \ell_\infty\}, \\
\left(\bar{N}, q\right)^{\beta} &= \{a \in \omega : b \in \ell_1 \text{ and } d \in c\}, \\
\left((\bar{N}, q)_\infty\right)^{\beta} &= \{a \in \omega : b \in \ell_1 \text{ and } d \in c_0\}, \\
\left((\bar{N}, q)_0\right)^{\gamma} = (\bar{N}, q)^{\gamma} = \left((\bar{N}, q)_\infty\right)^{\gamma} &= \{a \in \omega : b \in \ell_1 \text{ and } d \in \ell_\infty\}.
\end{aligned}$$

(b) ([57, Theorem 6] for X_p^{β} $(1 < p \le \infty)$) If $v = e$ and $u = (1/(n+1))_{n=1}^{\infty}$, then we put

$$b_k = (k+1)(a_k - a_{k+1}) \text{ and } d_k = (k+1)a_k \text{ for } k = 0, 1, \ldots,$$

and obtain from Corollary 2.10

$$\begin{aligned}
X_p^{\alpha} &= \left\{a \in \omega : d \in \ell_q\right\} \ (1 \le p \le \infty), \\
X_p^{\beta} = X_p^{\gamma} &= \left\{a \in \omega : b \in \ell_q \text{ and } d \in \ell_\infty\right\}, \ (1 \le p < \infty) \\
X_\infty^{\beta} = C_\infty^{\beta} &= \{a \in \omega : b \in \ell_1 \text{ and } d \in c_0\}, \\
C_0^{\beta} &= \{a \in \omega : b \in \ell_1 \text{ and } d \in \ell_\infty\}, \\
C^{\beta} &= \{a \in \omega : b \in \ell_1 \text{ and } d \in c\}
\end{aligned}$$

and

$$X_\infty^{\gamma} = C_\infty^{\gamma} = \{a \in \omega : b \in \ell_1 \text{ and } d \in c_0\}.$$

(c) ([65, Theorem 7.3.5 (ii), (v), (vii) and (v)] for the β- and γ-duals)
Let $cs_0 = (c_0)_\Sigma$. Then $cs_0^{\alpha} = \ell_1$. Furthermore, $bs^{\alpha} = \ell_1$, and so $cs^{\alpha} = \ell_1$; $bs^{\alpha} = bv_0 = bv \cap c_0$; $cs^{\beta} = bv$, since $bv \subset c$; $bs^{\gamma} = bv \cap \ell_\infty = bv$ and $cs^{\gamma} = bv$.

2.5 The α- and β-duals of $X_{\Delta^{(m)}}$

In this section, we study the α-, β- and γ-duals of the matrix domains of the difference operator. The situation is more complicated than in the case of the duals of matrix domains of the sum operator in Sect. 2.4.

We start, however, with an almost trivial result.

Proposition 2.3 *Let a be a sequence, T be a triangle and S be its inverse. We define the matrix* $B = B(T; a) = (b_{nk})_{n,k=0}^{\infty}$ *by*

$$B_n = a_n \cdot S_n \text{ for all } n, \text{ that is, } b_{nk} = a_n s_{nk} \text{ for all } n, k = 0, 1, \dots.$$

Then we have for all subsets X and Y of ω,

$$a \in M(X_T, Y) \text{ if and only if } B \in (X, Y). \tag{2.43}$$

Proof Since $z \in Z = X_T$ if and only if $x = Tz \in X$, and $z = Sx$, it follows that

$$a_n z_n = a_n(S_n x) = a_n \sum_{k=0}^{n} s_{nk} x_k = \sum_{k=0}^{n} a_n s_{nk} x_k = B_n x \text{ for } n = 0, 1, \dots, \tag{2.44}$$

and the statement in (2.43) is an immediate consequence. □

We apply Proposition 2.3 to determine the α-dual of the set $bv = (\ell_1)_\Delta$.

Example 2.12 Let $T = \Delta$. Then we have $S = \Sigma$ (Example 2.3), and so the matrix B of Proposition 2.3 is defined by

$$B_n = a_n e^{[n]}, \text{ that is, } b_{nk} = \begin{cases} a_n & (0 \le k \le n) \\ 0 & (k > n) \end{cases} \quad (n = 0, 1, \dots).$$

Therefore, we have by (2.43)

$$a \in bv^\alpha = ((\ell_1)_\Delta)^\alpha \text{ if and only if } B \in (\ell_1, \ell_1),$$

and by **19.** in Theorem 1.23, this is the case if and only if the condition in (19.1) is satisfied, that is, if and only if

$$\sup_k \sum_{n=0}^{\infty} |b_{nk}| = \sup_k \sum_{n=k}^{\infty} |a_n| < \infty,$$

hence $a \in \ell_1$. Therefore, we have

$$bv^\alpha = \ell_1. \tag{2.45}$$

Now we determine the β- and γ-duals of the set bv.

Example 2.13 Using the notations of Example 2.12, we obtain from (2.44),

$$\sum_{k=0}^{n} a_k z_k = \sum_{k=0}^{n}\sum_{j=0}^{k} b_{kj} x_j = \sum_{j=0}^{n}\left(\sum_{k=j}^{n} b_{jk}\right) x_j$$

for all $z \in X_T$, that is, for all $x = Sz \in X$. We write $C = (c_{nk})_{n,k=0}^{\infty}$ for the matrix with

$$c_{nk} = \begin{cases} \sum_{j=k}^{n} b_{jk} & (0 \le k \le n) \\ 0 & (k > n) \end{cases} \quad (n = 0, 1, \dots),$$

and obtain

$$a \in M(X_T, Y_\Sigma) \text{ if and only if } C \in (X, Y). \tag{2.46}$$

If $T = \Delta$, then we have

$$c_{nk} = \begin{cases} \sum_{j=k}^{n} a_j & (0 \le k \le n) \\ 0 & (k > n) \end{cases} \quad (n = 0, 1, \dots)$$

and (2.46) yields for $X = \ell_1$ and $Y = c$

$$a \in bv^\beta = M(bv, cs) = M\left((\ell_1)_\Delta, c_\Sigma\right) \text{ if and only if } C \in (\ell_1, c).$$

By **9.** in Theorem 1.23, this is the case if and only if the conditions in (4.1) and (7.1) are satisfied, that is, if and only if

$$\sup_{n,k} |c_{nk}| < \infty \text{ and } \lim_{n\to\infty} c_{nk} = \sum_{j=k}^{\infty} a_j \text{ exists for each } k.$$

Obviously these conditions are equivalent to $a \in cs$ and consequently

$$bv^\beta = cs \text{ ([65, Theorem 7.3.5 (iii)])}. \tag{2.47}$$

If $Y = \ell_\infty$ then we similarly obtain

$$a \in bv^\gamma = M(bv, bs) = M\left((\ell_1)_\Delta, (\ell_\infty)_\Sigma\right) \text{ if and only if } C \in (\ell_1, \ell_\infty).$$

By **4.** in Theorem 1.23, this is the case if and only if the condition in (4.1) is satisfied, that is, if and only if

$$\sup_{n,k}|c_{nk}| = \sup_{k,n\geq k}\left|\sum_{j=k}^{n} a_j\right| < \infty.$$

Obviously this condition is equivalent to $a \in bs$, and so

$$bv^{\gamma} = bs \text{ ([65, Theorem 7.3.5 (iv)])}. \tag{2.48}$$

Remark 2.4 The determination of $(X_T)^{\alpha}$ for arbitrary FK spaces X and triangles T is equivalent to $B(X, \ell_1)$ by (2.43), and by (1.21) in Theorem 1.13, this is equivalent to

$$\begin{aligned}\sup_{\substack{N \subset \mathbb{N}_0 \\ N \text{ finite}}}\left\|\left(\sum_{n\in N} b_{nk}\right)_{k=0}^{\infty}\right\|_{X,\delta}^{*} &= \sup_{\substack{N \subset \mathbb{N}_0 \\ N \text{ finite}}}\left(\sup_{x\in\overline{B_\delta}(0)}\left|\sum_{k=0}^{\infty}\left(\sum_{n\in N} b_{nk}\right)x_k\right|\right)\\ &= \sup_{\substack{N \subset \mathbb{N}_0 \\ N \text{ finite}}}\left(\sup_{x\in\overline{B_\delta}(0)}\left|\sum_{k=0}^{\infty}\left(\sum_{n\in N, n\geq k} a_n s_{nk}\right)x_k\right|\right) < \infty.\end{aligned}$$

This condition is not particularly practicable, in general, for instance, we obtain the following condition in the simple case of $X = \ell_\infty$ and $T = \Delta$:

$$\sup_{\substack{N \subset \mathbb{N}_0 \\ N \text{ finite}}}\left(\sum_{k=0}^{\infty}\left|\sum_{n\in N, n\geq k} a_n\right|\right) < \infty.$$

Therefore, we would want to find more applicable conditions. But it seems there is no general approach in the case of α-duals.

Let $m \in M$, $u \in \mathcal{U}$ and $\Delta^{(m)}$ be the triangle of the mth-order difference operator (Example 2.3). We define the sets

$$X(u\Delta^{(m)}) = \left\{x \in \omega : u \cdot \Delta^{(m)}x \in X\right\} \text{ for } X = c_0, c, \ell_\infty.$$

For $m = 1$, these sets reduce to the generalized difference spaces

$$c_0(u\Delta),\ \ c(u\Delta) \text{ and } \ell_\infty(u\Delta)$$

of Definition 2.3.

The following result holds.

Theorem 2.14 *Let* $m \in \mathbb{N}$, $u \in \mathcal{U}$ *and*

$$M^{\alpha,(m)}(u) = \left(\Sigma^{(m)} 1/|u|\right)^{-1} * \ell_1$$
$$= \left\{ a \in \omega : \sum_{k=0}^{\infty} |a_k| \sum_{j=0}^{k} \binom{m+k-j-1}{k-j} \frac{1}{|u_j|} < \infty \right\}.$$

Then we have

$$(X(u\Delta))^{\alpha} = M^{\alpha,(m)}(u) \text{ for } X = c_0, c, \ell_{\infty}. \tag{2.49}$$

Proof We write $M = M^{\alpha,(m)}(u)$, for short.

(i) First we show

$$M \subset \left(\ell_{\infty}(u\Delta^{(m)})\right)^{\alpha}. \tag{2.50}$$

Let $a \in M$ and $x \in \ell_{\infty}(u\Delta^{(m)})$ be given. Then we have

$$\sum_{k=0}^{\infty} |a_k| \Sigma_k^{(m)} 1/|u| < \infty, \tag{2.51}$$

and also $z = u \cdot (\Delta^{(m)} x) \in \ell_{\infty}$, that is, there exists a constant C such that $|z_k| \leq C$ for all $k = 0, 1, \ldots$. Since $x = \Sigma^{(m)}(z/u)$ (Example 2.3), it follows that

$$|a_k x_k| = \left| a_k \Sigma_k^{(m)}(z/u) \right| \leq |a_k| \sum_{j=0}^{k} \binom{m+k-j-1}{k-j} \frac{|z_j|}{|u_j|}$$
$$\leq C \cdot |a_k| \sum_{j=0}^{k} \binom{m+k-j-1}{k-j} \frac{1}{|u_j|} = C \cdot |a_k| \Sigma_k^{(m)}(1/|u|)$$

for $k = 0, 1, \ldots$, and so by (2.51)

$$\sum_{k=0}^{\infty} |a_k x_k| \leq C \cdot \sum_{k=0}^{\infty} |a_k| \Sigma_k^{(m)}(1/|u|) < \infty.$$

Thus, we have shown (2.50).

(ii) Now we show

$$\left(c_0(u\Delta^{(m)})\right)^{\alpha} \subset M. \tag{2.52}$$

We assume $a \notin M$, that is, $\sum_{k=0}^{\infty} |a_k| \Sigma_k^{(m)}(1/|u|) = \infty$, and consequently we can determine a strictly increasing sequence $(k(s))_{s=0}^{\infty}$ of integers with $k(0) = 0$ such that

$$\sum_{k=k(s)}^{k(s+1)-1} |a_k| \Sigma_k^{(m)}(1/|u|) \geq s+1 \text{ for } s = 0, 1, \ldots.$$

We define the sequence x by

$$x_k = \sum_{\ell=0}^{s-1} \frac{1}{l+1} \sum_{j=k(\ell)}^{k(\ell+1)-1} \binom{m+k-j-1}{k-j} \frac{1}{|u_j|} + \frac{1}{s+1} \sum_{j=k(s)}^{k} \binom{m+k-j-1}{k-j} \frac{1}{|u_j|}$$

for $k(s) \leq k \leq k(s+1) - 1$ $(s = 0, 1, \ldots)$.
Writing y for the sequence with

$$y_k = \frac{1}{(s+1)|u_k|} \text{ for } k(s) \leq k \leq k(s+1) - 1 \ (s = 0, 1, \ldots),$$

we obtain for k with $k(s) \leq k \leq k(s+1) - 1$

$$\begin{aligned}
\Sigma_k^{(m)} y &= \sum_{j=0}^{k} \binom{m+k-j-1}{k-j} y_j \\
&= \sum_{l=0}^{s-1} \sum_{j=k(\ell)}^{k(\ell+1)-1} \binom{m+k-j-1}{k-j} y_j + \sum_{j=k(s)}^{k} \binom{m+k-j-1}{k-j} y_j \\
&= \sum_{l=0}^{s-1} \frac{1}{\ell+1} \sum_{j=k(\ell)}^{k(\ell+1)-1} \binom{m+k-j-1}{k-j} \frac{1}{|u_j|} \\
&\qquad + \frac{1}{s+1} \sum_{j=k(s)}^{k} \binom{m+k-j-1}{k-j} \frac{1}{|u_j|} = x_k,
\end{aligned}$$

so $x = \Sigma^{(m)} y$. This implies $\Delta^{(m)} x = y$, that is, $u \cdot (\Delta^{(m)} x) = u \cdot y \in c_0$. Thus, we have $x \in c_0(u\Delta^{(m)})$. But, on the other hand, for each s,

$$\begin{aligned}
\sum_{k=k(s)}^{k(s+1)} |a_k x_k| &\geq \sum_{k=k(s)}^{k(s+1)} |a_k| \sum_{j=0}^{k} \binom{n+k-j-1}{k-j} \frac{1}{(s+1)|u_j|} \\
&= \frac{1}{s+1} \sum_{k=k(s)}^{k(s+1)} |a_k| \Sigma_k^{(m)}(1/|u|) \geq 1,
\end{aligned}$$

and so $a \notin (c_0(u\Delta^m)$. Thus, we have shown (2.52).
Finally, (2.49) follows from (2.50), (2.52) and the fact that

$$(\ell_\infty(u\Delta))^\alpha \subset (c(u\Delta))^\alpha \subset (c_0(u\Delta))^\alpha .$$

□

We write $\mathbf{n+1} = (n+1)_{n=0}^{\infty}$.

Remark 2.5 We obtain in the special case of $u = e$ and $m = 1$

$$(c_0(\Delta))^{\alpha} = (c(\Delta))^{\alpha} = (\ell_{\infty}(\Delta))^{\alpha} = (\mathbf{n+1})^{-1} * \ell_1 \\ = \left\{ a \in \omega : \sum_{k=0}^{\infty} (k+1)|a_k| < \infty \right\}.$$

Now we determine the β-duals of $c_0(u\Delta^{(m)})$, $c(u\Delta^{(m)})$ and $\ell_{\infty}(u\Delta^{(m)})$.

Theorem 2.15 *Let $m \in \mathbb{N}$ and $u \in \mathcal{U}$. Then, we have*

(a) $a \in (c_0(u\Delta^{(m)}))^{\beta}$ *if and only if*

$$\sum_{k=0}^{\infty} \left| \frac{1}{u_k} \sum_{j=k}^{\infty} \binom{m+j-k-1}{j-k} a_j \right| < \infty \tag{2.53}$$

and

$$\sup_n \sum_{k=0}^{n} \left| \frac{1}{u_k} \sum_{j=n}^{\infty} \binom{m+j-k-1}{j-k} a_j \right| < \infty; \tag{2.54}$$

moreover, if $a \in (c_0(u\Delta^{(m)}))^{\beta}$, *then*

$$\sum_{k=0}^{\infty} a_k x_k = \sum_{k=0}^{\infty} \left(\sum_{j=k}^{\infty} \binom{m+j-k-1}{j-k} a_j \right) \left(\Delta_k^{(m)} x \right); \tag{2.55}$$

(b) $a \in (c(u\Delta^{(m)}))^{\beta}$ *if and only if the conditions in* (2.53), (2.54) *and*

$$a \in \left(\Sigma^{(m)} 1/u \right)^{-1} * cs \tag{2.56}$$

hold;

(c) $a \in (\ell_{\infty}(u\Delta^{(m)}))^{\beta}$ *if and only if the condition in* (2.53) *holds and*

$$\lim_{n\to\infty} \sum_{k=0}^{n} \left| \frac{1}{u_k} \sum_{j=n}^{\infty} \binom{m+j-k-1}{j-k} a_j \right| = 0; \tag{2.57}$$

moreover, if $a \in (\ell_{\infty}(u\Delta^{(m)}))^{\beta}$ *then the identity in* (2.55) *holds for all* $x \in \ell_{\infty}(u\Delta^{(m)})$.

Proof Given $a \in \omega$, we write $b^{n,(m)} = b^{n,(m)}(a)$, $C^{(m)} = (c_{nk}^{(m)})_{n,k=0}^{\infty}$ and $D^{(m)} = (d_{nk}^{(m)})_{n,k=0}^{\infty}$ for the sequence and the matrices with

$$b_k^{n,(m)} = \begin{cases} \sum_{j=k}^{n} \binom{m+j-k-1}{j-k} a_j & (0 \le k \le n) \\ 0 & (k > n) \end{cases} \quad (n = 0, 1, \dots)$$

and $C_n^{(m)} = b^{n,(m)}$ and $D_n^{(m)} = (1/u) \cdot C_n^{(m)}$ for $n = 0, 1, \dots$, that is,

$$d_{nk}^{(m)} = \begin{cases} \frac{1}{u_k} \sum_{j=k}^{n} \binom{m+j-k-1}{j-k} a_j & (0 \le k \le n) \\ 0 & (k > n) \end{cases} \quad (n = 0, 1, \dots).$$

We note that, by Proposition 2.3, $a \in (X(u\Delta^{(m)}))^{\beta}$ if and only if $C^{(m)} \in (u^{-1} * X, c)$ and this is the case by Proposition 2.1 if and only if

$$D \in (X, c). \tag{2.58}$$

(a) First we show Part (a).

(a.i) First we show that if $a \in (c_0(u\Delta^{(m)}))^{\beta}$ then the conditions in (2.53) and (2.54) are satisfied.
Let $a \in (c_0(u\Delta^{(m)}))^{\beta}$ be given. Then we have $D^{(m)} \in (c_0, c)$ by (2.58) and by **12.** in Theorem 1.23 this is the case if and only if the conditions in (1.1) and (11.2) hold, that is,

$$K_D = \sup_n \sum_{k=0}^{\infty} |d_{nk}^{(m)}| = \sup_n \sum_{k=0}^{n} \frac{|b_k^{n,(m)}|}{|u_k|} < \infty \tag{2.59}$$

and

$$\beta_k^{(m)} = \lim_{n \to \infty} d_{nk}^{(m)} = \lim_{n \to \infty} \frac{b_k^{n,(m)}}{u_k} = \frac{1}{u_k} \sum_{j=k}^{\infty} \binom{m+j-k-1}{j-k} a_j \tag{2.60}$$

exists for each $k \in \mathbb{N}_0$. Let $n \in \mathbb{N}_0$ be given. We obtain from (2.59)

$$\sum_{k=0}^{n} \frac{|b_k^{\ell,(m)}|}{|u_k|} \le \sum_{k=0}^{\ell} \frac{|b_k^{\ell,(m)}|}{|u_k|} \le K_D \text{ for all } \ell \ge n,$$

hence, by (2.60)

$$\sum_{k=0}^{n} |\beta_k^{(m)}| = \lim_{\ell \to \infty} \sum_{k=0}^{n} \frac{|b_k^{\ell,(m)}|}{|u_k|} \le K_D.$$

Since n was arbitrary, we obtain (2.53) and this means $(\beta_k^{(m)})_{k=0}^{\infty} \in \ell_1 = c_0^{\beta}$. We observe that by (2.60)

$$b_k^{(m)} = \lim_{n \to \infty} b_k^{n,(m)} = u_k \beta_k^{(m)} \text{ exists for each } k,$$

and so we have $b^{(m)} \in (u^{-1} * c_0)^{\beta}$.
Now we define the matrix $W^{(m)} = (w_{nk}^{(m)})_{n,k=0}^{\infty}$ by $W_n^{(m)} = b_n^{(m)} e^{[n]}$ for $n = 0, 1, \dots$, that is,

$$w_{nk}^{(m)} = \begin{cases} \sum_{j=n}^{\infty} \binom{m+j-k-1}{j-1} a_j & (0 \le k \le n) \\ 0 & (k > n) \end{cases} \quad (n = 0, 1, \dots).$$

Let $x \in c_0(u\Delta^{(m)})$. Then we have $y = \Delta^m x \in u^{-1} * c_0$, and interchanging the order of summation and noting that

$$b_n^{n-1,(m)} = \sum_{j=n}^{n-1} \binom{m+j-k-1}{j-k} a_j = 0 \text{ for all } n = 0, 1, \dots,$$

we obtain

$$\begin{aligned}
\sum_{k=0}^{n-1} a_k x_k &= \sum_{k=0}^{n-1} a_k \Sigma_k^{(m)} y = \sum_{k=0}^{n-1} a_k \sum_{j=0}^{k} \binom{m+k-j-1}{k-j} y_j \\
&= \sum_{j=0}^{n-1} \left(\sum_{k=j}^{n-1} \binom{m+k-j-1}{k-j} a_k \right) y_j \\
&= \sum_{k=0}^{n-1} \left(\sum_{j=k}^{n-1} \binom{m+j-k-1}{j-k} a_j \right) y_k \\
&= \sum_{k=0}^{n-1} b_k^{n-1,(m)} y_k = \sum_{k=0}^{n} b_k^{n-1,(m)} y_k \\
&= \sum_{k=0}^{n} \left(\sum_{j=k}^{\infty} \binom{m+j-k-1}{j-k} a_j - \sum_{j=n}^{\infty} \binom{m+j-k-1}{j-k} a_j \right) y_k \\
&= \sum_{k=0}^{n} \left(b_k^{(m)} - w_{nk}^{(m)} \right) y_k = \sum_{k=0}^{n} b_k^{(m)} y_k - W_n^{(m)} y,
\end{aligned}$$

that is,

$$\sum_{k=0}^{n-1} a_k x_k = \sum_{k=0}^{n} b_k^{(m)} y_k - W_n^{(m)} y \text{ for } n = 0, 1, \dots. \tag{2.61}$$

Now since $a \in (c_0(u\Delta^{(m)})^\beta$ and $b^{(m)} \in (u^{-1} * c_0)^\beta$, it follows from (2.61) that $W^{(m)} \in (u^{-1} * c_0, c)$, and so $V^{(m)} \in (c_0, c)$ where the matrix $V^{(m)} = (v_{nk}^{(m)})_{n,k=0}^\infty$ is given by $v_{nk}^{(m)} = w_{nk}^{(m)}/u_k$ for $n, k = 0, 1, \dots$. It follows from (1.1) in Part **12.** of Theorem 1.23 that $V^{(m)} \in (c_0, c)$ implies

$$\begin{aligned} \sup_n \sum_{k=0}^{\infty} |v_{nk}^{(m)}| &= \sup_n \sum_{k=0}^{n} \frac{|w_{nk}^{(m)}|}{|u_k|} \\ &= \sup_n \sum_{k=0}^{n} \left| \frac{1}{u_k} \sum_{j=n}^{\infty} \binom{m+j-k-1}{j-k} a_j \right| < \infty \end{aligned} \tag{2.62}$$

which is (2.54).

Thus, we have shown that $a \in (c_0(u\Delta^{(m)}))^\beta$ implies the conditions in (2.53) and (2.54).

This completes the proof of Part (a.i).

(a.ii) Now we show that $a \in (c_0(u\Delta^{(m)}))^\beta$ implies (2.55).

It follows from (2.60) and the definition of the matrix $V^{(m)}$ that

$$\lim_{n\to\infty} v_{nk}^{(m)} = \lim_{n\to\infty} \frac{w_{nk}^{(m)}}{u_k} = \lim_{n\to\infty} \frac{1}{u_k} \sum_{j=n}^{\infty} \binom{m+j-k-1}{j-k} a_j = 0 \tag{2.63}$$

for each k. Therefore, if $a \in (c_0(u\Delta^{(m)}))^\beta$ then we have $b^{(m)} \in (u^{-1} * c_0)^\beta$ and (2.62) and (2.63) imply $W^{(m)} \in (u^{-1} * c_0, c_0)$ by (1.1) and (7.1) in **7.** of Theorem 1.23. Thus, (2.55) follows from (2.61).

This completes the proof of Part (a.ii).

(a.iii) Now we show that (2.53) and (2.54) imply $a \in (c_0(u\Delta^{(m)}))^\beta$.

We assume that the conditions in (2.53) and (2.54) are satisfied. Then again (2.63) holds and this and the condition in (2.54) together imply $W^{(m)} \in (u^{-1} * c_0, c_0)$ by (1.1) and (7.1) in **7.** of Theorem 1.23. Furthermore, the condition in (2.53) obviously implies $b^{(m)} \in (u^{-1} * c_0)^\beta$. Now it follows from (2.61) that $a \cdot x \in cs$ for all $x \in c_0(u\Delta^{(m)})$, that is, $a \in (c_0(u\Delta^{(m)}))^\beta$.

This completes the proof of Part (a.iii).

Thus, we have shown Part (a).

(b) Now we show Part (b).

(b.i) First, we show that if $a \in (c(u\Delta^{(m)}))^\beta$, then the conditions in (2.53), (2.54) and (2.56) are satisfied.

Let $a \in (c(u\Delta^{(m)}))^\beta$ be given. Then we have $a \in (c_0(u\Delta^{(m)}))^\beta$ and the conditions in (2.53) and (2.54) are satisfied by Part (a). Furthermore, we have seen in Part (a.i) of the proof of Part (a) that $a \in (c_0(u\Delta^{(m)}))^\beta$ implies $b^{(m)} \in (u^{-1} * c_0)^\beta$ and $W^{(m)} \in (u^{-1} * c_0, c_0)$. Now let $x \in c(u\Delta^{(m)})$. Then there exists a complex number ξ such that $u \cdot \Delta^{(m)}x - \xi e \in c_0$. We put $y = x - \xi\Sigma^{(m)}(1/u)$. Then $\Delta^{(m)}y = \Delta^{(m)}x - \xi(1/u)$ and $u \cdot \Delta^{(m)}y = u \cdot \Delta^{(m)}x - \xi e \in c_0$, that is, $y \in c_0(u\Delta^{(m)})$, and we obtain as in (2.61) with $z = \Delta^{(m)}y \in u^{-1} * c_0$ for $n = 0, 1, \dots$

$$\begin{aligned}\sum_{k=0}^{n-1} a_k x_k &= \sum_{k=0}^{n-1} a_k y_k + \xi \sum_{k=0}^{n-1} a_k \Sigma_k^{(m)}(1/u)\\ &= \sum_{k=0}^{n} b_k^{(m)} z_k - W_n^{(m)} z + \xi \sum_{k=0}^{n-1} a_k \Sigma_k^{(m)}(1/u).\end{aligned} \tag{2.64}$$

Now, by (2.64), $a \in (c_0(u\Delta^{(m)}))^\beta$, $b^{(m)} \in (u^{-1} * c_0)^\beta$ and $W^{(m)} \in (u^{-1} * c_0, c_0)$ imply $a \in (\Sigma^{(m)}(1/u))^{-1} * cs$ which is (2.56).
Thus, we have shown that $a \in (c(u\Delta^{(m)}))^\beta$ implies that the conditions in (2.53), (2.54) and (2.56) are satisfied.
This completes the proof of Part (b.i).

(b.ii) Now we show that the conditions in (2.53), (2.54) and (2.56) imply $a \in (c(u\Delta^{(m)}))^\beta$.
We assume that the conditions in (2.53), (2.54) and (2.56) are satisfied. Then the conditions in (2.53) and (2.54) together imply $W^{(m)} \in (u^{-1} * c_0, c_0)$ and $b^{(m)} \in (u^{-1} * c_0)$ as in Part (a.ii) of the proof of Part (a). Finally, it follows from (2.64) and the condition in (2.56) that $a \cdot x \in cs$ for all $x \in c(u\Delta^{(m)})$, that is, $a \in (c(u\Delta^{(m)}))^\beta$.
This completes the proof of Part (b.ii).

Thus, we have shown Part (b).

(c) Now we show Part (c).

(c.i) First, we show that if $a \in (\ell_\infty(u\Delta^{(m)}))^\beta$ then the conditions in (2.53) and (2.57) are satisfied.
We assume that $a \in (\ell_\infty(u\Delta^{(m)}))^\beta$. Then it follows that $a \in (c_0(u\Delta^{(m)}))^\beta$, and so $b^{(m)} \in (u^{-1} * c_0)^\beta$ as in the proof of Part (b.i), but obviously $(u^{-1} * c_0)^\beta = (u^{-1} * \ell_\infty)^\beta$. Now, since $a \in (\ell_\infty(u\Delta^{(m)}))^\beta$ and $b^{(m)} \in (u^{-1} * \ell_\infty)^\beta$, it follows from (2.61) and (2.63) that $W^{(m)} \in (u^{-1} * \ell_\infty, c_0)$. By (6.1) in **6.** of Theorem 1.23, $W^{(m)} \in (u^{-1} * \ell_\infty, c_0)$ implies

$$\lim_{n\to\infty} \sum_{k=0}^{\infty} \left| \frac{w_{nk}^{(m)}}{u_k} \right| = \lim_{n\to\infty} \sum_{k=0}^{n} \left| \frac{1}{u_k} \sum_{j=n}^{\infty} \binom{m+j-k-1}{j-k} a_j \right| = 0,$$

which is (2.57).

Thus, we have shown that $a \in (\ell_\infty(u\Delta^{(m)}))^\beta$ implies that the conditions in (2.53) and (2.57) are satisfied.

This completes the proof of Part (c.i).

(c.ii) Now we show that if $a \in (\ell_\infty(u\Delta^{(m)}))^\beta$ then (2.55) holds for all $x \in \ell_\infty(u\Delta^{(m)})$.

If $a \in (\ell_\infty(u\Delta^{(m)}))^\beta$, then, as we have seen in the proof of Part (c.i), $b^{(m)} \in (u^{-1} * \ell_\infty)^\beta$ and $W^{(m)} \in (u^{-1} * \ell_\infty, c_0)$. Therefore, it follows from (2.61) that (2.55) holds for all $x \in \ell_\infty(u\Delta^{(m)})$.

This completes the proof of Part (c.ii).

(c.iii) Finally, we show that if the conditions in (2.53) and (2.57) are satisfied, then $a \in (\ell_\infty(u\Delta^{(m)}))^\beta$.

We assume that the conditions in (2.53) and (2.57) are satisfied. Then, as before, (2.53) implies $b^{(m)}/u \in \ell_1 = \ell_\infty^\beta$, that is, $b^{(m)} \in (u^{-1} * \ell_\infty)^\beta$, and (2.57) is

$$\lim_{n\to\infty} \sum_{k=0}^{n} \frac{|w_{nk}^{(m)}|}{|u_k|} = 0$$

which implies $W^{(m)} \in (u^{-1} * \ell_\infty, c_0)$ by (6.1) in **6.** of Theorem 1.23. Now it follows from (2.61) that $a \cdot x \in cs$ for all $x \in \ell_\infty(u\Delta^{(m)})$, hence $a \in (\ell_\infty(u\Delta^{(m)}))^\beta$.

Thus, we have shown that the conditions in (2.53) and (2.57) imply $a \in (\ell_\infty(u\Delta^{(m)}))^\beta$.

This completes the proof of Part (c.iii).

This completes the proof of Part (c).

□

Now we consider the special case $m = 1$ and $u = e$ of Theorem 2.15. We need the following result.

Lemma 2.1 *(a)* ([38, Lemma 1]) *If* $b = (b_k)_{k=0}^\infty \in cs$ *then*

$$\lim_{n\to\infty} (n+1) \sum_{k=n}^{\infty} \frac{c_k}{k+1} = 0. \tag{2.65}$$

(b) ([38, Corollary]) *If* $a \in (\mathbf{n}+\mathbf{1})^{-1} * cs$ *then* $R \in (\mathbf{n}+\mathbf{1})^{-1} * c_0$ *where* $R = (R_k)_{k=0}^\infty$ *is the sequence with*

$$R_k = \sum_{j=k}^{\infty} a_j \text{ for } k = 0, 1, \ldots.$$

Proof (a) We assume $b \in cs$. Let $\varepsilon > 0$ be given. Then there exists $n_0 \in \mathbb{N}_0$ such that

$$\left|\sum_{j=n}^{n+k} b_j\right| < \frac{\varepsilon}{2} \text{ for all } n \geq n_0 \text{ and for all } k \in \mathbb{N}_0.$$

Let $n \geq m$ and $m \in \mathbb{N}_0$ be given. Then we obtain by Abel's summation by parts

$$\sum_{k=n}^{n+m} \frac{b_k}{k+1} = \sum_{k=0}^{m} \frac{b_{n+k}}{n+k+1}$$
$$= \sum_{k=0}^{m-1} \left(\frac{1}{n+k+1} - \frac{1}{n+k+2}\right) \sum_{j=0}^{k} b_{n+j} + \frac{1}{n+m+1} \sum_{k=0}^{m} b_{n+k},$$

hence

$$\left|\sum_{k=n}^{n+m} \frac{b_k}{k+1}\right| \leq \sum_{k=0}^{m-1} \left(\frac{1}{n+k+1} - \frac{1}{n+k+2}\right) \left|\sum_{j=n}^{n+k} b_j\right| + \frac{1}{n+m+1} \left|\sum_{k=n}^{n+m} b_k\right|$$
$$< \left(\frac{1}{n+1} - \frac{1}{n+m+1} + \frac{1}{n+m+1}\right) \cdot \frac{\varepsilon}{2} = \frac{\varepsilon}{2(n+1)}.$$

This shows that $b/(\mathbf{n}+\mathbf{1}) \in cs$ and

$$\left|\sum_{k=n}^{\infty} \frac{b_k}{k+1}\right| \leq \frac{\varepsilon}{2(n+1)} < \frac{\varepsilon}{n+1} \text{ for all } n \geq n_0,$$

that is,

$$\lim_{n\to\infty} \left|(n+1) \sum_{k=n}^{\infty} \frac{b_k}{k+1}\right| = 0.$$

Thus, we have shown that $b \in cs$ implies (2.65). This completes the proof of Part (a).

(b) We apply Part (a) with $b_k = (k+1)a_k$ for all $k = 0, 1, \ldots$. □

Corollary 2.11 ([38, Theorem 2 (a) and Corollary 3] for Parts (b) and (c) *We have*

(a) $a \in (c_0(\Delta))^\beta$ *if and only if*

$$\sum_{k=0}^{\infty} \left|\sum_{j=k}^{\infty} a_j\right| < \infty \text{ and } \sup_n \left(n \left|\sum_{j=n}^{\infty} a_j\right|\right) < \infty; \tag{2.66}$$

(b) $a \in (\ell_\infty(\Delta))^\beta$ *if and only if*

$$\sum_{k=0}^{\infty}\left|\sum_{j=k}^{\infty} a_j\right| < \infty \text{ and } \lim_{n\to\infty}\left(n\left|\sum_{j=n}^{\infty} a_j\right|\right) = 0; \tag{2.67}$$

(c) $(c(\Delta))^\beta = (\ell_\infty(\Delta))^\beta$.

Proof (a) It is easy to see that the conditions in (2.66) are the special cases of $m = 1$ and $u = e$ of the conditions (2.53) and (2.54), and so Part (a) follows from Part (a) of Theorem 2.15.

(b) It is easy to see that the conditions in (2.67) are the special cases of $m = 1$ and $u = e$ of the conditions (2.53) and (2.57), and so Part (b) follows from Part (c) of Theorem 2.15.

(c) First, $c(\Delta) \subset \ell_\infty(\Delta)$ implies $(\ell_\infty(\Delta))^\beta \subset (c(\Delta))^\beta$.
Conversely, if $a \in (c(\Delta))^\beta$ then it follows from Part (b) of Theorem 2.15 that the first condition in (2.67) holds and $a \in (\mathbf{n}+\mathbf{1})^{-1} * cs$ by (2.56). Now $a \in (\mathbf{n}+\mathbf{1})^{-1} * cs$ implies $(\mathbf{n}+\mathbf{1}) \cdot R \in c_0$ by Part (b) of Lemma 2.1, and this is the second condition in (2.67). Thus $a \in (\ell_\infty(\Delta))^\beta$ by Part (b).

□

Remark 2.6 Writing $R = (r_{nk})_{n,k=0}^\infty = \Sigma^t$ for the transpose of the matrix Σ, that is,

$$r_{nk} = \begin{cases} 1 & (k \geq n) \\ 0 & (0 \leq k < n) \end{cases} \quad (n = 0, 1, \dots),$$

we obtain from Corollary 2.11

$$(c_0(\Delta))^\beta = \left(\ell_1 \cap (\mathbf{n}+\mathbf{1})^{-1} * \ell_\infty\right)_R$$

and

$$(c(\Delta))^\beta = (\ell_\infty(\Delta))^\beta = \left(\ell_1 \cap (\mathbf{n}+\mathbf{1})^{-1} * c_0\right)_R.$$

2.6 The β-duals of Matrix Domains of Triangles in FK Spaces

In this section, we establish a general result which reduces the determination of the matrix domain of a triangle T in an FK space X to the determination of the β-dual of X and the characterization of the class (X, c_0).

Throughout T will always be a triangle, S be its inverse and $R = S^t$ be the transpose of S.

We will frequently use the following result.

Proposition 2.4 ([37, Theorem 1]) *Let X and Y be subsets of ω. Then we have*

$$A \in (X, Y_T) \text{ if and only if } T \cdot A \in (X, Y).$$

Proof (i) First we show that $A \in (X, Y_T)$ implies $T \cdot A \in (X, Y)$.
Let $A \in (X, Y_T)$. Then we have $A_n \in X^\beta$ for all n, and so $(T \cdot A)_n = \sum_{j=0}^n t_{nj} A_j \in X^\beta$, since the β-dual of any subset of ω obviously is a linear space.
Let $x \in X$ be given. Since $A_j \in X^\beta$ for all j, the series $\sum_{k=0}^\infty a_{jk}x_k$ converge for all j, and so

$$\begin{aligned}(T \cdot A)_n x &= \sum_{k=0}^\infty (T \cdot A)_{nk} x_k = \sum_{k=0}^\infty \left(\sum_{j=0}^n t_{nj} a_{jk} \right) x_k \\ &= \sum_{j=0}^n t_{nj} \left(\sum_{k=0}^\infty a_{jk} x_k \right) = \sum_{j=0}^n t_{nj} A_j x \\ &= T_n(Ax) \text{ for } n = 0, 1, \ldots,\end{aligned}$$

that is, $(T \cdot A)x = T(Ax)$. Since $A \in (X, Y_T)$, that is, $T(Ax) \in Y$ for all $x \in X$, it follows that $(T \cdot A) \in Y$ for all $x \in X$, hence $T \cdot A \in (X, Y)$.
Thus, we have shown that $A \in (X, Y_T)$ implies $T \cdot A \in (X, Y)$.

(ii) Now we show that $T \cdot A \in (X, Y)$ implies $A \in (X, Y_T)$.
We assume $T \cdot A \in (X, Y) = (X, (Y_T)_S)$, since matrix multiplication of triangles is associative. It follows by Part (i) that $B = S \cdot (T \cdot A) \in (X, Y_T)$. Since S and T are triangles, it follows that $B = S \cdot (T \cdot A) = (S \cdot T) \cdot A = A \in (X, Y)$.
Thus, we have shown that $T \cdot A \in (X, Y)$ implies $A \in (X, Y_T)$.

□

We also need the following result.

Lemma 2.2 ([43, Lemma 3.1]) *Let X be an FK space with AK. We write $R = S^t$ for the transpose of S. Then we have*

$$(X_T)^\beta \subset (X^\beta)_R. \tag{2.68}$$

Proof We write $Z = X_T$ and assume $a \in Z^\beta$. Then we have $B \in (X, cs)$ by Proposition 2.3, where B_n is the matrix with the rows $B_n = a_n S_n$ for $n = 0, 1, \ldots$. We write $C = \Sigma \cdot B$, that is, the matrix C has the entries

$$c_{nk} = \begin{cases} \sum_{j=k}^n a_j s_{jk} & (0 \le k \le n) \\ & (k > n) \end{cases} \quad (n = 0, 1, \ldots).$$

Then $C \in (X, c)$ by Proposition 2.4, since $cs = c_\Sigma$. Since X is an FK space with AK, it follows from (1.16) in Theorem 1.9 and Theorem 1.11 that $C \in (X, c)$ if and only if

$$\sup_n \|C_n\|^*_{X,\delta} < \infty \text{ for some } \delta > 0 \tag{2.69}$$

and

$$R_k a = \lim_{n\to\infty} c_{nk} = \sum_{j=k}^{\infty} a_j s_{jk} \text{ exists for each } k. \tag{2.70}$$

By (2.69), there exists a constant K such that

$$|C_n x| = \left|\sum_{k=0}^{n} c_{nk} x_k\right| \leq K \text{ for all } n \text{ and all } x \in \overline{B}_\delta(0). \tag{2.71}$$

Let $x \in X$ be given and $\rho = \delta/2$. Since $\overline{B}_\delta(0)$ is absorbing by Remark 1.3, and X has AK, there are a real $\lambda > 0$ and a non-negative integer m_0 such that $y^{[m]} = \lambda^{-1} \cdot x^{[m]} \in \overline{B}_\rho(0)$ for all $m \geq m_0$. Let $m \geq m_0$ be given. Then we have for all $n \geq m$ by (2.71)

$$\left|\sum_{k=0}^{m} c_{nk} x_k\right| = \lambda \left|\sum_{k=0}^{m} c_{nk} y_k^{[m]}\right| = \lambda \left|C_n y^{[m]}\right| \leq \lambda \cdot K,$$

and so by (2.70)

$$\left|\sum_{k=0}^{m} (R_k a) x_k\right| \leq \lambda \cdot \lim_{n\to\infty} \left|C_n y^{[m]}\right| \leq \lambda \cdot K.$$

Since $m \geq m_0$ was arbitrary, it follows that $(Ra) \cdot x \in bs$, that is, $Ra \in x^\gamma$, and since $x \in X$ was arbitrary, we have $Ra \in \bigcap_{x\in X} x^\gamma = X^\gamma$. Finally, since X has AK, we obtain $X^\beta = X^\gamma$ by Part (c) of Theorem 1.16. Therefore, we have $Ra \in X^\beta$, that is, $a \in (X^\beta)_R$.

Thus, we have shown the inclusion in (2.68). □

Now we reduce the determination of the β-dual of a triangle in X to the determination of the β-dual of X itself and the characterization of the class (X, c_0).

Theorem 2.16 ([43, Theorem 3.2]) *Let X be an FK space with AK. Then we have $a \in (X_T)^\beta$ if and only if*

$$a \in \left(X^\beta\right)_R \text{ and } W \in (X, c_0), \tag{2.72}$$

where the matrix $W = (w_{nk})_{n,k=0}^\infty$ is defined by

$$w_{nk} = \begin{cases} \sum\limits_{j=n}^{\infty} a_j s_{jk} & (0 \le k \le n) \\ 0 & (k > n) \end{cases} \quad (n = 0, 1, \dots).$$

Moreover, if $a \in (X_T)^\beta$ then we have

$$\sum_{k=0}^{\infty} a_k z_k = \sum_{k=0}^{\infty} (R_k a)(T_k z) \text{ for all } z \in Z = X_T. \tag{2.73}$$

Proof We write $Z = X_T$.

(i) First we show that $a \in Z^\beta$ implies that the conditions in (2.72) hold. We assume that $a \in Z^\beta$. Then it follows by Lemma 2.2 that $Ra \in X^\beta$. Hence the series $\sum_{j=n}^{\infty} a_j s_{jk}$ converge for all n and k, that is, the matrix W is defined. Furthermore, we have

$$\begin{aligned}\sum_{k=0}^{n} (R_k a)(T_k z) - \sum_{k=0}^{n} w_{nk} T_k z &= \sum_{k=0}^{n} (R_k a - w_{nk})\, T_k z \\ &= \sum_{k=0}^{n} \left(\sum_{j=k}^{\infty} a_j s_{jk} - \sum_{j=n}^{\infty} a_j s_{jk} \right) T_k z) \\ &= \sum_{k=0}^{n} \left(\sum_{j=k}^{n-1} a_j s_{jk} \right) T_k z = \sum_{j=0}^{n-1} \left(a_j \sum_{k=0}^{j+1} s_{jk} T_k z \right) \\ &= \sum_{j=0}^{n-1} a_j S_j(Tz) = \sum_{j=0}^{n-1} a_j z_j,\end{aligned}$$

that is,

$$\sum_{j=0}^{n-1} a_j z_j = \sum_{k=0}^{n} (R_k a)(T_k z) - W_n(Tz) \text{ for all } n \text{ and all } z \in Z. \tag{2.74}$$

Let $x \in X$ be given. Then we have $z = Sx \in Z$, and so $a \in z^\beta$ and $a \in (x^\beta)_R$. This implies $Wx = W(Tz) \in c$ by (2.74). Since $x \in X$ was arbitrary, we have $W \in (X, c) \subset (X, \ell_\infty)$. Furthermore, since $R_k a = \sum_{j=l}^{\infty} a_j s_{jk}$ exists for each k, we have

$$\lim_{n\to\infty} w_{nk} = \lim_{n\to\infty} \sum_{j=n}^{\infty} a_j s_{jk} = 0 \text{ for each } k. \tag{2.75}$$

Now $W \in (X, \ell_\infty)$ and (2.75) imply $W(X, c_0)$ by Theorem 1.11.
This completes the proof of (i).

(ii) Now we show that if $a \in Z^\beta$ then (2.73) holds.
Let $a \in Z^\beta$. Then it follows by (i) that the conditions (2.72) hold, and so (2.73) follows from (2.74).

(iii) Finally, we show that the conditions in (2.72) imply $a \in Z^\beta$.
We assume that the conditions in (2.72) are satisfied. If $z \in Z$ then $x = Tz \in X$, and so $a \cdot z \in cs$ by (2.74), that is, $a \in Z^\beta$.

□

Corollary 2.12 ([40, Proposition 3.1]) *Let X be an FK space with AK. Then we have $a \in (X_T)^\beta$ if and only if*

$$a \in \left(X^\beta\right)_R \text{ and } W \in (X, \ell_\infty). \tag{2.76}$$

Moreover, if $a \in (X_T)^\beta$ then the identity in (2.73) holds.

Proof First we assume $a \in (X_T)^\beta$. Then it follows from (2.72) in Theorem 2.16 that $a \in (X^\beta)_R$ and $W \in (X, c_0)$, but $(X, c_0) \subset (X, \ell_\infty)$.
Conversely, we assume $a \in (X^\beta)_R$ and $W \in (X, \ell_\infty)$. It follows from $Ra \in X^\beta$ that w_{mk} exists for each m and k and $\lim_{m\to\infty} w_{mk} = 0$ for each k. Since X has AK, this and $W \in (X, \ell_\infty)$ together imply $W \in (X, c_0)$ by Theorem 1.11. Finally, $a \in (X^\beta)_R$ and $W \in (X, c_0)$ together imply $a \in (X_T)^\beta$ by Theorem 2.16. □

The result of Theorem 2.16 can be extended to the spaces ℓ_∞ and c.

Corollary 2.13 ([43, Remark 3.3] for (a) and (b))

(a) The statement of Theorem 2.16 also holds when $X = \ell_\infty$.
(b) We have $a \in (c_T)^\beta$ if and only if $a \in (\ell_1)_R$ and $W \in (c, c)$. Moreover, if $a \in (c_T)^\beta$ then we have for all $z \in c_T$

$$\sum_{k=0}^{\infty} a_k z_k = \sum_{k=0}^{\infty} (R_k a)(T_k z) - \xi\alpha, \text{ where } \xi = \lim_{k\to\infty} T_k z \text{ and } \alpha = \lim_{n\to\infty} \sum_{k=0}^{n} w_{nk}. \tag{2.77}$$

(c) We have

$$\|a\|^*_{c_T} = \|Ra\|_1 + |\alpha| \text{ for all } a \in (c_T)^\beta. \tag{2.78}$$

Proof (i) First we show that if $X = c$ or $X = \ell_\infty$ then $a \in (X_T)^\beta$ if and only if $a \in (\ell_1)_R$ and $W \in (X, c)$.
Let $X = c$ or $X = \ell_\infty$. Then $X \supset c_0$ implies $X_T \supset (c_0)_T$, hence $(X_T)^\beta \subset ((c_0)_T)^\beta$. Since c_0 is a BK space with AK, it follows from Lemma 2.2 that $a \in (X_T)^\beta \subset ((c_0)_T)^\beta$ implies $a \in (c_0^\beta)_R = (\ell_1)_R = (X^\beta)_R$. We also obtain $W \in (X, c)$ from (2.74).
Conversely, if $a \in (X^\beta)_R$ and $W \in (X, c)$ then it follows from (2.74) that $a \in (X_T)^\beta$.

(a) Now let $X = \ell_\infty$. We have to show that $W \in (\ell_\infty, c)$ implies $W \in (\ell_\infty, c_0)$. If $W \in (\ell_\infty, c)$ then it follows from (11.1) in Part **11.** of Theorem 1.23 that

$$\sum_{k=0}^{n} |w_{nk}| \text{ is uniformly convergent in } n. \tag{2.79}$$

But, as before, in Part (i) of the proof of Theorem 2.16, we also have (2.75) and this and (2.79) imply

$$\lim_{n\to\infty} \sum_{k=0}^{n} |w_{nk}| = 0.$$

From this we obtain $W \in (\ell_\infty, c_0)$ by (6.1) in Part **6.** of Theorem 1.23. Thus, we have proved Part (a).

(b) It remains to show that $a \in (c_T)^\beta$ implies (2.77). Let $a \in (c_T)^\beta$ and $z \in c_T$ be given. Then we have $x = Tz \in c$ and $\xi = \lim_{k\to\infty} x_k$ exists. Hence, there is $x^{(0)} \in c_0$ such that $x = x^{(0)} + \xi \cdot e$. We put $z^{(0)} = Sx^{(0)}$. Then it follows that $z^{(0)} \in (c_0)_T$ and $z = Sx = S(x^{(0)} + \xi \cdot e) = z^{(0)} + \xi \cdot Se$, and we obtain as in (2.74)

$$\begin{aligned}\sum_{k=0}^{n-1} a_k z_k &= \sum_{k=0}^{n} (R_k a)(T_k z) - W_n\left(T(z^{(0)} + \xi \cdot Se)\right)\\ &= \sum_{k=0}^{n} (R_k a)(T_k z) - W_n(Tz^{(0)}) - \xi \cdot W_n e \text{ for all } n.\end{aligned}$$

The first term in the last equality converges as $n \to \infty$, since $Ra \in \ell_1$ by Part (i). The second term in the last equality tends to zero as $n \to \infty$, since $a \in (c_T)^\beta \subset ((c_0)_T)^\beta$ implies $W \in (c_0, c_0)$ by Theorem 2.16. Finally, we also have $W \in (c, c)$ by Part (i), and this implies by (13.1) in Part **13.** of Theorem 1.23 that $\alpha = \lim_{n\to\infty} W_n e$ exists. Now the identity in (2.77) follows.

(c) If $a \in (c_T)$, then it follows from Part (b) that $Ra \in \ell_1$ and (2.77) holds. Since $z \in c_T$ if and only if $x = Tz \in c$ and $\|z\|_{c_T} = \|x\|_\infty$, the right-hand side of (2.78) defines a functional $f \in c^*$ with its norm $\|f\|$ given by the right-hand side in (2.78) by (1.24) in Example 1.11.

□

Example 2.14 Let $1 < p < \infty$ and $q = p/(p-1)$. We write $bv^p = (\ell_p)_\Delta$. Then we have $a \in (bv^p)^\beta$ if and only if

$$\sum_{k=0}^{\infty} \left|\sum_{j=k}^{\infty} a_j\right|^q < \infty \text{ and } \sup_n \left(n^{1/q} \cdot \left|\sum_{k=n}^{\infty} a_k\right|\right) < \infty. \tag{2.80}$$

Moreover, if $a \in (bv^p)^\beta$ then

$$\|a\|^*_{bv^p} = \left(\sum_{k=0}^\infty \left|\sum_{j=k}^\infty a_j\right|^q\right)^{1/q} \quad \text{for all } a \in (bv^p)^\beta. \tag{2.81}$$

Proof (i) First we show that $a \in (bv^p)^\beta$ if and only if the conditions in (2.80) hold.

Since ℓ_p $(1 < p < \infty)$ is a BK space with AK by Part (c) of Example 1.4, we can apply Theorem 2.16 with $T = \Delta$, $S = \Sigma$ and $R = \Sigma^t$, that is, $r_{nk} = 1$ for $k \geq n$ and $r_{nk} = 0$ for $0 \leq k < n$ $(n = 0, 1, \ldots)$. We have by (2.72) that $a \in (bv^p)^\beta$ if and only if $Ra \in \ell_p^\beta = \ell_q$, that is,

$$\sum_{k=0}^\infty \left|\sum_{j=k}^\infty a_j\right|^q < \infty,$$

which is the first condition in (2.80), and $W \in (\ell_p, c_0)$, where $W = (w_{nk})_{n,k=0}^\infty$ is the matrix with

$$w_{nk} = \begin{cases} \sum\limits_{j=n}^\infty a_j & (0 \leq k \leq n) \\ 0 & (k > n) \end{cases} \quad (n = 0, 1, \ldots).$$

Now we have $W \in (\ell_p, c_0)$ by (5.1) and (7.1) in Part **10.** of Theorem 1.23, if and only if

$$\sup_n \left(\sum_{k=0}^\infty |w_{nk}|^q\right)^{1/p} = \sup_n \left((n+1)^{1/q} \cdot \left|\sum_{j=n}^\infty a_j\right|\right) < \infty,$$

which is the second identity in (2.80), and

$$\lim_{n\to\infty} w_{nk} = \lim_{n\to\infty} \sum_{j=n}^\infty a_j = 0,$$

which is redundant, since the series $\sum_{k=0}^\infty a_k$ converges.

Thus, we have shown that $a \in (bv^p)^\beta$ if and only if the conditions in (2.80) hold.

This completes the proof of Part (i).

(ii) Now we show that if $a \in (bv^p)^\beta$ then the identity in (2.81) holds.

Let $a \in (bv^p)^\beta$ be given. We observe that $x \in bv^p$ if and only if $y = \Delta x \in \ell_p$ and $\|x\|_{bv^p} = \|\Delta x\|_{\ell_p} = \|y\|_p$. Then we have by (2.73) in Theorem 2.16

$$\sum_{k=0}^{\infty} a_k x_k = \sum_{k=0}^{\infty} (R_k a) y_k.$$

Now $\|x\|_{bv^p} = \|y\|_p$ implies $\|a\|^*_{bv^p} = \|Ra\|^*_{\ell_p}$ and (2.81) follows from the fact that ℓ_p^* and ℓ_q are norm isomorphic.
This completes the proof of Part (ii).

□

Remark 2.7 (a) We have by Example 2.14

$$(bv^p)^\beta = \left(\ell_q \cap \left((n^{1/q})_{n=0}^{\infty}\right)^{-1} * \ell_\infty\right)_R.$$

(b) We observe that neither $\ell_q \subset ((n^{1/q})_{n=0}^{\infty})^{-1} * \ell_\infty$ nor $((n^{1/q})_{n=0}^{\infty})^{-1} * \ell_\infty \subset \ell_q$. To see this we define the sequences y and $\tilde{y}$ by

$$y_k = \begin{cases} \dfrac{1}{\nu+1} & (k = 2^\nu) \\ 0 & (k \neq 2^\nu) \end{cases} \quad (\nu = 0, 1, \dots) \text{ and } \tilde{y}_k = \frac{1}{(k+1)^{1/q}} \quad (k = 0, 1, \dots).$$

Then we have $y \in \ell_q \setminus [((n^{1/q})_{n=0}^{\infty})^{-1} * \ell_\infty]$ and $\tilde{y} \in [((n^{1/q})_{n=0}^{\infty})^{-1} * \ell_\infty] \setminus \ell_q$, since

$$\sum_{k=0}^{\infty} |y_k|^q = \sum_{\nu=0}^{\infty} \frac{1}{(\nu+1)^q} < \infty, \text{ but } |y_{2^\nu}|\,(2^\nu)^{1/q} = \frac{2^{\nu/q}}{\nu+1} \to \infty \ (\nu \to \infty),$$

and

$$k^{1/q}|\tilde{y}_k| = \left(\frac{k}{k+1}\right)^{1/q} \le 1 \text{ for } k = 0, 1, \dots, \text{ but } \sum_{k=0}^{\infty} |\tilde{y}_k|^q = \sum_{k=0}^{\infty} \frac{1}{k+1} = \infty.$$

Remark 2.8 We have seen at the beginning of the Proof of Lemma 2.2 that $a \in (X_T)^\beta$ if and only if $C \in (X, c)$ where the matrix C is defined by

$$c_{nk} = \begin{cases} \sum_{j=k}^{n} a_j s_{jk} & (0 \le k \le n) \\ 0 & (k > n) \end{cases} \quad (n = 0, 1, \dots).$$

(a) Therefore it follows from Theorem 2.16 or Part (a) of Corollary 2.13 that if X is an FK space with AK or $X = \ell_\infty$ then $C \in (X, c)$ if and only if

$$Ra \in X^\beta \text{ and } W \in (X, c_0).$$

(b) It follows from Part (b) of Corollary 2.13 that $C \in (c, c)$ if and only if

$$Ra \in \ell_1 \text{ and } W \in (c, c).$$

We close this section with an application.

Corollary 2.14 *We consider the* Euler sequence spaces e_p^r, e_0^r, e_c^r *of Part (a) of* Example *2.8 introduced in* [2, 5]. *Then we have*

(a) ([5, Theorem 4.5]) $a \in (e_1^r)^\beta$ *if and only if*

$$\sup_{n,k} \left| \sum_{j=k}^{n} \binom{j}{k} (r-1)^{j-k} r^{-j} a_j \right| < \infty \tag{2.82}$$

and

$$\sum_{j=k}^{\infty} \binom{j}{k} (r-1)^{j-k} r^{-j} a_j \textit{ converges for each } k; \tag{2.83}$$

(b) ([5, Theorem 4.5]) $a \in (e_p^r)^\beta$ *for* $1 < p < \infty$ *and* $q = p/(p-1)$ *if and only if the condition in (2.83) holds and*

$$\sup_n \left(\sum_{k=0}^{\infty} \left| \sum_{j=k}^{n} \binom{j}{k} (r-1)^{j-k} r^{-j} a_j \right|^q \right) < \infty; \tag{2.84}$$

(c) ([2, Theorem 4.2]) $a \in (e_0^r)^\beta$ *if and only if the condition in (2.83) holds and*

$$\sup_n \left(\sum_{k=0}^{\infty} \left| \sum_{j=k}^{n} \binom{j}{k} (r-1)^{j-k} r^{-j} a_j \right| \right) < \infty; \tag{2.85}$$

(d) ([2, Theorem 4.5]) $a \in (e_c^r)^\beta$ *if and only if the conditions in (2.83) and (2.85) hold and*

$$\lim_{n\to\infty} \left(\sum_{k=0}^{n} \sum_{j=k}^{n} \binom{j}{k} (r-1)^{j-k} r^{-j} a_j \right) \textit{ exists;} \tag{2.86}$$

(e) $a \in (e_\infty^r)^\beta$ *if and only if*

$$\sum_{n=0}^{\infty} \left| \sum_{k=n}^{\infty} \binom{k}{n} (r-1)^{k-n} r^{-k} a_k \right| < \infty \tag{2.87}$$

and

$$\lim_{m\to\infty} \sum_{n=0}^{m} \left| \sum_{k=m}^{\infty} \binom{k}{n} (r-1)^{k-n} r^{-k} a_k \right| = 0. \tag{2.88}$$

Proof (i) We verify that the matrix S defined by (2.20) is the inverse matrix of the Euler matrix $T = E^r$.

($i.\alpha$) In the proof of Part (i), we need the identity

$$\binom{n}{j}\binom{j}{k} = \binom{n}{k}\binom{n-k}{j-k} \text{ for } 0 \le k \le j \le n. \tag{2.89}$$

We have for $0 \le k \le j \le n$

$$\begin{aligned}\binom{n}{k}\binom{n-k}{j-k} &= \frac{n\cdots(n-k+1)}{k!}\cdot\frac{(n-k)\cdots(n-k-(j-k)+1)}{(j-k)!}\\ &= \frac{n\cdots(n-j+1)}{j!}\cdot\frac{j!}{k!(j-k)!}\\ &= \binom{n}{j}\binom{j}{k},\end{aligned}$$

that is, we have shown (2.89).

We write $y = E^r x$, that is,

$$y_j = E_j^r x = \sum_{k=0}^{j}\binom{j}{k}(1-r)^{j-k}r^k x_k \text{ for } n = 0, 1, \ldots.$$

Using (2.20) and (2.89), we obtain

$$\begin{aligned}S_n y &= \sum_{j=0}^{n} s_{nj}y_j = \sum_{j=0}^{n}\binom{n}{j}(r-1)^{n-j}r^{-n}\sum_{k=0}^{j}\binom{j}{k}(1-r)^{j-k}r^k x_k\\ &= \sum_{k=0}^{n} r^{k-n}x_k\sum_{j=k}^{n}\binom{n}{j}\binom{j}{k}(-1)^{n-j}(1-r)^{n-k}\\ &= \sum_{k=0}^{n}\binom{n}{k}\left(\frac{1-r}{r}\right)^{n-k}x_k\sum_{j=k}^{n}\binom{n-k}{j-k}(-1)^{n-j}\\ &= \sum_{k=0}^{n}\binom{n}{k}\left(\frac{1-r}{r}\right)^{n-k}x_k\sum_{j=0}^{n-k}\binom{n-k}{j}(-1)^{n-k-j}.\end{aligned}$$

Since

$$\sum_{j=0}^{n-k}\binom{n-k}{j}(-1)^{n-k-j} = (1-1)^{n-k} = \begin{cases}1 & (k=n)\\ 0 & (k \ne n),\end{cases}$$

it follows that

$$S_n y = \binom{n}{n}\left(\frac{1-r}{r}\right)^{n-n} x_n = x_n \text{ for } n = 0, 1, \ldots .$$

Thus, we have shown $S(E^r x) = x$. Since matrix multiplication of triangles is associative it follows that S is the inverse matrix of the Euler matrix E^r.
This completes the proof of Part (i).

To prove Parts (a) to (d) we apply Remark 2.8 to obtain $a \in X^\beta$ if and only if $C \in (X, c)$ in the corresponding cases, where

$$c_{nk} = \begin{cases} \sum_{j=k}^{n} s_{jk} a_j = \sum_{j=k}^{n} \binom{j}{k}(r-1)^{j-k} r^{-j} a_j & (0 \le k \le n) \\ 0 & (k > n) \end{cases} \quad (n = 0, 1, \ldots). \tag{2.90}$$

(a) We have $C \in (\ell_1, c)$ by (4.1) and (11.2) in Part **11.** of Theorem 1.23 and by (2.90) if and only if

$$\sup_{n,k} |c_{nk}| = \sup_{n,k} \left| \sum_{j=k}^{n} \binom{j}{k} (r-1)^{j-k} r^{-j} a_j \right| < \infty,$$

which is the condition in (2.82), and

$$\lim_{n \to \infty} c_{nk} = \sum_{j=k}^{\infty} \binom{j}{k} (r-1)^{j-k} r^{-j} a_j \text{ exists for each } k,$$

which is the condition in (2.83).
This completes the proof of Part (a).

(b) We have $C \in (\ell_p, c)$ for $1 < p < \infty$ by (5.1) and (11.2) of Part **15.** of Theorem 1.23, and by (2.90) if and only if

$$\sup_n \sum_{k=0}^{\infty} |c_{nk}|^q = \sup_n \sum_{k=0}^{n} \left| \sum_{j=k}^{n} \binom{j}{k} (r-1)^{j-k} r^{-j} a_j \right|^q < \infty,$$

which is the condition in (2.84), and (11.2) is the condition in (2.83) as in Part (a) of the proof.
This completes the proof of Part (b).

(c) We have $C \in (c_0, c)$ by (1.1) and (11.2) of Part **12.** of Theorem 1.23, and by (2.90) if and only if

$$\sup_n \sum_{k=0}^{\infty} |c_{nk}| = \sup_n \sum_{k=0}^{n} \left| \sum_{j=k}^{n} \binom{j}{k} (r-1)^{j-k} r^{-j} a_j \right| < \infty,$$

which is the condition in (2.85), and (11.2) is the condition in (2.83) as in Part (a) of the proof.
This completes the proof of Part (c).

(d) By Theorem 1.12, we have $C \in (c, c)$ if and only if $C \in (c_0, c)$ which yields the conditions in (2.83) and (2.85) by Part (c), and $Ce \in c$, that is,

$$\lim_{n\to\infty} \sum_{k=0}^{\infty} c_{nk} = \lim_{n\to\infty} \left(\sum_{k=0}^{n} \sum_{j=k}^{n} \binom{j}{k} (r-1)^{j-k} r^{-j} a_j \right) \text{ exists,}$$

which is the condition in (2.86).
This completes the proof of Part (d).

(e) Now we apply Theorem 2.16 and Part (a) of Corollary 2.13. We have $a \in (e_\infty^r)^\beta$ if and only if $R \in \ell_\infty^\beta = \ell_1$, which is the condition in (2.85), and $W \in (\ell_\infty, c_0)$, that is, by (6.1) of Part **6.** of Theorem 1.23,

$$\lim_{m\to\infty} \sum_{k=0}^{\infty} |w_{mk}| = \lim_{m\to\infty} \sum_{n=0}^{m} \left| \sum_{k=m}^{\infty} \binom{k}{n} (r-1)^{k-n} r^{-k} a_k \right| = 0,$$

which is (2.88).
This completes the proof of Part (e).

□

References

1. Alp, Z., Ilkhan, M.: On the difference sequence space $\ell_p(\hat{T}^q)$. Math. Sci. Appl. E-Notes **7**(2), 161–173 (2019)
2. Altay, B., Başar, F.: Some Euler sequence spaces on non–absolute type. Ukr. Math. J. **57**(1) (2005)
3. Altay, B., Başar, F.: Some paranormed sequence spaces of non-absolute type derived by weighted mean. J. Math. Anal. Appl. **319**(2), 494–508 (2006)
4. Altay, B., Başar, F.: Generalization of the sequence space $\ell(p)$ derived by weighted mean. J. Math. Anal. Appl. **330**(1), 174–185 (2007)
5. Altay, B., Başar, F., Mursaleen, M.: On the Euler sequence spaces which include ℓ_p and ℓ_∞ I. Inform. Sci. **176**, 1450–1462 (2006)
6. Başarir, M., Öztürk, M.: On the Riesz difference sequence space. Rend. Circ. Mat. Palermo **2**(57), 377–389 (2008)
7. Candan, M.: Domain of the double sequential band matrix in the classical sequence spaces. J. Inequal. Appl. **2012**, 281 (2012). http://www.journalofinequalitiesandapplications.com/content/2012/1/281

8. Candan, M.: Almost convergence and double sequential band matrix. Acta Math. Sci. Ser. B Engl. Ed. **34**(2), 354–366 (2014)
9. Candan, M.: Domain of the double sequential band matrix in the spaces of convergent and null sequences. J. Inequal. Appl. **2014**, 18: Adv. Differ. Equ. (2014)
10. Aydın, Ç.: Isomorphic sequence spaces and infinite matrices. PhD thesis, Inönü University, Malatya, Turkey (2001)
11. Aydın, Ç., Başar, F.: On the new sequence spaces which include the spaces c_0 and c. Hokkaido Math. J. **33**, 338–398 (2004)
12. Aydın, Ç., Başar, F.: Some new difference sequence spaces. Appl. Math. Comput. **157**, 677–693 (2004)
13. Cooke, R.C.: Infinite Matrices. MacMillan and Co. Ltd, London (1950)
14. Şengönül, M., Başar, F.: Some new Cesàro sequence spaces of nonabsolute type which include the spaces c_0 and c. Soochow J. Math. **31**(1), 107–119 (2005)
15. Şimşek, N., Karakaya, V.: Structure and some geometric properties of generalized Cesàro sequence space. Int. J. Contemp. Math. Sci. **3**(8), 389–399 (2008)
16. de Malafosse, B., Rakočević, V.: Applications of measure of noncompactness in operators on the spaces $s_\alpha, s_\alpha^\circ, s_\alpha^{(c)}, \ell_\alpha^p$. J. Math. Anal. Appl. **323**, 131–145 (2006)
17. Demiriz, S., Ilkhan, M., Kara, E.E.: Almost convergence and Euler totient matrix. Math. Sci. Appl. E-Notes (2020). https://doi.org/10.1007/s43034-019-00041-0
18. Djolović, I.: Compact operators on the spaces $a_0^r(\Delta)$ and $a_c^r(\Delta)$. J. Math. Anal. Appl. **318**, 658–666 (2006)
19. Djolović, I., Malkowsky, E.: A note on compact operators on matrix domains. J. Math. Anal. Appl. **340**, 291–303 (2008)
20. Erfanmanesh, S., Foroutannia, D.: Some new semi-normed spaces of non-absolute type and matrix transformations. Proc. Inst. Appl. Math. **4**(2), 96–108 (2015)
21. Et, M.: On some difference sequence spaces. Turkish J. Math. **17**, 18–24 (1993)
22. Foroutannia, D.: On the block sequence space $\ell_p(e)$ and related matrix transformations. Turk. J. Math. **39**, 830–841 (2015)
23. Grosse-Erdmann, K.-G.: The structure of sequence spaces of Maddox. Can. J. Math. **44**(2), 298–307 (1992)
24. Grosse-Erdmann, K.-G.: Matrix transformations between the sequence spaces of Maddox. J. Math. Anal. Appl. **180**, 223–238 (1993)
25. Grosse-Erdmann, K.-G.: The Blocking Technique, Weighted Mean Operators and Hardy's Inequality. Springer Lecture Notes in Mathematics, vol. 1679 (1999)
26. Ilkhan, M., Demiriz, S., Kara, E.E.: A new paranormed sequence space defined by Euler totient matrix. Karaelmas Sci. Eng. J. **9**(2), 277–282 (2019)
27. Jarrah, A.M., Malkowsky, E.: BK spaces, bases and linear operators. Rend Circ. Mat. Palermo, Ser. II, Suppl. **52**, 177–191 (1998)
28. Kara, E.E., Başarır, M., Mursaleen, M.: Compactness of matrix operators on some sequence spaces derived by Fibonacci numbers. Kragujevac J. of Math. **39**(2), 217–230 (2015)
29. Karakaya, V.: Some geometric properties of sequence spaces involving lacunary sequence. J. Inequal. Appl., Article ID 81028 (2007).https://doi.org/10.1155/2007/81028
30. Karakaya, V., Altun, M.: Fine spectra of upper triangular double-band matrices. J. Comput. Appl. Math. **234**, 1387–1394 (2010)
31. Karakaya, V., Altun, M.: On some geometric properties of a new paranormed sequence space. J. Funct. Spaces Appl., Article ID 685382, 8 (2014)
32. Kirişci, M.: The application domain of infinite matrices on classical sequence spaces. arXiv:1611.06138v1 [math.FA], 18 Nov 2016
33. Kirişci, M.: The sequence space bv and some applications. Math. Aeterna **4**(3), 207–223 (2014)
34. Kirişci, M.: Riesz type integrated and differentiated sequence spaces. Bull. Math. Anal. Appl. **7**(2), 14–27 (2015)
35. Kirişci, M., Başar, F.: Some new sequence spaces derived by the domain of generalized difference matrix. Comput. Math. Appl. **60**(5), 1299–1309 (2010)
36. Kızmaz, H.: On certain sequence spaces. Canad. Math. Bull. **24**(2), 169–176 (1981)

37. Malkowsky, E.: Linear operators in certain BK spaces. Bolyai Soc. Math. Stud. **5**, 259–273 (1996)
38. Malkowsky, E.: A note on the Köthe-Toeplitz duals of generalized sets of bounded and convergent difference sequences. J. Analysis **4**, 81–91 (1996)
39. Malkowsky, E.: Linear operators between some matrix domains. Rend. Circ. Mat. Palermo, Serie II, Suppl. **68**, 641–655 (2002)
40. Malkowsky, E., Nergiz, H.: Matrix transformations and compact operators on spaces of strongly Cesàro summable and bounded sequences of order α. Contemp. Anal. Appl. Math. (CAAM) **3**(2), 263–279 (2015)
41. Malkowsky, E., Özger, F.: A note on some sequence spaces of weighted means. Filomat **26**(3), 511–518 (2012)
42. Malkowsky, E., Parashar, S.D.: Matrix transformations in spaces of bounded and convergent difference sequences of order m. Analysis **17**, 187–196 (1997)
43. Malkowsky, E., Rakočević, V.: On matrix domains of triangles. Appl. Math. Comput. **189**, 1146–1163 (2007)
44. Malkowsky, E., Rakočević, V.: Advanced Functional Analysis. CRC Press, Taylor & Francis Group, Boca Raton, London, New York (2019)
45. Mohiuddine, S.A., Alotaibi, A.: Weighted almost convergence and related infinite matrices. J. Inequal. Appl. **2018**(1), 15 (2018)
46. Mursaleen, M.: Generalized spaces of difference sequences. J. Math. Anal. Appl. **203**(3), 738–745 (1996)
47. Mursaleen, M.: On some geometric properties of a sequence space related to ℓ_p. Bull. Australian Math. Soc. **67**(2), 343–347 (2003)
48. Mursaleen, M., Başar, F., Altay, B.: On the euler sequence spaces which include the spaces ℓ_p and ℓ_∞ II. Nonlinear Anal. **65**(3), 707–717 (2006)
49. Mursaleen, M., Noman, A.K.: Some new sequence spaces derived by the domain of generalized difference matrix. Math. Comput. Modelling **52**(3–4), 603–617 (2010)
50. Mursaleen, M., Noman, A.K.: On generalized means and some related sequence spaces. Comput. Math. Appl. **61**, 988–999 (2011)
51. Mursaleen, M., Noman, A.K.: On some new sequence spaces of non-absolute type related to the spaces ℓ_p and ℓ_∞ II. Math. Commun. **16**, 383–398 (2011)
52. Natarajan, P.N.: Cauchy multiplication of (M, λ_n) summable series. Adv. Dev. Math. Sci. **3**(12), 39–46 (2012)
53. Natarajan, P.N.: On the (M, λ_n) method of summability. Analysis (München) **33**(2), 51–56 (2013)
54. Natarajan, P.N.: A product theorem for the Euler and Natarajan methods of summability. Analysis (München) **33**(2), 189–195 (2013)
55. Natarajan, P.N.: New properties of the Natarajan method of summability. Comment. Math. **55**(1), 9–15 (2015)
56. Natarajan, P.N.: Classical Summability Theory. Springer Nature Singapore Pte Ltd. (2017)
57. Ng, P.-N., Lee, P.-Y.: Cesàro sequence spaces of non-absolute type. Commentat. Math, XX (1978)
58. Polat, H., Başar, F.: Some Euler spaces of difference sequences of order m. Acta Math. Sci. Ser. B Engl. Ed. **27B**(2), 254–266 (2007)
59. Başarır, M., Kara, E.E.: On compact operators on the Riesz $B(m)$–difference sequence space. Iran J. Sci. Technol. Trans. A Sci. **35**(4), 279–285 (2011)
60. Roopaefi, H., Foroutannia, D.: A new sequence space and norm of certain matrix operators on this space. Sahand Commun. Math. Anal **3**(1), 1–12 (2016)
61. Sanhan, W., Suantai, S.: Some geometric properties of Cesàro sequence space. Kyungpook Math. J. **43**, 191–197 (2003)
62. Sönmez, A., Başar, F.: Generalized difference spaces of non-absolute type of convergent and null sequences. Abstr. Appl. Anal. 20 (2012). https://doi.org/10.1155/2012/435076

63. Talebi, G.: On multipliers of matrix domains. J. Inequal. Appl. **2018**, 296 (2018). https://doi.org/10.1186/s13660-018-1887-4
64. Tuğ, O., Rakočević, V., Malkowsky, E.: On the domain of the four–dimensional sequential band matrix in some double sequence spaces. Mathematics **8** (2020). https://doi.org/10.3390/math8050789
65. Wilansky, A.: Summability Through Functional Analysis. Mathematical Studies, vol. 85. North-Holland, Amsterdam (1984)

Chapter 3
Operators Between Matrix Domains

In this chapter, we apply the results of the previous chapters to characterize matrix transformations on the spaces of generalized weighted means and on matrix domains of triangles in BK spaces. We also establish estimates or identities for the Hausdorff measure of noncompactness of matrix transformations from arbitrary BK spaces with AK into c, c_0 and ℓ_1, and also from the matrix domains of an arbitrary triangle in ℓ_p, c and c_0 into c, c_0 and ℓ_1. Furthermore we determine the classes of compact operators between the spaces just mentioned. Finally we establish the representations of the general bounded linear operators from c into itself and from the space bv^+ of sequences of bounded variation into c, and the determination of the classes of compact operators between them.

Throughout this chapter, T will always be a triangle, S be its inverse and $R = S^t$ be the transpose of S.

Section 3.5 is related to the results recently published in [7].

We also list some more recent publications concerning measures of noncompactness and its applications such as [1, 4, 5, 8–11, 16–19].

3.1 Matrix Transformations on $W(u, v; X)$

In this section, we establish a result that characterizes the classes $(W(u, c; X), Y)$ of matrix transformations, where the sets $W(u, c; X)$ are the spaces of generalized weighted means of Definition 2.4 in Sect. 2.4.

Throughout, let $u, v \in \mathcal{U}$.

Theorem 3.1 ([12, Theorem 2.6]) *Let X and Y be subets of ω and X be normal.*

(a) Then we have $A \in (X_\Sigma, Y)$ if and only if

$$A_n \in M(X, c_0) \text{ for all } n = 0, 1, \dots \tag{3.1}$$

B. de Malafosse et al., *Operators Between Sequence Spaces and Applications*,
https://doi.org/10.1007/978-981-15-9742-8_3

and

$$B \in (X, Y) \text{ where } B_n = \Delta^+ A_n \text{ for } n = 0, 1, \ldots. \tag{3.2}$$

(b) *Then we have* $A \in (W(u, v; X), Y)$ *if and only if*

$$A_n = (1/(u \cdot v))^{-1} * M(X, c_0) \text{ for } n = 0, 1, \ldots \tag{3.3}$$

and

$$\hat{B} \in (X, Y) \text{ where } \hat{B}_n = (1/u) \cdot \Delta^+(A_n/v) \text{ for } n = 0, 1, \ldots. \tag{3.4}$$

Proof (a) We write $Z = X_\Sigma$.

(i) First we show that $A \in (Z, Y)$ implies that the conditions in (3.1) and (3.2) hold.
We assume $A \in (Z, Y)$. Then we have $A_n \in Z^\beta$ for $n = 0, 1, \ldots$, hence $A_n \in M(X, c_0)$ by the first part of (2.30) in Corollary 2.7. Let $x \in X$ be given. Then $z = \Delta x \in Z$ and $x = \Sigma z$, and it follows from $A_n \in Z^\beta$ and (2.31) in Corollary 2.7 that

$$B_n x = A_n z \text{ for } n = 0, 1, \ldots, \text{ that is, } Bx = Az. \tag{3.5}$$

Therefore, $Az \in Y$ implies $Bx \in Y$.
Thus we have shown that $A \in (Z, Y)$ implies that the conditions in (3.1) and (3.2) hold.
This completes the proof of Part (i).

(ii) Now we show that if the conditions in (3.1) and (3.2) are satisfed, then $A \in (Z, Y)$.
We assume that the conditions in (3.1) and (3.2) are satisfied. First we have $B_n \in X^\beta$ for $n = 0, 1, \ldots$ by (3.2) which means $A_n \in (X^\beta)_{\Delta^+}$ by the definition of the matrix B. Now this and $A_n \in M(X, c_0)$ in (3.1) together imply $A_n \in Z^\beta$ for $n = 0, 1, \ldots$ by the first part of (2.30) in Corollary 2.7. Again it follows that (3.5), and therefore $Az \in Y$ for all $z \in Z$.
Thus we have shown that if the conditions in (3.1) and (3.2) are satisfed, then $A \in (Z, Y)$.
This completes the proof of Part (ii).

(b) Since, by Proposition 2.1, $A \in (u^{-1} * X, Y)$ if and only if $D \in (X, Y)$ where $D_n = A_n/u$ for $n = 1, 2, \ldots$, Part (b) is an immediate consequence of Part (a). □

We apply Theorem 3.1 to characterize matrix transformations from bs and cs into ℓ_∞, c_0 and c.

Corollary 3.1 *Then the necessary and sufficient conditions for* $A \in (X, Y)$ *can be read from the following table:*

From / To	bs	cs
ℓ_∞	**1.** ([20, 8.4.5C])	**2.** ([20, 8.4.5B])
c_0	**3.** ([20, 8.5.6E])	**4.** ([20, 8.4.5B])
c	**5.** ([20, 8.5.6D])	**6.** ([20, 8.4.5B])

where

1. *(1.1), (1.2) where*
  (1.1) $\sup_n \sum_{k=0}^{\infty} |a_{nk} - a_{n,k-1}| < \infty$
  (1.2) $\lim_{k\to\infty} a_{nk} = 0$ *for each n*

2. *(1.1)*

3. *(1.2), (3.1), (3.2) where*
  (3.1) $\sum_{k=0}^{\infty} |a_{nk} - a_{n,k+1}|$ *converges uniformly in n*
  (3.2) $\lim_{n\to\infty} a_{nk} = 0$ *for each k*

4. *(1.1), (3.2)*

5. *(1.2), (3.1), (5.1) where*
  (5.1) $\lim_{n\to\infty} a_{nk} = \alpha_k$ *exists for each k*

6. *(1.1), (5.1).*

Proof **1.** Since $X = \ell_\infty$ obviously is normal, we have by (3.1) and (3.2) in Theorem 3.1 that $A \in (bv, \ell_\infty)$ if and only if $A_n \in M(\ell_\infty, c_0)$ for all n, that is, $A_n \in c_0$ by Part (b) of Proposition 2.2, which is the condition in (1.2). Also $B = \Delta^+ A_n \in (\ell_\infty, \ell_\infty)$, that is, by (1.1) of Part **1.** of Theorem 1.23

$$\sup_n \sum_{k=0}^{\infty} |b_{nk}| = \sup_n \sum_{k=0}^{\infty} |a_{nk} - a_{n,k+1}| < \infty,$$

which, together with (1.2), obviously is equivalent to the condition in (1.1). This completes the proof of Part 1.

2. Since $X = cs$ is a BK space with AK by Part (d) of Example 2.7, $Z = \ell_1$ is a BK space with AK by Part (c) of Example 1.4 with $Z^\beta = \ell_\infty^\beta = \ell_1$ by Example 1.8, and $cs^\beta = bv$ by Part (a) of Corollary 2.8, we obtain from Theorem 1.22 that $A \in (cs, \ell_\infty)$ if and only if $D = A^t \in (\ell_1, bv)$. This is the case by Proposition 2.4 if and only if $C = \Delta \cdot D \in (\ell_1, \ell_1)$, that is, by (19.1) of Part **19.** of Theorem 1.23, if and only if

$$\sup_k \sum_{n=0}^{\infty} |c_{nk}| < \infty. \tag{3.6}$$

Since

$$c_{nk} = (\Delta \cdot D)_{nk} = d_{nk} - d_{n,k-1} = a_{kn} - a_{k,n-1} \text{ for all } n \text{ and } k,$$

we obtain from (3.6) that

$$\sup_k \sum_{n=0}^{\infty} |a_{kn} - a_{k,n-1}| < \infty,$$

which is the condition in (1.1).
This completes the proof of Part 2..

5. (i) First we show the sufficiency of the conditions in 5.
We assume that the conditions in (1.2), (5.1) and (3.2) are satisfied. First we note that the condition in (1.2) is equivalent to $A_n \in M(\ell_\infty, c_0)$ for all n by Part (b) of Proposition 2.2, which is the condition in (3.1) of Part (a) of Theorem 3.1. Also the condition in (5.1) obviously implies

$$\lim_{n\to\infty} b_{nk} = \lim_{n\to\infty} (a_{nk} - a_{n,k+1}) \text{ exists for each } k$$

and this and the condition in (3.2) together imply $B \in (\ell_\infty, c)$ by the conditions (11.1) and (11.2) of Part **11.** of Theorem 1.23. Now $B \in (\ell_\infty, c)$ is the condition in (3.2) which together with the condition in (3.1) implies $A \in (bs, c)$ by Part (a) of Theorem 3.1.
Thus we have shown the sufficiency of the conditions in (1.2), (5.1) and (3.2).

(ii) Now we show the necessity of the conditions in 5..
We assume $A \in (bs, c)$. Since ℓ_∞ is a normal set it follows as in the proof of Part 1. that $A_n \in M(\ell_\infty, c_0)$ which is the condition in (1.2), and $B \in (\ell_\infty, c)$ which, by the conditions in (11.1) and (11.2) of Part **11.** of Theorem 1.23, yields the condition in (3.1). Finally, since $e^{(k)} \in bs$ for all k, we obtain $Ae^{(k)} \in c$ for all k which is the condition in (5.1). Thus we have shown that $A \in (bs, c)$ implies that the conditions in 5. hold.

3. The proof of Part 3. is analogous to that of Part 5.

4. Since cs has AK it follows by Theorem 1.11 that $A \in (cs, c_0)$ if and only $A \in (cs, \ell_\infty)$ and $Ae^{(k)} \in c$ for all k, which is the condition in (3.2); $A \in (cs, \ell_\infty)$ is to equivalent the condition in (1.1) by Part 2.

6. The proof of Part 6. is very similar to that of Part 4.

□

Now we establish the characterizations of the classes (bs, bs), (cs, bs), (bs, cs) and (cs, cs).

Corollary 3.2 *The necessary and sufficient conditions for $A \in (X, Y)$, when $X \in \{bs, cs\}$ and $Y \in \{\ell_\infty, c_0, c\}$, can be read from the following table:*

From / *To*	bs	cs
bs	**1.** ([20, 8.4.6C])	**2.** ([20, 8.4.6B])
cs	**3.**	**4.** ([20, 8.4.6B])

where

1. *(1.1), (1.2) where*

(1.1) $\sup_n \sum_{k=0}^{\infty} |\sum_{j=0}^{n} (a_{jk} - a_{j,k-1}| < \infty$

(1.2) $\lim_{k\to\infty} a_{nk} = 0$ *for each* n

2. *(1.1)*

3. *(1.2), (3.1), (3.2) where*

(3.1) $\sum_{k=0}^{\infty} |\sum_{j=0}^{n} (a_{jk} - a_{j,k+1})|$ *converges uniformly in* n

(3.2) $\sum_{j=0}^{\infty} a_{jk}$ *exists for each* k

4. *(1.1), (3.2).*

Proof We write $C = \Sigma \cdot A$, that is,

$$c_{nk} = \sum_{j=0}^{n} a_{jk} \text{ for } n, k = 0, 1, \ldots.$$

Applying Proposition 2.4, we obtain Parts **1.**, **2.**, **3.** and **4.** by replacing the entries of the matrix A in the conditions in Parts 1., 2., 5. and 6. of Corollary 3.1 by those of the matrix C and observing that the condition

$$\lim_{k\to\infty} c_{nk} = \lim_{k\to\infty} \sum_{j=0}^{n} a_{jk} = 0 \text{ for each } n$$

obviously is equivalent to the condition

$$\lim_{k\to\infty} a_n = 0 \text{ for each } n,$$

which is the condition in (1.2). □

Remark 3.1 By [20, 8.5.9], $A \in (bs, cs)$ if and only if

$$\lim_{n\to\infty} \sum_{k=0}^{\infty} \left| \sum_{j=n}^{\infty} (a_{jk} - a_{j,k+1}) \right| = 0 \text{ and (1.2) in Corollary 3.2.} \tag{3.7}$$

It can be shown that the conditions in (3.7) are equivalent to those in (1.2), (3.1) and (3.2) in Part **3.** of Corollary 3.2.

Now we characterize matrix transformations from the spaces of generalized weighted means into the classical sequence spaces and into spaces of generalized weighted means.

We write $\sup_N$, $\sup_K$ and $\sup_{k,N}$ for the suprema taken over all finite subsets N and K of $\mathbb{N}_0$ and over all non-negative integers k and all finite subsets N of $\mathbb{N}_0$, respectively.

Corollary 3.3 ([12, Theorem 3.3]) *The necessary and sufficient conditions for* $A \in (W(u,v;X),Y)$, *when* $X \in \{\ell_p\ (1 \le p \le \infty), c_0, c\}$ *and* $Y \in \{\ell_\infty, c_0, c,\}$ *and* $Y = \ell_r\ (1 < r < \infty)$ *for* $X = \ell_1$ *can be read from the following table:*

To / From	ℓ_∞	c_0	c	ℓ_1	$\ell_r\ (1 < r < \infty)$
$W(u,v;\ell_p)\ (1 \le p < \infty)$	**1.**	**2.**	**3.**	**4.**	**17.** *only* $p = 1$
$W(u,v;c_0)$	**5.**	**6.**	**7.**	**8.**	**18.**
$W(u,v;c)$	**9.**	**10.**	**11.**	**12.**	**15.**
$W(u,v;\ell_\infty)$	**13.**	**14.**	**15.**	**16.**	**20.**

where

1. *(1.1),(1.2) where*

(1.1) $\sup_k \sum_{k=0}^{\infty} |a_{nk}/u_k v_k| < \infty$ *for all n*

(1.2) $\begin{cases} \sup_n \sum_{k=0}^{\infty} |(1/u_k)(a_{nk}/v_k - a_{n,k+1}/v_{k+1})|^q < \infty & (1 < p < \infty) \\ \sup_{n,k} |(1/u_k)(a_{nk}/v_k - a_{n,k+1}/v_{k+1})| < \infty & (p = \infty) \end{cases}$

2. *(1.1),(1.2), (2.1) where*

(2.1) $\lim_{n\to\infty}((1/u_k)(a_{nk}/v_k - a_{n,k+1}/v_{k+1})) = 0$ *for each k*

3. *(1.1),(1.2), (3.1) where*

(3.1) $\lim_{n\to\infty}((1/u_k)(a_{nk}/v_k - a_{n,k+1}/v_{k+1})) = \alpha_k$ *for each k*

4. *(1.1), (4.1) where*

(4.1) $\begin{cases} \sup_N \sum_{k=0}^{\infty} |(1/u_k) \sum_{n\in N}(a_{nk}/v_k - a_{n,k+1}/v_{k+1})|^q < \infty \\ \qquad (1 < p < \infty) \\ \sup_k |1/u_k| \sum\limits_{n=0}^{\infty} |a_{nk}/v_k - a_{n,k+1}/v_{k+1}| < \infty \quad (p = 1) \end{cases}$

5. *(1.1), (5.1) where*

(5.1) $\sup_n \sum_{k=0}^{\infty} |(1/u_k)(a_{nk}/v_k - a_{n,k+1}/v_{k+1})| < \infty$

6. *(1.1),(2.1),(5.1)*

7. *(1.1),(3.1),(5.1)*

8. *(1.1), (8.1) where*

(8.1) $\sup_N \sum_{k=0}^{\infty} |(1/u_k) \sum_{n\in N}(a_{nk}/v_k - a_{n,k+1}/v_{k+1})| < \infty$

9. *(5.1), (9.1), (9.2) where*

(9.1) $\lim_{k\to\infty} a_{nk}/(u_k v_k) = \beta_n$ *for all n*

(9.2) $\sup_n \beta_n < \infty$

10. *(2.1),(5.1), (9.1),(9.2), (10.1) where*

(10.1) $\lim_{n\to\infty} \sum_{k=0}^{\infty} a_{nk}/v_k(1/u_k - 1/u_{k-1}) = 0$

11. *(3.1),(5.1), (9.1),(9.2), (11.1) where*

(11.1) $\lim_{n\to\infty} \sum_{k=0}^{\infty} a_{nk}/v_k(1/u_k - 1/u_{k-1}) = \gamma$

12. *(8.1),(9.1), (12.1) where*

(12.1) $\sum_{n=0}^{\infty} |\beta_n| < \infty$

13. *(5.1), (13.1) where*
(13.1) $\lim_{k\to\infty} a_{nk}/(u_k v_k) = 0$ *for all n*

14. *(2.1),(13.1) (14.1) where*
(14.1) $\sum_{k=0}^{\infty} |(1/u_k)(a_{nk}/v_k - a_{n,k+1}/v_{k+1})$ *is uniformly convergent in n*

15. *(3.1),(13.1), (14.1)*

16. *(8.1),(13.1)*

17. *(1.1), (17.1) where*
(17.1) $\sup_k \sum_{n=0}^{\infty} |(1/u_k)(a_{nk}/v_k - a_{n,k+1}/v_{k+1}|^r < \infty$

18. *(1.1), (18.1) where*
(18.1) $\sup_K \sum_{n=0}^{\infty} |\sum_{k\in K}(1/u_k)(a_{nk}/v_k - a_{n,k+1}/v_{k+1}|^r < \infty$

19. *(9.1),(18.1) (19.1) where*
(19.1) $\sum_{n=0}^{\infty} |\beta_n|^r < \infty$

20. *(13.1),(18.1).*

Proof The characterizations of the classes $(W(u, v; X), Y)$ in the first, second and third rows of the table follow from Part (b) of Theorem 3.1. The conditions for $A_n/(u \cdot v) \in M(X, c_0)$ in (3.3) are given in Parts (d) and (a) of Proposition 2.2; for $X = \ell_p$ $(1 \le p < \infty)$ and $X = c_0$, which accounts for (1.1) in Parts **1.**, **2.**, **3.**, **4.**, **17.**, **5.**, **6.**, **7.**, **8.**, and **18.**, (c) of Proposition 2.2 for $X = \ell_\infty$, which accounts for (13.1) in Parts **13.**, **14.**, **15.**, **16.** and **20.**. Furthermore, the conditions for $B \in (X, Y)$ in (3.4) of Part (b) of Theorem 3.1 are $B \in (c_0, \ell_1), (\ell_\infty, \ell_1)$ which yields (8.1) in Parts **8.** and **16.** by (16.1) in Parts **16.** and **17.** of Theorem 1.23; the conditions for $B \in (\ell_1, \ell_\infty)$ and $B \in (\ell_p, \ell_\infty)$ for $1 < p < \infty$ in (4.1) and (5.1) in Parts **4.** and **5.** of Theorem 1.23 yield the condition in (1.2) in Part **1.**; for $B \in (\ell_p, c_0)$ and $B \in (\ell_p, c)$ we have to add the conditions $Be^{(k)} \in c_0$ and $Be^{(k)} \in c$ for all $k \in \mathbb{N}_0$ by Theorem 1.11, which account for the conditions (2.1) and (3.1) in Parts **2.** and **3.**; the conditions for $B \in (\ell_p, \ell_1)$ in (19.1) for $p = 1$ and (20.1) for $1 < p < \infty$ in Parts **19.** and **20.** of Theorem 1.23 yield the conditions in (4.1) of Part **4.**; the condition for $B \in (\ell_1, \ell_r)$ for $(1 < r < \infty)$ in (24.1) in Part **24.** of Theorem 1.23 accounts for the condition in (17.1) of Part **17.**; the condition for $B \in (c_0, \ell_\infty)$ and $B \in (\ell_\infty, \ell_\infty)$ in (1.1) in Parts **1.** and **2.** of Theorem 1.23 accounts for the condition (5.1) in Parts **5.** and **13.**; for $B \in (c_0, c_0)$ and $B \in (c_0, c)$ we have to add the conditions $Be^{(k)} \in c_0$ and $Be^{(k)} \in c$ for all $k \in \mathbb{N}_0$ by Theorem 1.11, which account for the conditions (2.1) and (3.1) in Parts **6.** and **7.**; the condition for $B \in (c_0, \ell_r)$ and $B \in (\ell_\infty, \ell_r)$ $(1 < r < \infty)$ in (21.1) in Parts **21.** and **22.** of Theorem 1.23 accounts for the condition (18.1) in Parts **18.** and **20.**; the conditions for $B \in (\ell_\infty, c_0)$ in Part (i) of Remark 1.6 and in (7.1) in Part **7.** of Theorem 1.23 account for (14.1) and (2.1) in Part **14.**; the conditions for $B \in (\ell_\infty, c)$ in (11.1) and (11.2) in Part **11.** account for (14.1) and (3.1) in Part **15.**.

Now we prove the characterizations of the classes (Z, Y) where $Z = W(u, v; c)$ and $Y = \ell_\infty, \ell_1, \ell_r$ for $1 < r < \infty$, that is, Parts **9.**, **12.** and **19.**. First we assume $A \in (Z, Y)$. Then we have $A_n \in Z^\beta$ for all n and so $A_n \in Z^\beta \subset (1/(u \cdot v))^{-1} * c$ by

Corollary 2.10, which is the condition in (9.1). Furthermore $(Z, Y) \subset (W(u, v; c_0), Y)$ implies that the conditions in (5.1), (8.1) or (18.1) hold for $Y = \ell_\infty$, $Y = \ell_1$ or $Y = \ell_r$, respectively, and moreover $A \in (W(u, v; Y), c_0)$ implies $B \in (c_0, Y)$ by (3.4) in Part (b) of Theorem 3.1. But $(c_0, Y) = (c, Y) = (\ell_\infty, Y)$ for $Y = \ell_\infty$ by Parts **1.**, **2.** and **3.** of Theorem 1.23, **16.**, **17.** and **18.** for $Y = \ell_1$, and **21.**, **22.** and **23.** for $Y = \ell_r (1 < r < \infty)$. Let $z \in Z$ be given. Then we have $x = u \cdot \Sigma(v \cdot z) \in c$, hence $\lim_{k\to\infty} = \xi$ for some $\xi \in \mathbb{C}$, and so $A_n z = B_n x + \xi\beta_n$ for all n by (2.42) and (9.1). Finally, $Az \in Y$ and $Bx \in Y$ together imply $(\beta_n)_{n=0}^\infty \in Y$, that is, (9.2), (12.1) or (19.1) for $Y = \ell_\infty, \ell_1, \ell_r$. Conversely, we assume that the conditions in Parts **9.**, **12.** and **19.** hold. The conditions in (9.1) and (5.1), or (8.1), or (18.1) yield $A_n \in Z^\beta$ for all n by Corollary 2.9 (in the case of (18.1), we apply the inequality in (1.20) of Lemma 1.1). Again we have $A_n z = B_n x + \xi\beta_n$ for all n, all $z \in Z$ and all $x = Tz \in c$ where $\xi = \lim_{k\to\infty} x_k$. By the conditions in (5.1), or (8.1), or (18.1), we have $B \in (c, Y)$, for $Y = \ell_\infty$ by (1.1) in Part **3.** of Theorem 1.23, Parts **1.**, **2.** and **3.**, for $Y = \ell_1$ by (16.1) in Part **18.** of Theorem 1.23, and for $Y = \ell_r$ $(1 < r < \infty)$ by (21.1) in Part **23.** of Theorem 1.23. Also the conditions in (9.2), or (12.1), or (19.1) imply $(\beta_n)_{n=0}^\infty \in Y$, hence $Az \in Y$ for all $x \in Z$.

To prove the conditons in Parts **10.** and **11.**, we apply [13, Theorem 1 (c)] and [12, Theorem 2.2] and have to add the conditions $Aw^{(k)} \in Y$ for $Y = c_0, c$ where $w^{(-1)} = (1/v) \cdot \Delta(e/u)$ and $w^{(k)} = (1/v) \cdot \Delta(e^{(k)}/u)$ for $k \in \mathbb{N}_0$. The condition $Aw^{(-1)} \in Y$ yields the conditions in (10.1) and (11.1), and the conditions $Aw^{(k)} \in Y$ for $k \in \mathbb{N}_0$ yield the conditions in (2.1) and (3.1). □

Remark 3.2 Applying Proposition 2.4, we obtain the characterizations of the classes $(W(u, v; X), W(u', v', Y))$ from those for $(W(u, v; X), Y)$ in Corollary 3.1 by replacing a_{nk} by $c_{nk} = u'_n \sum_{j=0}^n a_{nj} v_{j'}$ $(n, k = 0, 1, \dots)$.

3.2 Matrix Transformations on X_T

In this section, we reduce the characterisations of the classes (X_T, Y) to those of the classes (X, Y) and (X, c_0), when T is an arbitrary triangle and X is an arbitrary FK space with AK or $X \in \{\ell_\infty, c\}$.

Theorem 3.2 ([14, Theorem 3.4]) *Let X be an FK space with AK, Y be an arbitrary subset of ω. Then we have $A \in (X_T, Y)$ if and only if*

$$\hat{A} \in (X, Y) \text{ and } W^{(A_n)} \in (X, c_0) \text{ for } n = 0, 1, \dots, \tag{3.8}$$

where $\hat{A}_n = RA_n$ for $n = 0, 1, \dots$ and the triangles $W^{(A_n)} = (w_{mk}^{(A_n)})_{m,k=0}^\infty$ are defined by

$$w_{mk}^{(A_n)} = \begin{cases} \sum_{j=m}^\infty a_{nj} s_{jk} & (0 \le k \le m) \\ 0 & (k > m) \end{cases} \quad (m = 0, 1, \dots).$$

Moreover, if $A \in (X_T, Y)$ *then*

$$Az = \hat{A}(Tz) \text{ for all } z \in Z = X_T. \tag{3.9}$$

Proof We write $W^{(n)} = W^{(A_n)}$, for short.

(i) First we show that $A \in (X_T, Y)$ implies the conditions in (3.8) and (3.9). We assume $A \in (X_T, Y)$. Then it follows that $A_n \in Z^\beta$, hence, by (2.72) in Theorem 2.16, $\hat{A}_n \in X^\beta$ and $W^{(n)} \in (X, c_0)$ for all n, which is the second condition in (3.8). Let $x \in X$ be given, hence $z = Sx \in Z$. Since $A_n \in Z^\beta$ implies $A_n z = \hat{A}_n(Tz) = \hat{A}_n x$ for all n by (2.73) in Theorem 2.16, and $Az \in Y$ for all $z \in Z$ implies $\hat{A}x = Az \in Y$, we have $\hat{A} \in (X, Y)$. Moreover (3.9) holds.

(ii) Now we show that the conditions in (3.8) imply $A \in (Z, Y)$. We assume $\hat{A} \in (X, Y)$ and $W^{(n)} \in (X, c_0)$ for all n. Then we have $\hat{A}_n \in X^\beta$ for all n, and this and $W^{(n)} \in (X, c_0)$ together imply $A_n \in Z^\beta$ for all n by (2.72) in Theorem 2.16. Now let $z \in Z$ be given, hence $x = Tz \in X$. Again we have $A_n z = \hat{A}_n x$ for all n by (2.73) in Theorem 2.16, and $\hat{A}x \in Y$ for all $x \in X$ implies

$$Az = \hat{A}x \in Y.$$

Hence we have $A \in (X, Y)$.

□

Remark 3.3 Similarly as in Corollary 2.12, the condition $W^{(A_n)} \in (X, c_0)$ in (3.8) of Theorem 3.2 may be replaced by $W^{(A_n)} \in (X, \ell_\infty)$.

The result of Theorem 3.2 can be extended to matrix transformations on the spaces $(\ell_\infty)_T$ and c_T.

Corollary 3.4 ([14, Remark 3.5])

(a) The statement of Theorem *3.2 also holds for* $X = \ell_\infty$.
(b) Let $Y \subset \omega$ *be a linear space. Then we have* $A \in (c_T, Y)$ *if and only if*

$$\hat{A} \in (c_0, Y), \quad W^{(A_n)} \in (c, c) \text{ for all } n, \tag{3.10}$$

and

$$\hat{A}e - \left(\alpha^{(n)}\right)_{n=0}^\infty \in Y, \text{ where } \alpha^{(n)} = \lim_{m\to\infty} \sum_{k=0}^m w_{mk}^{(A_n)} \ (n = 0, 1, \ldots). \tag{3.11}$$

Moreover, if $A \in (c_T, Y)$ *then we have*

$$Az = \hat{A}(Tz) - \xi\left(\alpha^{(n)}\right)_{n=0}^\infty \text{ for all } z \in c_T, \text{ where } \xi = \lim_{k\to\infty} T_k z. \tag{3.12}$$

Proof (a) Part (a) is obvious from Part (a) of Corollary 2.13.
(b)

(i) First we show that $A \in (c_T, Y)$ implies the conditions in (3.10)–(3.12).
First we assume $A \in (c_T, Y)$. Since $c_T \supset (c_0)_T$, it follows that $A \in ((c_0)_T, Y)$ and so, by the first condition in (3.9) of Theorem 3.9, $\hat{A} \in (c_0, Y)$ which is the first condition in (3.10). Also, by Part (b) of Corollary 2.13, $A_n \in (c_T)^\beta$ for all n implies $W^{(A_n)} \in (c, c)$ for all n, which is the second condition in (3.10). Furthermore we obtain (3.11) from (2.77) in Part (b) of Corollary 2.13. Moreover, if $A \in (c_T, Y)$ then (3.11) follows from (2.77).
This completes the proof of (i).

(ii) Now we show that the conditions in (3.10) and (3.11) imply $A \in (c_T, Y)$.
We assume that the conditions in (3.10) and (3.11) are satisfied. Then $\hat{A}_n = RA_n \in c_0^\beta = \ell_1$ and $W^{(A_n)} \in (c, c)$ for all n, the conditions in (3.10), together imply $A_n \in (c_T)^\beta$ for all n by Part (b) of Corollary 2.13. Let $z \in c_T$ be given. Then we have $x = Tz \in c$. We put $x^{(0)} = x - \xi \cdot e$ where $\xi = \lim_{k\to\infty} x_k$. Then we have $x^{(0)} \in c_0$ and, by (2.77) in Part (b) of Corollary 2.13

$$Az = \hat{A}(Tz) - \xi \left(\alpha^{(n)}\right)_{n=0}^\infty = \hat{A}x^{(0)} + \xi \cdot \left(\hat{A}e - \left(\alpha^{(n)}\right)_{n=0}^\infty\right) \in Y,$$

since $\hat{A} \in (c_0, Y)$, $\hat{A}e - (\alpha^{(n)})_{n=0}^\infty \in Y$ and Y is a linear space.
This completes the proof of Part (ii).

This completes the proof of Part (b). □

Now we establish an relation between the operator norm of the matrix operators of $A \in (X_T, Y)$ and $\hat{A} \in (X, Y)$.

Theorem 3.3 ([14, Theorem 3.6]) *Let X and Y be BK spaces and X have AK or $X = \ell_\infty$. If $A \in (X_T, Y)$ then we have*

$$\|L_A\| = \|L_{\hat{A}}\|, \tag{3.13}$$

where $L_A(z) = Az$ for all $z \in Z = X_T$ and $L_{\hat{A}}(x) = \hat{A}(x)$ for all $x \in X$ (Part (b) of Theorem 1.10).

Proof We assume $A \in (X_T, Y)$. Since X is a BK space, so is $Z = X_T$ with the norm $\|\cdot\|_Z = \|T(\cdot)\|_X$ by Theorem 2.8. This also means that $x \in B_X(0, 1)$ if and only if $z = Sx \in B_Z(0, 1)$. It follows by Theorem 1.4 that $L_A \in \mathcal{B}(Z, Y)$, and so $L_{\hat{A}} \in \mathcal{B}(X, Y)$ by Theorem 3.2. We have by (3.9) in Theorem 3.2 or in Part (a) of Corollary 3.4 for $X = \ell_\infty$

$$\begin{aligned}\left\|L_{\hat{A}}\right\| &= \sup_{x\in B_X(0,1)} \left\|L_{\hat{A}}(x)\right\| = \sup_{x\in B_X(0,1)} \left\|\hat{A}x\right\| \\ &= \sup_{z\in B_Z(0,1)} \|Az\| = \sup_{z\in B_Z(0,1)} \|L_A(x)\| = \|L_A\|,\end{aligned}$$

which implies (3.13). □

We obtain the next result as a corollary of Theorem 3.13.

Corollary 3.5 ([6, Theorem 2.8]) *Let $X = \ell_p$ $(1 \le p \le \infty)$ or $X = c_0$, and q be the conjugate number of p.*

(a) Let $Y \in \{c_0, c, \ell_\infty\}$. If $A \in (X_T, Y)$, then we put

$$\|A\|_{(X_T,\infty)} = \sup_n \left\|\hat{A}_n\right\|_q = \begin{cases} \sup_n \sum_{k=0}^\infty |\hat{a}_{nk}| & (X \in \{c_0, \ell_\infty\}) \\ \sup_n \left(\sum_{k=0}^\infty |\hat{a}_{nk}^q|\right)^{1/q} & (X \in \ell_p \text{ for } 1 < p < \infty) \\ \sup_{n,k} |\hat{a}_{nk}| & (X = \ell_1). \end{cases}$$

Then we have

$$\|L_A\| = \|A\|_{(X_T,\infty)}. \tag{3.14}$$

(b) Let $Y = \ell_1$ and $\sup_N$ denote the supremum taken over all finite subsets N of $\mathbb{N}_0$. If $A \in (X_T, \ell_1)$ and $\hat{b}^{(N)} = (\hat{b}_k^{(N)})_{k=0}^\infty$ denotes the sequence with $\hat{b}_k^{(N)} = \sum_{n\in N} \hat{a}_{nk}$ $(k = 0, 1, \dots)$ for any finite subset N of $\mathbb{N}_0$, then we put

$$\|A\|_{(X_T,1)} = \sup_N \left\|\hat{b}^{(N)}\right\|_q$$
$$= \begin{cases} \sup_N \sum_{k=0}^\infty \left|\sum_{n\in N} \hat{a}_{nk}\right| & (X \in \{c_0, \ell_\infty\} \\ \sup_N \left(\sum_{k=0}^\infty \left|\sum_{n\in N} \hat{a}_{nk}\right|^q\right)^{1/q} & (X = \ell_p \text{ for } 1 < p < \infty) \end{cases}$$

and

$$\|A\|_{((\ell_1)_T,\ell_1))} = \sup_k \left\|\hat{A}^k\right\|_1 = \sup_k \left\|\left(\hat{a}_{nk}\right)_{n=0}^\infty\right\|_1 = \sup_k \sum_{n=0}^\infty |\hat{a}_{nk}|.$$

If $X = \ell_1$ then

$$\|L_A\| = \|A\|_{((\ell_1)_T,1)} \tag{3.15}$$

holds; otherwise we have

$$\|A\|_{(X_T,1)} \le \|L_A\| \le 4 \cdot \|A\|_{(X_T,1)}. \tag{3.16}$$

Proof Let $A \in (X_T, Y)$. Since X is a BK space with AK or $X = \ell_\infty$, $A \in (X_T, Y)$ implies $\hat{A} \in (X, Y)$ by Theorem 3.3 and Part (a) of Corollary 3.4 for $X = \ell_\infty$, and (3.9) in Theorem 3.2 holds in each case.

(a) If $Y \in \{c_0, c, \ell_\infty\}$, then we have $\|L_A\| = \|L_{\hat{A}}\| = \sup_n \|\hat{A}_n\|^*$ by (1.18) and (1.19) in Part (c) of Theorem 1.10 and Theorems 1.11 and 2.8. Now (3.14) follows from the definition of the norm $\|\cdot\|_{(X,\ell_\infty)}$ and the fact that $\|\cdot\|_X^* = \|\cdot\|_q$ by Example 1.11.
(b) Let $Y = \ell_1$. The cases $X = \ell_p$ $(1 < p \leq \infty)$ and $X = c_0$ are proved in exactly the same way as in Part (a) of the proof by applying (1.22) in Remark 1.4.
Finally, let $X = \ell_1$. Again we have $\hat{A} \in (\ell_1, \ell_1)$ and (3.13) in Theorem 3.3 holds. Since $L_{(\hat{A})} \in \mathcal{B}(\ell_1, \ell_1)$ by Part (b) of Theorem 1.10, it easily follows that

$$\left\|L_{\hat{A}}(x)\right\|_1 \leq \sum_{n=0}^{\infty}\sum_{k=0}^{\infty} |\hat{a}_{nk}x_k| \leq \sup_k \left\|\hat{A}^k\right\|_1 \cdot \|x\|_1 \leq \|A\|_{((\ell_1)_T,1)} \cdot \|x\|_1,$$

hence $\|L_{\hat{A}}\| \leq \|A\|_{((\ell_1)_T,1)}$.
We also have for each k

$$\left\|L_{\hat{A}}(e^{(k)})\right\|_1 = \left\|\hat{A}^k\right\|_1 \leq \|L_{\hat{A}}\|,$$

hence $\|A\|_{((\ell_1)_T,1)} = \sup_k \|\hat{A}^k\|_1 \leq \|L_{\hat{A}}\|$.
Thus we have shown (3.15).

□

Because of (3.12) in Part (b) of Corollary 3.4 the case $X = c$ has to be treated separately for the estimates of the norm of operators given by matrix transformations from c_T into any of the spaces c_0, c, ℓ_∞ and ℓ_1.

Theorem 3.4 ([6, Theorem 2.9])

(a) *Let $A \in (c_T, Y)$ where $Y \in \{c_0, c, \ell_\infty\}$. Then we have*

$$\|L_A\| = \|A\|_{(c_T,\infty)} = \sup_n \left(\sum_{k=0}^{\infty} |\hat{a}_{n,k}| + |\alpha^{(n)}|\right) \text{ with } \alpha^{(n)} \text{ from 3.11.} \quad (3.17)$$

(b) *Let $A \in (c_T, \ell_1)$. Then we have*

$$\|A\|_{(c_T,1)} = \sup_N \left(\sum_{k=0}^{\infty}\left|\sum_{n\in N} \hat{a}_{nk}\right| + \left|\sum_{n\in N}\alpha^{(n)}\right|\right) \leq \|L_A\| \leq 4 \cdot \|A\|_{(c_T,1)}. \quad (3.18)$$

Proof We write $L = L_A$, for short, and assume $A \in (c_T, Y)$ for $Y \in \{c_0, c, \ell_\infty, \ell_1\}$. Then $A_n \in (c_T)^\beta$ for all n.

(a) Let $Y \in \{c_0, c, \ell_\infty\}$. First $A \in (c_T, Y)$ implies $\hat{A} \in (c_0, Y)$ by the first part of (3.10) in Part (b) of Corollary 3.4 and so we have $\sup_n \sum_{k=0}^{\infty} |\hat{a}_{nk}| < \infty$ by (1.1) in **1.–3.** of Theorem 1.23, that is, $\hat{A}e \in \ell_\infty$. Since also $\hat{A}e - (\alpha^{(n)})_{n=0}^{\infty} \in Y \subset \ell_\infty$ by (3.11) in Part (b) of Corollary 3.4, we obtain $(\alpha^{(n)})_{n=0}^{\infty} \in \ell_\infty$. Therefore the third term in (3.17) is defined and finite. Since c_T is a BK space by Theorem 2.8 it follows from (1.18) and (1.19) in Part (c) of Theorems 1.10, 1.11 and 2.8 (as in Part (a) of the proof of Corollary 3.5) that

$$\|L\| = \sup_n \|A_n\|^*_{c_T}. \tag{3.19}$$

Also $A_n \in (c_T)^\beta$ for $n = 0, 1, \dots$ implies by (1.24) in Example 1.11

$$\|A_n\|^*_{c_T} = \|RA_n\|_1 + |\alpha^{(n)}| = \sum_{k=0}^\infty |\hat{a}_{nk}| + |\alpha^{(n)}| \text{ for all } n = 0, 1 \dots. \tag{3.20}$$

Now (3.17) follows from (3.19) and (3.20).
(b) Now let $Y = \ell_1$. Since c_T is a BK space, it follows by (1.21) in Theorem 1.13 and (1.22) in Remark 1.4 that

$$\|A\|_{(c_T,1)} = \sup_N \left\| \sum_{n\in N} A_n \right\|^*_{c_T} \le \|L\| \le 4 \cdot \|A\|_{(c_T,1)}. \tag{3.21}$$

Let N be a finite subset of $\mathbb{N}_0$ and the sequence $b^{(N)} = \sum_{n\in N} A_n$ defined as the sequence $\hat{b}^{(N)}$ in Part (b) of Corollary 3.5 with the entries $\hat{a}_{nk}$ of $\hat{A}$ replaced by the entries (a_{nk}) of A. Since $A_n \in (c_T)^\beta$ for all n, it follows by (2.68) in Lemma 2.2 that the series $R_k A_n$, and consequently $R_k b^{(N)}$ converge for all k, and we obtain

$$\begin{aligned} R_k b^{(N)} &= \sum_{j=k}^\infty s_{jk} b_j^{(N)} = \sum_{j=k}^\infty s_{jk} \sum_{n\in N} a_{nj} = \sum_{n\in N} \sum_{j=k}^\infty s_{jk} a_{nj} \\ &= \sum_{n\in N} R_k A_n = \sum_{n\in N} \hat{a}_{nk} = \hat{b}_k^{(N)} \end{aligned}$$

for all k, that is,

$$Rb^{(N)} = \hat{b}^{(N)},$$

$$w_{mk}^{(b^{(N)})} = \begin{cases} \sum_{j=m}^\infty s_{jk} b_j^{(N)} = \sum_{n\in N} \left(\sum_{j=m}^\infty s_{jk} a_{nj} \right) & (0 \le k \le m) \\ 0 & (k > m) \end{cases} \quad (m = 0, 1, \dots). \tag{3.22}$$

Hence $w_{mk}^{(b^{(N)})} = \sum_{n\in N} w_{mk}^{(A_n)}$ for all m and k, and

$$\beta^{(N)} = \lim_{m\to\infty} \sum_{k=0}^m w_{mk}^{(b^{(N)})} = \lim_{m\to\infty} \sum_{n\in N} \sum_{k=0}^m w_{mk}^{(A_n)} = \sum_{n\in N} \alpha^{(n)}. \tag{3.23}$$

Furthermore, $A \in (c_T, \ell_1)$ implies $\hat{A} \in (c_0, \ell_1)$ and $\hat{A}e - (\alpha^{(n)})_{n=0}^{\infty} \in \ell_1$ by the first part of (3.10) and by (3.10) in Part (b) of Corollary 3.4. It follows from (1.21) in Theorem 1.13, the fact that $\|\cdot\|_{c_0}^{*} = \ell_1$ by Example 1.11 and from (3.22) that

$$M_1 = \sup_N \sum_{k=0}^{\infty} \left| \sum_{n \in N} \hat{a}_{nk} \right| = \sup_N \|Rb^{(N)}\|_1 = \sup_N \left\|\hat{b}^{(N)}\right\|_1 < \infty.$$

Furthermore, $\hat{A}e - (\alpha^{(n)})_{n=0}^{\infty} \in \ell_1$ and (3.23) yield, if we put $M_2 = \|\hat{A}e - (\alpha^{(n)})_{n=0}^{\infty}\|_1$,

$$\begin{aligned} |\beta^{(N)}| &= \left| \sum_{n \in N} \alpha^{(n)} \right| \le \left| \sum_{n \in N} \left(\hat{A}_n e - \alpha^{(n)} \right) \right| + \left| \sum_{n \in N} \hat{A}_n e \right| \\ &\le M_2 + \left| \sum_{n \in N} \sum_{k=0}^{\infty} \hat{a}_{nk} \right| \le M_2 + \left\| \hat{b}^{(N)} \right\|_1 \le M_1 + M_2 < \infty. \end{aligned}$$

This implies that $\|A\|_{(c_T,1)}$ is defined and finite.
Finally, $A_n \in (c_T)^{\beta}$ for all n implies $b^{(N)} \in (c_T)^{\beta}$ for all finite subsets N of $\mathbb{N}_0$, and so we obtain from (2.78) in Part (c) of Corollary 2.78, (3.22) and (3.23) that

$$\left\|\hat{b}^{(N)}\right\|_{c_T}^{*} = \|Rb^{(N)}\|_1 + |\beta^{(N)}| = \sum_{k=0}^{\infty} \left| \sum_{n \in N} \hat{a}_{nk} \right| + \left| \sum_{n \in N} \alpha^{(n)} \right|.$$

Now (3.18) follows from (3.21).

□

In view of Remark 2.8 and Theorem 3.2, we obtain

Theorem 3.5 ([14, Theorem 3.9]) *Let X be an FK space with AK amd Y be an arbitrary subset of ω. Then we have $A \in (X_T, Y)$ if and only if*

$$\hat{A} \in (X, Y) \text{ and } V^{(A_n)} \in (X, c) \text{ for } n = 0, 1, \ldots, \tag{3.24}$$

where the matrices $\hat{A} = (\hat{a}_{nk})_{n,k=0}^{\infty}$ and $V^{(A_n)} = (v_{mk}^{(A_n)})_{m,k=0}^{\infty}$ $(n = 0, 1, \ldots)$ are defined by $\hat{A}_n = RA_n$ and

$$v_{mk}^{(A_n)} = \begin{cases} \sum_{j=k}^{m} s_{jk} a_{nj} & (0 \le k \le n) \\ 0 & (k > n) \end{cases} \quad (m = 0, 1, \ldots).$$

Proof First we assume that the conditions in (3.24) are satisfied. By Remark 2.8, $V^{(A_n)} \in (X, c)$ implies $W^{(A_n)} \in (X, c_0)$, and this and $\hat{A} \in (X, Y)$ together imply $A \in (X_T, Y)$ by Theorem 3.2.

Conversely, if $A \in (X_T, Y)$ then the first condition in (3.24) holds by Theorem 3.2; also $A_n \in (X_T)^\beta$ implies $V^{(A_n)} \in (X, c)$ by Remark 2.8. □

Remark 3.4 (a) ([14, Remark 3.10 (a)]) The statement of Theorem 3.5 also holds for $X = \ell_\infty$ by Part (a) of Remark 2.8.
(b) ([14, Remark 3.10 (c)]) Let Y be a linear subspace of ω. Then it follows by Part (b) of Remark 2.8 and Part (b) of Corollary 3.4 that $A \in (c_T, Y)$ if and only if

$$\hat{A} \in (c_0, Y), \ \ V^{(A_n)} \in (c, c) \text{ for all } n \text{ and } \hat{A}e - \left(\alpha^{(n)}\right)_{n=0}^{\infty} \in Y. \tag{3.25}$$

We apply Theorem 3.5 to obtain the following result Part (b) is due to *K. Zeller* in [21].

Example 3.1 We have

(a) $A \in (bv, \ell_\infty)$ if and only if

$$\sup_{n,k} \left| \sum_{j=k}^{\infty} a_{nj} \right| < \infty; \tag{3.26}$$

(b) $A \in (bv, c)$ if and only if (3.26) holds and

$$\alpha = \lim_{n \to \infty} \sum_{k=0}^{\infty} a_{nk} \text{ exists} \tag{3.27}$$

and

$$\alpha_k = \lim_{n \to \infty} a_{nk} \text{ exists for each } k; \tag{3.28}$$

(c) $A \in (bv, c_0)$ if and only if (3.26) holds and

$$\lim_{n \to \infty} \sum_{k=0}^{\infty} a_{nk} = 0 \tag{3.29}$$

and

$$\lim_{n \to \infty} a_{nk} = 0 \text{ for each } k. \tag{3.30}$$

Proof Here we have $T = \Delta$ and $bv = (\ell_1)_\Delta$. Since ℓ_1 is a BK space with AK, we can apply Theorem 3.5. We obtain

$$\hat{a}_{nk} = \sum_{j=0}^{\infty} r_{kj} a_{nj} = \sum_{j=k}^{\infty} a_{nj} \ (n, k = 0, 1, \ldots)$$

and

$$v_{mk}^{(n)} = \begin{cases} \sum_{j=k}^{m} a_{nj} & (0 \le k \le m) \\ 0 & (k > m) \end{cases} \quad (m = 0, 1, \dots)$$

(a) We have by Theorem 3.5 $A \in (bv, \ell_\infty)$ if and only if $\hat{A} \in (\ell_1, \ell_1)$, which is the case by (4.1) in Part **4.** of Theorem 1.23 is the case if and only if

$$\sup_{n,k} |\hat{a}_{nk}| = \sup_{n,k} \left| \sum_{j=k}^{\infty} a_{nj} \right| < \infty,$$

which is (3.26), and also $V^{(A_n)} \in (\ell_1, c)$, which is the case by (4.1) and (7.1) in Part **7.** of Theorem 1.23 if and only if

$$\sup_{m,k} |v_{mk}^{(A_n)}| = \sup_{m \ge k,k} \left| \sum_{j=k}^{m} a_{nj} \right| < \infty \tag{3.31}$$

and

$$\lim_{m\to\infty} v_{mk}^{(A_n)} = \lim_{m\to\infty} \sum_{j=k}^{m} a_{nj} = \sum_{j=k}^{\infty} a_{nj} = \beta_k^{(n)} \text{ exists for all } n \text{ and } k. \tag{3.32}$$

Obviously the conditions in (3.31) and (3.32) are redundant.
So we have $A \in (bv, \ell_\infty)$ if and only if the condition in (3.26) is satisfied.
(b) We have by Theorem 3.5 $A \in (bv, c)$ if and only if $\hat{A} \in (\ell_1, c)$, which is the case by (4.1) and (11.2) in Part **14.** of Theorem 1.23 if and only if (3.26) holds and

$$\lim_{n\to\infty} \hat{a}_{nk} = \lim_{n\to\infty} \sum_{j=k}^{\infty} a_{nj} = \hat{\alpha}_k \text{ exists for each } k, \tag{3.33}$$

and also $V^{(A_n)} \in (\ell_1, c)$, which is redundant as in Part (a) of the proof.
So we have $A \in (bv, c)$ if and only if the conditions in (3.26) and (3.33) hold.
We show that the condition in (3.33) is equivalent to those in (3.27) and (3.28).
If $\hat{\alpha}_k$ in (3.33) exists for each k, then, in particular, $\hat{\alpha}_0$ exists, which is the condition in (3.27); we also have for each fixed $k \in \mathbb{N}_0$

$$\alpha_k = \lim_{n\to\infty} a_{nk} = \lim_{n\to\infty} \left(\sum_{j=k}^{\infty} a_{nj} - \sum_{j=k+1}^{\infty} a_{nj} \right) = \hat{\alpha}_k - \hat{\alpha}_{k+1},$$

that is, (3.28) holds.

Conversely if the limits α and α_k for $k \in \mathbb{N}_0$ exist in (3.27) and (3.28), then $\lim_{n\to\infty}\sum_{j=0}^{k} a_{nj} = \sum_{j=0}^{k}\alpha_j$ for each k, and the limits

$$\hat{\alpha}_k = \lim_{n\to\infty}\sum_{j=k}^{\infty} a_{nj} = \lim_{n\to\infty}\left(\sum_{j=0}^{\infty} a_{nj} - \sum_{j=1}^{k-1} a_{nj}\right) = \hat{\alpha}_0 - \sum_{j=0}^{k-1}\alpha_j \text{ exist for all } k.$$

(c) The proof of Part (c) is similar to Part (b) of the proof.

□

Remark 3.5 It is easy to see that the condition in (3.26) of Example 3.1 is equivalent to the conditions

$$\sup_n\left|\sum_{k=0}^{\infty} a_{nk}\right| < \infty \text{ and } \sup_{n,m}\left|\sum_{k=0}^{m} a_{nk}\right| < \infty. \tag{3.34}$$

Now we apply Theorems 3.5 and 3.2, Remark 3.4 and Corollary 3.4 to characterize some matrix transformations on the spaces e_p^r for $1 \le p \le \infty$, e_0^r and e_c^r.

It follows from (2.20) in Corollary 2.14 that the matrices $\hat{A}$, $V^{(n)} = V^{(A_n)}$ and $W^{(n)} = W^{(A_n)} (n = 0, 1, \dots)$ of Theorems 3.5 and 3.2 are given by

$$\hat{a}_{nk} = R_k A_n = \sum_{j=k}^{\infty}\binom{j}{k}(r-1)^{j-k}r^{-j}a_{nj} \quad (n, k = 0, 1, \dots), \tag{3.35}$$

$$\begin{aligned} v_{mk}^{(n)} &= \sum_{j=k}^{m} a_{nj}s_{jk} \\ &= \begin{cases} \sum_{j=k}^{m}\binom{j}{k}(r-1)^{j-k}r^{-j}a_{nj} & (0 \le k \le m) \\ 0 & (k > m) \end{cases} \quad (m = 0, 1, \dots), \end{aligned} \tag{3.36}$$

and

$$\begin{aligned} w_{mk}^{(n)} &= \sum_{j=m}^{\infty} a_{nj}s_{jk} \\ &= \begin{cases} \sum_{j=m}^{\infty}\binom{j}{k}(r-1)^{j-k}r^{-j}a_{nj} & (0 \le k \le m) \\ 0 & (k > m) \end{cases} \quad (m = 0, 1, \dots). \end{aligned} \tag{3.37}$$

Corollary 3.6 *We have*

(a) ([2, Theorem 2.2 (i)]) $A \in (e_1^r, \ell_\infty)$ *if and only if*

$$\sup_{n,k} \left| \sum_{j=k}^{\infty} \binom{j}{k} (r-1)^{j-k} r^{-j} a_{nj} \right| < \infty; \tag{3.38}$$

$A \in (e_1^r, c_0)$ *if and only if the condition in* (3.38) *holds and*

$$\lim_{n\to\infty} \left(\sum_{j=k}^{\infty} \binom{j}{k} (r-1)^{j-k} r^{-j} a_{nj} \right) = 0 \text{ for each } k; \tag{3.39}$$

$A \in (e_1^r, c)$ *if and only if the condition in* (3.38) *holds and*

$$\hat{\alpha}_k = \lim_{n\to\infty} \left(\sum_{j=k}^{\infty} \binom{j}{k} (r-1)^{j-k} r^{-j} a_{nj} \right) \text{ exists for each } k; \tag{3.40}$$

(b) ([2, Theorem 2.2 (ii)]) $A \in (e_p^r, \ell_\infty)$ *for* $1 < p < \infty$ *and* $q = p/)p-1)$ *if and only if*

$$\sup_n \sum_{k=0}^{\infty} \left| \sum_{j=k}^{\infty} \binom{j}{k} (r-1)^{j-k} r^{-j} a_{nj} \right|^q < \infty; \tag{3.41}$$

and

$$\sup_m \sum_{k=0}^{m} \left| \sum_{j=k}^{m} \binom{j}{k} (r-1)^{j-k} r^{-j} a_{nj} \right|^q < \infty \text{ for all } n; \tag{3.42}$$

$A \in (e_p^r, c_0)$ *if and only if the conditions in* (3.41), (3.42) *and* (3.39) *hold;*
$A \in (e_p^r, c)$ *if and only if the conditions in* (3.41), (3.42) *and* (3.40) *hold;*

(c) $A \in (e_\infty^r, \ell_\infty)$ *if and only if*

$$\sup_n \sum_{k=0}^{\infty} \left| \sum_{j=k}^{\infty} \binom{j}{k} (r-1)^{j-k} r^{-j} a_{nj} \right| < \infty; \tag{3.43}$$

and

$$\lim_{m\to\infty} \sum_{k=0}^{m} \left| \sum_{j=m}^{\infty} \binom{j}{k} (r-1)^{j-k} r^{-j} a_{nj} \right| = 0 \text{ for all } n; \tag{3.44}$$

$A \in (e_\infty^r, c_0)$ *if and only if the condition in* (3.44) *holds and*

$$\lim_{n\to\infty} \sum_{k=0}^{\infty} \left| \sum_{j=k}^{\infty} \binom{j}{k} (r-1)^{j-k} r^{-j} a_{nj} \right| = 0; \tag{3.45}$$

$A \in (e^r_\infty, c)$ *if and only if the conditions in* (3.40), (3.44) *and* (3.45) *hold and*

$$\sum_{k=0}^{\infty}\left|\sum_{j=k}^{\infty}\binom{j}{k}(r-1)^{j-k}r^{-j}a_{nj}\right| \text{ converges uniformly in } n. \tag{3.46}$$

Proof In the proof of Parts (a) and (b), we use the conditions in (3.24) of Theorem 3.5, and in the proof of Part (c), the conditions in (3.8) of Theorem 3.2 for the characterizations of the respective classes of matrix transformations.

(a) The conditions for $A \in (e^r_1, \ell_\infty)$ are $\hat{A} \in (\ell_1, \ell_\infty)$, which is the condition in (3.38) by (4.1) in Part **4.** of Theorem 1.23 and (3.35), and $V^{(n)} \in (\ell_1, c)$ for all n which is equivalent to the conditions in (3.39) and

$$\lim_{m\to\infty} v^{(n)}_{mk} \text{ exists for each } k \tag{3.47}$$

by (4.1) and (11.2) in Part **14.** of Theorem 1.23 and (3.36); the condition in (3.47) is redundant in view of (3.35) and (3.38).
The conditions for $A \in (e^r_1, c_0)$ are $\hat{A} \in (\ell_1, c_0)$, which means by Theorem 1.11 that we have to add the condition $\lim_{n\to\infty}\hat{a}_{nk} = 0$ for each k, that is, the condition in (3.39), to the condition in (3.38) for $\hat{A} \in (\ell_1, \ell_\infty)$, and again $V^{(n)} \in (\ell_1, c)$ for all n.
For $A \in (e^r_1, c)$ we have to replace the condition in (3.39) for $\hat{A} \in (\ell_1, c_0)$ by $\hat{\alpha}_k = \lim_{n\to\infty}\hat{a}_{nk}$ exists for each k, which is the condition in (3.40), for $\hat{A} \in (\ell_1, c)$.

(b) The proof of Part (b) is very similar to that of Part (b). Now we obtain the conditions in (3.41) and (3.42) for $\hat{A} \in (\ell_r, \ell_\infty)$ and $V^{(n)} \in (\ell_r, c)$ $(n = 0, 1, \dots)$ from the condition in (5.1) in Parts **5.** and **15.** of Theorem 1.23; the condition in (3.47) that comes from (11.2) in Part **15.** of Theorem 1.23 again is redundant.
(c) Now we use the conditions in (3.8) of Theorem 3.2 and (3.37) for the entries of the matrices $W^{(n)}$ $(n = 0, 1, \dots)$.
The conditions for $A \in (e^r_\infty, \ell_\infty)$ are $\hat{A} \in (\ell_1, \ell_1)$, which is the condition in (3.43) by (1.1) in Part **1.** of Theorem 1.23, and $W^{(n)} \in (\ell_\infty, c_0)$ $(n = 0, 1, \dots)$, which, by (6.1) in Part **6.** of Theorem 1.23, is equivalent to $\lim_{m\to\infty}\sum_{k=0}^{m}|w^{(n)}_{mk}| = 0$ $(n = 0, 1,)$, that is, to the condition in (3.44). The conditions for $A \in (e^r_\infty, c_0)$ are $\hat{A} \in (\ell_\infty, c_0)$, which, by (6.1) in Part **6.** of Theorem 1.23, is equivalent to $\lim_{m\to\infty}\sum_{k=0}^{\infty}|\hat{a}_{mk}| = 0$, that is, to the condition in (3.45), and $W^{(n)} \in (\ell_\infty, c_0)$ which again is equivalent to the condition in (3.44). The conditions for $A \in (e^r_\infty, c)$ are $W^{(n)} \in (\ell_\infty, c_0)$, that is, (3.44) as before, and $\hat{A} \in (\ell_\infty, c)$ which, by (11.1) and (11.2) in Part **11.** of Theorem 1.23, is equivalent to $\sum_{k=0}^{\infty}|\hat{a}_{nk}|$ converges uniformly in n, that is, the condition in (3.46), and $\lim_{n\to\infty}\hat{a}_{nk} = \hat{a}_k$ exists for each k, which is the condition in (3.40).

□

Finally we characterize matrix transformations on the matrix domains of Δ in the classical sequence spaces. Now we have $T = \Delta$ and $S = \Sigma$, hence the matrices $\hat{A}$ and $W^{(n)}$ $(n = 0, 1, \dots)$ of Theorem 3.2 are given by

$$\hat{a}_{nk} = R_k A_n = \sum_{j=k}^{\infty} a_{nj} \quad (n, k = 0, 1, \dots), \tag{3.48}$$

and

$$w_{mk}^{(n)} = \sum_{j=m}^{\infty} a_{nj} s_{jk} = \begin{cases} \sum_{j=m}^{\infty} a_{nj} & (0 \le k \le m) \\ 0 & (k > m) \end{cases} \quad (m = 0, 1, \dots). \quad \square \tag{3.49}$$

We need the following results the first of which is a generalization of Theorem 1.12.

Proposition 3.1 *Let X be an FK space, $b \notin X$ be a sequence, $X_1 = X \oplus b = \{x_1 = x + \lambda b : x \in X,\ \lambda \in \mathbb{C}\}$, and Y be a linear subspace of ω. Then $A \in (X_1, Y)$ if and only if $A \in (X, Y)$ and $Ab \in Y$.*

Proof First, we assume $A \in (X_1, Y)$. Then $X \subset X_1$ implies $A \in (X, Y)$, and $b \in X_1$ implies $Ab \in Y$.

Conversely, we assume $A \in (X, Y)$ and $Ab \in Y$. Let $x_1 \in X_1$ be given. Then there are $x \in X$ and $\lambda \in \mathbb{C}$ such that $x_1 = x + \lambda b$, and it follows that $Ax_1 = A(x + \lambda b) = Ax + \lambda Ab \in Y$. □

Proposition 3.2 *Let Y be a linear subspace of ω. Then we have $A \in (c(\Delta), Y)$ if and only if*

$$\hat{A} \in (c_0, Y), \tag{3.50}$$

$$\sup_m (m+1) \left| \sum_{j=m}^{\infty} a_{nj} \right| < \infty \textit{ for } n = 0, 1, \dots \tag{3.51}$$

and

$$A(\Sigma e) \in Y. \tag{3.52}$$

Proof Since $x \in c(\Delta)$ if and only if $x^{(0)} = x - \xi \Sigma e \in c_0(\Delta)$, applying Proposition 3.1 with $X = c_0(\Delta)$, $X_1 = c(\Delta)$ and $b = \Sigma e$, we obtain $A \in (c(\Delta), Y)$ if and only if $A \in (c_0, \Delta)$ and $A(\Sigma e) \in Y$ which is the condition in (3.52). Furthermore, we have, by Theorem 3.2, $A \in (c_0, \Delta)$ if and only if $\hat{A} \in (c_0, Y)$, which is the condition in (3.50), and $W^{(n)} \in (c_0, c_0)$ for all n, which by the conditions (1.1) (7.1) in Part **7.** of Theorem 1.23 and by (3.49) is equivalent to

$$\sup_m \sum_{k=0}^{\infty} |w_{mk}^{(n)}| = \sup_m \sum_{k=0}^{m} \left| \sum_{j=m}^{\infty} a_{nj} \right| = \sup_m (m+1) \left| \sum_{j=m}^{\infty} a_{nj} \right| < \infty \text{ for } n = 0, 1, \ldots,$$

which is the condition in (3.51), and

$$\lim_{m \to \infty} w_{mk}^{(n)} = \lim_{m \to \infty} \sum_{j=m}^{\infty} a_{nj} = 0 \text{ for } n = 0, 1, \ldots, \tag{3.53}$$

which is redundant. □

Corollary 3.7 *Let* $1 < p, r < \infty$, $q = p/(p-1)$ *and* $s = r/(r-1)$. *Then the necessary and sufficient conditions for* $A \in (X, Y)$ *can be read from the following table:*

From / *To*	$\ell_\infty(\Delta)$	$c_0(\Delta)$	$c(\Delta)$	bv	bv^p
ℓ_∞	**1.**	**2.**	**3.**	**4.**	**5.**
c_0	**6.**	**7.**	**8.**	**9.**	**10.**
c	**11.**	**12.**	**13.**	**14.**	**15.**
ℓ_1	**16.**	**17.**	**18.**	**19.**	**20.**
ℓ_r	**21.**	**22.**	**23.**	**24.**	**unknown**

where

1. *where*

(1.1) $\sup_n \sum_{k=0}^{\infty} |\sum_{j=k}^{\infty} a_{nj}| < \infty$

(1.2) $\lim_{m\to\infty} (m+1)|\sum_{j=m}^{\infty} a_{nj}| = 0$ *for all* n

2. *(1.1), (2.1) where*

(2.1) $\sup_m (m+1)|\sum_{j=m}^{\infty} a_{nj}| < \infty$ *for all* n

3. *(1.1), (2.1), (3.1) where*

(3.1) $\sup_n |\sum_{k=0}^{\infty} (k+1) a_{nk}| < \infty$ *for all* n

4. *(4.1) where*

(4.1) $\sup_{n,k} |\sum_{j=k}^{\infty} a_{nj}| < \infty$

5. *(5.1),(5.2) where*

(5.1) $\sup_n \sum_{k=0}^{\infty} |\sum_{j=k}^{\infty} a_{nk}|^q < \infty$

(5.2) $\sup_m (m+1)|\sum_{j=m}^{\infty} a_{nj}|^s < \infty$ *for all n*

6. *(1.2), (6.1) where*

(6.1) $\lim_{n\to\infty} \sum_{k=0}^{\infty} |\sum_{j=k}^{\infty} a_{nk}| = 0$

7. *(1.1), (2.1), (7.1) where*

(7.1) $\lim_{n\to\infty} \sum_{j=k}^{\infty} a_{nj} = 0$ *for every k*

8. *(1.1), (2.1), (7.1), (8.1) where*

(8.1) $\lim_{n\to\infty} \sum_{k=0}^{\infty} (k+1)a_{nk} = 0$

9. *(4.1), (7.1)*

10. *(5.1), (5.2), (7.1)*

11. *(1.2), (11.1), (11.2) where*

(11.1) $\sum_{k=0}^{\infty} |\sum_{j=k}^{\infty} a_{nj}|$ *converges uniformly in n*

(11.2) $\lim_{n\to\infty} \sum_{j=k}^{\infty} a_{nk} = \hat{\alpha}_k$ *exists for every k*

12. *(1.1), (2.1), (11.2)*

13. *(1.1), (2.1), (11.2), (13.1) where*

(13.1) $\lim_{n\to\infty} \sum_{k=0}^{\infty} (k+1)a_{nk} = \gamma$ *exists*

14. *(4.1), (11.2)*

15. *(5.1), (5.2), (11.2)*

16. *(1.2), (16.1) where*

(16.1) $\sup_{\substack{N \subset \mathbb{N}_0 \\ N \text{ finite}}} (\sum_{k=0}^{\infty} |\sum_{n\in N} \sum_{j=k}^{\infty} a_{nj}|) < \infty$

17. *(2.1),(16.1)*

18. *(2.1), (16.1), (18.1) where*

(18.1) $\sum_{n=1}^{\infty} |\sum_{k=0}^{\infty} (k+1)a_{nk}| < \infty$

19. *(19.1) where*

(19.1) $\sup_k \sum_{n=0}^{\infty} |\sum_{j=k}^{\infty} a_{nj}| < \infty$

20. *(5.2) (20.1) where*

(20.1) $\sup_{\substack{N \subset \mathbb{N}_0 \\ N \text{ finite}}} (\sum_{k=0}^{\infty} |\sum_{n\in N} \sum_{j=k}^{n} a_{nj}|^q) < \infty$

21. *(1.2), (21.1) where*

(21.1) $\sup_{\substack{K \subset \mathbb{N}_0 \\ K \text{ finite}}} \sum_{n=0}^{\infty} |\sum_{k\in K} \sum_{j=k}^{\infty} a_{nk}|^r < \infty$

22. *(2.1), (21.1)*

23. *(2.1), (21.1), (23.1) where*

(23.1) $\sum_{n=0}^{\infty} |\sum_{k=0}^{\infty} |(k+1)a_{nk}|^r) < \infty$

24. *(24.1) where*

(24.1) $\sup_k \sum_{n=0}^{\infty} |\sum_{j=k}^{\infty} a_{nj}|^r < \infty.$

Proof (i) First we show that if X is an FK space with AK and Y is any subset of ω, then $A \in (X_T, Y)$ if and only if

$$\hat{A} \in (X, Y) \text{ and } W^{(n)} \in (X, \ell_\infty) \text{ for } n = 0, 1, \dots. \tag{3.54}$$

We assume $A \in (X_T, Y)$. Then, by the condition in (3.8) of Theorem 3.8, $\hat{A} \in (X, Y)$ and $W^{(n)} \in (X, \ell_\infty)$ for $n = 0, 1, \cdots$, and these conditions clearly imply those in (3.54), since $(X, c_0) \subset (X, \ell_\infty)$.
Conversely, we assume that the conditions in (3.54) hold. Then the series $R_k A_n = \sum_{j=k}^\infty s_{jk} a_{nj}$ converge for all n and k, hence

$$\lim_{m\to\infty} w_{mk}^{(n)} = \lim_{m\to\infty} \sum_{j=m}^\infty s_{jk} a_{nj} = 0 \text{ for each } k \text{ and all } n.$$

This condition and $W^{(n)} \in (X, \ell_\infty)$ $(n = 0, 1, \dots)$ together imply $W^{(n)} \in (X, c_0)$ $(n = 0, 1, \dots)$. So the conditions in (3.8) of Theorem 3.8 are satisfied and consequently we have $A \in (X_T, Y)$.
This completes the proof of Part (i)

(ii) We prove Parts **1.**, **6.**, **11.**, **16.** and **21.** By Part (a) of Remark 3.4, we have $A \in (\ell_\infty, Y)$ if and only if $\hat{A} \in (\ell_\infty, Y)$ and $W^{(n)} \in (\ell_\infty, c_0)$ $(n = 0, 1, \dots)$.
For $Y = \ell_\infty$, the condition $A \in (\ell_\infty, Y)$ yields (1.1) in Part **1.** by (1.1) in Part **1.** of Theorem 1.23; for $Y = c_0$, the condition in $A \in (\ell_\infty, Y)$ yields (6.1) in Part **6.** by (6.1) in Part **6.** of Theorem 1.23; for $Y = c$, the condition in $A \in (\ell_\infty, Y)$ yields (11.1) and (11.2) in Part **11.** by (11.1) and (11.2) in Part **11.** of Theorem 1.23; for $Y = \ell_1$, the condition in $A \in (\ell_\infty, Y)$ yields (16.1) in Part **16.** by (16.1) in Part **16.** of Theorem 1.23; for $Y = \ell_r$ $(1 < r < \infty)$, the condition in $A \in (\ell_\infty, Y)$ yields (21.1) in Part **21.** by (21.1) in Part **21.** of Theorem 1.23. Also the condition $W^{(n)} \in (\ell_\infty, c_0)$ yields (1.2) in Parts **1.**, **6.**, **11.**, **16.** and **21.** by (6.1) in Part **6.** of Theorem 1.23.
This completes the proof of Part (ii).

(iii) We prove Parts **2.**, **7.**, **12.** **17.** and **22.**.
Since $(c_0, Y) = (\ell_\infty, Y)$ for $Y = \ell_\infty, \ell_1, \ell_r$ $(1 < r < \infty)$ by Parts **2.**, **17.** and **22.** of Theorem 1.23, we obtain (1.1) in Part **2.**, (16.1) in Part **17.** and (21.1) in Part **22.**. For $Y = c_0, c$, the condition $\hat{A} \in (c_0, Y)$ yields $\hat{A} \in (c_0, \ell_\infty)$ and $Ae^{(k)} \in Y$ for all k, that is, (1.1) and (7.1) in Part **7.**, and (1.1) and (11.2) in Part **12.**. Also the second condition in (3.54), that is, $W^{(n)} \in (c_0, \ell_\infty)$ for $n = 0, 1, \dots$ yields (2.1) in Parts **2.**, **7.**, **12.** **17.** and **22.** by (1.1) in Part **2.** of Theorem 1.23 This completes the proof of Part (iii).

(iv) We prove Parts **3.**, **8.**, **13.** **18.** and **23.**.
By Proposition 3.1, we have to add the condition

$$A(\Sigma e) = \left(\sum_{k=0}^\infty (k+1) a_{nk} \right)_{n=0}^\infty \in Y$$

to the conditions for $A \in (c_0, Y)$ in each case. This yields (3.1), (8.1), (13.1), (18.1) and (23.1) in Parts **3.**, **8.**, **13. 18.** and **23.**.
This completes the proof of Part (iv).

(v) We prove Parts **4.**, **9.**, **14. 19.** and **24.**.
For $Y = \ell_\infty$, the condition $A \in (\ell_1, Y)$ yields (4.1) in Part **4.** by (4.1) in Part **4.** of Theorem 1.23; for $Y = c_0, c$ we have to add the condition $Ae^{(k)} \in c_0, c$ for each k, that is, (7.1) in Part **9.**, and (11.2) in Part **14.**; for $Y = \ell_1$, the condition $A \in (\ell_1, Y)$ yields (19.1) in Part **19.** by (19.1) in Part **19.** of Theorem 1.23; for $Y = \ell_r$, the condition $A \in (\ell_1, Y)$ yields (24.1) in Part **24.** by (24.1) in Part **24.** of Theorem 1.23.
The condition $W^{(n)} \in (\ell_1, \ell_\infty)$ for $n = 0, 1, \dots$ in (3.54) is

$$\sup_{m \geq k, k} \left| \sum_{j=m}^{\infty} a_{nj} \right| < \infty \text{ for all } n \tag{3.55}$$

by (4.1) in Part **4.** of Theorem 1.23. It is obvious that the condition in (3.55) is contained in (4.1) in Parts **4.**, **9.** and **14.**, in (19.1) and (24.1) of Parts **19.** and **24.**.
This completes the proof of Part (v).

(vi) Finally, we prove Parts **5.**, **10.**, **15.** and **20.**. For $Y = \ell_\infty$, the condition $A \in (\ell_r, Y)$ $(1 < r < \infty)$ yields (5.1) in Part **5.** by (5.1) in Part **5.** of Theorem 1.23; for $Y = c_0, c$ we have to add the condition $Ae^{(k)} \in c_0, c$ for each k, that is, (7.1) in Part **10.**, and (11.2) in Part **15.**; For $Y = \ell_1$, the condition $A \in (\ell_r, Y)$ $(1 < r < \infty)$ yields (20.1) in Part **20.** by (20.1) in Part **20.**. of Theorem 1.23. Also the $W^{(n)} \in (\ell_r, \ell_\infty)$ $(1 < r < \infty; s = r/(r-1))$ for $n = 0, 1, \dots$ in (3.54) is

$$\begin{aligned} \sup_m \sum_{k=0}^{\infty} |w_{mk}^{(n)}|^s &= \sup_m \sum_{j=m}^{\infty} \left| \sum_{j=m}^{\infty} a_{nj} \right|^s \\ &= \sup_m (m+1) \left| \sum_{j=m} a_{nj} \right|^s < \infty \text{ for all } n, \end{aligned} \tag{3.56}$$

by (5.1) in Part **5.** of Theorem 1.23, and (3.56) obviously is (5.2) in Parts **5.**, **10.**, **15.** and **20.**.
This completes the proof of Part (vi). □

3.3 Compact Matrix Operators

In this section, we establish some identities or estimates for the Hausdorff measure of noncompactness of matrix operators from BK with AK into the spaces c_0, c and ℓ_1. This is achieved by applying Theorems 1.10, 1.27 and 1.29. The identities or estimates and Theorem 1.30 yield the characterizations of the corresponding classes of compact matrix operators.

We need the following concept and result. A norm $\|\cdot\|$ on a sequence space X is said to be *monotonous*, if $x, \tilde{x} \in X$ with $|x_k| \leq |\tilde{x}_k|$ for all k implies $\|x\| \leq \|\tilde{x}\|$. Part (a) of the following lemma generalizes Example 1.14.

Lemma 3.1 ([15, Lemma 9.8.1]) *(a) Let X be a monotonous BK space with AK, $\mathcal{P}_n : X \to X$ be the projector onto the linear span of $\{e^{(1)}, e^{(2)}, \dots, e^{(n)}\}$ and $\mathcal{R}_n = I - \mathcal{P}_n$, where I is the identity on X. Then we have*

$$L = \lim_{n\to\infty} \|\mathcal{R}_n\| = 1 \tag{3.57}$$

and

$$\chi(Q) = \lim_{n\to\infty}\left(\sup_{x\in Q}\|\mathcal{R}_n(x)\|\right) \text{ for all } Q \in \mathcal{M}_X \tag{3.58}$$

(b) Let $\mathcal{P}_n : c \to c$ be the projector onto the linear span of $\{e, e^{(1)}, e^2, \dots, e^{(n)}\}$ and $\mathcal{R}_n = I - \mathcal{P}_n$. Then

$$L = \lim_{n\to\infty} \|\mathcal{R}_n\| = 2. \tag{3.59}$$

and the estimate in (1.48) *of* Part (b) *in* Example 1.14 *holds.*

Proof It is clear that

$$\mu_n(Q) = \sup_{x\in Q}\|\mathcal{R}_n(x)\| < \infty \text{ for all } n \text{ and all } Q \in \mathcal{M}_X \text{ or } Q \in \mathcal{M}_c.$$

(a) Since X is a monotonous BK space with AK, we obtain for all n and all $x \in X$

$$\|\mathcal{R}_n(x)\| = \|x - x^{[n]}\| \geq \|x - x^{[n+1]}\| = \|\mathcal{R}_{n+1}(x)\|. \tag{3.60}$$

Let $n \in \mathbb{N}_0$ and $\varepsilon > 0$ be given. Then there exists a sequence $x^{(0)} \in Q$ such that $\|\mathcal{R}_n(x^{(0)})\| \geq \mu_{n+1}(Q) - \varepsilon$ and it follows from (3.60) that

$$\mu_n(Q) \geq \|\mathcal{R}_n(x^{(0)})\| \geq \|\mathcal{R}_{n+1}(x^{(0)})\| \geq \mu_{n+1}(Q) - \varepsilon. \tag{3.61}$$

Since $n \in \mathbb{N}_0$ and $\varepsilon > 0$ were arbitrary, we have $\mu_n(Q) \geq \mu_{n+1}(Q) \geq 0$ for all n, and so $\lim_{n\to\infty}\mu_n(Q)$ exists. This shows that the limit in (3.58) exists. Furthermore, since the norm $\|\cdot\|$ is monotonous, we have $\|\mathcal{R}_n(x)\| = \|x - x^{[n]}\| \leq \|x\|$ for all $x \in X$ and all n, hence

$$\|\mathcal{R}_n\| < 1 \text{ for all } n. \tag{3.62}$$

To prove the converse inequality, given $n \in \mathbb{N}_0$, we have $\|\mathcal{R}_n(e^{(n+1)})\| = \|e^{(n+1)}\| \neq 0$, and consequently

$$\|\mathcal{R}_n\| \geq 1 \text{ for all } n. \tag{3.63}$$

Finally (3.62) and (3.63) imply (3.57) and the equality in (3.58) follows from Remark 1.12.

(b) (This is Part (b) of Example 1.14.)
The sequence $(e, e^{(0)}, e^{(1)}, \dots)$ is a Schauder basis of the space c by Example 1.6 and every sequence $x = (x_k)_{k=0}^{\infty} \in c$ has a unique representation

$$x = \xi\, e + \sum_{k=0}^{\infty}(x_k - \xi)e^{(k)} \text{ where } \xi = \xi(x) = \lim_{k\to\infty} x_k$$

by (1.14) in Example 1.6. Then we have

$$\mathcal{R}_n(x) = \sum_{k=n+1}^{\infty} (x_k - \xi)e^{(k)} \text{ for all } n \in \mathbb{N}.$$

Since $|x_k - \xi| \leq \|x\|_\infty + |\xi| \leq 2 \cdot \|x\|_\infty$ for all k and all $x \in c$, we obtain

$$\|\mathcal{R}_n(x)\|_\infty \leq \sup_{k\geq n+1} |x_k - \xi| \leq 2 \cdot \|x\|_\infty,$$

that is,

$$\|\mathcal{R}_n\| \leq 2 \text{ for all } n \in \mathbb{N}. \tag{3.64}$$

On the other hand, if $n \in \mathbb{N}$ is given, then we have for $x = -e + 2 \cdot e^{(n+1)}$

$$\xi = -1, \ \|x\|_\infty = 1 \text{ and } \mathcal{R}_n(x) = 2 \cdot e^{(n+1)}, \text{ that is, } \|\mathcal{R}_n\| \geq 2.$$

This and (3.64) together imply $\|\mathcal{R}_n\| = 2$ for all $n \in \mathbb{N}$, and so

$$a = \lim_{n\to\infty} \|\mathcal{R}_n\| = 2.$$

Similarly as in the proof of Part (a) it can be shown that

$$\lim_{n\to\infty}\left(\sup_{x\in Q} \|\mathcal{R}_n(x)\|_\infty\right)$$

exists, and so (1.48) follows from (1.44) and (1.45) in Theorem 1.27. □

Example 3.2 The Hausdorff measure of noncompactness of any operator $L \in \mathcal{B}(\ell_1)$ is given by

$$\|L\|_\chi = \lim_{m\to\infty}\left(\sup_k \sum_{n=m}^{\infty} |a_{nk}|\right), \tag{3.65}$$

where $A = (a_{nk})_{n,k=0}^{\infty}$ is the matrix that represents L (Part (b) of Theorem 1.10).

Proof We write $\overline{B}_{\ell_1}$ for the closed unit ball in ℓ_1 and $A^{<m>}$ for the matrix with the rows $A_n^{<m>} = 0$ for $n \le m$ and $A_n^{<m>} = A_n$ for $n \ge m+1$. Then we obviously have $(\mathcal{R}_m \circ L)(x) = A^{<m>}(x)$ for all $x \in \ell_1$ and we obtain by (1.52) in Theorem 1.29, Remark 1.12, (3.58) and (3.57) in Lemma 3.1, and (19.1) in **19.** of Theorem 1.23

$$\|L\|_\chi = \chi\left(L(\overline{B}_{\ell_1})\right) = \lim_{m\to\infty}\left(\sup_{\|x\|=1} \|(\mathcal{R}_m \circ L)(x)\|\right) = \lim_{m\to\infty} \|A^{<m>}\|$$

$$= \lim_{m\to\infty}\left(\sup_k \sum_{n=m+1}^{\infty} |a_{nk}|\right),$$

that is, (3.65) holds. □

Remark 3.6 It follows from (3.65) and (1.54) in Theorem 1.30 that $L \in \mathcal{K}(\ell_1)$ if and only if

$$\lim_{m\to\infty}\left(\sup_k \sum_{n=m}^{\infty} |a_{nk}|\right) = 0. \tag{3.66}$$

Theorem 3.6 ([15, Theorem 9.8.4]) *Let X and Y be BK spaces, X have AK, $L \in \mathcal{B}(X, Y)$ and $A \in (X, Y)$ be the matrix that represents L* (Part (b) *of* Theorem 1.10).

(a) *If $Y = c$, then we have*

$$\frac{1}{2}\cdot \lim_{m\to\infty}\left(\sup_{n\ge m} \left\|A_n - (\alpha_k)_{k=0}^{\infty}\right\|_X^*\right) \le \|L\|_\chi$$

$$\le \lim_{m\to\infty}\left(\sup_{n\ge m} \left\|A_n - (\alpha_k)_{k=0}^{\infty}\right\|_X^*\right), \tag{3.67}$$

where

$$\alpha_k = \lim_{n\to\infty} a_{nk} \text{ for all } k. \tag{3.68}$$

(b) *If $Y = c_0$, then we have*

$$\|L\|_\chi = \lim_{m\to\infty}\left(\sup_{n\ge m} \|A_n\|_X^*\right). \tag{3.69}$$

(c) *If* $Y = \ell_1$, *then we have*

$$\lim_{m\to\infty}\left(\sup_{N_m}\left\|\sum_{n\in N_m} A_n\right\|_X^*\right) \le \|L\|_X \le 4\cdot\lim_{m\to\infty}\left(\sup_{N_m}\left\|\sum_{n\in N_m} A_n\right\|_X^*\right), \tag{3.70}$$

where the supremum is taken over all finite subsets of integers $\ge m$.

Proof We write $\|\cdot\| = \|\cdot\|_X^*$ and $\|A\| = \|A\|_{(X,\ell_\infty)} = \sup_n \|A_n\|$, for short.

(a) Let $A = (a_{nk})_{n,k=0}^\infty \in (X, c)$. Then $\|A\| < \infty$, α_k in (3.68) exists for all k by (11.2) in **13.** of Theorem 1.23, and $\|L\| = \|A\|$ by (1.19) in Theorem 1.10.

(i) We show $(\alpha_k)_{k=0}^\infty \in X^\beta$.
Let $x \in X$ be given. Since X has AK, there exists a positive constant C such that $\|x^{[m]}\| \le C\|x\|$ for all $m \in \mathbb{N}_0$ and it follows that

$$\left|\sum_{k=0}^m a_{nk}x_k\right| = \left|A_n x^{[m]}\right| \le C\|A_n\|^*\|x\| \le C\|A\|\cdot\|x\| \text{ for all } n \text{ and } m,$$

hence by (3.68)

$$\left|\sum_{k=0}^m \alpha_k x_k\right| = \lim_{n\to\infty}\left|\sum_{k=0}^m a_{nk}x_k\right| \le C\|A\|\cdot\|x\| \text{ for all } m. \tag{3.71}$$

This implies $(\alpha_k x_k)_{k=0}^\infty \in bs$, and since $x \in X$ was arbitrary, we conclude $(\alpha_k)_{k=0}^\infty \in X^\gamma$, and $X^\gamma = X^\beta$ by Theorem 1.16 (c), since X has AK and so AD.

Also $(\alpha_k)_{k=0}^\infty \in X^\beta$ implies $\|(\alpha_k)_{k=0}^\infty\|^* < \infty$ by Theorem 1.14.

(ii) Now we show

$$\lim_{n\to\infty} A_n x = \sum_{k=0}^\infty \alpha_k x_k \text{ for all } x \in X. \tag{3.72}$$

Let $x \in X$ and $\varepsilon > 0$ be given. Since X has AK, there exists $k_0 \in \mathbb{N}_0$ such that

$$\|x - x^{[k_0]}\| \le \frac{\varepsilon}{2(M+1)}, \text{ where } M = \|A\| + \|(\alpha_k)_{k=0}^\infty\|^*. \tag{3.73}$$

It also follows from (3.68) that there exists $n_0 \in \mathbb{N}_0$ such that

$$\left|\sum_{k=0}^{k_0}(a_{nk} - \alpha_k)x_k\right| < \frac{\varepsilon}{2} \text{ for all } n \ge n_0. \tag{3.74}$$

Let $n \ge n_0$ be given. Then it follows from (3.73) and (3.74) that

$$\left|A_n x - \sum_{k=0}^{\infty} \alpha_k x_k\right| \le \left|\sum_{k=0}^{k_0} (a_{nk} - \alpha_k) x_k\right| + \left|\sum_{k=k_0+1}^{\infty} (a_{nk} - \alpha_k)\right|$$

$$< \frac{\varepsilon}{2} + \|A_n - (\alpha_k)_{k=0}^{\infty}\|^* \|x - x^{[k_0]}\| < \frac{\varepsilon}{2} + \frac{\varepsilon}{2} = \varepsilon.$$

Thus we have shown (3.72).

(iii) Now we show the inequalities in (3.67).
Let $y = (y_n)_{n=0}^{\infty} \in c$ be given. Then, by Example 1.6, y has unique representation $y = \eta e + \sum_{n=0}^{\infty} (y_n - \eta) e^{(n)}$, where $\eta = \lim_{n\to\infty} y_n$. We obtain $\mathcal{R}_m y = \sum_{n=m+1}^{\infty} (y_n - \eta) e^{(n)}$ for all m. Writing $y_n = A_n x$ for $n = 0, 1, \ldots$ and $B = (b_{nk})_{n,k=0}^{\infty}$ for the matrix with $b_{nk} = a_{nk} - \alpha_k$ for all n and k, we obtain from (3.72)

$$\|\mathcal{R}_m(Ax)\| = \sup_{n\ge m+1} |y_n - \eta| = \sup_{n\ge m+1} \left|A_n x - \sum_{k=0}^{\infty} \alpha_k x_k\right| = \sup_{n\ge m+1} |B_n x|,$$

whence

$$\sup_{x\in \overline{B}_X} \|\mathcal{R}_m(Ax)\| = \sup_{n\ge m+1} \|B_n\|^* \text{ for all } m.$$

Now the inequalities in (3.67) follow from (1.52) in Theorem 1.29, (3.58) and (3.59) in Lemma 3.1, and (1.45) in Theorem 1.27.

Thus we have shown Part (a).
(b) Part (b) follows from (a) with $\alpha_k = 0$ and $L = \lim_{m\to\infty} \|\mathcal{R}_m\| = 1$.
(c) We note that we have by Theorem 1.10 (d) and the fact that $\mathcal{R}_m$ is given by the matrix $A^{<m+1>}$

$$\sup_{N_{m+1}} \left\|\sum_{n\in N_{m+1}} A_n\right\| = \sup_{\substack{N\subset \mathbb{N}_0 \\ N \text{ finite}}} \left\|\sum_{n\in N} A_n^{<m+1>}\right\|_X^* \le \|\mathcal{R}_m \circ L\|_X^* =$$

$$\|A^{<m+1>}\|_{(X,\ell_1)} \le 4 \cdot \sup_{\substack{N\subset \mathbb{N}_0 \\ N \text{ finite}}} \left\|\sum_{n\in N} A_n^{<m+1>}\right\|_X^* = 4 \cdot \sup_{N_{m+1}} \left\|\sum_{n\in N_{m+1}} A_n\right\|. \quad (3.75)$$

Now the inequalities in (3.70) follow from (1.52) in Theorem 1.29, (3.58) and (3.57) in Lemma 3.1, and Remark 1.13.

□

Theorem 3.7 ([14, Theorem 4.2]) *Let X be a complete linear metric sequence space with a translation invariant metric, T be a triangle, and χ_T and χ denote the Hausdorff measures of noncompactness on X_T and X, respectively. Then we have*

$$\chi_T(Q) = \chi(T(Q)) \text{ for all } Q \in \mathcal{M}_{X_T}. \quad (3.76)$$

Proof We write $Z = X_T$, $B(x,r)$ and $B_T(z,r)$ for the open balls of radius r and X and Z, centred at x and z, and observe that $Q \in \mathcal{M}_Z$ if and only if $T(Q) \in \mathcal{M}_X$ by the definition of the metric on Z (Theorem 2.8).

(i) First we show that

$$\chi_T(Q) \leq \chi(T(Q)) \text{ for all } Q \in \mathcal{M}_Z. \tag{3.77}$$

We assume that $r = \chi_T(Q) > \chi(T(Q)) = s$ for some $Q \in \mathcal{M}_Z$. Then there are a real ε with $s < \varepsilon < t, x_1, x_2, \dots, x_n \in X$ and $r_1, r_2, \dots, r_n < \varepsilon$ such that

$$T(Q) \subset \bigcup_{k=1}^{\infty} B(x_k, r_k)$$

by the definition of $\chi_T(Q)$ in Part (b) of Definition 1.8. Let $v \in Q$ be given. We put $u = Tv \in X$, and so there are $z_j \in Z$ with $x_j = Tz_j$ and $r_j < \varepsilon$ such that $u \in B(x_j, r_j)$, that is,

$$\begin{aligned} d(u, x_j) &= d(Tv, Tz_j) = d(Tv - Tz_j, 0) = d(T(v - z_j), 0) = d_T(v - z_j, 0) \\ &= d_T(v, z_j) < r_j, \end{aligned}$$

hence

$$v \in B_T(x_j, r_j) \subset \bigcup_{k=1}^{n} B_T(z_k, r_k).$$

Since $v \in Q$ was arbirtary, we have

$$Q \subset \bigcup_{k=1}^{n} B_T(z_k, r_k),$$

and so $\chi_T(Q) \leq \varepsilon < t$, which is a contradiction to $\chi_T(Q) = t$. Therefore (3.77) must hold.

(ii) Now we show

$$\chi_T(Q) \geq \chi(T(Q)) \text{ for all } Q \in \mathcal{M}_Z. \tag{3.78}$$

Applying with X and Z replaced by Z and $Z_S = X$, respectively, where $S = T^{-1}$, we obtain

$$\chi(T(Q)) = (\chi_T)_S(T(Q)) \leq \chi_T(S(T(Q))) = \chi_T(Q) \text{ for all } Q \in \mathcal{M}_Z,$$

that is, (3.78) is satisfied.

Finally (3.77) and (3.78) imply (3.76). □

Now we apply the results above to establish identities or estimates of the Hausdorff measure of noncompactness of continuous linear operators L_A on matrix domains of triangles in the classical sequence spaces. We write N_r ($r \in \mathbb{N}_0$) for subsets of $\mathbb{N}_0$ with elements that are greater or equal to r, and $\sup_{N_r}$ for the supremum taken overall finite sets N_r.

Corollary 3.8 ([6, Corollary 3.6]) *Let $1 \le p \le \infty$ and q be the conjugate number of p.*

(a) If $A \in ((\ell_p)_T, c_0)$ or $A \in ((c_0)_T, c_0)$, then we have

$$\|L_A\|_X = \lim_{r\to\infty}\left(\sup_{n\ge r}\left\|\hat{A}_n\right\|_q\right) = \begin{cases} \lim_{r\to\infty}\left(\sup_{n\ge r}\left(\sum_{k=0}^{\infty}|\hat{a}_{nk}|\right)\right) & (p=\infty \text{ or } X=c_0) \\ \lim_{r\to\infty}\left(\sup_{n\ge r}\left(\sum_{k=0}^{\infty}|\hat{a}_{nk}|^q\right)^{1/q}\right) & (1<p<\infty) \\ \lim_{r\to\infty}\left(\sup_{n\ge r,k\ge 0}|\hat{a}_{nk}|\right) & (p=1). \end{cases} \tag{3.79}$$

(b) If $A \in ((\ell_p)_T, \ell_1)$ for $1 < p \le \infty$ or $A \in ((c_0)_T, \ell_1)$, then we have

$$\lim_{r\to\infty}\left(\sup_{N_r}\left\|\sum_{n\in N_r}\hat{A}_n\right\|_q\right) \le \|L_A\|_X \le 4\cdot\lim_{r\to\infty}\left(\sup_{N_r}\left\|\sum_{n\in N_r}\hat{A}_n\right\|_q\right). \tag{3.80}$$

If $A \in ((\ell_1)_T, \ell_1)$, then we have

$$\lim_{r\to\infty}\left(\sup_k\left(\sum_{n=r}^{\infty}|\hat{a}_{nk}|\right)\right). \tag{3.81}$$

(c) If $A \in ((\ell_p)_T, c)$ or $A \in ((c_0)_T, c)$, then we have

$$\frac{1}{2}\cdot\lim_{r\to\infty}\left(\sup_{n\ge r}\left\|\hat{A}_n-\hat{\alpha}\right\|_q\right) \le \|L_A\|_X \le \lim_{r\to\infty}\left(\sup_{n\ge r}\left\|\hat{A}_n-\hat{\alpha}\right\|_q\right) \tag{3.82}$$
$$\text{where } \hat{\alpha}=(\hat{\alpha})_{k=0}^{\infty} \text{ with } \hat{\alpha}_k=\lim_{n\to\infty}\hat{a}_{nk} \text{ for every } k.$$

Proof (a) This follows from Part (a) of Lemma 3.1, (1.52) in Theorem 1.29 and (3.58) in Part (a) in Corollary 3.5.

(b) This follows from (3.58) in Part (a) of Lemma 3.1, (1.52) in Theorem 1.29 and Part (b) in Corollary 3.5.

(c) This follows from (3.58) in Part (a) of Lemma 3.1, Part (a) Theorem 3.6 and (3.58) in Part (a) in Corollary 3.5.

□

Example 3.3 If $T = \Delta$ and $A \in (bv^p, c) = (\ell_p(\Delta), c)$ for $1 \le p \le \infty$, then

$$\hat{A}_n = \left(\sum_{j=k}^{\infty} a_{nj}\right)_{k=0}^{\infty} \quad \text{for all } n \text{ and } \hat{\alpha}_k = \lim_{n\to\infty} \hat{a}_{nk} = \lim_{n\to\infty}\left(\sum_{j=k}^{\infty} a_{nj}\right) \text{ for all } k \tag{3.83}$$

and we obtain from (3.82) in Part (c) of Corollary 3.8

$$\frac{1}{2}\cdot \lim_{r\to\infty}\left(\sup_{n\ge r}\left\|\left(\sum_{j=k}^{\infty} a_{nj} - \lim_{n\to\infty}\left(\sum_{j=k}^{\infty} a_{nj}\right)\right)\right\|_q\right) \le \|L_A\|_\chi \le \lim_{r\to\infty}\left(\sup_{n\ge r}\left\|\left(\sum_{j=k}^{\infty} a_{nj} - \lim_{n\to\infty}\left(\sum_{j=k}^{\infty} a_{nj}\right)\right)\right\|_q\right) \tag{3.84}$$

with $\|\cdot\|_q$ as in (3.79) in Part (a) of Corollary 3.8.

Example 3.4 Now we determine identities or estimates for the Hausdorff measure of noncompactness of L_A when $A \in (X, c_0)$ and $A \in (X, c)$ where $X = e_p^r$ for $1 \le p \le \infty$ or $X = e_0^r$ for the Euler sequence spaces e_p^r and e_0^r of Part (a) in Example 2.8.

Now it follows from (2.21) in Part (a) of Example 2.8 that

$$\hat{a}_{nk} = R_k A_n = \sum_{j=k}^{\infty} \binom{j}{k} (r-1)^{j-k} r^j a_{nj} \text{ for } n, k = 0, 1, \dots \tag{3.85}$$

and

$$\hat{\alpha}_k = \lim_{n\to\infty} \hat{a}_{nk} \text{ for } k = 0, 1, \dots. \tag{3.86}$$

Then the identities for $\|L_A\|_\chi$ when $A \in (e_p^r, c_0)$ for $1 \le p \le \infty$ and $A \in (e_0^r, c_0)$ are given by (3.79) in Part (a) of Corollary 3.8 with $\hat{a}_{nk}$ from (3.85) and the estimates for $\|L_A\|_\chi$ when $A \in (e_p^r, c)$ for $1 \le p \le \infty$ and $A \in (e_0^r, c)$ are given by (3.82) in Part (c) of Corollary 3.8 with $\hat{a}_{nk}$ from (3.85) and $\hat{\alpha}_k$ from (3.86).

Now we establish estimates and an identity for $\|L_A\|_\chi$ when $A \in (c_T, Y)$, where $Y \in \{c_0, \ell_1, \ell_1\}$.

Theorem 3.8 ([6, Theorem 3.7])

(a) *Let $A \in (c_T, c)$,*

$$\hat{\alpha}_n = \lim_{n\to\infty} \hat{a}_{nk} \textit{ for } k = 0, 1, \dots, \tag{3.87}$$

$$\gamma_n = \lim_{m\to\infty} \sum_{k=0}^{m} w_{mk}^{(A_n)} \textit{ for } n = 0, 1, \dots \tag{3.88}$$

and

$$\beta = \lim_{n\to\infty}\left(\sum_{k=0}^{\infty} \hat{a}_{nk} - \gamma_n\right). \tag{3.89}$$

Then we have

$$\frac{1}{2}\cdot \lim_{r\to\infty}\left(\sup_{n\geq r}\left(\sum_{k=0}^{\infty} |\hat{a}_{nk} - \hat{\alpha}_k| + \left|\sum_{k=0}^{\infty} \hat{\alpha}_k - \beta - \gamma_n\right|\right)\right)$$
$$\leq \|L_A\|_{\chi} \leq \lim_{r\to\infty}\left(\sup_{n\geq r}\left(\sum_{k=0}^{\infty} |\hat{a}_{nk} - \hat{\alpha}_k| + \left|\sum_{k=0}^{\infty} \hat{\alpha}_k - \beta - \gamma_n\right|\right)\right). \tag{3.90}$$

(b) *Let $A \in (c_T, c_0)$. Then we have*

$$\|L_A\|_{\chi} = \lim_{r\to\infty}\left(\sup_{n\geq r}\left(\sum_{k=0}^{\infty} |\hat{a}_{nk}| + |\gamma_n|\right)\right). \tag{3.91}$$

(c) *Let $A \in (c_T, \ell_1)$. Then we have*

$$\lim_{r\to\infty}\left(\sup_{N_r}\left(\sum_{k=0}^{\infty}\left|\sum_{n\in N_r} \hat{a}_{nk}\right| + \left|\sum_{n\in N_r} \gamma_n\right|\right)\right)$$
$$\leq \|L_A\|_{\chi} \leq 4\cdot \lim_{r\to\infty}\left(\sup_{N_r}\left(\sum_{k=0}^{\infty}\left|\sum_{n\in N_r} \hat{a}_{nk}\right| + \left|\sum_{n\in N_r} \gamma_n\right|\right)\right). \tag{3.92}$$

Proof (a) Let $A \in (c_T, c)$. Then it follows by (3.10) and (3.11) in Part (b) of Corollary 3.4 that $\hat{A} \in (c_0, c)$, which implies that the limits α_k in (3.87) exist for all k, and by Part (i) of the proof of Theorem 3.6, $(\hat{\alpha}_k)_{k=0}^{\infty} \in c_0^{\beta} = \ell_1 \subset cs$, $W^{(A_n)} \in (c, c)$ for all n, which implies that the limits γ_n in (3.88) exist for all n, and $\hat{A}e - (\gamma_n)_{n=0}^{\infty} \in c$, which implies that the limit β in (3.89) exists.
Let $z \in c_T$ be given and $\xi_T = \lim_{k\to\infty} T_k z$. Then we have by (3.12) in Part (b) of Corollary 3.4

$$y_n = A_n z = \hat{A}_n(Tz) - \xi_T\gamma_n = \hat{A}_n(Tz - \xi_T e) + \xi_T(\hat{A}_n e - \gamma_n) \text{ for all } n. \tag{3.93}$$

First, $\hat{A} \in (c_0, c)$ and $Tz - \xi_T e \in c_0$ together imply by (3.72) in Part (ii) of the proof of Theorem 3.6

$$\hat{\eta}_0 = \lim_{n\to\infty} \hat{A}_n(Tz - \xi_T e) = \sum_{k=0}^{\infty} \hat{\alpha}_k(T_k z - \xi_T) = \sum_{k=0}^{\infty} \hat{\alpha}_k T_k z - \xi_T \sum_{k=0}^{\infty} \hat{\alpha}_k,$$

and so by (3.93) and (3.89)

$$\hat{\eta} = \lim_{n\to\infty} y_n = \hat{\eta}_0 - \xi_T\beta.$$

Therefore we have

$$\begin{aligned} y_n - \hat{\eta} &= \sum_{k=0}^{\infty} \hat{a}_{nk} T_k z - \xi_T \gamma_n - (\hat{\eta}_0 + \xi_T \beta) \\ &= \sum_{k=0}^{\infty} (\hat{a}_{nk} - \hat{\alpha}_k) T_k z + \xi_T \left(\sum_{k=0}^{\infty} \hat{\alpha}_k - \beta - \gamma_n \right). \end{aligned}$$

Finally, since $\|z\|_{c_T} = \|Tz\|_\infty$, we obtain from (3.17) in Part (a) of Theorem 3.4

$$\sup_{z \in S_{c_T}} \|\mathcal{R}_{r-1}(Az)\|_\infty = \sup_{n \geq r} \left(\sum_{k=0}^{\infty} |\hat{a}_{nk} - \hat{\alpha}_k| + \left| \sum_{k=0}^{\infty} \hat{\alpha}_k - \beta - \gamma_n \right| \right)$$

and (3.90) follows.

(b) Let $A \in (c_T, c_0)$. Then we obtain as in Part (a) of the proof $\hat{\alpha}_k = 0$ for all k, $\beta = 0$ and then $y_n = \hat{A}(Tz)$ $(n = 0, 1, \dots)$ for all $z \in c_T$. Therefore it follows that

$$\sup_{z \in S_{c_T}} \|\mathcal{R}_{r-1}(Az)\|_\infty = \sup_{n \geq r} \left(\sum_{k=0}^{\infty} |\hat{a}_{nk}| + |\gamma_n| \right) \tag{3.94}$$

and (3.91) follows.

(c) Let $A \in (c_T, \ell_1)$. As in Part (b) of the proof, we obtain (3.94), and (3.92) follows from (3.18) in Part (b) of Theorem 3.4

□

Remark 3.7 If $A \in (X_T, Y)$, then we can easily obtain necessary and sufficient conditions for $L_A \in \mathcal{K}(c_T, c)$ from our results above and (1.54) in Theorem 1.30. For instance, if $A \in (c_T, c)$, then $L_A \in \mathcal{K}(c_T, c)$ if and only if

$$\lim_{r \to \infty} \left(\sup_{n \geq r} \left(\sum_{k=0}^{\infty} |\hat{a}_{nk} - \hat{\alpha}_k| + \left| \sum_{k=0}^{\infty} \hat{\alpha}_k - \beta - \gamma_n \right| \right) \right) = 0$$

by (3.90) in Part (a) of Theorem 3.8.

Example 3.5 Let $T = \Delta$ and $A \in (X_T, Y)$ for $X, Y \in \{c_0, c\}$, $\hat{A}_n$ $n = 0, 1, \dots$ and $\hat{\alpha}_k$ $(k = 0, 1, \dots)$ be defined by (3.83) in Example 3.3, and

$$\gamma_n = \lim_{m \to \infty} \sum_{k=0}^{m} w_{mk}^{(A_n)} = \lim_{m \to \infty} \left(\sum_{k=0}^{m} \sum_{j=m}^{\infty} a_{nj} \right) \text{ for all } n \tag{3.95}$$

and

$$\beta = \lim_{n\to\infty}\left(\sum_{k=0}^{\infty}\hat{a}_{nk} - \gamma_n\right) = \lim_{n\to\infty}\left(\sum_{k=0}^{\infty}\sum_{j=k}^{\infty} a_{nj} - \lim_{m\to\infty}\left(\sum_{k=0}^{m}\sum_{j=m}^{\infty} a_{nj}\right)\right). \tag{3.96}$$

Then the identities or estimates for $\|L_A\|_\chi$ can be read from the following table

From / To	$c_0(\Delta)$	$c(\Delta)$
c_0	**1.**	**2.**
c	**3.**	**4.**

where

1. (1.1), that is, the case $X = c_0$ of (3.79) in Part (a) of Corollary 3.8

2. (2.1), that is, (3.91) in Part (b) of Theorem 3.8

3. (3.1), that is, (3.83) with $q = 1$ in Example 3.3 (by Part (c) of Corollary 3.8)

4. (4.1), that is, (3.90) in Part (a) of Theorem 3.8.

3.4 The Class $\mathcal{K}(c)$

The vital assumption in Theorem 3.6 is that the initial space has AK. Hence it can not be applied in the case of $= \mathcal{B}(c, c)$.

The next result gives the representation of the general operator $L \in \mathcal{B}(c)$, a formula for its norm $\|L\|$, an estimate for its Hausdorff measure of noncompactness $\|L\|_\chi$ and a formula for the limit of the sequence $L(x)$.

Theorem 3.9 ([15, Theorem 9.9.1]) *We have $L \in \mathcal{B}(c)$ if and only if there exist a sequence $b \in \ell_\infty$ and a matrix $A = (a_{nk})_{n,k=0}^{\infty} \in (c_0, c)$ such that*

$$L(x) = b\xi + Ax \text{ for all } x \in c, \text{ where } \xi = \lim_{k\to\infty} x_k, \tag{3.97}$$

$$a_{nk} = L_n(e^{(k)}),\ b_n = L_n(e) - \sum_{k=0}^{\infty} a_{nk} \text{ for all } n \text{ and } k, \tag{3.98}$$

$$\beta = \lim_{n\to\infty}\left(b_n + \sum_{k=0}^{\infty} a_{nk}\right) \text{ exists} \tag{3.99}$$

and

$$\|L\| = \sup_n\left(|b_n| + \sum_{k=0}^{\infty}|a_{nk}|\right). \tag{3.100}$$

Moreover, if $L \in \mathcal{B}(c)$, then we have

$$\frac{1}{2} \cdot \limsup_{n\to\infty} \left(\left| b_n - \beta + \sum_{k=0}^{\infty} \alpha_k \right| + \left\| \tilde{A}_n \right\|_1 \right) \leq \|L\|_\chi \leq \limsup_{n\to\infty} \left(\left| b_n - \beta + \sum_{k=0}^{\infty} \alpha_k \right| + \left\| \tilde{A}_n \right\|_1 \right), \tag{3.101}$$

where

$$\alpha_k = \lim_{n\to\infty} a_{nk} \text{ for all } k \in \mathbb{N}_0 \tag{3.102}$$

and $\tilde{A} = (\tilde{a}_{nk})_{n,k=0}^{\infty}$ *is the matrix with* $\tilde{a}_{nk} = a_{nk} - \alpha_k$ *for all* n *and* k*; we also have for all* $x \in c$

$$\eta = \lim_{n\to\infty} L_n(x) = \xi\beta + \sum_{k=0}^{\infty} \alpha_k(x_k - \xi) = \left(\beta - \sum_{k=0}^{\infty} \alpha_k\right)\xi + \sum_{k=0}^{\infty} \alpha_k x_k. \tag{3.103}$$

Proof (i) First we assume that $L \in \mathcal{B}(c)$ and show that L has the given representation and satisfies (3.100).
We assume $L \in \mathcal{B}(c)$.
We write $L_n = P_n \circ L$ for $n = 0, 1, \dots$ where $P_n : c \to \mathbb{C}$ is the nth coordinate with $P_n(x) = x_n$ for all $x = (x_k)_{k=0}^{\infty} \in c$. Since c is a BK space, we have $L_n \in c^*$ for all $n \in \mathbb{N}$, that is, by Example 1.11 (with $\chi(f)$ and a replaced by b_n and A_n, respectively)

$$L_n(x) = b_n \cdot \lim_{k\to\infty} x_k + \sum_{k=0}^{\infty} a_{nk}x_k \text{ for all } x \in c \tag{3.104}$$

with b_n and a_{nk} from (3.98); we also have by Example 1.11

$$\|L_n\| = |b_n| + \sum_{k=0}^{\infty} |a_{nk}| \text{ for all } n. \tag{3.105}$$

Now (3.104) yields the representation of the operator L in (3.97).
Furthermore, since $L(x^{(0)}) = Ax^{(0)}$ for all $x^{(0)} \in c_0$, we have $A \in (c_0, c)$ and so

$$\|A\| = \sup_n \sum_{k=0}^{\infty} |a_{nk}| < \infty$$

by (1.1) in Theorem 1.23 **12.**. Also $L(e) = b + Ae$ by (3.97), and so $L(e) \in c$ yields (3.99), and we obtain

$$\|b\|_\infty \leq \|L(e)\|_\infty + \|A\| < \infty,$$

that is, $b \in \ell_\infty$. Consequently, we have

$$\sup_n \|L_n\| = \sup_n \left(|b_n| + \sum_{k=0}^{\infty} |a_{nk}| \right) < \infty.$$

Since $|\lim_{k\to\infty} x_k| \leq \sup_k |x_k| = \|x\|_\infty$ for all $x \in c$, we obtain by (3.99)

$$\begin{aligned}\|L(x)\|_\infty &= \sup_n \left| b_n \cdot \lim_{k\to\infty} x_k + \sum_{k=0}^{\infty} a_{nk}x_k \right| \\ &\leq \left[\sup_n \left(|b_n| + \sum_{k=0}^{\infty} |a_{nk}| \right) \right] \cdot \|x\|_\infty = \sup_n \|L_n\| \cdot \|x\|_\infty,\end{aligned}$$

hence $\|L\| \leq \sup_n \|L_n\|$. We also have

$$|L_n(x)| \leq \|L(x)\|_\infty \leq \|L\| \text{ for all } x \in S_c \text{ and all } n,$$

that is, $\sup_n \|L_n\| \leq \|L\|$, and we have shown (3.100).
This completes Part (i) of the proof.

(ii) Now we show that if L has the given representation then $L \in \mathcal{B}(c)$.
We assume $A \in (c_0, c)$ and $b \in \ell_\infty$ and that the conditions in (3.97), (3.99) and (3.100) are satisfied. Then we obtain from $A \in (c_0, c)$ by (1.1) in Theorem 1.23 **12.** that $\|A\| < \infty$, and this and $b \in \ell_\infty$ imply $\|L\| < \infty$ by (3.100), hence $L \in \mathcal{B}(c, \ell_\infty)$. Finally, let $x \in c$ be given and $\xi = \lim_{k\to\infty} x_k$. Then $x - \xi \cdot e \in c_0$, so by (3.97)

$$L_n(x) = b_n \cdot \xi + \sum_{k=0}^{\infty} a_{nk}x_k = \left(b_n + \sum_{k=0}^{\infty} a_{nk} \right) \cdot \xi + A_n(x - \xi \cdot e) \text{ for all } n,$$

and it follows from (3.99) and $A \in (c_0, c)$ that $\lim_{n\to\infty} L_n(x)$ exists. Since $x \in c$ was arbitrary, we have $L \in \mathcal{B}(c)$.
This completes Part (ii) of the proof.

(iii) Now we show that if $L \in \mathcal{B}(c)$ then $\|L\|_\chi$ satisfies the inequalities in (3.101).
We assume $L \in \mathcal{B}(c)$.
Let $x \in c$ be given, $\xi = \lim_{k\to\infty} x_k$ and $y = L(x)$. Then we have by Part (i) of the proof

$$y = b \cdot \xi + Ax, \text{ where } A \in (c_0, c) \text{ and } b \in \ell_\infty,$$

and we note that the limits α_k in (3.102) exist for all k by (11.2) in Theorem 1.23 **12.**, and $(\alpha_k)_{k=0}^{\infty} \in \ell_1$ by Part (a) (i) of the proof of Theorem 3.6. So we can write

$$y_n = b_n \cdot \xi + A_n x = \xi \cdot \left(b_n + \sum_{k=0}^{\infty} a_{nk} \right) + A_n(x - \xi \cdot e) \text{ for all } n \in \mathbb{N}. \tag{3.106}$$

Since $A \in (c_0, c)$ it follows from (3.72) in Part (a) (ii) of the proof of Theorem 3.6 that

$$\lim_{n\to\infty} A_n(x - \xi \cdot e) = \sum_{k=0}^{\infty} \alpha_k(x_k - \xi) = \sum_{k=0}^{\infty} \alpha_k x_k - \xi \sum_{k=0}^{\infty} \alpha_k. \tag{3.107}$$

So we obtain by (3.106), (3.99) and (3.107)

$$\begin{aligned} \eta = \lim_{n\to\infty} y_n &= \lim_{n\to\infty} \left[\xi \left(b_n + \sum_{k=0}^{\infty} a_{nk} \right) + A_n(x - \xi \cdot e) \right] \\ &= \xi \cdot \lim_{n\to\infty} \left(b_n + \sum_{k=0}^{\infty} a_{nk} \right) + \lim_{n\to\infty} A_n(x - \xi \cdot e) \\ &= \xi \cdot \beta + \sum_{k=0}^{\infty} \alpha_k x_k - \xi \sum_{k=0}^{\infty} \alpha_k = \xi \left(\beta - \sum_{k=0}^{\infty} \alpha_k \right) + \sum_{k=0}^{\infty} \alpha_k x_k, \end{aligned} \tag{3.108}$$

that is, we have shown (3.103).
For each m, we have

$$\mathcal{R}_m(y) = \sum_{n=m+1}^{\infty} (y_n - \eta) e^{(n)} \text{ for } y \in c \text{ and } \eta = \lim_{n\to\infty} y_n.$$

Writing

$$f_n^{(m)}(x) = (\mathcal{R}_m(L(x)))_n,$$

we obtain for $n \geq m+1$ by (3.106) and (3.108)

$$\begin{aligned} f_n^{(m)}(x) = y_n - \eta &= \xi \cdot b_n + A_n(x) - \left[\xi \left(\beta - \sum_{k=0}^{\infty} \alpha_k \right) + \sum_{k=0}^{\infty} \alpha_k x_k \right] \\ &= \xi \cdot \left(b_n - \beta + \sum_{k=0}^{\infty} \alpha_k \right) + \sum_{k=0}^{\infty} (a_{nk} - \alpha_k) x_k. \end{aligned}$$

Since $f_n^{(m)} \in c^*$, we have by Example 1.11

$$\| f_n^{(m)} \| = \left| b_n - \beta + \sum_{k=0}^{\infty} \alpha_k \right| + \sum_{k=0}^{\infty} |a_{nk} - \alpha_k|,$$

and it follows that

$$\sup_{x \in S_c} \|\mathcal{R}_m(L(x))\|_\infty = \sup_{n \geq m+1} \|f_n^{(m)}\|$$
$$= \sup_{n \geq m+1} \left(\left| b_n - \beta + \sum_{k=0}^{\infty} \alpha_k \right| + \sum_{k=0}^{\infty} |a_{nk} - \alpha_k| \right).$$

Now the inequalities in (3.101) follow from (1.52) in Theorem 1.29, (3.58) and (3.59) in Lemma 3.1, and (1.47) in Remark 1.11.

□

We obtain the following result for $L \in \mathcal{B}(c, c_0)$ as in Theorem 3.9 with $\beta = \alpha_k = 0$ for all k and by replacing the factor $1/2$ in (3.101) by !:

Corollary 3.9 ([15, Corollary 9.9.2]) *We have $L \in \mathcal{B}(c, c_0)$ if and only if there exist a sequence $b \in \ell_\infty$ and a matrix $A = (a_{nk})_{n,k=0}^{\infty} \in (c_0, c)$ such that* (3.97) *holds, where the sequence b and the matrix A are given by* (3.98)*, and the norm of L satisfies* (3.100).

Moreover, if $L \in \mathcal{B}(c, c_0)$, then we have

$$\|L\|_\chi = \limsup_{n \to \infty} \left(|b_n| + \|A_n\|_1\right). \tag{3.109}$$

Now we give the estimates or identities for the Hausdorff measures of noncompactness of bounded linear operators between the spaces c_0 and c.

Theorem 3.10 [15, Theorem 9.9.3]) *Let $L \in \mathcal{B}(X, Y)$ where X and Y are any of the spaces c_0 and c. Then the estimates or identities for the Hausdorff measure of the operators L can be obtained from the following table*

From / *To*	c_0	c
c_0	**1.**	**2.**
c	**3.**	**4.**

where

1. $\|L\|_\chi = \limsup_{n\to\infty} \left(\sum_{k=0}^{\infty} |a_{nk}| \right)$

2. $\|L\|_\chi = \limsup_{n\to\infty} \left(|b_n| + \sum_{k=0}^{\infty} |a_{nk}| \right)$

3. $\frac{1}{2} \limsup_{n\to\infty} \left(\sum_{k=0}^{\infty} |a_{nk} - \alpha_k| \right) \leq \|L\|_\chi \leq \limsup_{n\to\infty} \left(\sum_{k=0}^{\infty} |a_{nk} - \alpha_k| \right)$

4. $\frac{1}{2} \limsup_{n\to\infty} \left(\left| b_n - \beta + \sum_{k=0}^{\infty} \alpha_k \right| + \sum_{k=0}^{\infty} |a_{nk} - \alpha_k| \right) \leq \|L\|_\chi$

$$\leq \limsup_{n\to\infty} \left(\left| b_n - \beta + \sum_{k=0}^{\infty} \alpha_k \right| + \sum_{k=0}^{\infty} |a_{nk} - \alpha_k| \right).$$

Proof **1.** and **3.** The conditions follow from Theorem 3.6 (b) for **1.** and (a) for **3.**, and (1.47) in Remark 1.11 and the fact that $\| \cdot \|_X^* = \| \cdot \|_1$ for $X = c_0$ and $X = c$ by Example 1.11.

2. and **4.** The conditions follow from (3.109) in Corollary 3.9 and from (3.102) in Theorem 3.9 (b). □

Applying (1.54) in Theorem 1.30 we obtain the following characterizations for compact operators between the spaces c_0 and c.

Corollary 3.10 ([15, Corollary 9.9.4]) *Let $L \in \mathcal{B}(X, Y)$ where X and Y are any of the spaces c_0 and c. Then the necessary and sufficient conditions for $L \in \mathcal{K}(X, Y)$ can be obtained from the following table*

From / *To*	c_0	c
c_0	**1.**	**2.**
c	**3.**	**4.**

where

1. $\|L\|_\chi = \lim_{n\to\infty} \left(\sum_{k=0}^{\infty} |a_{nk}| \right) = 0$

2. $\|L\|_\chi = \lim_{n\to\infty} \left(|b_n| + \sum_{k=0}^{\infty} |a_{nk}| \right) = 0$

3. $\lim_{n\to\infty} \left(\sum_{k=0}^{\infty} |a_{nk} - \alpha_k| \right) = 0$

4. $\lim_{n\to\infty} \left(\left| b_n - \beta + \sum_{k=0}^{\infty} \alpha_k \right| + \sum_{k=0}^{\infty} |a_{nk} - \alpha_k| \right) = 0.$

The following example will show that a regular matrix cannot be compact; this is a well-known result by *Cohen* and *Dunford* ([3]).

Example 3.6 ([15, Example 9.9.5]) Let $A \in (c, c)$. It follows from the condition in **4.** in Corollary 3.10 with $b_n = 0$ for all n and $\beta = \lim_{n\to\infty} \sum_{k=0}^{\infty} a_{nk}$ that L_A is compact if and only if

$$\lim_{n\to\infty}\left(\left|\sum_{k=0}^{\infty}\alpha_k - \beta\right| + \sum_{k=0}^{\infty}|a_{nk} - \alpha_k|\right) = 0. \tag{3.110}$$

If A is regular, then $\alpha_k = 0$ for $k = 0, 1, \dots$ and $\beta = 1$, and so

$$\left|\sum_{k=0}^{\infty}\alpha_k - \beta\right| + \sum_{k=0}^{\infty}|a_{nk} - \alpha_k| = 1 + \sum_{k=0}^{\infty}|a_{nk}| \geq 1 \text{ for all } n.$$

Thus a regular matrix cannot satisfy (3.110) and consequently cannot be compact.

3.5 Compact Operators on the Space bv^+

In this section, we define and study the spaces bv_0^+ and bv^+, characterize the classes (X, Y) where $X \in \{bv_0^+, bv^+\}$ and $Y \in \{\ell_\infty, c, c_0\}$, establish the representations of the general operators in the classes $\mathcal{B}(bv^+, Y)$ where $Y \in \{c_0, c, \ell_\infty\}$, determine the classes $\mathcal{B}(bv^+, Y)$, give estimates for the Hausdorff measures of noncompactness $\|L\|_\chi$ for $L \in \mathcal{B}(bv^+, c_0)$ and $L \in \mathcal{B}(bv^+, c)$ and characterize the classes $\mathcal{K}(bv^+, c_0)$ and $\mathcal{K}(bv^+, c)$.

Definition 3.1 We define the Matrix $\Delta^+ = (\delta^+_{nk})_{n,k=0}^{\infty}$ of the *forward differences* by

$$\delta^+_{nk} = \begin{cases} 1 & (k = n) \\ -1 & (k = n+1) \\ 0 & (k \neq n, n+1) \end{cases} \quad (n = 0, 1, \dots)$$

and define the sets

$$bv^+ = (\ell_1)_{\Delta^+} = \left\{x \in \omega : \sum_{k=0}^{\infty}|x_k - x_{k+1}| < \infty\right\} \text{ and } bv_0^+ = bv^+ \cap c_0.$$

The definition of bv^+ is very similar to that of bv, but Δ^+ is not a triangle, so we cannot apply the previous results for matrix domains of triangles.

Remark 3.8 We have $bv^+ \subset c$.

Proof Let $x \in bv^+$ and $\varepsilon > 0$ be given. Then there exsits $n_0 \in \mathbb{N}_0$ such that

$$\sum_{k=n}^{\infty} |x_k - x_{k+1}| < \varepsilon \text{ for all } n \geq n_0,$$

hence

$$|x_n - x_m| \leq \sum_{k=n}^{m-1} |x_k - x_{k+1}| < \varepsilon \text{ for all } m > n \geq n_0.$$

Thus $x = (x_k)_{k=0}^{\infty}$ is a Cauchy sequence of complex numbers, hence convergent. □

We need the following general result.

Theorem 3.11 ([20, Theorem 4.5.1]) *Let* $(X, (p_n))$ *and* $(Y, (q_n))$ *be* FH *spaces (in* H*) (*Definition 1.2 *and the notation introduced after* Theorem 2.1*),* $X \cap Y = \{0\}$ *and*

$$Z = X + Y = \{x + y : x \in X, \ y \in Y\}.$$

Then Z *is an* FH *space and* X *and* Y *are closed in* Z.

If X *and* Y *are* BH *spaces, then* Z *is a* BH *space.*

Proof Each $z \in Z$ can be written uniquely as $z = x + y$, where $x \in X$ and $y \in Y$. We define

$$r_n(z) = p_n(z) + q_n(y) \text{ for each } n.$$

(i) First we show that $(Z, (r_n))$ is complete.
If z is a Cauchy sequence in $(Z, (r_n))$, then x and y are Cauchy sequences in $(X, (p_n))$ and $(Y, (q_n))$, respectively, hence convergent by the completeness of X and Y. So there are $s \in X$ and $t \in Y$ such that $x \to s$ and $y \to t$. We put $u = s + t$. Then we have

$$r_n(z - u) = p_n(x - s) + q_n(y - t) \to 0 \text{ for all } n,$$

hence $z \to u$ in $(Z, (r_n))$.
Thus we have shown that $(Z, (r_n))$ is complete.

(ii) Now we show that $(Z, (r_n))$ is an FH space.
We assume $z \to 0$ in $(Z, (p_n))$. Let $z = x + y$ with $x \in X$ and $y \in Y$. Then $z \to 0$ implies

$$p_n(x) \leq p_n(x) + q_n(y) = r_n(z) \text{ for all } n,$$

hence $x \to 0$ in $(X, (p_n))$ and so $x \to 0$ in H, since X is an FH space. Similarly we obtain $y \to 0$ in $(Y, (q_n))$; and so $z = x + y \to 0$ in H.
Thus we have shown that $(Z, (r_n))$ is an FH space.

(iii) Now we show that X is closed in Z.
For $x \in X$, we have

$$r_n(x) = p_n(x) + q_n(0) = p_n(x) \text{ for all } n.$$

Hence X has the relative topology of Z, and consequently X is closed in Z by Theorem 1.5.

(iv) The last statement is trivial.

□

Example 3.7 ([20, Example 4.5.5]) Let X be an FH space (in H), $y \in H \setminus X$ and $Y = span(\{y\})$. Then we write $X \oplus y = X + Y$.
By Theorem 3.11, $X \oplus y$ is an FH space and X is a closed subspace.

Lemma 3.2 *We put*

$$\|x\|_{bv^+} = \left|\lim_{k\to\infty} x_k\right| + \sum_{k=0}^{\infty} |x_k - x_{k+1}| \textit{ for all } x = (x_k)_{k=0}^{\infty} \in bv^+. \tag{3.111}$$

Then we have

(a) bv_0^+ *is a* BK *space with* AK*;*
(b) bv^+ *is a* BK *space.*

Proof We note that since $bv^+ \subset c$ by Remark 3.8, $\|\cdot\|_{bv^+}$ in (3.111) is defined on bv^+.

(i) We show that bv_0^+ and ℓ_1 are norm isomorphic.
We define the map T on bv_0^+ by

$$y = Tx = (x_k - x_{k+1})_{k=0}^{\infty} \text{ for all } x = (x_k)_{k=0}^{\infty} \in bv_0^+.$$

Then obviously $T \in \mathcal{L}(bv_0^+, \ell_1)$.
If $Tx = 0$, then $x_k = x_{k+1}$ for all all k, and $x \in c_0$ implies $x = 0$. This shows that T is one to one.
If $y \in \ell_1$ is given, we put

$$x = \left(\sum_{k=n}^{\infty} y_k\right)_{n=0}^{\infty}.$$

Since $\ell_1 \subset cs$, we have $x \in c_0$; also

$$\sum_{n=0}^{\infty} |x_n - x_{n+1}| = \sum_{n=0}^{\infty} \left|\sum_{k=n}^{\infty} y_k - \sum_{k=n+1}^{\infty} y_k\right| = \sum_{n=0}^{\infty} |y_n| < \infty,$$

hence $x \in c_0 \cap bv^+ = bv_0^+$ and

$$Tx = (x_n - x_{n+1})_{n=0}^{\infty} = \left(\sum_{k=n}^{\infty} y_k - \sum_{n=n+1}^{\infty} y_k\right)_{n=0}^{\infty} = y.$$

This shows that $T : bv_0^+ \to \ell_1$ is also onto.
Consequently bv_0^+ and ℓ_1 are isomorphic.
Finally

$$\|x\|_{bv^+} = \sum_{k=0}^{\infty} |x_k - x_{k+1}| = \sum_{k=0}^{\infty} \left|(Tx)_{k=0}^{\infty}\right| = \|Tx\|_1 \text{ for all } x \in bv_0^+.$$

Thus we have shown that bv_0^+ is a Banach space with $\|\cdot\|_{bv^+}$.

(ii) Now we show that bv_0^+ is a BK space with AK.
Let $x \in bv_0^+$ be given. Since $x \in c_0$ and $\Delta^+ x \in \ell_1 \subset cs$, it follows that

$$x_n = \sum_{k=n}^{m}(x_k - x_{k+1}) + x_{m+1} \to \sum_{k=n}^{\infty}(x_k - x_{k+1}) \ (m \to \infty),$$

hence

$$x_n \leq \|x\|_{bv^+} \text{ for all } n,$$

and so bv_0^+ is a BK space. Furthermore we have

$$\left\|x - x^{[m]}\right\|_{bv^+} = |x_{m+1}| + \sum_{k=m+1}^{\infty} |x_k - x_{k+1}| \to 0 \ (m \to \infty),$$

hence bv_0^+ has AK.

(iii) Finally, since $bv^+ = bv_0^+ \oplus e$, bv^+ is a BK space by Example 3.7.

□

Lemma 3.3 *We have*

$$(bv_0^+)^\beta = bs \ \text{([20 Theorem 7.3.5 (ii)])} \tag{3.112}$$

and

$$(bv^+)^\beta = cs \ \text{([20 Theorem 7.3.5 (iii)])}. \tag{3.113}$$

Proof Let $a, x \in \omega$ and $m \in \mathbb{N}_0$ be given. Then we have

$$\sum_{k=0}^{m}(x_k - x_{k+1})\sum_{j=0}^{k} a_j = \sum_{k=0}^{m} x_k \sum_{j=0}^{k} a_j - \sum_{k=0}^{m} x_{k+1} \sum_{j=0}^{k} a_j$$
$$= a_0x_0 + \sum_{k=1}^{m} x_k \sum_{j=0}^{k} a_j - \sum_{k=1}^{m} x_k \sum_{j=0}^{k-1} a_j - x_{m+1} \sum_{j=0}^{m} a_j$$
$$= a_0x_0 + \sum_{k=1}^{m} a_k x_k - x_{m+1} \sum_{j=0}^{m} a_j,$$

that is,

$$\sum_{k=0}^{m} a_k x_k = \sum_{k=0}^{m}(x_k - x_{k+1})\sum_{j=0}^{k-1} a_j + x_{m+1}\sum_{j=0}^{m} a_j \text{ for all } m. \tag{3.114}$$

(i) First we show

$$bs \subset (bv_0^+)^\beta \tag{3.115}$$

and

$$cs \subset (bv^+)^\beta. \tag{3.116}$$

First we observe that the first term on the right-hand side of (3.114) converges as $m \to \infty$ for all $x \in bv^+ \supset bv_0^+$ and all $a \in bs \supset cs$, since

$$\sum_{k=0}^{\infty} |x_k - x_{k+1}| \left|\sum_{j=0}^{k-1} a_j\right| \le \|a\|_{bs} \cdot \|x\|_{bv^+}. \tag{3.117}$$

Furthermore, if $a \in bs$ and $x \in bv_0^+ \subset c_0$, then the second term on the right-hand side of (3.114) converges to 0 as $m \to \infty$, so we have established the inclusion in (3.115); we also note that in this case

$$\left|\sum_{k=0}^{\infty} a_k x_k\right| \le \|a\|_{bs} \cdot \|x\|_{bv^+}. \tag{3.118}$$

Finally, if $a \in cs$ and $x \in bv^+ \subset c$ the second term on the right-hand side of (3.114) converges as $m \to \infty$, so we have established the inclusion in (3.116).

(ii) Conversely we show

$$(bv_0^+)^\beta \subset bs. \tag{3.119}$$

Let $a \in (bv_0^+)^\beta$. Then $f(x) = \sum_{k=0}^{\infty} a_k x_k$ for all $x \in X$ defines a functional $f \in (bv_0^+)^*$ by Theorem 1.3. If $m \in \mathbb{N}_0$ is given, then we have for $x = e^{[m]} \in bv_0^+$ that $\|e^{[m]}\|_{bv^+} = 1$ and

$$\left|\sum_{k=0}^{m} a_k\right| \leq \|f\| \cdot \|e^{[m]}\|_{bv^+} \leq \|f\| < \infty,$$

and since $m \in \mathbb{N}_0$ was arbitrary, we obtain

$$\|a\|_{bs} \leq \|f\| < \infty, \tag{3.120}$$

hence $a \in bs$.
Thus we have shown the inclusion in (3.119).

(iii) If $a \in (bv^+)^\beta$, then we obtain, since $e \in bv^+$, $a = a \cdot e \in cs$, and we have shown

$$(bv^+)^\beta \subset cs. \tag{3.121}$$

Now (3.112) is an immediate consequence of (3.115) and (3.119), and (3.113) is an immediate consequence of (3.116) and (3.121). □

The following result is useful for the representation of continuous linear functionals on bv^+.

Lemma 3.4 *Let X be a BK space and $Y = b \oplus X$, where $b \in \omega \setminus X$. Then $f \in Y^*$ if and only if there exist $c \in \mathbb{C}$ and $g \in X^*$ such that*

$$f(y) = \lambda \cdot c + g(x) \text{ for all } y = \lambda b + x \in Y, \text{ where } x \in X. \tag{3.122}$$

Moreover

$$\frac{1}{2} \cdot (|c| + \|g\|) \leq \|f\| \leq |c| + \|g\|. \tag{3.123}$$

Proof We note that Y is a BK space with

$$\|y\|_Y = |\lambda| + \|x\|_X \text{ for } y = \lambda \cdot b + x$$

by Example 3.7.

(i) We show: If there exist $c \in \mathbb{C}$ and $g \in X^*$ such that (3.122) holds then $f \in Y^*$.

(α) We show that f is linear (trivial).
Let $\mu \in \mathbb{C}$ and $y, \tilde{y} \in Y$. Then there are unique $\lambda, \tilde{\lambda} \in \mathbb{C}$ and $x, \tilde{x} \in X$ such that $y = \lambda \cdot b + x$ and $\tilde{\lambda} \cdot b + \tilde{x}$. Then we have

$$\mu y + \tilde{y} = \mu(\lambda \cdot b + x) + \tilde{\lambda} \cdot b + \tilde{x} = (\mu\lambda + \tilde{\lambda}) \cdot b + \mu x + \tilde{x}$$

and it follows from (3.122) and the linearity of g that

$$\begin{aligned} f(\mu y + \tilde{y}) &= (\mu\lambda + \tilde{\lambda}) \cdot c + g(\mu x + \tilde{x}) \\ &= \mu(\lambda \cdot c + g(x)) + \tilde{\lambda} \cdot c + g(\tilde{x}) = \mu f(y) + f(\tilde{y}), \end{aligned}$$

that is, f is linear.

(β) We show f is continuous.
Since $g \in X^*$, we obtain for $y = \lambda \cdot b + x$

$$\begin{aligned}|f(y)| &= |\lambda \cdot c + g(x)| \leq |\lambda| \cdot |c| + \|g\| \cdot \|x\|_X \\ &\leq (|c| + \|g\|) \cdot (|\lambda| + \|x\|_X) \\ &= (|c| + \|g\|) \cdot \|y\|_Y,\end{aligned}$$

that is,

$$\|f\| \leq |c| + \|g\|. \tag{3.124}$$

Hence we have $f \in Y^*$.

(ii) We show that if $f \in Y^*$ then there exist $c \in \mathbb{C}$ and $g \in X^*$ such that (1) holds. Let $f \in Y^*$. We put $c = f(b)$ and $g = f|_X$, the restriction of f on X. Then clearly $g \in X^*$ and we obtain for all $y = \lambda \cdot b + x \in Y$

$$f(y) = f(\lambda \cdot b + x) = \lambda \cdot f(b) + f(x) = \lambda \cdot c + g(x),$$

that is, f has the representation in (3.122).

(iii) It remains to show the left-hand side of (3.123).
Let $\varepsilon > 0$ be given. Then there exists $x_0 \in S_X$, where S_X denotes the unit sphere in X, such that $g(x_0) \geq \|g\| - \varepsilon$. We put $y_0 =\text{sgn}(c) \cdot b + g(x_0)$ with $c \neq 0$ and obtain

$$|f(y_0)| = |\,|c| + g(x_0)| \geq |c| + \|g\| - \varepsilon$$

and

$$\|y_0\| = |\,|\text{sgn}(c)| + 1| = 2.$$

Since $\varepsilon > 0$ was arbitrary, we obtain

$$\|f\| \geq \frac{1}{2} \cdot (|c| + \|g\|).$$

□

Theorem 3.12 *(a) The space $(bv_0^+)^*$ is norm isomorphic to bs.*

(b) We have $f \in bv^*$ if and only if there exist $c \in \mathbb{C}$ and a sequence $a = (a_k)_{k=0}^\infty \in$ bs such that

$$f(y) = \eta \cdot c + \sum_{k=0}^\infty a_k(y_k - \xi) \textit{ for all } y = (y_k)_{k=0}^\infty \in bv, \textit{ where } \eta = \lim_{k\to\infty} y_k; \tag{3.125}$$

moreover, if $f \in bv^*$ then

$$\frac{1}{2} \cdot (|c| + \|a\|_{bs}) \leq \|f\| \leq |c| + \|a\|_{bs}. \tag{3.126}$$

Proof (a) Since bv_0^+ is a BK space with AK by Part (a) of Lemma 3.2, $(bv_0^+)^*$ and $(bv_0^+)^\beta$ are isomorphic by Theorem 1.14, and $(bv_0^+)^\beta = bs$ by (3.112) in Lemma 3.3.
Furthermore if $f(x) = \sum_{k=0}^{\infty} a_k x_k$ $(x \in bv_0^+)$ for some sequence $a \in bs$, then it follows from (3.118) and (3.120) in the proof of Lemma 3.3 that

$$\|f\| = \|a\|_{bs}. \tag{3.127}$$

(b) We note that $(bv^+, \|\cdot\|_{bv^+})$ is a BK space with

$$\|y\|_{bv^+} = |\xi| + \sum_{k=0}^{\infty} |y_k - y_{k+1}| \text{ for all } y = (y_k)_{k=0}^{\infty} \in bv^+, \text{ where } \eta = \lim_{k\to\infty} y_k.$$

and $bv^+ = bv_0^+ \oplus e$ by Part (b) of Lemma 3.2. Applying Lemma 3.4 with $b = e$ and $\lambda = \eta$, we obtain by (3.122) $f \in (bv^+)^*$ if and only if, writing $x = y - \eta e = \sum_{k=0}^{\infty}(y_k - \eta)e^{(k)} \in bv_0^+$

$$f(y) = \eta \cdot c + g(x), \text{ where } c = f(e) \text{ and } g \in (bv_0^+)^*.$$

Furthermore we have by Part (a) that $g \in (bv_0^+)^*$ if and only if there exists a sequence $a \in bs$ such that

$$g(x) = \sum_{k=0}^{\infty} a_k x_k \text{ for all } x \in bv_0^+,$$

hence we have established the representation in (3.125).
Moreover, since $\|g\| = \|a\|_{bs}$, we obtain the inequalities in (3.126) from (3.123).

□

Theorem 3.13 *Let $X \in \{bv_0^+, bv^+\}$ and $Y \in \{\ell_\infty, c, c_0, \ell_1\}$. Then the necessary and sufficient conditions for $A \in (X, Y)$ can be read from the following table*

From / *To*	bv_0^+	bv^+
ℓ_∞	**1.**	**2.**
c_0	**3.**	**4.**
c	**5.**	**6.**
ℓ_1	**7.**	**8.**

where

1. (1.1), *where* (1.1) $\|A\|_{(bv^+,\infty)} = \sup_{n,k} \left| \sum_{j=0}^{k} a_{nj} \right| < \infty$

2. (1.1) *and* (2.1), *where* (2.1) $\sup_n \left| \sum_{k=0}^{\infty} a_{nk} \right| < \infty$

3. (1.1) *and* (3.1), *where* (3.1) $\lim_{n\to\infty} a_{nk} = 0$ *for each* k

4. (1.1), (2.1) *and* (4.1), *where* (4.1) $\lim_{n\to\infty} \sum_{k=0}^{\infty} a_{nk} = 0$

5. (1.1) *and* (5.1), *where* (5.1) $\alpha_k = \lim_{n\to\infty} a_{nk}$ *exsists for each* k

6. (1.1), (2.1) *and* (6.1), *where* (6.1) $\alpha = \lim_{n\to\infty} \sum_{k=0}^{\infty} a_{nk}$ *exists*

7. (7.1), *where* (7.1) $\|A\|_{(bv^+,1)} = \sup_N \sup_k \left| \sum_{j=0}^{k} \sum_{n\in N} a_{nj} \right| < \infty$

8. (7.1) *and* (8.1), *where* (8.1) $\sum_{n=0}^{\infty} \left| \sum_{k=0}^{\infty} a_{nk} \right| < \infty.$

Moreover if $A \in (X, Y)$ *for* $X \in \{bv_0^+, bv^+\}$ *and* $Y \in \{\ell_\infty, c, c_0\}$, *then*

$$\|L_A\| = \|A\|_{(bv^+,\infty)} \tag{3.128}$$

and if $A \in (X, \ell_1)$, *then*

$$\|A\|_{(bv^+,1)} \leq \|L_A\| = 4 \cdot \|A\|_{(bv^+,1)}. \tag{3.129}$$

Proof **1.** and **7.** Since bv_0^+ is a BK space by Part (a) in Lemma 3.2, it follows by (1.18) in Part (b) of Theorem 1.10 that $A \in (bv_0^+, \ell_\infty)$ if and only if $\sup_n \|A_n\|^*_{bv_0^+} < \infty$, and by (1.21) in Theorem 1.13 that $A \in (bv_0^+, \ell_1)$ if and only if

$$\sup_N \|\sum_{n\in N} A_n\|^*_{bv_0^+} < \infty.$$

Since $\|\cdot\|^*_{bv_0^+} = \|\cdot\|_{bs}$ by (3.112) in Lemma 3.3, these conditions yield (1.1) in **1.** and (7.1) in **7.**, respectively.

3. and **5.** Since bv_0^+ has AK by Part (a) in Lemma 3.2, Theorem 1.11 yields the conditions $Ae^{(k)} \in c_0$ for each k, that is, (3.1) in **3.**, and $Ae^{(k)} \in c$ for each k, that is, (5.1) in **5.**.

2., **4.**, **6.** and **8.** Since $bv^+ = bv_0^+ \oplus e$, Theorem 1.12 yields the conditions $Ae \in \ell_\infty$, that is, (2.1) in **2.**, $Ae \in c_0$, that is, (4.1) in **4.**, $Ae \in c$, that is, (6.1) in **6.** and $Ae \in \ell_1$, that is, (8.1) in **8.**.

Finally, the conditions in (3.128) and (3.129) follow from (1.19) in Part (b) of Theorem 1.10 and (1.22) in Remark 1.4, respectively. □

Remark 3.9 It is easy to see that the conditions (1.1) in **1.** and (2.1) in **2.** Theorem 3.11 are equivalvent to

$$\sup_{n,k}\left|\sum_{j=k}^{\infty} a_{nj}\right| < \infty. \tag{3.130}$$

Theorem 3.14 *(a) We have $L \in \mathcal{B}(bv^+, \ell_\infty)$ if and only if there exists a matrix $A \in (bv_0^+, \ell_\infty)$ and a sequence $b \in \ell_\infty$ such that*

$$L(x) = b \cdot \xi + Ax^{(0)} \text{ for all } x \in bv^+, \tag{3.131}$$

where $\xi = \lim_{k\to\infty} x_k$ and $x^{(0)} = x - \xi \cdot e \in bv_0^+$.

(b) We have $L \in \mathcal{B}(bv^+, c)$ if and only if there exists a matrix $A \in (bv_0^+, c)$ and a sequence $b \in c$ such that (3.131) *holds.*

(c) We have $L \in \mathcal{B}(bv^+, c_0)$ if and only if there exists a matrix $A \in (bv_0^+, c_0)$ and a sequence $b \in c_0$ with such that (3.131) *holds.*

(d) If $L \in \mathcal{B}(bv^+, Y)$, where $Y \in \{\ell_\infty, c, c_0\}$, then

$$\frac{1}{2}\cdot\sup_n\left(|b_n| + \|A_n\|_{bs}\right) \le \|L\| \le \sup_n\left(|b_n| + \|A_n\|_{bs}\right). \tag{3.132}$$

Proof (a)(a.i) First we assume $L \in \mathcal{B}(bv^+, \ell_\infty)$.
Trivially $L \in \mathcal{B}(bv^+, \ell_\infty)$ implies $L \in \mathcal{B}(bv_0^+, \ell_\infty)$ and since bv_0^+ has AK by Part (a) of Lemma 3.2, there exists a matrix $A \in (bv_0^+, \ell_\infty)$ by Part (b) of Theorem 1.10 such that $L(x^{(0)}) = Ax^{(0)}$ for all $x^{(0)} \in bv_0^+$. Now, for $x \in bv^+$, there exists $x^{(0)} \in bv_0^+$ such that $x = x^{(0)} + \xi \cdot e \in bv^+$ where $\xi = \lim_{k\to\infty} x_k$ and we obtain

$$L(x) = L(x^{(0)} + \xi \cdot e) = L(x^{(0)}) + \xi \cdot L(e),$$

$L(x^{(0)}) = Ax^{(0)} \in \ell_\infty$ and $L(e) \in \ell_\infty$. Putting $= L(e)$ we obtain $b \in \ell_\infty$ and (3.131) is satisfied.

(a.ii) Conversely, we assume that there exist a matrix $A \in (bv_0^+, \ell_\infty)$ and a sequence $b \in \ell_\infty$ such that (3.131) holds.
Clearly L maps bv^+ into ℓ_∞ and is linear. Also $A \in (bv_0^+, \ell_\infty)$ implies $A_n = (a_{nk})_{k=0}^\infty \in (bv_0^+)^\beta$ for all n, and $(bv_0^+)^\beta = bs$ by (3.112) in Lemma 3.3. Now we obtain from (3.131) for each n

$$L_n(x) = b_n \cdot \xi + A_n(x^{(0)}) \text{ for all } x = x^{(0)} + \xi \cdot e \in bv^+,$$

and so $L_n \in (bv^+)^*$ for all n by Part (b) of Theorem 3.12.

(b) (b.i) First we assume $L \in \mathcal{B}(bv^+, c)$.
This implies $L \in \mathcal{B}(bv^+, \ell_\infty)$ and by Part (a), there exist a matrix $A \in (bv_0^+, \ell_\infty)$ and a sequence $b \in \ell_\infty$ such that (3.131) holds. It follows that

$$L(e^{(k)}) = Ae^{(k)} = (a_{nk})_{k=0}^\infty \in c \text{ for each } k,$$

hence the limits

$$\alpha_k = \lim_{n\to\infty} a_{nk} \text{ exist for all } k. \tag{3.133}$$

Now $A \in (bv_0^+, \ell_\infty)$ and (3.133) together imply $A \in (bv_0^+, c)$ by **5.** in Theorem 3.11. Furthermore, we have $L(e) = b \in c$.

(b.ii) Conversely, we assume that there exist a matrix $A \in (bv_0^+, c)$ and a sequence $b \in c$ such that (3.131) holds. Then we have $L \in \mathcal{B}(bv^+, \ell_\infty)$ by Part (a). Let $x \in bv^+$ be given. Then $x = x^{(0)} + \xi \cdot e$ with $x^{(0)} \in bv_0^+$ and $\xi = \lim_{k\to\infty} x_k$ and $L(x) = b\xi + Ax^{(0)} \in c$.

(c) The proof is similar to that in Part (b).

(d) In each case, $L \in \mathcal{B}(bv^+, Y)$ implies $L_n \in (bv^+)^*$ for $n = 0, 1, \dots$, and it follows from the right-hand side of (3.126) in Part (b) of Theorem 3.12 that

$$\|L_n\| \le |b_n| + \|A_n\|_{bs} \text{ for each } n.$$

Since $b \in \ell_\infty$ and $A \in (bv^+, \ell_\infty)$ implies $\sup_n \|A_n\|_{bs} < \infty$ by **1.** in Theorem 3.11, we obtain

$$\|L(x)\|_\infty = \sup_n |L_n(x)| \le \sup_n \|L_n\| \cdot \|x\|_{bv^+} \le \sup_n (|b_n| + \|A_n\|_{bs}) \cdot \|x\|_{bv^+},$$

hence

$$\|L\| \le \sup_n (|b_n| + \|A_n\|_{bs}),$$

which is the second inequality in (3.132). Also by the left-hand side of (3.126) in Part (b) of Theorem 3.12, given $\varepsilon > 0$, for each $n \in \mathbb{N}_0$ there exists $x^{(n)} \in S_{bv^+}$ such that

$$\left|L_n(x^{(n)})\right| \ge \frac{1}{2} \cdot (|b_n| + \|A_n\|_{bs}) - \varepsilon,$$

whence

$$\|L\| \ge \frac{1}{2} \cdot (|b_n| + \|A_n\|_{bs}) - \varepsilon.$$

Since $\varepsilon > 0$ was arbitrary, we have

$$\|L\| \ge \frac{1}{2} \cdot (|b_n| + \|A_n\|_{bs}),$$

which is the first inequality in (3.132).

□

Now we are able to establish an estimate for the Hausdorff neasure of noncompactness of an arbitrary operator $L \in \mathcal{B}(bv^+, c)$. We use the notations of Theorem 3.14.

Theorem 3.15 *Let $L \in \mathcal{B}(bv^+, c)$. Then we have*

$$\frac{1}{4} \cdot \limsup_{n\to\infty} \left(|b_n - \beta| + \sup_m \left| \sum_{k=0}^{m} a_{nk} - \alpha_k \right| \right) \leq \|L\|_\chi \leq$$
$$\leq \limsup_{n\to\infty} \left(|b_n - \beta| + \sup_m \left| \sum_{k=0}^{m} a_{nk} - \alpha_k \right| \right), \tag{3.134}$$

where

$$\alpha_k = \lim_{n\to\infty} a_{nk} \textit{ for all } k \textit{ and } \beta = \lim_{n\to\infty} b_n. \tag{3.135}$$

Proof Let $L \in \mathcal{B}(bv^+, c)$. Then $L(x) \in c$ for all $x \in bv^+$, hence by (1.14) in Example 1.6

$$L(x) = \eta(x) \cdot e + \sum_{m=0}^{\infty} (L_m(x) - \eta(x))\, e^{(m)}, \tag{3.136}$$

where $\eta(x) = \lim_{j\to\infty} L_j(x)$. Also, by Part (b) of Theorem 3.14, there exists a matrix $A \in (bv_0^+, c)$ and a sequence $b \in c$ such that

$$L(x) = b \cdot \xi + Ax^{(0)} \text{ for all } x = x^0 + \xi \cdot e, \tag{3.137}$$

where $x^{(0)} \in bv_0^+$ and $\xi = \lim_{k\to\infty} x_k$. We observe that the limits β and α_k ($k = 0, 1, \dots$) in (3.135) exist, and

$$b = L(e) \text{ and } a_{nk} = L_n(e^{(k)}) \text{ for all } n \text{ and } k.$$

We write $Q^{(r)} = \mathcal{R}_r \circ L$ for all r and obtain by (3.136)

$$Q^{(r)}(x) = \mathcal{R}_r(L(x)) = \sum_{m=r+1}^{\infty} (L_m(x) - \eta(x)) e^{(m)} \text{ for each } r. \tag{3.138}$$

On the other hand by (3.137), for each r there exists a sequence $d^{(r)} = Q^{(r)}(e) \in c$ and a matrix $C^{(r)} = (c_{nk}^{(r)} \in (bv_0^+, c)$ with $c_{nk}^{(r)} = Q_n^{(r)}(e^{(k)})$ for all n and k such that

$$Q^{(r)}(x) = d^{(r)} \cdot \xi + C^{(r)} x^{(0)}. \tag{3.139}$$

Now it follows from (3.139) and (3.138) that

$$d^{(r)} = Q^{(r)}(e) = \sum_{m=r+1}^{\infty} (L_m(e) - \eta(e))\, e^{(m)} = \sum_{m=r+1}^{\infty} \left(b_m - \lim_{j\to\infty} b_j \right) e^{(m)}$$
$$= \sum_{m=r+1}^{\infty} (b_m - \beta)\, e^{(m)},$$

that is,

$$d_n^{(r)} = \begin{cases} 0 & (n \leq r) \\ b_n - \beta & (n \geq r+1), \end{cases}$$

and

$$c_{nk}^{(r)} = Q_n^{(r)}(e^{(k)}) = \sum_{m=r+1}^{\infty} \left(L_n(e^{(k)}) - \eta(e^{(k)})\right) e_n^{(m)} = \sum_{m=r+1}^{\infty} (a_{nk} - \alpha_k)\, e_n^{(m)},$$

hence

$$c_{nk}^{(r)} = \begin{cases} 0 & (n \leq r) \\ a_{nk} - \alpha_k & (n \geq r+1). \end{cases}$$

Since $Q^{(r)} \in \mathcal{B}(bv_0^+, c)$ for each r, we obtain by (3.132) in Part (d) of Theorem 3.14

$$\frac{1}{2} \cdot \sup_{n \geq r+1} \left(|b_n - \beta| + \sup_m \left| \sum_{k=0}^{m} a_{nk} - \alpha_k \right| \right) \leq \| Q^{(r)} \| \leq$$
$$\leq \sup_{n \geq r+1} \left(|b_n - \beta| + \sup_m \left| \sum_{k=0}^{m} a_{nk} - \alpha_k \right| \right).$$

Now the inequalities in (3.134) follow from (1.52) in Theorem 1.29, (3.59) in Lemma 3.1 and (1.47) in Remark 1.11. □

Now the characterization of $\mathcal{K}(bv^+, c)$ is an immediate consequence of (3.134) in Theorem 3.15 and (1.54) in Theorem 1.30.

Corollary 3.11 *Let $L \in \mathcal{B}(bv^+, c)$. Then we have $L \in \mathcal{K}(bv^+, c)$ if and only if*

$$\lim_{n \to \infty} \left(|b_n - \beta| + \sup_m \left| \sum_{k=0}^{m} a_{nk} - \alpha_k \right| \right) = 0.$$

References

1. Alotaibi, A., Mursaleen, M., Alamiri, B.A.S., Mohiuddine, S.A.: Compact operators on some Fibonacci difference sequence spaces. J. Inequal. Appl. **2015**(1), 1–8 (2015)
2. Altay, B., Başar, F., Mursaleen, M.: On the Euler sequence spaces which include ℓ_p and ℓ_∞ I. Inf. Sci. **176**, 1450–1462 (2006)
3. Cohen, L.W., Dunford, N.: Transformations on sequence spaces. Duke Math. J. **3**(4), 689–701 (1937)
4. Demiriz, S.: Applications of measures of noncompactness to the infinite system of differential equations in bv_p spaces. Electron. J. Math. Anal. Appl. **5**(1), 313–320 (2017)
5. Demiriz, S., Kara, E.E.: On compact operators on some sequence spaces related to matrix $B(r, s, t)$. Thai J. Math. **14**(3), 651–666 (2016)

6. Djolović, I., Malkowsky, E.: A note on compact operators on matrix domains. J. Math. Anal. Appl. **340**, 291–303 (2008)
7. Djolović, I., Petković, K., Malkowsky, E.: Matrix mappings and general bounded linear operators on the space bv. Math. Slovaca **68**(2), 405–414 (2018)
8. Ilkhan, M., Şimşek, Kara, E.E.: A new regular infinite matrix defined by Jordan totient function and its matrix domain in ℓ_p. Math. Methods. Appl. Sci. (2020). https://doi.org/10.1002/mma.6501
9. Ilkhan, M., Kara, E.E.: A new Banach space defined by Euler totient matrix operator. Oper. Matrices **13**(2), 527–544 (2019)
10. Kara, E.E., Ilkhan, M., Şimşek, N.: A study on certain sequence spaces using Jordan totient function. In: 8th International Eurasian Conference on Mathematical Sciences and Applications, Baku Azerbaijan (2019)
11. Kara, E.E., Başarır, M.: On compact operators and some Euler $B(m)$-difference sequence spaces and compact operators. J. Math. Anal. Appl. **379**(2), 499–511 (2011)
12. Malkowsky, E.: Linear operators between some matrix domains. Rend. Circ. Mat. Palermo, Serie II, Suppl. **68**, 641–655 (2002)
13. Malkowsky, E., Rakočević, V.: The measure of noncompactness of linear operators between certain sequence spaces. Acta Sci. Math. (Szeged) **64**, 151–170 (1998)
14. Malkowsky, E., Rakočević, V.: On matrix domains of triangles. Appl. Math. Comput. **189**, 1146–1163 (2007)
15. Malkowsky, E., Rakočević, V.: Advanced Functional Analysis. CRC Press, Taylor & Francis Group, Boca Raton, London, New York (2019)
16. Mursaleen, M., Mohiuddine, S.A.: Applications of measures of non-compactness to the infinite system of differential equations in ℓ_p spaces. Nonlinear Anal. **74**(4), 2111–2115 (2012)
17. Mursaleen, M., Noman, A.K.: Compactness by the Hausdorff measure of noncompactness. Nonlinear Anal. **73**(8), 2541–2557 (2010)
18. Başarır, M., Kara, E.E.: On compact operators on the Riesz $B(m)$–difference sequence space. Iran J. Sci. Technol. Trans. A Sci. **35**(4), 279–285 (2011)
19. Başarır, M., Kara, E.E.: On the B-difference sequence space derived by generalized weighted mean and compact operators. J. Math. Anal. Appl. **379**(2), 499–511 (2011)
20. Wilansky, A.: Summability through Functional Analysis, Mathematical Studies, vol. 85. North–Holland, Amsterdam (1984)
21. Zeller, K.: Faktorfolgen bei Limitierungsverfahren. Math. Z. **56**, 134–151 (1952)

Chapter 4
Computations in Sequence Spaces and Applications to Statistical Convergence

In this chapter, we define elementary sets that are used to simplify some sequence spaces, in particular, we are interested in applications of the theory of infinite matrices in various topics of mathematics. Since we know the characterizations of the classes (X, Y) when X and Y are any of the sets ℓ_∞, c or c_0 (Theorem 1.23), it is interesting to deal with the sets s_τ, s_τ^0 and $s_\tau^{(c)}$ introduced in Example 2.2, where $\tau = (\tau_n)_{n=1}^\infty$ is a sequence with $\tau_n > 0$ for all n; these may also be considered as the matrix domain in ℓ_∞, c and c_0 of the diagonal matrix with the sequence τ on its diagonal.

So we obtain characterizations of the sets (X_1, Y_1) where $X_1 \in \{s_\tau, s_\tau^0, s_\tau^{(c)}\}$ and $Y_1 \in \{s_\nu, s_\nu^0, s_\nu^{(c)}\}$ where τ and μ are real sequences with positive terms. We also obtain simplifications of some sets of sequences, such as $w_\tau^p(\lambda)$, $w_\tau^{\circ p}(\lambda)$, $w_\tau^{\bullet p}(\lambda)$ and $c_\tau^{+.p}(\lambda, \mu)$ with $p > 0$. These sets generalize in a certain sense the sets w_∞^p, w^p, c_0^p, c_∞^p and c^p. So we may do calculations on these sets and extend some well-known results.

In particular, we consider the sum and the product of some linear spaces of sequences, and present many computations involving the sets $w_0(\lambda)$ and $w_\infty(\lambda)$, which are generalizations of w_0 and w_∞, to obtain new results on *statistical convergence*. This notion was first introduced by Steinhaus in 1949 [35], and studied by several authors such as Fast [21], Fridy [24] and Connor [1], and more recently Mursaleen and Mohhiuddine [33], Mursaleen and Edely [32], Patterson and Savaş [34], and de Malafosse and Rakočević [19]. Then Fridy and Orhan [23] introduced the notion of lacunary statistical convergence which can be considered as a special case of statistical convergence. The notion of statistical convergence was also used for Hardy's *Tauberian theorem for Cesàro means* [25]. It was shown by Fridy and Khan [22] that the hypothesis of the convergence of the arithmetic means of the sequence x can be replaced by the weaker assumption of its statistical convergence. Here, we extend Hardy's *Tauberian theorem for Cesàro means* to the cases where the arithmetic means are successively replaced by the *operator of weighted means* $\overline{N}_q$ and the generalized arithmetic means $C(\lambda)$.

B. de Malafosse et al., *Operators Between Sequence Spaces and Applications*,
https://doi.org/10.1007/978-981-15-9742-8_4

4.1 On Strong τ-Summability

Let $\mathcal{U}^+ \subset \omega$ be the set of all sequences $u = (u_n)_{n=1}^\infty$ with $u_n > 0$ for all n. We recall that $\mathcal{U}$ is the set of all nonzero sequences (beginning of Sect. 2.2). Then, for given sequence $u = (u_n)_{n=1}^\infty \in \omega$, we define the infinite diagonal matrix D_u with $(D_u)_{nn} = u_n$ for all n. Let E be any subset of ω and $u \in \mathcal{U}$, using Wilansky's notations [36] we write

$$\left(\frac{1}{u}\right)^{-1} * E = D_u E = \left\{y = (y_n)_{n=1}^\infty \in \omega : \frac{y}{u} = \left(\frac{y_n}{u_n}\right)_{n=1}^\infty \in E\right\}.$$

We note that we can also write

$$E_\tau = D_\tau E$$

for $\tau \in \mathcal{U}^+$ and for any subset E of ω.

We recall the definition in Example 2.2 of the following sets for any sequence $\tau \in \mathcal{U}^+$:

$$s_\tau = \left(\frac{1}{\tau}\right)^{-1} * \ell_\infty = \left\{x \in \omega : \left(\frac{x_n}{\tau_n}\right)_{n=1}^\infty \in \ell_\infty\right\},$$
$$s_\tau^0 = \left(\frac{1}{\tau}\right)^{-1} * c_0 = \left\{x \in \omega : \left(\frac{x_n}{\tau_n}\right)_{n=1}^\infty \in c_0\right\}$$

and

$$s_\tau^{(c)} = \left(\frac{1}{\tau}\right)^{-1} * c = \left\{x_n \in \omega : \left(\frac{x_n}{\tau_n}\right)_{n=1}^\infty \in c\right\}.$$

It was also noted in Example 2.2 that each of the sets s_τ, s_τ^0 and $s_\tau^{(c)}$ is a BK space normed by

$$\|x\|_{s_\tau} = \sup_n \left(\frac{|x_n|}{\tau_n}\right)$$

and s_τ^0 has AK.

Now let $\tau, \nu \in \mathcal{U}^+$. We write $S_{\tau,\nu}$ for the class of infinite matrices $A = (a_{nk})_{n,k=1}^\infty$ such that

$$\|A\|_{S_{\tau,\nu}} = \sup_n \left(\frac{1}{\nu_n}\sum_{k=0}^\infty |a_{nk}|\tau_k\right) < \infty. \tag{4.1}$$

The set $S_{\tau,\nu}$ is a Banach space with the norm $\|A\|_{S_{\tau,\nu}}$. It was noted in Proposition 2.1 that

$$A \in (s_\tau, s_\nu) \text{ if and only if } A \in S_{\tau,\nu}.$$

So we can write $(s_\tau, s_\nu) = S_{\tau,\nu}$. In this way, we may state the next extension.

Lemma 4.1 *Let $\tau, \nu \in \mathcal{U}^+$ and $E, F \subset \omega$. Then we have*

$$A \in (E_\tau, F_\nu) \text{ if anf only if } D_{1/\nu} A D_\tau = \left(\frac{1}{\nu_n} a_{nk} \tau_k\right)_{n,k=1}^{\infty} \in (E, F).$$

Now by Parts **1.–3.** of Theorem 1.23, we have $(s_1, s_1) = (c_0, s_1) = (c, s_1)$, and so by Lemma 4.1

$$(s_\tau, s_\nu) = (s_\tau^0, s_\nu) = (s_\tau^{(c)}, s_\nu) = S_{\tau,\nu}.$$

When $s_\tau = s_\nu$ we obtain the *Banach algebra with identity* $S_{\tau,\nu} = S_\tau$, (see [2, 3, 5, 6, 8]) normed by $\|A\|_{S_\tau} = \|A\|_{S_{\tau,\tau}}$. We also have

$$A \in (s_\tau, s_\tau) \text{ of and only if } A \in S_\tau.$$

If $\|I - A\|_{S_\tau} < 1$, we say that $A \in \Gamma_\tau$. The set S_τ is a *Banach algebra with identity*. We have the useful result: if $A \in \Gamma_\tau$ then A is bijective from s_τ into itself. If $\tau = (r^n)_{n=1}^\infty$, we write $\Gamma_r, S_r, s_r, s_r^0$ and $s_r^{(c)}$ for $\Gamma_\tau, S_\tau, s_\tau, s_\tau^0$ and $s_\tau^{(c)}$, respectively [10]. When $r = 1$, we obtain $s_1 = \ell_\infty$, $s_1^0 = c_0$, $s_r^{(c)} = c$ and $S_1 = S_e$. We have by **1.–3.** of Theorem 1.23

$$(s_1, s_1) = (c_0, s_1) = (c, s_1) = S_1.$$

For any subset E of ω, we write, $AE = \{y \in \omega : y = Ax \text{ for some } x \in E\}$. Then the matrix domain of A in $F \subset \omega$ is denoted by $F(A)$ or $F_A = \{x \in \omega : Ax \in F\}$. So for any given $a \in \mathcal{U}$ we have $F_a = F_{D_{1/a}}$.

First, we need to recall some results.

Lemma 4.2 (Parts (i) and (ii) of Proposition 1.1) *Let E and F be arbitrary subsets of ω and let $u \in \mathcal{U}$. Then we have*

(i) $M(E, F) \subset M(\widetilde{E}, F)$ *for all* $\widetilde{E} \subset E$,
(ii) $M(E, F) \subset M(E, \widetilde{F})$ *for all* $F \subset \widetilde{F}$.

Lemma 4.3 ([29, Lemma 3.1, p. 648] and [31, Example 1.28, p. 157]) *We have*

(i) $M(c_0, F) = \ell_\infty$ *for* $F = c_0$, *or* c, *or* ℓ_∞;
(ii) $M(\ell_\infty, G) = \ell_\infty$ *for* $G = c_0$, *or* c;
(iii) $M(c, c) = c$ *and* $M(\ell_\infty, c) = c_0$.

We deduce the next corollary from Lemma 4.1.

Corollary 4.1 *Let $\alpha, \beta \in \mathcal{U}^+$. Then we have*

(i) $M(s_\alpha^0, H) = s_{\beta/\alpha}$ *for* $H = s_\beta^0$, *or* $s_\beta^{(c)}$, *or* s_β;
(ii) $M(K, s_\beta) = s_{\beta/\alpha}$ *if* $K = s_\alpha^0, s_\alpha^{(c)}$;
(iii) $M(s_\alpha, s_\beta^{(c)}) = s_{\beta/\alpha}^0$ *and* $M(s_\alpha^{(c)}, s_\beta^{(c)}) = s_{\beta/\alpha}^{(c)}$.

Proof This result follows from Lemma 4.1. Indeed, for any given sets $E, F \subset \omega$, we have $a \in M(D_\alpha * E, D_\beta * F)$ if and only if we successively have

$$D_a \in (D_\alpha * E, D_\beta * F),\;\; D_{a\alpha/\beta} \in (E, F) \text{ and } a\alpha/\beta \in M(E, F).$$

□

4.2 Sum and Product of Spaces of the Form s_ξ, s_ξ^0, or $s_\xi^{(c)}$

In the following, we use the next two properties, where we write $\mathcal{U}_1^+$ (resp. $c(1)$) for the set of all sequences ξ that satisfy $0 < \xi_n \leq 1$ for all n, (resp. that satisfy $\xi_n \to 1$ $(n \to \infty)$). For any linear space of sequences χ, we consider the conditions

$$\chi \subset \chi(D_\xi) \text{ for all } \xi \in \mathcal{U}_1^+ \tag{4.2}$$

and

$$\chi \subset \chi(D_\xi) \text{ for all } \xi \in c(1). \tag{4.3}$$

Lemma 4.4 ([16, Proposition 5.1]) *Let E be a linear space of sequences that satisfies condition (4.2). Then we have*

$$E_{\alpha+\beta} = E_\alpha + E_\beta \textit{ for all } \alpha, \beta \in \mathcal{U}^+. \tag{4.4}$$

Proof Let $y = (\alpha + \beta)z = \alpha z + \beta z$ with $z \in E$. Then we have $y \in E_\alpha + E_\beta$ and $E_{\alpha+\beta} \subset E_\alpha + E_\beta$. Now let $y \in E_\alpha + E_\beta$. Then there are $u, v \in E$, such that

$$y = \alpha u + \beta v = (\alpha + \beta)\left(\frac{\alpha}{\alpha+\beta}u + \frac{\beta}{\alpha+\beta}v\right),$$

since E is a linear space by the condition in (4.2) we have $y \in E_{\alpha+\beta}$ and $E_\alpha + E_\beta \subset E_{\alpha+\beta}$. So we have shown $E_{\alpha+\beta} = E_\alpha + E_\beta$. □

Remark 4.1 The condition in (4.2) is true when E is either of the sets c_0 or ℓ_∞. This implies $s_\alpha + s_\beta = s_{\alpha+\beta}$ and $s_\alpha^0 + s_\beta^0 = s_{\alpha+\beta}^0$ for $\alpha, \beta \in \mathcal{U}^+$.

Remark 4.2 We note that for any linear space of sequences E we have $E_{\alpha+\beta} \subset E_\alpha + E_\beta$ for all $\alpha, \beta \in \mathcal{U}^+$. So if $y \in E$ implies $\alpha y \in E$ for all y and for all $\alpha \in \mathcal{U}_1^+$, then the condition in (4.4) is equivalent to $E_\alpha + E_\beta \subset E_{\alpha+\beta}$.

Now we state some results concerning the *sum* of particular sequence spaces.

Theorem 4.1 ([11, Theorem 4, p. 293]) *Let $\alpha, \beta \in \mathcal{U}^+$. Then we have*

(i) (a) $s_\alpha = s_\beta$ *if and only if there are* $K_1, K_2 > 0$ *such that*

$$K_1 \le \frac{\beta_n}{\alpha_n} \le K_2 \text{ for all } n;$$

(b) $s_\alpha + s_\beta = s_{\alpha+\beta} = s_{\max\{\alpha,\beta\}}$;
(c) $s_\alpha + s_\beta = s_\alpha$ *if and only if* $\beta/\alpha \in \ell_\infty$.
(ii) (a) $s_\alpha^0 \subset s_\beta^0$ *if and only if* $\alpha/\beta \in \ell_\infty$;
(b) $s_\alpha^0 = s_\beta^0$ *if and only if* $s_\alpha = s_\beta$;
(c) $s_\alpha^0 + s_\beta^0 = s_{\alpha+\beta}^0$;
(d) $s_\alpha^0 + s_\beta^0 = s_\alpha^0$ *if and only if* $\beta/\alpha \in \ell_\infty$;
(e) $s_\alpha^{(c)} \subset s_\beta^{(c)}$ *if and only if* $\alpha/\beta \in c$;
(f) *The condition* $\alpha_n/\beta_n \to l \neq 0$ *for some* $l \in \mathbb{C}$ *is equivalent to*

$$s_\alpha^{(c)} = s_\beta^{(c)},$$

and if $\alpha_n/\beta_n \to l \neq 0$, *then*

$$s_\alpha = s_\beta,\ s_\alpha^0 = s_\beta^0 \text{ and } s_\alpha^{(c)} = s_\beta^{(c)}.$$

(iii) (a) $s_{\alpha+\beta}^{(c)} \subset s_\alpha^{(c)} + s_\beta^{(c)}$;
(b) *The condition* $\alpha/(\alpha+\beta) \in c$ *is equivalent to*

$$s_\alpha^{(c)} + s_\beta^{(c)} = s_{\alpha+\beta}^{(c)};$$

(c) *The condition* $\beta/\alpha \in c$ *is equivalent to* $s_\alpha^{(c)} + s_\beta^{(c)} = s_{\alpha+\beta}^{(c)} = s_\alpha^{(c)}$;
(iv) $s_\alpha^0 + s_\beta^{(c)} = s_\beta^{(c)}$ *is equivalent to* $\alpha/\beta \in \ell_\infty$, *and the condition* $\beta/\alpha \in c_0$ *is equivalent to* $s_\alpha^0 + s_\beta^{(c)} = s_\alpha^0$;
(v) (a) $s_\alpha + s_\beta^0 = s_\alpha$ *is equivalent to* $\beta/\alpha \in \ell_\infty$;
(b) $s_\alpha + s_\beta^0 = s_\beta^0$ *is equivalent to* $\alpha/\beta \in c_0$;
(vi) (a) $s_\alpha^{(c)} + s_\beta = s_\alpha^{(c)}$ *is equivalent to* $\beta/\alpha \in c_0$;
(b) $s_\alpha^{(c)} + s_\beta = s_\beta$ *is equivalent to* $\alpha/\beta \in \ell_\infty$;
(c) $s_\alpha^{(c)} + s_\beta^{(c)} = s_\alpha^{(c)}$ *is equivalent to* $\beta/\alpha \in c$.

Proof (i)

(a) follows from [9, Proposition 1, p. 244].
(b) follows from Remark 4.1. Then from the inequalities

$$\max\{\alpha_n, \beta_n\} \le \alpha_n + \beta_n \le 2\max\{\alpha_n, \beta_n\} \text{ for all } n$$

we obtain by (i) (a) the identity $s_{\max(\alpha,\beta)} = s_{\alpha+\beta}$.
This concludes the proof of Part (i) (b).

(c) We have $s_\alpha + s_\beta = s_\alpha$ if and only if $s_\beta \subset s_\alpha$, that is, $\beta/\alpha \in M(s_1, s_1) = \ell_\infty$.

(ii)

(a) follows from the equivalence of $s_\alpha^0 \subset s_\beta^0$ and $\alpha/\beta \in M(c_0, c_0) = \ell_\infty$.

(b) We deduce from (ii) (a) that $s_\alpha^0 = s_\beta^0$ is equivalent to α/β and $\beta/\alpha \in \ell_\infty$, that is, $s_\alpha = s_\beta$.

(c) follows from Remark 4.1.

(d) The identity $s_\alpha^0 + s_\beta^0 = s_\alpha^0$ is equivalent to $s_{\alpha+\beta}^0 = s_\alpha^0$ and to $s_{\alpha+\beta} = s_\alpha + s_\beta = s_\alpha$ and we conclude using (i) (c).

(e) We have $s_\alpha^{(c)} \subset s_\beta^{(c)}$ if and only if

$$\alpha/\beta \in M(c, c) = c$$

and we have shown Part (ii) (e).

(f) follows from the equivalence of $s_\alpha^{(c)} = s_\beta^{(c)}$ and $\alpha/\beta, \beta/\alpha \in c$, that is,

$$\frac{\alpha_n}{\beta_n} \to l \ (n \to \infty) \text{ for some scalar } l \neq 0. \tag{4.5}$$

Furthermore, the condition in (4.5) implies α/β and $\beta/\alpha \in \ell_\infty$. This means $s_\alpha = s_\beta$ and we conclude by (ii) (b) and the first part of (ii) (f).

(iii)

(a) For any given $x \in s_{\alpha+\beta}^{(c)}$ there is $\varphi \in c$ such that

$$x = (\alpha + \beta)\varphi = \alpha\varphi + \beta\varphi \in s_\alpha^{(c)} + s_\beta^{(c)},$$

which implies

$$s_{\alpha+\beta}^{(c)} \subset s_\alpha^{(c)} + s_\beta^{(c)}.$$

(b) *Necessity.* Let $\alpha/(\alpha + \beta) \in c$. Then we have $\beta/(\alpha + \beta) = 1 - \alpha/(\alpha + \beta) \in c$. So we obtain $s_\alpha^{(c)} \subset s_{\alpha+\beta}^{(c)}$ and $s_\beta^{(c)} \subset s_{\alpha+\beta}^{(c)}$ and $s_\alpha^{(c)} + s_\beta^{(c)} \subset s_{\alpha+\beta}^{(c)}$ and we conclude by (iii) (a)

$$s_{\alpha+\beta}^{(c)} = s_\alpha^{(c)} + s_\beta^{(c)}.$$

Sufficiency. Conversely, if $s_{\alpha+\beta}^{(c)} = s_\alpha^{(c)} + s_\beta^{(c)}$, then we have

$$\alpha \in s_\alpha^{(c)} + s_\beta^{(c)}$$

and $\alpha \in s_{\alpha+\beta}^{(c)}$ which imply $\alpha/(\alpha + \beta) \in c$.
This concludes the proof of (b).

(c) By (ii) (f), the identity $s_{\alpha+\beta}^{(c)} = s_\alpha^{(c)}$ is equivalent to

$$\frac{\alpha_n + \beta_n}{\alpha_n} = 1 + \frac{\beta_n}{\alpha_n} \to l \neq 0 \ (n \to \infty),$$

that is, $\beta/\alpha \in c$. We conclude, since the condition $\beta/\alpha \in M(c, c) = c$ is equivalent to $s_\alpha^{(c)} + s_\beta^{(c)} = s_\alpha^{(c)}$.

(iv) The condition $s_\alpha^0 + s_\beta^{(c)} = s_\beta^{(c)}$ is equivalent to $s_\alpha^0 \subset s_\beta^{(c)}$. Then the inclusion $s_\alpha^0 \subset s_\beta^{(c)}$ is equivalent to $\alpha/\beta \in M(c_0, c) = \ell_\infty$, and we conclude $s_\alpha^0 + s_\beta^{(c)} = s_\beta^{(c)}$ if and only if $\alpha/\beta \in \ell_\infty$. Similarly, $s_\alpha^0 + s_\beta^{(c)} = s_\alpha^0$ is equivalent to $s_\beta^{(c)} \subset s_\alpha^0$, and since $\beta/\alpha \in M(c, c_0) = c_0$ we conclude $s_\alpha^0 + s_\beta^{(c)} = s_\alpha^0$ is equivalent to $\beta/\alpha \in c_0$.
(v)

(a) follows from the equivalence of $s_\alpha + s_\beta^0 = s_\alpha$ and $\beta/\alpha \in M(c_0, \ell_\infty) = \ell_\infty$.

Parts (v) (b) and (vi) can be obtained in the similar way. □

As a direct consequence of the preceding results, we obtain the following corollary.

Corollary 4.2 ([11, Corollary 5, p. 296]) *The following conditions are equivalent:*

(i) $\beta/\alpha \in \ell_\infty$;
(ii) $s_\alpha^0 + s_\beta^0 = s_\alpha^0$;
(iii) $s_\alpha + s_\beta = s_\alpha$;
(vi) $s_\alpha + s_\beta^{(c)} = s_\alpha$.

Remark 4.3 It can easily be shown that if $\beta = e$, and α is defined by $\alpha_{2n} = 1$ and $\alpha_{2n+1} = n$ for all n, then we have $s_\alpha^{(c)} + s_1 \neq s_1$ and $\neq s_\alpha^{(c)}$.

We easily deduce from the preceding results the following proposition.

Proposition 4.1 ([11, Proposition 12, p. 306]) *Let* $\alpha, \beta, \gamma \in \mathcal{U}^+$. *Then we have*

(i) $S_{\alpha+\beta,\gamma} = S_{\alpha,\gamma} \bigcap S_{\beta,\gamma} = S_{\max\{\alpha,\beta\},\gamma}$.
(ii) $(E + F, s_\gamma) = S_{\alpha+\beta,\gamma} = S_{\max\{\alpha,\beta\},\gamma}$ *for* $E = s_\alpha$, s_α^0, *or* $s_\alpha^{(c)}$ *and* $F = s_\beta$, s_β^0, *or* $s_\beta^{(c)}$.

Now we deal with some properties of the *product* $E * F$ of certain subsets E and F of ω. These results generalize some of those given in [9].
For any given sets of sequences E and F, we write

$$E * F = \{xy = (x_n y_n)_{n=1}^\infty \in \omega : x \in E \text{ and } y \in F\}. \tag{4.6}$$

We state the following results without proof.

Proposition 4.2 ([11, Proposition 7, p. 298]) *Let* $\alpha, \beta, \gamma \in \mathcal{U}^+$. *Then we have*

(i) $s_\alpha * s_\beta = s_\alpha * s_\beta^{(c)} = s_{\alpha\beta}$;
(ii) $s_\alpha * s_\beta^0 = s_\alpha^0 * s_\beta^0 = s_\alpha^{(c)} * s_\beta^0 = s_{\alpha\beta}^0$;
(iii) $s_\alpha^{(c)} * s_\beta^{(c)} = s_{\alpha\beta}^{(c)}$;
(iv) $s_\alpha^{(c)} * s_\beta = s_{\alpha\beta}$;
(v) *Let* E *be any of the sets* s_α *or* s_α^0 *or* $s_\alpha^{(c)}$. *Then* $E * s_\beta^0 = s_\gamma^0$ *if and only if there are* $K_1, K_2 > 0$ *such that*

$$K_1 \gamma_n \le \alpha_n \beta_n \le K_2 \gamma_n \text{ for all } n;$$

(vi) (a) $s_\alpha * s_\beta = s_\alpha * s_\gamma$ *if and only if* $s_\beta = s_\gamma$ *and*
(b) $s_\alpha^0 * s_\beta^0 = s_\alpha^0 * s_\gamma^0$ *if and only if* $s_\beta^0 = s_\gamma^0$.

4.3 Properties of the Sequence $C(\tau)\tau$

Since we intend to simplify certain sets of sequences, in particular, those of τ-strongly summable sequences, we need to study the sequence $C(\tau)\tau$.

First, we deal with the operators represented by $C(\lambda)$ and $\Delta(\lambda)$. We define $C(\lambda)$ for $\lambda = (\lambda_n)_{n=1}^\infty \in \mathcal{U}$ by

$$(C(\lambda))_{nk} = \begin{cases} \dfrac{1}{\lambda_n} & \text{if } k \le n \\ 0 & \text{otherwise.} \end{cases} \tag{4.7}$$

It can easily be shown that the matrix $\Delta(\lambda)$ defined by

$$(\Delta(\lambda))_{nk} = \begin{cases} \lambda_n & \text{if } k = n \\ -\lambda_{n-1} & \text{if } k = n-1 \text{ and } n \ge 1 \\ 0 & \text{otherwise} \end{cases} \tag{4.8}$$

is the inverse of the triangle $C(\lambda)$, see [13]. If $\lambda = e$, we obtain the operator of the first difference represented by $\Delta(e) = \Delta$. In the following, we use the well-known notation $\Sigma = C(e)$. Note that $\Delta = \Sigma^{-1}$ and $\Delta, \Sigma \in S_R$ for any real $R > 1$. We let $\Delta(\lambda)^T = \Delta(\lambda)^+$ and $C(\lambda)^T = C(\lambda)^+$. Then we have $\Delta^+(\lambda) = D_\lambda \Delta^+$. Now we consider the following sets defined in [6, 7] where cs is the set of all convergent series and we write $\tau^\bullet = (\tau_{n-1}/\tau_n)_{n=1}^\infty$ for $\tau \in \mathcal{U}^+$.

$$\widehat{C_1} = \left\{ \tau \in \mathcal{U}^+ : \frac{1}{\tau_n} \left(\sum_{k=1}^{n} \tau_k \right) = O(1) \; (n \to \infty) \right\}, \tag{4.9}$$

$$\widehat{C} = \left\{ \tau \in \mathcal{U}^+ : \left(\frac{1}{\tau_n} \left(\sum_{k=1}^{n} \tau_k \right) \right)_{n=1}^{\infty} \in c \right\},$$

$$\widehat{C_1^+} = \left\{ \tau \in \mathcal{U}^+ \bigcap cs : \frac{1}{\tau_n} \left(\sum_{k=n}^{\infty} \tau_k \right) = O(1) \; (n \to \infty) \right\},$$

$$\Gamma = \left\{ \tau \in \mathcal{U}^+ : \overline{\lim_{n \to \infty}} \tau_n^\bullet < 1 \right\} \text{ and}$$

$$\Gamma^+ = \left\{ \tau \in \mathcal{U}^+ : \overline{\lim_{n \to \infty}} \left(\frac{\tau_{n+1}}{\tau_n} \right) < 1 \right\}.$$

Note that $\tau \in \Gamma^+$ if and only if $1/\tau \in \Gamma$. We will see in Proposition 4.3 that $\tau_n \to \infty$ $(n \to \infty)$ if $\tau \in \widehat{C_1}$. We also have $\tau \in \Gamma$ if and only if there is an integer $q \geq 1$ such that

$$\gamma_q(\tau) = \sup_{n \geq q+1} \tau_n^\bullet < 1. \tag{4.10}$$

In the following, we show $\widehat{C} = \widehat{\Gamma}$, where

$$\widehat{\Gamma} = \left\{ \tau \in \mathcal{U}^+ : \lim_{n \to \infty} \tau_n^\bullet < 1 \right\}.$$

For this we consider the set $\mathcal{B}(s_\tau^{(c)}) = \mathcal{B}(s_\tau^{(c)}, s_\tau^{(c)})$ of all bounded linear operators mapping from $s_\tau^{(c)}$ into itself. We recall that since $s_\tau^{(c)}$ is a Banach space with the norm $\|\cdot\|_{s_\tau}$, and the set $\mathcal{B}(s_\tau^{(c)})$ of all linear operators $A : s_\tau^{(c)} \to s_\tau^{(c)}$ normed by

$$\|A\|_{\mathcal{B}(s_\tau^{(c)})} = \sup_{x \neq 0} \left(\frac{\|Ax\|_{s_\tau}}{\|x\|_{s_\tau}} \right)$$

is the Banach algebra of all bounded linear operators that map $s_\tau^{(c)}$ into itself, see [10].

Lemma 4.5 ([14, Proposition 2.2. p. 88]) We have $\hat{\Gamma} \subset \hat{C}$.

Proof Assume $\tau \in \widehat{\Gamma}$ and let $\Delta_\tau = D_{1/\tau} \Delta D_\tau$. The nonzero entries of Δ_τ are defined by $(\Delta_\tau)_{nn} = 1$ for all n and $(\Delta_\tau)_{n,n-1} = -\tau_n^\bullet$ for all n. Then from the characterization of (c, c), the condition $\Delta_\tau \in (c, c)$ is equivalent to $(\tau_n^\bullet)_{n=1}^\infty \in c$. Now we show that Δ is invertible in $\mathcal{B}(s_\tau^{(c)})$. We consider the matrix

$$\Sigma^{(\eta)} = \begin{pmatrix} \left(\Delta^{(\eta)}\right)^{-1} & 0 \\ 0 & 1 \end{pmatrix},$$

for any given integer $\eta \geq 1$, where $\Delta^{(\eta)}$ is the finite matrix whose entries are those of the η first rows and of the η first columns of Δ. We obtain $\Sigma^{(\eta)} \Delta = (a_{nk})_{n,k=1}^\infty$

with $a_{nn} = 1$ for all n, $a_{n,n-1} = -1$ for all $n \geq \eta$ and $a_{nk} = 0$ otherwise. We deduce

$$\|I - \Sigma^{(\eta)}\Delta\|_{\mathcal{B}(s_\tau)} = \|I - \Sigma^{(\eta)}\Delta\|_{S_\tau} = \sup_{n \geq \eta} \tau_n^\bullet.$$

So we obtain

$$\overline{\lim_{n\to\infty}} \tau_n^\bullet = \lim_{n\to\infty} \tau_n^\bullet < 1 \text{ and } \|I - \Sigma^{(\eta)}\Delta\|_{\mathcal{B}(s_\tau)} < 1.$$

We conclude $\Sigma^{(\eta)}\Delta$ is invertible in $\mathcal{B}(s_\tau^{(c)})$ and $\Delta = (\Sigma^{(\eta)})^{-1}\Sigma^{(\eta)}\Delta$ is bijective from $s_\tau^{(c)}$ into itself. Finally, since Δ is bijective from $s_\tau^{(c)}$ into itself, we have $\Sigma = \Delta^{-1} \in (s_\tau^{(c)}, s_\tau^{(c)})$ and $D_{1/\tau}\Sigma D_\tau \in (c, c)$. This implies $\tau \in \widehat{C}$ and we have shown $\widehat{\Gamma} \subset \widehat{C}$. □

Remark 4.4 We note that $\Gamma \neq \widehat{C_1}$. Indeed, we consider the sequence τ defined by

$$\tau_n = \begin{cases} \zeta^k & \text{if } n = 2k \\ \zeta^k & \text{if } n = 2k+1, \end{cases}$$

where $\zeta > 1$ is a real number. It can easily be seen that $\tau \in \widehat{C_1}$ and $\overline{\lim_{n\to\infty}} \tau_n^\bullet = 1$.

We obtain the following results in which we let $[C(\tau)\tau]_n = (\sum_{k=1}^n \tau_k)/\tau_n$.

Proposition 4.3 ([8, Proposition 2.1. p. 1656], [18]) *Let $\tau \in \mathcal{U}^+$. Then we have*

(i)

$$\tau_n^\bullet \to 0 \textit{ if and only if } [C(\tau)\tau]_n \to 1 \ (n \to \infty).$$

(ii)

$$[C(\tau)\tau]_n \to l \textit{ if and only if } \tau_n^\bullet \to 1 - \frac{1}{l} \ (n \to \infty) \textit{ for some scalar } l > 0.$$

(iii)

$$\tau \in \widehat{C_1} \textit{ implies } \tau_n \geq K\gamma^n \textit{ for some } K > 0 \textit{ and } \gamma > 1 \textit{ for all } n. \tag{4.11}$$

(iv) The condition $\tau \in \Gamma$ implies that $\tau \in \widehat{C_1}$ and there exists a real $b > 0$ such that

$$[C(\tau)\tau]_n \leq \frac{1}{1 - \gamma_q(\tau)} + b\left[\gamma_q(\tau)\right]^n \textit{ for } n \geq q+1. \tag{4.12}$$

(v) We have $\Gamma^+ \subset \widehat{C_1^+}$.

Proof (i) We assume $\tau_n^\bullet \to 0 \ (n \to \infty)$. Then there is an integer N such that

$$\tau_n^\bullet \leq \frac{1}{2} \text{ for all } n \geq N+1. \tag{4.13}$$

So there exists a real $K > 0$ such that $\tau_n \geq K2^n$ for all n and

$$\frac{\tau_k}{\tau_n} = \tau^\bullet_{k+1} \cdots \tau^\bullet_n \leq \left(\frac{1}{2}\right)^{n-k} \text{ for } N \leq k \leq n-1. \tag{4.14}$$

Then we have

$$\begin{aligned} \frac{1}{\tau_n}\left(\sum_{k=1}^{n-1} \tau_k\right) &= \frac{1}{\tau_n}\left(\sum_{k=1}^{N-1} \tau_k\right) + \sum_{k=N}^{n-1} \frac{\tau_k}{\tau_n} \\ &\leq \frac{1}{K2^n}\left(\sum_{k=1}^{N-1} \tau_k\right) + \sum_{k=N}^{n-1} \left(\frac{1}{2}\right)^{n-k}, \end{aligned} \tag{4.15}$$

and since

$$\sum_{k=N}^{n-1} (1/2)^{n-k} = 1 - (1/2)^{n-N} \to 1 \ (n \to \infty),$$

we deduce

$$\frac{\sum_{k=1}^{n-1} \tau_k}{\tau_n} = O(1) \ (n \to \infty)$$

and

$$([C(\tau)\tau]_n)_{n=1}^{\infty} \in \ell_\infty.$$

From the identities

$$\begin{aligned} [C(\tau)\tau]_n &= \frac{\tau_1 + \cdots + \tau_{n-1}}{\tau_{n-1}} \tau^\bullet_n + 1 \\ &= [C(\tau)\tau]_{n-1} \, \tau^\bullet_n + 1 \text{ for all } n \geq 1, \end{aligned} \tag{4.16}$$

we obtain

$$[C(\tau)\tau]_n \to 1 \ (n \to \infty).$$

This proves the necessity.

Conversely, assume $[C(\tau)\tau]_n \to 1 \ (n \to \infty)$. By the identities given in (4.16), we have

$$\tau^\bullet_n = \frac{[C(\tau)\tau]_n - 1}{[C(\tau)\tau]_{n-1}}, \tag{4.17}$$

and we conclude $\tau^\bullet_n \to 0 \ (n \to \infty)$.

(ii) The necessity is a direct consequence of the identity in (4.17) and $\hat{C} \subset \hat{\Gamma}$. Conversely, we assume $\tau^\bullet_n \to 1 - 1/l \ (n \to \infty)$. Let $L = 1 - 1/l$ and $\sigma_n = \sum_{k=1}^n \tau_k$ and note that $l \geq 1$, since $\sigma_n/\tau_n = 1 + \sigma_{n-1}/\tau_n \geq 1$ for all n. By Lemma

4.5 we have that $\widehat{C} = \widehat{\Gamma}$, so we can write $\sigma_n/\tau_n \to l_1$ $(n \to \infty)$ for some scalar l_1, and we must show that $l_1 = l$. We have for every $n > 2$

$$\tau_n^{\bullet} = \frac{\sigma_{n-1} - \sigma_{n-2}}{\tau_n} = \frac{\sigma_{n-1}}{\tau_{n-1}}\tau_n^{\bullet} - \frac{\sigma_{n-2}}{\tau_{n-2}}\frac{\tau_{n-2}}{\tau_{n-1}}\tau_n^{\bullet}$$

and

$$\frac{\sigma_{n-1} - \sigma_{n-2}}{\tau_n} \to l_1 L - l_1 L^2 = L \ (n \to \infty).$$

If $L \neq 0$ then we have $l_1 = 1/\left(1 - L\right)$ and since $L = 1 - 1/l$, we conclude

$$l_1 = \frac{1}{1 - \left(1 - \frac{1}{l}\right)} = l.$$

If $L = 0$ then we have $l = 1$ and

$$\frac{\sigma_n}{\tau_n} = \frac{\sigma_{n-1}}{\tau_{n-1}}\tau_n^{\bullet} + 1 \to 1 \ (n \to \infty).$$

(iii) For a real $M > 1$ we have

$$[C(\tau)\tau]_n = \frac{\sigma_n}{\sigma_n - \sigma_{n-1}} \leq M \text{ for all } n \geq 2.$$

So we have

$$\sigma_n \geq \frac{M}{M-1}\sigma_{n-1}$$

and

$$\sigma_n \geq \left(\frac{M}{M-1}\right)^{n-1} \tau_1 \text{ for all } n.$$

Then from the inequalities

$$\frac{\tau_1}{\tau_n}\left(\frac{M}{M-1}\right)^{n-1} \leq [C(\tau)\tau]_n = \frac{\sigma_n}{\tau_n} \leq M,$$

we conclude

$$\tau_n \geq K\gamma^n \text{ for all } n,$$

with

$$K = (M-1)\tau_1/M^2 \text{ and } \gamma = M/(M-1) > 1.$$

(iv) If $\tau \in \Gamma$ there is an integer $q \geq 1$ for which

$$\tau_k^\bullet \le \chi < 1 \text{ for } k \ge q+1 \text{ with } \chi = \gamma_q(\tau).$$

So there is a real $M' > 0$ for which

$$\tau_n \ge \frac{M'}{\chi^n} \text{ for all } n \ge q+1. \tag{4.18}$$

Writing $\sigma_{nq} = (\sum_{k=1}^{q} \tau_k)/\tau_n$ and $d_n = [C(\tau)\tau]_n - \sigma_{nq}$, we obtain

$$d_n = \frac{1}{\tau_n}\left(\sum_{k=q+1}^{n} \tau_k\right) = 1 + \sum_{j=q+1}^{n-1}\left(\prod_{k=1}^{n-j} \tau_{n-k+1}^\bullet\right) \le \sum_{j=q+1}^{n} \chi^{n-j} \le \frac{1}{1-\chi}.$$

We obtain by the condition in (4.18)

$$\sigma_{nq} \le \frac{1}{M'}\chi^n\left(\sum_{k=1}^{q} \tau_k\right).$$

We conclude

$$[C(\tau)\tau]_n \le a + b\chi^n$$

with

$$a = \frac{1}{1-\chi} \text{ and } b = \frac{1}{M'}\left(\sum_{k=1}^{q} \tau_k\right).$$

(v) If $\tau \in \Gamma^+$, there are $\chi' \in (0,1)$ and an integer $q' \ge 1$ such that

$$\tau_k/\tau_{k-1} \le \chi' \text{ for } k \ge q'.$$

Then for $n \ge q'$, we have

$$\frac{1}{\tau_n}\left(\sum_{k=n}^{\infty} \tau_k\right) = \sum_{k=n}^{\infty}\left(\frac{\tau_k}{\tau_n}\right) \le 1 + \sum_{k=n+1}^{\infty}\prod_{i=0}^{k-n-1}\left(\frac{\tau_{k-i}}{\tau_{k-i-1}}\right) \le \sum_{k=n}^{\infty} \chi'^{k-n} = O(1)$$
$$(n \to \infty).$$

This completes the proof. □

Now we state a result where we let $c^\bullet = \{\tau \in \mathcal{U}^+ : \tau^\bullet \in c\}$.

Lemma 4.6 *We have*

$$\widehat{C_1} \cap c^\bullet = \Gamma. \tag{4.19}$$

Proof Since we have $\Gamma \subset \widehat{C_1} \cap c^\bullet$, it is enough to show $\widehat{C_1} \cap c^\bullet \subset \Gamma$. For this, we assume that there is $\tau \in \widehat{C_1} \cap c^\bullet$ such that $\tau \notin \Gamma$. Then we have $\lim_{n\to\infty} \tau_n^\bullet \ge 1$. So for any given $\varepsilon > 0$, there is an integer $q > 0$ such that $\tau_n^\bullet \geq 1 - \varepsilon$ for all $n \geq q + 1$ and

$$[C(\tau)\tau]_{2q} \geq \frac{1}{\tau_{2q}} \left(\sum_{k=q}^{2q} \tau_k \right) \geq \sum_{k=q}^{2q-1} \left(\tau_{k+1}^\bullet \dots \tau_{2q}^\bullet \right) + 1$$
$$\geq (1-\varepsilon)^q + \dots + (1-\varepsilon) + 1 \geq \frac{1-(1-\varepsilon)^{q+1}}{\varepsilon}.$$

Then we have

$$\frac{1-(1-\varepsilon)^{q+1}}{\varepsilon} \sim \frac{1-[1-(q+1)\varepsilon]}{\varepsilon} \sim q+1 \ (\varepsilon \to 0)$$

and $([C(\tau)\tau]_{2q})_q \notin \ell_\infty$ which implies $\tau \notin \widehat{C_1}$. This leads to a contradiction and we have shown the lemma. □

4.4 Some Properties of the Sets $s_\tau(\Delta)$, $s_\tau^0(\Delta)$ and $s_\tau^{(c)}(\Delta)$

In this section, we consider Δ as an operator from X into itself where X is any of the sets s_τ, s_τ^0 or $s_\tau^{(c)}$. Then we obtain necessary and sufficient conditions for $\Delta \in (X, X)$ to be bijective. In this way, we have the following results.

Theorem 4.2 ([6, Theorem 2.6. p. 1789]) *Let $\tau \in \mathcal{U}^+$. Then we have*

(i) $s_\tau(\Delta) = s_\tau$ *if and only if* $\tau \in \widehat{C_1}$;
(ii) $s_\tau^0(\Delta) = s_\tau^0$ *if and only if* $\tau \in \widehat{C_1}$;
(iii) $s_\tau^{(c)}(\Delta) = s_\tau^{(c)}$ *if and only if* $\tau \in \widehat{\Gamma}$;
(iv) $\Delta_\tau = D_{1/\tau} \Delta D_\tau$ *is bijective from c into itself with* $\lim x = \Delta_\tau - \lim x$, *if and only if*

$$\tau_n^\bullet = \frac{\tau_{n-1}}{\tau_n} \to 0.$$

Proof (i) We have $s_\tau(\Delta) = s_\tau$ if and only if $\Delta, \Sigma \in (s_\tau, s_\tau)$. This means that $\Delta, \Sigma \in S_\tau$, that is,

$$\|\Delta\|_{S_\tau} = \sup_n (1 + \tau_n^\bullet) < \infty \text{ and } \|\Sigma\|_{S_\tau} = \sup_n [C(\tau)\tau]_n < \infty.$$

Since

$$0 < \tau_n^\bullet \leq [C(\tau)\tau]_n \text{ for all } n,$$

we deduce $\Delta, \Sigma \in S_\tau$ if and only if $\|\Sigma\|_{S_\tau} < \infty$, that is, $\tau \in \widehat{C_1}$.
(ii) The condition $s_\tau^0(\Delta) = s_\tau^0$ is equivalent to

$$\Delta_\tau = D_{1/\tau}\Delta D_\tau \text{ and } \Sigma_\tau = D_{1/\tau}\Sigma D_\tau \in (c_0, c_0).$$

Since

$$(\Sigma_\tau)_{nk} = \begin{cases} \dfrac{\tau_k}{\tau_n} & \text{if } k \leq n \\ 0 & \text{if } k > n, \end{cases}$$

the condition $s_\tau^0(\Delta) = s_\tau^0$ is equivalent to $s_\tau(\Delta) = s_\tau$ and $\tau_k/\tau_n = o(1)$ $(n \to \infty)$ for all k. Part (i) and by Part (iii) of Proposition 4.3, the condition $s_\tau(\Delta) = s_\tau$ implies $\tau \in \widehat{C_1}$ and $\tau_n \to \infty$ $(n \to \infty)$.
So we have shown Part (ii).
(iii) As above we obtain $s_\tau^{(c)}(\Delta) = s_\tau^{(c)}$ if and only if Δ_τ and $\Sigma_\tau \in (c, c)$. We have $\Delta_\tau \in (c, c)$ if and only if $\tau^\bullet = (\tau_{n-1}/\tau_n)_{n=1}^\infty \in c$. Indeed, we have $\Delta_\tau \in S_1$ and

$$\sum_{k=1}^{n} (\Delta_\tau)_{nk} = 1 - \tau_n^\bullet$$

tends to a limit as $n \to \infty$. Then we have $\Sigma_\tau \in (c, c)$ if and only if each of the next conditions is satisfied, where

$$\Sigma_\tau \in S_1, \text{ that is, } \tau \in \widehat{C_1}; \tag{a}$$

$$\lim_{n\to\infty} \frac{\tau_k}{\tau_n} = 0 \text{ for all } k, \tag{b}$$

and

$$\tau \in \widehat{C}. \tag{c}$$

From Proposition 4.3, the condition in (c) implies $\lim_{n\to\infty} \tau_n = \infty$, so (c) implies (a) and (b). Finally, from Part (i) of Proposition 4.3, we conclude that $\tau \in \widehat{C}$ implies $\tau^\bullet \in c$. We have shown $s_\tau^{(c)}(\Delta) = s_\tau^{(c)}$ if and only if $\tau \in \widehat{C}$ and we conclude, since $\widehat{C} = \widehat{\Gamma}$ by Parts (i) and (ii) of Proposition 4.3.
This completes the proof of Part (iii).
(iv) We have $\Delta_\tau \in (c, c)$ and $\lim x = \Delta_\tau - \lim x$ if and only if $\lim_{n\to\infty} \tau_n^\bullet = 0$. We conclude using Part (iii), since $\lim_{n\to\infty} \tau_n^\bullet = 0$ implies $\tau \in \widehat{C}$. □

Remark 4.5 In Part (iv) of Theorem 4.2, we saw that $\Sigma_\tau \in (c, c)$ and $\lim x = \Sigma_\tau - \lim x$ if and only if $\tau_n^\bullet \to 0$ $(n \to \infty)$. Indeed, we must have $\sigma_{nk} = \tau_k/\tau_n = o(1)$ $(n \to \infty)$ for all k and

$$\lim_{n\to\infty} \left(\sum_{k=1}^{n} \sigma_{nk} \right) = \lim_{n\to\infty} \left(1 + \sum_{k=1}^{n-1} \frac{\tau_k}{\tau_n} \right) = 1;$$

and from Part (i) of Proposition 4.3, the previous property is satisfied if and only if $\tau_n^\bullet \to 0$ $(n \to \infty)$.

Remark 4.6 It can be seen that the condition $\tau^{\bullet} \in c$ does not imply $\tau \in \widehat{C_1}$. To see this, it is enough to notice that $C(e)e = (n)_{n=1}^{\infty} \notin c_0$.

4.5 The Spaces $w_\tau(\lambda)$, $\overset{\circ}{w}_\tau(\lambda)$ and $w_\tau^{\bullet}(\lambda)$

In this section, we establish some properties of the sets $w_\tau(\lambda)$, $\overset{\circ}{w}_\tau(\lambda)$, $w_\tau^{\bullet}(\lambda)$. We also consider the sum of sets of the form $w_\tau(\lambda)$, and show that under some conditions, the spaces $w_\tau(\lambda)$, $\overset{\circ}{w}_\tau(\lambda)$ and $w_\tau^{\bullet}(\lambda)$ can be written in the form s_ξ or s_ξ^0. We need to recall some definitions and properties. For every sequence $x = (x_n)_{n=1}^{\infty}$, we write $|x| = (|x_n|)_{n=1}^{\infty}$ and define

$$
\begin{aligned}
w_\tau(\lambda) &= \{x \in \omega : C(\lambda)(|x|) \in s_\tau\}, \\
\overset{\circ}{w}_\tau(\lambda) &= \left\{x \in \omega : C(\lambda)(|x|) \in s_\tau^0\right\}, \\
w_\tau^{\bullet}(\lambda) &= \left\{x \in \omega : x - le \in \overset{\circ}{w}_\tau(\lambda) \text{ for some } l \in \mathbb{C}\right\}.
\end{aligned}
$$

For instance, we see that

$$
w_\tau(\lambda) = \left\{ x = (x_n)_{n=1}^{\infty} \in \omega : \sup_n \left(\frac{1}{|\lambda_n| \tau_n} \sum_{k=1}^{n} |x_k| \right) < \infty \right\}.
$$

If there exist $A, B > 0$, such that $A \leq \tau_n \leq B$ for all n, we obtain the well-known spaces $w_\tau(\lambda) = w_\infty(\lambda)$, $\overset{\circ}{w}_\tau(\lambda) = w_0(\lambda)$ and $w_\tau^{\bullet}(\lambda) = w(\lambda)$. It was proved that if λ is a *strictly increasing sequence of reals tending to infinity* the sets $w_0(\lambda)$ and $w_\infty(\lambda)$ are BK spaces and $w_0(\lambda)$ has AK with respect to the norm

$$
\|x\| = \|C(\lambda)\,|x|\,\|_{s_1} = \sup_n \left(\frac{1}{\lambda_n} \sum_{k=1}^{n} |x_k| \right).
$$

We have the next result.

Theorem 4.3 ([7, Theorem 1, p. 18]) *Let $\tau, \lambda \in \mathcal{U}^+$. Then we have*

(i) We consider the following conditions:

(a) $\lim_{n\to\infty} (\tau\lambda)_n^{\bullet} = 0$;
(b) $s_\tau^{(c)}(C(\lambda)) = s_{\tau\lambda}^{(c)}$;
(c) $\tau\lambda \in \widehat{C_1}$;
(d) $w_\tau(\lambda) = s_{\tau\lambda}$;
(e) $\overset{\circ}{w}_\tau(\lambda) = s_{\tau\lambda}^0$;
(f) $w_\tau^{\bullet}(\lambda) = s_{\tau\lambda}^0$.

Then (a) is equivalent to (b), (c) is equivalent to (d), and (c) implies (e) and (f).

(ii) *If $\tau\lambda \in \widehat{C_1}$, then $w_\tau(\lambda)$, $\overset{\circ}{w}_\tau(\lambda)$ and $w_\tau^\bullet(\lambda)$ are BK spaces with respect to the norm*

$$\|x\|_{s_{\tau\lambda}} = \sup_n \left(\frac{|x_n|}{\tau_n\lambda_n}\right), \tag{4.20}$$

and $\overset{\circ}{w}_\tau(\lambda) = w_\tau^\bullet(\lambda)$ has AK.

Proof (i) We have

$$s_\tau^{(c)}(C(\lambda) = \Delta(\lambda)s_\tau^{(c)} = \Delta D_\lambda s_\tau^{(c)} = \Delta s_{\tau\lambda}^{(c)}.$$

This shows $s_\tau^{(c)}(C(\lambda)) = s_\tau^{(c)}$ is equivalent to $(s_{\tau\lambda}^{(c)})_\Delta = s_{\tau\lambda}^{(c)}$ and the equivalence of (a) and (b) follows from Part (iii) of Theorem 4.2.
Now we show the equivalence of (c) and (d).
First, we assume that (c) holds. Then we have

$$w_\tau(\lambda) = \{x : |x| \in \Delta(\lambda)s_\tau\}.$$

From the identity $\Delta(\lambda) = \Delta D_\lambda$, we obtain $\Delta(\lambda)s_\tau = \Delta s_{\tau\lambda}$. Now using (c) we see that Δ is bijective from $s_{\tau\lambda}$ into itself and $w_\tau(\lambda) = s_{\tau\lambda}$.
Conversely, assume $w_\tau(\lambda) = s_{\tau\lambda}$. Then $\tau\lambda \in s_{\tau\lambda}$ implies $C(\lambda)(\tau\lambda) \in s_\tau$, and since $D_{1/\tau}C(\lambda)(\tau\lambda) \in s_1 = \ell_\infty$ we conclude $C(\tau\lambda)(\tau\lambda) \in \ell_\infty$.
The proof that (c) implies (e) follows the same lines as that of (c) implies (d) by replacing $s_{\tau\lambda}$ by $s_{\tau\lambda}^0$.
Now we show that (c) implies (f). Let $x \in w_\tau^\bullet(\lambda)$. There is $l \in \mathbb{C}$ such that $C(\lambda)(|x - le|) \in s_\tau^0$. So we have

$$|x - le| \in \Delta(\lambda)s_\tau^0 = \Delta s_{\tau\lambda}^0,$$

and from Part (ii) of Theorem 4.2, we have $\Delta s_{\tau\lambda}^0 = s_{\tau\lambda}^0$. Now, since (c) holds we deduce from Part (iii) of Proposition 4.3 that $\tau_n\lambda_n \to \infty$ $(n \to \infty)$ and $le \in s_{\tau\lambda}^0$. We conclude $x \in w_\tau^\bullet(\lambda)$ if and only if $x \in le + s_{\tau\lambda}^0 = s_{\tau\lambda}^0$.
(ii) is a direct consequence of (i). □

Now we establish some conditions to obtain $w_\tau(\lambda) + w_\tau(\mu) = w_\tau(\lambda + \mu)$, $w_\tau(\lambda) + w_\nu(\lambda) = w_{\tau+\nu}(\lambda)$ or $w_{\tau\nu}(\lambda \cdot \mu) = w_\tau(\lambda) * w_\nu(\mu)$. We obtain the next result.

Lemma 4.7 ([9, Proposition 4, p. 251]) *Let $\tau, \nu \in \mathcal{U}^+$. Then we have*

(i) *The condition $s_\tau(\Delta) = s_\nu$ is equivalent to*

$$s_\tau = s_\nu \text{ and } \tau \in \widehat{C_1}.$$

(ii) (a) *For any given n, we have*

$$[C(\tau+\nu)(\tau+\nu)]_n \leq [C(\tau)\tau]_n + [C(\nu)\nu]_n \tag{4.21}$$

and

$$[C(\tau\nu)(\tau\nu)]_n \leq ([C(\tau)\tau]_n)([C(\nu)\nu]_n) \text{ for all } n.$$

(b) *If* $\tau, \nu \in \widehat{C_1}$, *then* $\tau+\nu, \tau\nu \in \widehat{C_1}$.

(iii) *We assume* $s_\tau = s_\nu$. *Then*

(a) $\tau, \nu \in \widehat{C_1}$ *if and only if* $\tau+\nu \in \widehat{C_1}$;
(b) $\tau \in \widehat{C_1}$ *if and only if* $\nu \in \widehat{C_1}$.

Proof (i) If Δ is bijective from s_τ into s_ν, then we have $\Delta \in (s_\tau, s_\nu)$ and $\Sigma = \Delta^{-1} \in (s_\nu, s_\tau)$. So $(\tau_{n-1}+\tau_n)/\nu_n = O(1)$ $(n \to \infty)$, and there is a real $M > 0$ such that

$$\frac{\nu_n}{\tau_n} \leq \frac{1}{\tau_n}\left(\sum_{k=1}^{n} \nu_k\right) \leq M \text{ for all } n.$$

Then $\tau_n = O(\nu_n)$ and $\nu_n = O(\tau_n)$ $(n \to \infty)$ and from Part (i) of Theorem 4.1, we conclude $s_\tau = s_\nu$ and $C(\tau)\tau \in \ell_\infty$.
The converse has been shown in Part (i) of Theorem 4.2.
(ii)

(a) The inequality in (4.21) is obvious, and

$$[C(\tau\nu)(\tau\nu)]_n = \frac{1}{\tau_n\nu_n}\left(\sum_{k=1}^{n} \tau_k\nu_k\right) \leq \left(\frac{1}{\tau_n}\sum_{k=1}^{n} \tau_k\right)\left(\frac{1}{\nu_n}\sum_{k=1}^{n} \nu_k\right)$$
$$\leq ([C(\tau)\tau]_n)([C(\nu)\nu]_n) \text{ for all } n.$$

(b) is a direct consequence of (a).

(iii)

(a) The necessity follows from (ii) (a).
Conversely, since $s_\tau = s_\nu$ there are $K_1, K_2 > 0$ such that

$$K_1 \leq \frac{\tau_n}{\nu_n} \leq K_2 \text{ for all } n. \tag{4.22}$$

So $\tau_n + \nu_n \leq (K_2+1)\nu_n$ and

$$\frac{1}{(K_2+1)\nu_n}\sum_{k=1}^{n}(\tau_k+\nu_k) \leq [C(\tau+\nu)(\tau+\nu)]_n \text{ for all } n.$$

From (4.22), we get $1/\nu_n \geq K_1/\tau_n$ and we conclude

$$K_1[C(\tau)\tau]_n + [C(\nu)\nu]_n \le (K_2+1)[C(\tau+\nu)(\tau+\nu)]_n \text{ for all } n$$

and $C(\tau)\tau$, $C(\nu)\nu \in \ell_\infty$.
So we have shown (a).

(b) Finally, since the condition in (4.22) implies

$$\frac{K_1}{K_2}[C(\nu)\nu]_n \le [C(\tau)\tau]_n \le \frac{K_2}{K_1}[C(\nu)\nu]_n \text{ for all } n,$$

we obtain (b).

This concludes the proof. □

We then have the next theorem.

Theorem 4.4 ([9, Theorem 1, p. 257]) *Let λ, μ, τ, $\nu \in \mathcal{U}^+$. Then we have*

(i) The conditions $\tau\lambda, \tau\mu \in \widehat{C_1}$ imply

$$w_\tau(\lambda+\mu) = w_\tau(\lambda) + w_\tau(\mu) = s_{\tau(\lambda+\mu)}. \tag{4.23}$$

(ii) The conditions $\tau\lambda, \nu\lambda \in \widehat{C_1}$ imply

$$w_{\tau+\nu}(\lambda) = w_\tau(\lambda) + w_\nu(\lambda) = s_{(\tau+\nu)\lambda}.$$

(iii) The conditions $\tau\lambda, \nu\mu \in \widehat{C_1}$ imply

$$w_{\tau\nu}(\lambda\cdot\mu) = w_\tau(\lambda) * w_\nu(\mu) = s_{\tau\nu\lambda\mu}.$$

(iv) (a) If $\tau\lambda, \tau\mu \in \Gamma$ then

$$w_\tau(\lambda+\mu) = w_\tau(\lambda) + w_\tau(\mu) = s_{\tau(\lambda+\mu)}.$$

(b) If $\tau\lambda, \nu\mu \in \Gamma$ then

$$w_{\tau\nu}(\lambda\cdot\mu) = w_\tau(\lambda) * w_\nu(\mu) = s_{\tau\nu\lambda\mu}.$$

Proof (i) Let $\tau\lambda, \tau\mu \in \widehat{C_1}$. By Part (i) of Theorem 4.3, we obtain $w_\tau(\lambda) = s_{\tau\lambda}$ and $w_\tau(\mu) = s_{\tau\mu}$. So we have

$$w_\tau(\lambda) + w_\tau(\mu) = s_{\tau(\lambda+\mu)},$$

and using Part (ii) of Lemma 4.7, we obtain $\tau(\lambda+\mu) \in \widehat{C_1}$. So $w_\tau(\lambda+\mu) = s_{\tau(\lambda+\mu)}$ and the condition in (4.23) holds.
(ii) We obtain from Part (i) of Theorem 4.3

$$w_\tau(\lambda) + w_\nu(\lambda) = s_{\tau\lambda} + s_{\nu\lambda} = s_{(\tau+\nu)\lambda}.$$

Furthermore, if $\tau\lambda, \nu\lambda \in \widehat{C_1}$, then we have

$$C((\tau+\nu)\lambda)((\tau+\nu)\lambda) \in \ell_\infty \text{ and } w_{\tau+\nu}(\lambda) = s_{(\tau+\nu)\lambda} = w_\tau(\lambda) + w_\nu(\lambda).$$

This completes the proof of Part (ii).
(iii) Let $\tau\lambda, \nu\mu \in \widehat{C_1}$. It follows from Part (i) of Theorem 4.3 and Part (ii) of Lemma 4.7 that $\tau\lambda\nu\mu \in \widehat{C_1}$ and $w_{\tau\nu}(\lambda\cdot\mu) = s_{\tau\nu\lambda\mu}$. We conclude

$$w_\tau(\lambda) * w_\nu(\mu) = s_{\tau\lambda} * s_{\nu\mu} = s_{\tau\nu\lambda\mu} = w_{\tau\nu}(\lambda\cdot\mu).$$

(iv) (a) follows from Part (i), since $\tau\lambda, \tau\mu \in \Gamma$ together imply $\tau\lambda, \tau\mu \in \widehat{C_1}$.
(iv) (b) can be obtained similarly using Part (iii). □

Remark 4.7 We note that $w_\tau(\lambda) * w_\nu(\mu) \subset w_{\tau\nu}(\lambda\cdot\mu)$ for all $\lambda, \mu \in \mathcal{U}^+$. Indeed, let $x = y\cdot z \in w_\tau(\lambda) * w_\nu(\mu)$ with $y = (y_n)_{n=1}^\infty$, $z = (z_n)_{n=1}^\infty$. We obtain $C(\lambda)(|y|) \in s_\tau$ and $C(\mu)(|z|) \in s_\nu$ and since

$$\frac{1}{\lambda_n\mu_n}\left(\sum_{k=1}^n |y_k z_k|\right) \le \left[\frac{1}{\lambda_n}\left(\sum_{k=1}^n |y_k|\right)\right]\left[\left(\frac{1}{\mu_n}\left(\sum_{k=1}^n |z_k|\right)\right)\right] = \tau_n\nu_n O(1)\ (n\to\infty),$$

we conclude $x \in w_{\tau\nu}(\lambda\mu)$.

Using some results given in Lemma 4.7, we obtain the following results.

Proposition 4.4 ([9, Proposition 6, p. 258]) *Let $\lambda, \mu, \tau, \nu \in \mathcal{U}^+$. Then we have*

(i) We assume $s_\tau = s_\nu$.

(a) The condition $\tau\lambda, \nu\lambda \in \widehat{C_1}$ is equivalent to

$$w_{\tau+\nu}(\lambda) = w_\tau(\lambda) + w_\nu(\lambda) = s_{\tau\lambda}.$$

(b) If $\lambda_n \sim \mu_n$ $(n\to\infty)$, then the condition $\tau\lambda, \nu\mu \in \widehat{C_1}$ is equivalent to

$$w_{\tau+\nu}(\mu) = w_\tau(\lambda) + w_\nu(\mu) = s_{\tau\lambda}.$$

(ii) If $\lambda_n \sim \mu_n$ $(n\to\infty)$ and $\tau\lambda, \nu\mu \in \widehat{C_1}$, then

$$w_{\tau+\nu}(\mu) = w_\tau(\lambda) + w_\nu(\mu) = s_{(\tau+\nu)\lambda}.$$

Proof (i)

(a) The necessity was shown in Theorem 4.4.
Conversely, assume $w_{\tau+\nu}(\lambda) = s_{\tau\lambda}$. The condition $s_\tau = s_\nu$ implies $s_{\tau\lambda} = s_{\nu\lambda}$ and $s_{\tau\lambda} = s_{(\tau+\nu)\lambda}$. So $w_{\tau+\nu}(\lambda) = s_{(\tau+\nu)\lambda}$ and $(\tau+\nu)\lambda \in \widehat{C_1}$. From Part (iii) (a) of Lemma 4.7, we deduce $\tau\lambda, \nu\lambda \in \widehat{C_1}$, so $w_\tau(\lambda) = s_{\tau\lambda}$, $w_\nu(\lambda) = s_{\nu\lambda}$ and $w_{\tau+\nu}(\mu) = w_\tau(\lambda) + w_\nu(\lambda) = s_{(\tau+\nu)\lambda}$.

(b) *Necessity*. The conditions $\tau\lambda, \nu\mu \in \widehat{C_1}$ together imply

$$w_\tau(\lambda) + w_\nu(\mu) = s_{\tau\lambda} + s_{\nu\mu},$$

and since $\lambda_n \sim \mu_n$ $(n \to \infty)$, we obtain $s_{\nu\mu} = s_{\nu\lambda}$. So we have

$$w_\tau(\lambda) + w_\nu(\mu) = w_{\tau+\nu}(\mu) = s_{(\tau+\nu)\lambda}$$

and we conclude since $s_\tau = s_\nu$ implies $s_{\tau\lambda} = s_{(\tau+\nu)\lambda}$.
Conversely, as above, the condition $s_\tau = s_\nu$ implies $s_{\tau\lambda} = s_{(\tau+\nu)\lambda}$, and the identity $w_{\tau+\nu}(\mu) = s_{(\tau+\nu)\lambda}$ implies $(\tau+\nu)\lambda \in \widehat{C_1}$, and using Part (ii) of Lemma 4.7, we deduce $\tau\lambda, \nu\lambda \in \widehat{C_1}$. Furthermore, the condition $\lambda_n \sim \mu_n$ as $n \to \infty$ implies $s_\lambda = s_\mu$ and $s_{\nu\lambda} = s_{\nu\mu}$. So, by Part (iii) (b) of Lemma 4.7, the condition $\nu\lambda \in \widehat{C_1}$ implies $\nu\mu \in \widehat{C_1}$.

This concludes the proof of Part (i).
(ii) The conditions $\tau\lambda, \nu\mu \in \widehat{C_1}$ imply $\tau\lambda + \nu\mu \in \widehat{C_1}$. Since $s_\lambda = s_\mu$, we have $s_{\tau\lambda} = s_{\tau\mu}$ and $s_{\tau\lambda+\nu\mu} = s_{(\tau+\nu)\mu}$. So using Part (iii) of Lemma 4.7, we have $\tau\lambda + \nu\mu \in \widehat{C_1}$ if and only if $(\tau+\nu)\lambda \in \widehat{C_1}$ holds. Finally, we obtain

$$w_\tau(\lambda) + w_\nu(\mu) = s_{\tau\lambda+\nu\mu} = s_{(\tau+\nu)\mu} = s_{\tau\lambda+\nu\mu}.$$

□

Remark 4.8 In Part (ii) of Proposition 4.4, we can replace the condition $\tau\lambda, \nu\mu \in \widehat{C_1}$ by $\tau\lambda, \nu\mu \in \Gamma$.

4.6 Matrix Transformations From $w_\tau(\lambda) + w_\nu(\mu)$ into s_γ

We obtain the following result by Proposition 4.4.

Proposition 4.5 ([9, Proposition 7, p. 262]) *Let $\tau, \nu, \gamma, \lambda, \mu \in \mathcal{U}^+$. We assume $\lambda_n \sim \mu_n$ $(n \to \infty)$ and $\tau\lambda, \nu\mu \in \widehat{C_1}$. Then we have*

(i) $A \in (w_\tau(\lambda) + w_\nu(\mu), s_\gamma)$ if and only if

$$\sup_n \left(\frac{1}{\gamma_n} \sum_{k=1}^{\infty} |a_{nk}| (\tau_k + \nu_k) \mu_k \right) < \infty; \tag{4.24}$$

(ii) $A \in (c + w_\tau(\lambda) + w_\nu(\mu), s_\gamma)$ if and only if (4.24) holds.

Proof (i) By Part (ii) of Proposition 4.4, we obtain $w_\tau(\lambda) + w_\nu(\mu) = s_{(\tau+\nu)\lambda}$ and

$$(w_\tau(\lambda) + w_\nu(\mu), s_\gamma) = S_{(\tau+\nu)\lambda,\gamma}.$$

(ii) By Part (ii) (b) of Lemma 4.7, we have $\tau\lambda + \nu\mu \in \widehat{C_1}$. Then the condition $\lambda_n \sim \mu_n$ $(n \to \infty)$ implies $s_{\tau\lambda} = s_{\tau\mu}$ and $s_{\tau\lambda+\nu\mu} = s_{(\tau+\nu)\mu}$, and by Part (iii) (b) of Lemma 4.7, we obtain $(\tau + \nu)\mu \in \widehat{C_1}$. So there is $M > 0$ such that

$$\frac{(\tau_1 + \nu_1)\mu_1}{(\tau_n + \nu_n)\mu_n} \leq [C((\tau + \nu)\mu)((\tau + \nu)\mu)]_n \leq M \text{ for all } n,$$

which implies $1/(\tau + \nu)\mu \in \ell_\infty$. Using Part (vi) (b) of Theorem 4.1, we deduce $c + w_\tau(\lambda) + w_\nu(\mu) = w_\tau(\lambda) + w_\nu(\mu)$ and we conclude applying Part (i). □

Example 4.1 ([9, Example 1, p. 262]) Under the hypotheses of the previous proposition it can easily be seen that

$$(w_\tau(\lambda) + w_\nu(\mu), (c + s_\tau) * s_\nu) = S_{(\tau+\nu)\lambda(\tau+e)\nu}.$$

Indeed, from Part (iv) of Proposition 4.2, we have $(c + s_\tau) * s_\nu = c * s_\nu + s_\tau * s_\nu = s_\nu + s_{\tau\nu}$ and as above we obtain $w_\tau(\lambda) + w_\nu(\mu) = s_{(\tau+\nu)\lambda}$. So we have shown

$$Ax + Ay \in (c + s_\tau) * s_\nu \text{ for all } x \in w_\tau(\lambda) \text{ and } y \in w_\nu(\mu)$$

if and only if

$$\sup_n \left(\frac{1}{(\tau_n + 1)\nu_n} \sum_{k=1}^{\infty} |a_{nk}|(\tau_k + \nu_k)\lambda_k \right) < \infty.$$

4.7 On the Sets $c_\tau(\lambda, \mu)$, $\overset{\circ}{c}_\tau(\lambda, \mu)$ and $c_\tau^\bullet(\lambda, \mu)$

In this section, we deal with spaces that generalize the sets $c_0(\lambda)$, $c(\lambda)$ and $c_\infty(\lambda)$. We will see that under some conditions, the spaces $c_\tau(\lambda, \mu)$, $\overset{\circ}{c}_\tau(\lambda, \mu)$ and $c_\tau^\bullet(\lambda, \mu)$ can be written in the form s_ξ, s_ξ^0, or $s_\xi^{(c)}$.

Let $\tau = (\tau_n)_{n=1}^\infty \in \mathcal{U}^+$ and let $\lambda \in \mathcal{U}$ and $\mu \in \omega$. We consider the set

$$c_\tau(\lambda, \mu) = (w_\tau(\lambda))_{\Delta(\mu)} = \{x \in \omega : \Delta(\mu)x \in w_\tau(\lambda)\}.$$

It is easy to see that

$$c_\tau(\lambda, \mu) = \{x \in \omega : C(\lambda)(|\Delta(\mu)x|) \in s_\tau\},$$

that is,

$$c_\tau(\lambda, \mu) = \left\{ x = (x_n)_{n=1}^\infty \in \omega : \sup_n \left(\frac{1}{|\lambda_n|\tau_n} \sum_{k=1}^{n} |\mu_k x_k - \mu_{k-1}x_{k-1}| \right) < \infty \right\},$$

with the convention $x_{-1} = 0$. Similarly, we define the following sets:

$$\overset{\circ}{c}_\tau(\lambda, \mu) = \left\{x \in \omega : C(\lambda)(|\Delta(\mu)x|) \in s_\tau^0\right\},$$
$$c_\tau^\bullet(\lambda, \mu) = \left\{x \in \omega : x - le \in \overset{\circ}{c}_\tau(\lambda\mu) \text{ for some } l \in \mathbb{C}\right\}.$$

We recall that for $\lambda = \mu$, we write $c_0(\lambda) = (w_0(\lambda))_{\Delta(\lambda)}$,

$$c(\lambda) = \{x \in \omega : x - le \in c_0(\lambda) \text{ for some } l \in \mathbb{C}\},$$

and $c_\infty(\lambda) = (w_\infty(\lambda))_{\Delta(\lambda)}$, see [28, 31]. It can easily be seen that

$$c_0(\lambda) = \overset{\circ}{c}_e(\lambda, \lambda),\ c_\infty(\lambda) = c_e(\lambda, \lambda) \text{ and } c(\lambda) = c_e^\bullet(\lambda, \lambda).$$

These sets of sequences are called *strongly convergent to* 0, *strongly convergent* and *strongly bounded*. If $\lambda \in \mathcal{U}^+$ is a sequence *strictly increasing to infinity*, $c(\lambda)$ is a Banach space with respect to

$$\|x\|_{c_\infty(\lambda)} = \sup_n \left(\frac{1}{\lambda_n} \sum_{k=1}^n |\lambda_k x_k - \lambda_{k-1} x_{k-1}|\right)$$

with the convention $x_0 = 0$. Each of the spaces $c_0(\lambda)$, $c(\lambda)$ and $c_\infty(\lambda)$ is a BK space with the previous norm (see [28]); $c_0(\lambda)$ has AK and every $x \in c(\lambda)$ has a unique representation given by

$$x = le + \sum_{k=1}^\infty (x_k - l)e^{(k)}, \tag{4.25}$$

where $x - le \in c_0(\lambda)$. The number l is called *the strong $c(\lambda)$-limit* of the sequence x. We obtain the next result.

Theorem 4.5 ([6, Theorem 4.2, p. 1798]) *Let $\tau, \lambda, \mu \in \mathcal{U}^+$.*

(i) We consider the following properties:

(a) $\tau\lambda \in \widehat{C_1}$,
(b) $c_\tau(\lambda, \mu) = s_{\tau\frac{\lambda}{\mu}}$,
(c) $\overset{\circ}{c}_\tau(\lambda, \mu) = s^0_{\tau\frac{\lambda}{\mu}}$,
(d) $c_\tau^\bullet(\lambda, \mu) = \{x \in \omega : x - le \in s^0_{\tau\frac{\lambda}{\mu}}$ *for some* $l \in \mathbb{C}\}$.

Then (a) is equivalent to (b), (a) implies (c) and (a) implies (d).

(ii) If $\tau\lambda \in \widehat{C_1}$, then $c_\tau(\lambda, \mu)$, $\overset{\circ}{c}_\tau(\lambda, \mu)$ and $c_\tau^\bullet(\lambda, \mu)$ are BK spaces with respect to the norm

$$\|x\|_{s_{\tau\frac{\lambda}{\mu}}} = \sup_n \left(\mu_n \frac{|x_n|}{\tau_n \lambda_n}\right);$$

$\overset{\circ}{c}_\tau(\lambda, \mu)$ has AK and every $x \in c_\tau^\bullet(\lambda, \mu)$ has a unique representation (4.25), where $x - le \in s^0_{\tau\lambda/\mu}$.

Proof (i) We show that (a) implies (b).
Let $x \in c_\tau(\lambda, \mu)$. We have $\Delta(\mu)x \in w_\tau(\lambda)$, which is equivalent to

$$x \in C(\mu)s_{\tau\lambda} = D_{1/\mu}\Sigma s_{\tau\lambda},$$

and by Part (i) of Theorem 4.2, each of the operators represented by Δ and $\Sigma = \Delta^{-1}$ is bijective from $s_{\tau\lambda}$ into itself. So we have $\Sigma s_{\tau\lambda} = s_{\tau\lambda}$ and $x \in D_{1/\mu}\Sigma s_{\tau\lambda} = s_{\tau\lambda/\mu}$. Hence (b) holds.
Now we show that (b) implies (a).
First we let

$$\widetilde{\tau}_{\lambda,\mu} = ((-1)^n \tau_n \lambda_n/\mu_n)_{n=1}^\infty.$$

We have $\widetilde{\tau}_{\lambda,\mu} \in s_{\tau\lambda/\mu} = c_\tau(\lambda, \mu)$. Since

$$\Delta(\mu) = \Delta D_\mu \text{ and } D_\mu \widetilde{\tau}_{\lambda,\mu} = ((-1)^n \tau_n \lambda_n)_{n=1}^\infty,$$

we obtain $|\Delta(\mu)\widetilde{\tau}_{\lambda,\mu}| = (\xi_n)_{n=1}^\infty$, with

$$\xi_n = \begin{cases} \lambda_1\tau_1 & \text{if } n = 1 \\ \lambda_{n-1}\tau_{n-1} + \lambda_n\tau_n & \text{if } n \geq 2. \end{cases}$$

We deduce $\Sigma|\Delta(\mu)\widetilde{\tau}_{\lambda,\mu}| \in s_{\tau\lambda}$. This means

$$C_n' = \frac{1}{\tau_n\lambda_n}\left(\lambda_1\tau_1 + \sum_{k=2}^n (\lambda_{k-1}\tau_{k-1} + \lambda_k\tau_k)\right) = O(1)(n \to \infty).$$

From the inequality

$$[C(\tau\lambda)(\tau\lambda)]_n \leq C_n' \text{ for all } n$$

we obtain (a).
The proof of (a) implies (c) follows the same lines as that of (a) implies (b) with s_τ replaced by s^0_τ.
Now we show that (a) implies (d). Let $x \in c_\tau^\bullet(\lambda, \mu)$. Then there exists $l \in \mathbb{C}$ such that

$$\Delta(\mu)(x - le) \in \overset{\circ}{w}_\tau(\lambda)$$

and since (c) implies Part (e) in Theorem 4.3, we have $\overset{\circ}{w}_\tau(\lambda) = s^0_{\tau\lambda}$. So we have

$$x - le \in C(\mu)s^0_{\tau\lambda} = D_{\frac{1}{\mu}}\Sigma s^0_{\tau\lambda},$$

and from Theorem 4.2, we have $\Sigma s^0_{\tau\lambda} = s^0_{\tau\lambda}$ and $D_{1/\mu}\Sigma s^0_{\tau\lambda} = s^0_{\tau\lambda/\mu}$. We conclude that $x \in c_\tau^\bullet(\lambda,\mu)$ if and only if

$$x \in le + s^0_{\tau\lambda/\mu} \text{ for some } l \in \mathbb{C}.$$

(ii) is a direct consequence of (i) and of the fact that for every $x \in c_\tau^\bullet(\lambda,\mu)$, we have

$$\left\| x - le - \sum_{k=1}^{N}(x_k - l)e^{(k)} \right\|_{s_{\tau\frac{\lambda}{\mu}}} = \sup_{n \geq N+1}\left(\mu_n \frac{|x_n - l|}{\tau_n \lambda_n}\right) = o(1)(N \to \infty).$$

This completes the proof. □

We immediately obtain the following Corollary.

Corollary 4.3 ([6, Theorem 4.3, pp. 1799–1800]) *Let $\tau, \lambda, \mu \in \mathcal{U}^+$. Then we have*

(i) If $\tau\lambda \in \widehat{C_1}$ and $\mu \in \ell_\infty$, then

$$c_\tau^\bullet(\lambda,\mu) = s^0_{\tau\frac{\lambda}{\mu}}. \tag{4.26}$$

(ii) We have $\lambda \in \Gamma$ implies $\lambda \in \widehat{C_1}$, and $\lambda \in \widehat{C_1}$ implies $c_0(\lambda) = s^0_\lambda$ and $c_\infty(\lambda) = s_\lambda$.

Proof (i) Since $\mu \in \ell_\infty$, by Part (iii) of Proposition 4.3, we deduce that there are $K > 0$ and $\gamma > 1$ such that

$$\frac{\tau_n \lambda_n}{\mu_n} \geq K\gamma^n \text{ for all } n.$$

So $le \in s^0_{\tau\lambda/\mu}$ and (4.26) holds.
(ii) follows from Theorem 4.5, since $\Gamma \subset \widehat{C_1}$. □

4.8 Sets of Sequences of the Form $[A_1, A_2]$

In this section, we consider spaces that generalize the sets given in Sect. 4.7 and establish necessary conditions to reduce them to the form s_ξ or s^0_ξ. In this section, we deal with the sets

$$[A_1(\lambda), A_2(\mu)] = \{x \in \omega : A_1(\lambda)(|A_2(\mu)x|) \in s_\tau\},$$

where A_1 and A_2 are of the form $C(\xi)$, $C^+(\xi)$, $\Delta(\xi)$ or $\Delta^+(\xi)$ and we give necessary conditions to obtain $[A_1(\lambda), A_2(\mu)]$ in the form s_γ.

Let $\lambda, \mu \in \mathcal{U}^+$. Throughout this section, we write

$$[A_1, A_2] = [A_1(\lambda), A_2(\mu)]$$

for any matrices

$$A_1(\lambda) \in \{\Delta(\lambda), \Delta^+(\lambda), C(\lambda), C^+(\lambda)\} \text{ and } A_2(\mu) \in \{\Delta(\mu), \Delta^+(\mu), C(\mu), C^+(\mu)\}$$

to simplify. So we have, for instance,

$$[C, \Delta] = \{x \in \omega : C(\lambda)(|\Delta(\mu)x|) \in s_\tau\} = (w_\tau(\lambda))_{\Delta(\mu)}, \text{etc.}$$

Now we consider the spaces $[C, C]$, $[C, \Delta]$, $[\Delta, C]$ and $[\Delta, \Delta]$. For the reader's convenience, we obtain the next identities, where $A_1(\lambda)$ and $A_2(\mu)$ are triangles and we use the convention $\mu_{-1} = 0$.

$$[C, C] = \left\{x \in \omega : \frac{1}{\lambda_n}\left(\sum_{m=1}^{n}\left|\frac{1}{\mu_m}\left(\sum_{k=1}^{m} x_k\right)\right|\right) = \tau_n O(1)\ (n \to \infty)\right\},$$

$$[C, \Delta] = \left\{x \in \omega : \frac{1}{\lambda_n}\left(\sum_{k=1}^{n}|\mu_k x_k - \mu_{k-1}x_{k-1}|\right) = \tau_n O(1)\ (n \to \infty)\right\},$$

$$[\Delta, C] = \left\{x \in \omega : -\lambda_{n-1}\left|\frac{1}{\mu_{n-1}}\left(\sum_{k=1}^{n-1} x_k\right)\right| + \lambda_n\left|\frac{1}{\mu_n}\left(\sum_{k=1}^{n} x_k\right)\right| = \tau_n O(1)\right.$$
$$\left.(n \to \infty)\right\}$$

and

$$[\Delta, \Delta] =$$
$$\{x \in \omega : -\lambda_{n-1}|\mu_{n-1}x_{n-1} - \mu_{n-2}x_{n-2}| + \lambda_n|\mu_n x_n - \mu_{n-1}x_{n-1}|$$
$$= \tau_n O(1)\ (n \to \infty)\}.$$

We note that for $\tau = e$ and $\lambda = \mu$, $[C, \Delta]$ is the well-known set of sequences that are strongly bounded, denoted $c_\infty(\lambda)$, see [28]. We obtain the following results from Proposition 4.3 and Theorem 4.5.

Theorem 4.6 ([8, Theorem 3.1, p. 1665]) *Let $\tau, \lambda, \mu \in \mathcal{U}^+$. Then we have*

(i) If $\tau\lambda, \tau\lambda\mu \in \Gamma$, then

$$[C, C] = s_{(\tau\lambda\mu)}.$$

(ii) If $\tau\lambda \in \Gamma$, then

$$[C, \Delta] = s_{(\tau \frac{\lambda}{\mu})}.$$

(iii) *If* $\tau, \tau\mu/\lambda \in \Gamma$, *then*

$$[\Delta, C] = s_{(\tau \frac{\mu}{\lambda})}.$$

(iv) *If* $\tau, \tau/\lambda \in \Gamma$, *then*

$$[\Delta, \Delta] = s_{(\frac{\tau}{\lambda\mu})}.$$

Proof (i) For any given x we have $C(\lambda)(|C(\mu)x|) \in s_\tau$ if and only if

$$C(\mu)x \in s_\tau(C(\lambda)).$$

Since $\tau\lambda \in \Gamma$ we have $s_\tau(C(\lambda)) = s_{(\tau\lambda)}$. So we obtain

$$x \in \Delta(\mu)s_{\tau\lambda}$$

and the condition $\tau\lambda\mu \in \Gamma$ implies $\Delta(\mu)s_{\tau\lambda} = s_{(\tau\lambda\mu)}$. So we have shown Part (i)
(ii) is a direct consequence of Theorem 4.5 (i).
(iii) Similarly, we have $\Delta(\lambda)(|C(\mu)x|) \in s_\tau$ if and only if

$$|C(\mu)x| \in s_\tau(\Delta(\lambda)) = C(\lambda)s_\tau.$$

Since $\tau \in \Gamma$, we obtain $C(\lambda s_\tau = D_{1/\lambda}\Sigma s_\tau = s_{(\tau/\lambda)}$. So we have for any given x

$$x \in \Delta(\mu)s_{(\frac{\tau}{\lambda})} = \Delta s_{(\frac{\tau\mu}{\lambda})}.$$

We conclude, since $\tau\mu/\lambda \in \Gamma$ implies $\Delta s_{(\tau\mu/\lambda)} = s_{(\tau\mu/\lambda)}$.
(iv) Since $\tau \in \Gamma$, we have as above $C(\lambda)s_\tau = s_{(\tau/\lambda)}$. So we obtain $\Delta(\lambda)(|\Delta(\mu)x|) \in s_\tau$ if and only if

$$\Delta(\mu)x \in C(\lambda)s_\tau = s_{(\frac{\tau}{\lambda})}.$$

Then we have

$$x \in C(\mu)s_{(\frac{\tau}{\lambda})} = s_{(\frac{\tau}{\lambda\mu})},$$

since $\tau/\lambda \in \Gamma$. So Part (iv) holds. This concludes the proof. □

Remark 4.9 We have by Theorem 4.5

$$\tau\lambda \in \widehat{C_1} \text{ if and only if } [C, \Delta] = s_{(\tau \frac{\lambda}{\mu})}.$$

We obtain as direct consequence of Lemma 4.6

Corollary 4.4 *Under the condition* $\tau\lambda \in c^\bullet$, *we have*

$$\tau\lambda \in \Gamma \textit{ if and only if } [C, \Delta] = s_{(\tau \frac{\lambda}{\mu})}.$$

Remark 4.10 We obtain the same results as in Theorem 4.6 for the set

$$[A_1, A_2]_0 = \left\{x \in \omega : A_1(\lambda)(|A_2(\mu)x|) \in s_\tau^0\right\},$$

replacing in each of the cases (i), (ii), (iii) and (iv) the set s_ξ by the set s_ξ^0.

Now we study the sets $[\Delta, \Delta^+]$, $[\Delta, C^+]$, $[C, \Delta^+]$, $[\Delta^+\Delta]$, $[\Delta^+, C]$, $[\Delta^+\Delta^+]$, $[C^+, C]$, $[C^+, \Delta]$, $[C^+, \Delta^+]$ and $[C^+, C^+]$. We immediately obtain the following from the definitions of the operators $\Delta(\xi)$, $\Delta^+(\eta)$, $C(\xi)$ and $C^+(\eta)$:

$$= \{x : \lambda_n|\mu_n(x_n - x_{n+1})| - \lambda_{n-1}|\mu_{n-1}(x_{n-1} - x_n)| = \tau_n O(1) \quad (n \to \infty)\},$$

$$[\Delta, C^+] = \left\{x : \lambda_n \left|\sum_{i=n}^{\infty} \frac{x_i}{\mu_i}\right| - \lambda_{n-1}\left|\sum_{i=n-1}^{\infty} \frac{x_i}{\mu_i}\right| = \tau_n O(1) \ (n \to \infty)\right\},$$

$$[\Delta^+, \Delta] = \{x : \lambda_n|\mu_n x_n - \mu_{n-1}x_{n-1}| - \lambda_n|\mu_{n+1}x_{n+1} - \mu_n x_n| = \tau_n O(1) \quad (n \to \infty)\},$$

$$[\Delta^+, C] = \left\{x : \lambda_n \left(\frac{1}{\mu_n}\left|\sum_{i=1}^{n} x_i\right| - \frac{1}{\mu_{n+1}}\left|\sum_{i=1}^{n+1} x_i\right|\right) = \tau_n O(1) \ (n \to \infty)\right\},$$

$$[\Delta^+, \Delta^+] = \{x : \lambda_n\mu_n|x_n - x_{n+1}| - \lambda_n\mu_{n+1}|x_{n+1} - x_{n+2}| = \tau_n O(1) \ (n \to \infty)\},$$

$$[C^+, C] = \left\{x : \sum_{k=n}^{\infty}\left(\frac{1}{\lambda_k}\left|\frac{1}{\mu_k}\sum_{i=1}^{k} x_i\right|\right) = \tau_n O(1) \ (n \to \infty)\right\},$$

$$[C^+, C^+] = \left\{x : \sum_{k=n}^{\infty}\left(\frac{1}{\lambda_k}\left|\sum_{i=k}^{\infty} \frac{x_i}{\mu_i}\right|\right) = \tau_n O(1) \ (n \to \infty)\right\}.$$

The spaces $[C^+, \Delta]$, $[C^+, \Delta^+]$ and $[C, \Delta^+]$ will be generalized in Sect. 4.9, where we will see that $[C^+, \Delta] = c_\tau^{+.1}(\lambda, \mu)$, $[C^+, \Delta^+] = c_\tau^{+1}(\lambda, \mu)$ and $[C, \Delta^+] = c_\tau^{;+1}(\lambda, \mu)$.

Given any sequence $\nu = (\nu_n)_{n=1}^{\infty} \in \mathcal{U}^+$ we write $\nu^- = (\nu_{n-1})_{n=1}^{\infty}$ with the convention $\nu_0 = 0$. We state the following result, in which we use the convention $\tau_n = \mu_n = \nu_n = 1$ for $n < 0$.

Theorem 4.7 ([8, Theorem 3.3, pp. 1666–1667])

(i) *Let $\tau \in \Gamma$. Then we have*

$$[\Delta, \Delta^+] = s_{\left(\frac{\tau}{\lambda\mu}\right)^-} \text{ if } \frac{\tau}{\lambda\mu} \in \Gamma, \tag{a}$$

$$[\Delta, C^+] = s_{\left(\tau\frac{\mu}{\lambda}\right)} \text{ if } \tau, \frac{\lambda}{\tau} \in \Gamma. \tag{b}$$

(ii) *If $\tau\lambda$ and $\tau\lambda/\mu \in \Gamma$, then*

$$[C, \Delta^+] = s_{(\tau \frac{\lambda}{\mu})^-}.$$

(iii) *If* $\tau/\lambda \in \Gamma$, *then*

$$[\Delta^+, \Delta] = s_{(\frac{\tau_{n-1}}{\mu_n \lambda_{n-1}})_n} = s_{(\frac{1}{\mu}(\frac{\tau}{\lambda})^-)}.$$

(iv) *If* $\tau/\lambda, \mu(\tau/\lambda)^- = (\mu_n \tau_{n-1}/\lambda_{n-1})_n \in \Gamma$, *then*

$$[\Delta^+, C] = s_{\mu(\frac{\tau}{\lambda})^-}.$$

(v) *If* $\tau/\lambda, (\tau/\lambda)^- \mu^{-1} = (\tau_{n-1}(\mu_n \lambda_{n-1})^{-1})_n \in \Gamma$, *then*

$$[\Delta^+, \Delta^+] = s_{(\frac{(\frac{\tau}{\lambda})^-}{\mu})^-} = s_{(\frac{\tau_{n-2}}{\lambda_{n-2}\mu_{n-1}})_n}.$$

(vi) *If* $1/\tau, \tau\lambda\mu \in \Gamma$, *then*

$$[C^+, C] = s_{(\tau\lambda\mu)}.$$

(vii) *If* $1/\tau, \tau\lambda \in \Gamma$, *then*

$$[C^+, \Delta] = s_{(\tau \frac{\lambda}{\mu})}.$$

(viii) *If* $1/\tau, \tau\lambda/\mu \in \Gamma$, *then*

$$[C^+, \Delta^+] = s_{(\tau \frac{\lambda}{\mu})^-}.$$

(ix) *If* $1/\tau, 1/\tau\lambda \in \Gamma$, *then*

$$[C^+, C^+] = s_{(\tau\lambda\mu)}.$$

Proof These results follow from the next properties. If $\alpha \in \mathcal{U}^+$, then we have $s_\alpha(\Delta^+) = s_{\alpha^-}(\Delta)$ and the condition $\alpha \in \Gamma$ implies $s_\alpha(\Delta^+) = s_{\alpha^-}$. Then it can easily be shown that if $\alpha \in \Gamma^+$ then we have $s_\alpha(\Sigma^+) = s_\alpha$. Indeed, the inclusion $s_\alpha(\Sigma^+) \subset s_\alpha$ is immediate, since $(\Sigma^+)^{-1} = \Delta^+ \in (s_\alpha, s_\alpha)$. Then by Proposition 4.3 (v), we have $\alpha \in \widehat{C_1^+}$ which implies $\Sigma^+ \in (s_\alpha, s_\alpha)$ and $s_\alpha \subset s_\alpha(\Sigma^+)$.

(i) (a) First, for any given x, the condition $\Delta(\lambda)(|\Delta^+(\mu)x|) \in s_\tau$ is equivalent to

$$|\Delta^+(\mu)x| \in s_\tau(\Delta(\lambda)).$$

Then we have $s_\tau(\Delta(\lambda)) = D_{1/\lambda} s_\tau(\Delta)$, since the condition $y \in s_\tau(\Delta(\lambda))$ means $\Delta D_\lambda y \in s_\tau$ and $y \in D_{1/\lambda} s_\tau(\Delta)$ for all y. Finally, the condition $\tau \in \Gamma$ implies $s_\tau(\Delta(\lambda)) = s_{(\tau/\lambda)}$, and using the identity $\Delta^+(\mu) = D_\mu \Delta^+$, we conclude $[\Delta, \Delta^+] = s_{(\frac{\tau}{\lambda\mu})^-}$ if $\tau/\lambda\mu \in \Gamma$.
So we have shown (i) (a).
(i) (b) We have $\Delta(\lambda)(|C^+(\mu)x|) \in s_\tau$ if and only if

$$|C^{+}(\mu)x| \in C(\lambda)s_{\tau} = D_{\frac{1}{\lambda}}\Sigma s_{\tau}.$$

Since $\tau \in \Gamma$, we have $\Sigma s_{\tau} = s_{\tau}$ and $D_{1/\lambda}\Sigma s_{\tau} = s_{(\tau/\lambda)}$. Then, for $\tau/\lambda \in \Gamma^{+}$, we have $x \in [\Delta, C^{+}]$ if and only if

$$x \in s_{(\tau/\lambda)}\left(C^{+}(\mu)\right).$$

Since $C^{+}(\mu) = \Sigma^{+}D_{1/\mu}$, we conclude $[\Delta, C^{+}] = s_{(\tau\frac{\mu}{\lambda})}$.
(ii) We have $C(\lambda)(|\Delta^{+}(\mu)|) \in s_{\tau}$ if and only if $|\Delta^{+}(\mu)x| \in \Delta(\lambda)s_{\tau} = \Delta s_{\tau\lambda}$. Since $\tau\lambda \in \Gamma$, we have $\Delta s_{\tau\lambda} = s_{\tau\lambda}$ and the condition $C(\lambda)(|\Delta^{+}(\mu)x|) \in s_{\tau}$ is equivalent to $|\Delta^{+}(\mu)x| \in s_{\tau\lambda}$. Now, by the identity $\Delta^{+}(\mu) = D_{\mu}\Delta^{+}$ and since $\tau\lambda/\mu \in \Gamma$, the condition $|\Delta^{+}(\mu)x| \in s_{\tau\lambda}$ is equivalent to $x \in s_{(\tau\lambda/\mu)}(\Delta^{+}) = s_{(\tau\lambda/\mu)^{-}}$. This concludes the proof of Part (ii).
(iii) Here, we have $\Delta^{+}(\lambda)(|\Delta(\mu)x|) \in s_{\tau}$ if and only if

$$|\Delta(\mu)x| \in s_{\tau}(\Delta^{+}(\lambda)) = s_{(\frac{\tau}{\lambda})^{-}},$$

since $\tau/\lambda \in \Gamma$. Thus, we obtain

$$x \in C(\mu)s_{(\frac{\tau}{\lambda})^{-}} = D_{\frac{1}{\mu}}\Sigma s_{(\frac{\tau}{\lambda})^{-}} = s_{\left(\frac{\tau_{n-1}}{\lambda_{n-1}\mu_{n}}\right)}$$

if $(\tau/\lambda)^{-} \in \Gamma$, that is, $\tau/\lambda \in \Gamma$.
(iv) If $\tau/\lambda \in \Gamma$, then we have $\Delta^{+}(\lambda)(|C(\mu)x|) \in s_{\tau}$ if and only if

$$|C(\mu)x| \in s_{\tau}\left(\Delta^{+}(\lambda)\right) = s_{(\frac{\tau}{\lambda})^{-}},$$

that is, $x \in \Delta(\mu)s_{(\tau/\lambda)^{-}}$. Since $\mu(\tau/\lambda)^{-} \in \Gamma$, we conclude $[\Delta^{+}, C] = s_{(\mu(\tau/\lambda)^{-})}$.
(v) We have

$$[\Delta^{+}, \Delta^{+}] = \left\{x : |\Delta^{+}(\mu)x| \in s_{\tau}(\Delta^{+}(\lambda))\right\},$$

and since $\tau/\lambda \in \Gamma$, we obtain

$$s_{\tau}(\Delta^{+}(\lambda)) = s_{(\frac{\tau}{\lambda})^{-}}.$$

So the condition $\tau/\lambda \in \Gamma$ implies

$$[\Delta^{+}, \Delta^{+}] = s_{(\frac{\tau}{\lambda})^{-}}(\Delta^{+}(\mu)).$$

Then the condition $(\tau/\lambda)^{-}\mu^{-1} \in \Gamma$ implies

$$s_{(\frac{\tau}{\lambda})^{-}}\left(\Delta^{+}(\mu)\right) = s_{\left(\frac{(\frac{\tau}{\lambda})^{-}}{\mu}\right)^{-}} = s_{\left(\frac{\tau_{n-2}}{\lambda_{n-2}\mu_{n-1}}\right)_{n}}.$$

This concludes the proof of (v).
(vi) Here we have $C^+(\lambda)(|C(\mu)x|) \in s_\tau$ if and only if $|C(\mu)x| \in s_\tau(C^+(\lambda))$ and since $\tau \in \Gamma^+$, we have $s_\tau(C^+(\lambda)) = s_{\tau\lambda}$. Then for $\tau\lambda\mu \in \Gamma$, we have $x \in [C^+, C]$ if and only if $x \in \Delta(\mu)s_{\tau\lambda} = s_{(\tau\lambda\mu)}$ and $[C^+, C] = s_{(\tau\lambda\mu)}$.
(vii) , (viii) and (ix) can be shown using similar arguments as those used above. □

Remark 4.11 In each of the conditions in Theorem 4.7, we have $s_\tau(A_1 A_2) = [A_1, A_2] = (s_\tau(A_1))_{A_2}$, for

$$A_1 \in \{\Delta(\lambda), \Delta^+(\lambda), C(\lambda), C^+(\lambda)\} \text{ and } A_2 \in \{\Delta(\mu), \Delta^+(\mu), C(\mu), C^+(\mu)\}.$$

For instance, we have

$$[\Delta, C] = \left\{x : \left(\frac{\lambda_n}{\mu_n} - \frac{\lambda_{n-1}}{\mu_{n-1}}\right)\sum_{i=1}^{n-1} x_i + \frac{\lambda_n}{\mu_n} x_n = \tau_n O(1) \ (n \to \infty)\right\} \text{ for } \frac{\tau\mu}{\lambda} \in \Gamma.$$

Under conditions similar to those given in Theorems 4.6 and 4.7, we may explicitly calculate the sets

$$[\Delta, \Delta] = \{x : -\lambda_{n-1}\mu_{n-2}x_{n-2} + \mu_{n-1}(\lambda_n + \lambda_{n-1})x_{n-1} - \lambda_n\mu_n x_n = \tau_n O(1) \\ (n \to \infty)\},$$

$$[\Delta, C^+] = \left\{x : \frac{\lambda_n}{\mu_n} x_n + (\lambda_n - \lambda_{n-1}) \sum_{m=n-1}^{\infty} \frac{x_m}{\mu_m} = \tau_n O(1) \ (n \to \infty)\right\},$$

$$[\Delta, \Delta^+] = \{x : -\lambda_{n-1}\mu_{n-1}x_{n-1} + (\mu_n\lambda_n + \mu_{n-1}\lambda_{n-1})x_n - \lambda_n\mu_n x_{n+1} = \tau_n O(1) \\ (n \to \infty)\}$$

and

$$[\Delta^+, \Delta] = \{x : -\lambda_n\mu_{n+1}x_{n-1} + 2\lambda_n\mu_n x_n - \lambda_n\mu_{n+1}x_{n+1} = \tau_n O(1) \ (n \to \infty)\}.$$

4.9 Extension of the Previous Results

Here we study the sets $X(\Delta(\mu)), X(\Delta^+(\mu))$ for $X \in \{s_\tau, s_\tau^0, s_\tau^{(c)}\}$, and the sets $w_\tau^p(\lambda)$, $\overset{\circ}{w}_\tau^{p}(\lambda)$, $w_\tau^{+p}(\lambda)$, $w_\tau^{+p}(\lambda)$ and $\overset{\circ}{w}_\tau^{+p}(\lambda)$ for $p > 0$.

First, we deal with the sets $X(\Delta(\mu)), X(\Delta^+(\mu))$ for $X \in \{s_\tau, s_\tau^0, s_\tau^{(c)}\}$, and the sets $w_\tau^p(\lambda)$, $\overset{\circ}{w}_\tau^{p}(\lambda)$, $w_\tau^{+p}(\lambda)$ and $\overset{\circ}{w}_\tau^{+p}(\lambda)$. We use some properties of $\Delta^+(\lambda)$ for $\lambda \in \mathcal{U}$. We see that for any given $\tau \in \mathcal{U}^+$, the condition

$$\frac{\lambda_n^\bullet}{\tau_n^\bullet} = O(1) \ (n \to \infty)$$

implies

$$\Delta^+(\lambda) = D_\lambda \Delta^+ \in (s_{(\tau/|\lambda|)}, s_\tau).$$

To state some new results we need the following lemmas.

Lemma 4.8 *If Δ^+ is bijective from s_τ into itself, then $\tau \in cs$.*

Proof We assume $\tau \notin cs$, that is, $\sum_{n=1}^{\infty} \tau_n = \infty$. We are led to deal with two cases.

1- $e \in \mathrm{Ker}\Delta^+ \bigcap s_\tau$. Then Δ^+ is not bijective from s_τ into itself.
2- $e \notin \mathrm{Ker}\Delta^+ \bigcap s_\tau$. Then $1/\tau \notin s_1$ and there is a strictly increasing sequence of integers $(n_i)_{i=1}^{\infty}$ such that $1/\tau_{n_i} \to \infty$ $(i \to \infty)$. We assume that the equation $\Delta^+ x = \tau$ has a solution $x = (x_{n,0})_{n=1}^{\infty}$ in s_τ. Then there is a unique scalar x_0 such that

$$x_{n,0} = x_0 - \sum_{k=1}^{n-1} \tau_k.$$

So we obtain

$$\frac{|x_{n_i,0}|}{\tau_{n_i}} = \left| \frac{1}{\tau_{n_i}} \left(x_0 - \sum_{k=1}^{n_i-1} \tau_k \right) \right| \to \infty \ (i \to \infty),$$

and $x \notin s_\tau$. So we are led to a contradiction.

We conclude that each of the properties $e \in \mathrm{Ker}\Delta^+ \bigcap s_\tau$ and $e \notin \mathrm{Ker}\Delta^+ \bigcap s_\tau$ is not satisfied and Δ^+ is not bijective from s_τ into itself. This completes the proof. □

We also need to state the following elementary result.

Lemma 4.9 *We have $\Sigma^+(\Delta^+ x) = x$ for all $x \in c_0$ and $\Delta^+(\Sigma^+ x) = x$ for all $x \in cs$.*

By Lemmas 4.8 and 4.9 we obtain the next proposition, (cf. [8, Theorem 2.7, p. 1659]).

Theorem 4.8 *Let $\tau \in \mathcal{U}^+$. The operator $\Delta^+ \in (s_\tau, s_\tau)$ is bijective if and only if $\tau \in \widehat{C_1^+}$.*

Now let

$$w_\tau^p(\lambda) = \left\{ x \in \omega : C(\lambda)\left(|x|^p\right) \in s_\tau \right\} \tag{4.27}$$

$$\overset{\circ}{w}{}_\tau^p(\lambda) = \left\{ x \in \omega : C(\lambda)\left(|x|^p\right) \in s_\tau^0 \right\} \tag{4.28}$$

$$w_\tau^{+p}(\lambda) = \left\{ x \in \omega : C^+(\lambda)\left(|x|^p\right) \in s_\tau \right\} \tag{4.29}$$

$$\overset{\circ}{w}{}_\tau^{+p}(\lambda) = \left\{ x \in \omega : C^+(\lambda)\left(|x|^p\right) \in s_\tau^0 \right\}, \tag{4.30}$$

see [7].

We obtain the following theorem from [7, Theorem 2, pp. 19–20].

Theorem 4.9 *Let $\tau \in \mathcal{U}^+$, $\lambda, \mu \in \mathcal{U}$ and $p > 0$. Then we have*

(i) (a) $s_\tau(\Delta(\mu)) = s_{(\tau/|\mu|)}$ *if and only if* $\tau \in \widehat{C_1}$;

(b) $s_\tau^0(\Delta(\mu)) = s^0_{(\tau/|\mu|)}$ *if and only if* $\tau \in \widehat{C_1}$;

(c) $s_\tau^{(c)}(\Delta(\mu)) = s^{(c)}_{(\tau/|\mu|)}$ *if and only if* $\tau \in \widehat{\Gamma}$;

(d) $s_\tau(\Delta^+(\mu)) = s_{(\tau/|\mu|)^-}$ *if and only if* $\tau/|\mu| \in \widehat{C_1}$.

(ii) (a) $s_\tau(\Sigma^+) = s_\tau$ *if and only if* $\tau \in \widehat{C_1^+}$;

(b) $s_\tau^0(\Sigma^+) = s_\tau^0$ *if and only if* $\tau \in \widehat{C_1^+}$;

(c) $\tau \in \widehat{C_1^+}$ *if and only if* $w_\tau^{+p}(\lambda) = s_{(\tau|\lambda|)^{1/p}}$;

(d) *if* $\tau \in \widehat{C_1^+}$, *then* $\overset{\circ}{w}{}_\tau^{+p}(\lambda) = s^0_{(\tau|\lambda|)^{1/p}}$;

(e) $\tau|\lambda| \in \widehat{C_1}$ *if and only if* $w_\tau^p(\lambda) = s_{(\tau|\lambda|)^{1/p}}$.

(f) *If* $\tau|\lambda| \in \widehat{C_1}$, *then* $\overset{\circ}{w}{}_\tau^{p}(\lambda) = s^0_{(\tau|\lambda|)^{1/p}}$.

Proof (i)

(a), (b), (c) can be shown as in Theorem 4.2.

(d) We write $\nu = \tau/|\mu|$ and we show

$$s_\tau(\Delta^+(\mu)) = s_{\nu^-}(\Delta). \tag{4.31}$$

We have $x \in s_\tau(\Delta^+(\mu))$ if and only if $D_\mu \Delta^+ x \in s_\tau$, that is, $\Delta^+ x \in s_\nu$. But the condition $\Delta^+ x \in s_\nu$ means $\Delta x \in s_{\nu^-}$. We conclude that (4.31) holds. Now by Theorem 4.2, we obtain $s_{\nu^-}(\Delta) = s_{\nu^-}$ if and only if $\nu^- \in \widehat{C_1}$, that is, $\nu \in \widehat{C_1}$. This concludes the proof of Part (i) (d).

(ii)

(a), (b) follow from the equivalence of each of the inclusions $s_\tau \subset s_\tau(\Sigma^+)$, or $s_\tau^0 \subset s_\tau^0(\Sigma^+)$ and the condition $\tau \in \widehat{C_1^+}$.

(c) Let $\tau \in \widehat{C_1^+}$. Since $C^+(\lambda) = \Sigma^+ D_{1/\lambda}$, we have

$$w_\tau^{+p}(\lambda) = \left\{x \in \omega : (\Sigma^+ D_{\frac{1}{\lambda}})(|x|^p) \in s_\tau\right\} = \left\{x : D_{\frac{1}{\lambda}}(|x|^p) \in s_\tau(\Sigma^+)\right\};$$

and since $\tau \in \widehat{C_1^+}$ implies $s_\tau(\Sigma^+) = s_\tau$, we conclude

$$w_\tau^{+p}(\lambda) = \left\{x \in \omega : |x|^p \in D_\lambda s_\tau = s_{\tau|\lambda|}\right\} = s_{(\tau|\lambda|)^{\frac{1}{p}}}.$$

Conversely, we have $(\tau|\lambda|)^{1/p} \in s_{(\tau|\lambda|)^{1/p}} = w_\tau^{+p}(\lambda)$. So we have

$$C^+(\lambda)\left[(\tau|\lambda|)^{\frac{1}{p}}\right]^p = \left(\sum_{k=n}^{\infty} \frac{\tau_k|\lambda_k|}{|\lambda_k|}\right)_{n=1}^{\infty} \in s_\tau,$$

that is, $\tau \in \widehat{C_1^+}$ and we have shown Part (ii) (c).
(d) We obtain Part (ii) (d) using similar arguments as above.
(e), (f) We only need to show Part (ii) (e), since the proof of Part (ii) (f) is similar.
We assume $\tau|\lambda| \in \widehat{C_1}$. Then we have

$$w_\tau^{\circ p}(\lambda) = \left\{x \in \omega : |x|^p \in \Delta(\lambda)s_\tau^0\right\}.$$

Since $\Delta(\lambda) = \Delta D_\lambda$, we obtain $\Delta(\lambda)s_\tau^0 = \Delta s_{\tau|\lambda|}^0$. Now, Part (i) (b), the condition $\tau|\lambda| \in \widehat{C_1}$ implies that Δ is bijective from $s_{\tau|\lambda|}^0$ into itself and $w_\tau^{\circ p}(\lambda) = s^0_{(\tau|\lambda|)^{1/p}}$. This concludes the proof. □

As a direct consequence of Theorem 4.9 we obtain the following results.

Corollary 4.5 ([7, Corollary 1, p. 23]) *Let $r > 0$ be any real. Then the next statements are equivalent, where (i) $r > 1$, (ii) $s_r(\Delta) = s_r$, (iii) $s_r^0(\Delta) = s_r^0$ and (iv) $s_r(\Delta^+) = s_r$.*

Proof Indeed, we see from Parts (i) and (ii) of Theorem 4.9 that it is enough to show that $\tau = (r^n)_{n=1}^\infty \in \widehat{C_1}$ if and only if $r > 1$. We have $(r^n)_{n=1}^\infty \in \widehat{C_1}$ if and only if $r \neq 1$ and

$$r^{-n}\left(\sum_{k=1}^n r^k\right) = \frac{1}{1-r}r^{-n+1} - \frac{r}{1-r} = O(1)\ (n \to \infty).$$

This implies $r > 1$ and concludes the proof. □

4.10 Sets of Sequences that are Strongly τ-Bounded With Index p

Now we consider the spaces $c_\tau^p(\lambda, \mu)$, $c_\tau^{\cdot +p}(\lambda, \mu)$, $c_\tau^{+\cdot p}(\lambda, \mu)$ and $c_\tau^{+p}(\lambda, \mu)$ that generalize the sets $[C, \Delta]$, $[C, \Delta^+]$, $[C^+, \Delta]$ and $[C^+, \Delta^+]$ studied in the previous section. These sets also generalize the well-known spaces of sequences $c_\infty^p(\lambda)$ and $c_0^p(\lambda)$ that are strongly bounded and convergent to naught with index p. They also generalize the sets $c_\infty(\lambda, \mu)$ and $c_0(\lambda, \mu)$ that are strongly τ-bounded and convergent to naught we studied in the previous section.

For given real $p > 0$ and $\lambda, \mu \in \mathcal{U}$, we write

$$\begin{aligned} c_\tau^p(\lambda, \mu) &= \left(w_\tau^p(\lambda)\right)_{\Delta(\mu)} = \left\{x : C(\lambda)(|\Delta(\mu)x|^p) \in s_\tau\right\}, \qquad (4.32)\\ c_\tau^{\cdot +p}(\lambda, \mu) &= \left(w_\tau^p(\lambda)\right)_{\Delta^+(\mu)} = \left\{x : C(\lambda)(|\Delta^+|(\mu)x|^p) \in s_\tau\right\},\\ c_\tau^{+\cdot p}(\lambda, \mu) &= \left(w_\tau^{+p}(\lambda)\right)_{\Delta(\mu)} = \left\{x : C^+(\lambda)(|\Delta(\mu)x|^p) \in s_\tau\right\},\\ c_\tau^{+p}(\lambda, \mu) &= \left(w_\tau^{+p}(\lambda)\right)_{\Delta^+(\mu)} = \left\{x : C^+(\lambda)(|\Delta^+(\mu)x|^p) \in s_\tau\right\}. \end{aligned}$$

When s_τ is replaced by s_τ^0 in the previous definitions, we write $\widetilde{c_\tau^p}(\lambda,\mu)$, $\widetilde{c_\tau^{\cdot+p}}(\lambda,\mu)$, $\widetilde{c_\tau^{+\cdot p}}(\lambda,\mu)$ and $\widetilde{c_\tau^{+p}}(\lambda,\mu)$, instead of $c_\tau^p(\lambda,\mu)$, $c_\tau^{\cdot+p}(\lambda,\mu)$, $c_\tau^{+\cdot p}(\lambda,\mu)$ and $c_\tau^{+p}(\lambda,\mu)$. For instance, it can easily be seen that

$$c_\tau^p(\lambda,\mu)=\left\{x=(x_n)_{n=1}^\infty:\sup_n\left[\frac{1}{|\lambda_n|\tau_n}\left(\sum_{k=1}^n|\mu_kx_k-\mu_{k-1}x_{k-1}|^p\right)\right]<\infty\right\},$$

$$c_\tau^{\cdot+p}(\lambda,\mu)=\left\{x=(x_n)_{n=1}^\infty:\sup_n\left[\frac{1}{|\lambda_n|\tau_n}\left(\sum_{k=1}^n|\mu_k(x_k-x_{k+1})|^p\right)\right]<\infty\right\},$$

$$c_\tau^{+\cdot p}(\lambda,\mu)=\left\{x=(x_n)_{n=1}^\infty:\sup_n\left[\frac{1}{\tau_n}\sum_{k=n}^\infty\left(\frac{1}{|\lambda_k|}|\mu_kx_k-\mu_{k-1}x_{k-1}|^p\right)\right]<\infty\right\},$$

$$\widetilde{c_\tau^{+p}}(\lambda,\mu)=\left\{x=(x_n)_{n=1}^\infty:\lim_{n\to\infty}\left[\frac{1}{\tau_n}\sum_{k=n}^\infty\left(\frac{1}{|\lambda_k|}|\mu_k(x_k-x_{k+1})|^p\right)\right]=0\right\}$$

with the convention $x_0=0$. We say that $c_\tau^p(\lambda,\mu)$ and $\widetilde{c_\tau^p}(\lambda,\mu)$ are the sets of sequences that are strongly τ-bounded and τ-convergent to 0 with index p. If $\lambda=\mu$ and $\tau=e$, then $c_\tau^p(\lambda,\mu)=c_\infty^p(\lambda)$ and $\widetilde{c_\tau^p}(\lambda,\mu)=c_0^p(\lambda)$ are the sets of sequences that are strongly bounded and strongly convergent to zero with index p.

Now we let

$$\zeta_p=\frac{(\tau|\lambda|)^{1/p}}{|\mu|}=\left(\frac{(\tau_n|\lambda_n|)^{1/p}}{|\mu_n|}\right)_{n=1}^\infty,\quad \zeta_p^-=\left(\frac{(\tau_{n-1}|\lambda_{n-1}|)^{1/p}}{|\mu_{n-1}|}\right)_{n=1}^\infty,$$

with

$$\frac{(\tau_{-1}|\lambda_{-1}|)^{1/p}}{|\mu_{-1}|}=1.$$

We obtain the following theorem from the previous results, (cf. [7, Theorem 4, p. 26]).

Theorem 4.10 *Let $\lambda,\mu\in\mathcal{U}$ and $\tau\in\mathcal{U}^+$. Then we have*

(i) If $\tau|\lambda|$, $(\tau|\lambda|)^{1/p}\in\widehat{C_1}$, then

$$c_\tau^p(\lambda,\mu)=s_{\zeta_p}\text{ and }\widetilde{c_\tau^p}(\lambda,\mu)=s_{\zeta_p}^0.$$

(ii) If $\tau|\lambda|$, $\zeta_p\in\widehat{C_1}$, then we have

$$c_\tau^{\cdot+p}(\lambda,\mu)=s_{\zeta_p^-}\text{ and }\widetilde{c_\tau^{\cdot+p}}(\lambda,\mu)=s_{\zeta_p^-}^0.$$

(iii) *Let* $\tau \in \widehat{C_1^+}$. *Then*

(a) $(\tau|\lambda|)^{1/p} \in \widehat{C_1}$ *implies*

$$c_\tau^{+\cdot p}(\lambda,\mu) = s_{\zeta_p} \text{ and } \widetilde{c_\tau^{+\cdot p}}(\lambda,\mu) = s^0_{\zeta_p}.$$

(b) *If* $\zeta_p \in \widehat{C_1}$, *then*

$$c_\tau^{+p}(\lambda\mu) = s_{\zeta_p^-} \text{ and } \widetilde{c_\tau^{+p}}(\lambda,\mu) = s^0_{\zeta_p^-}.$$

Proof (i) First, we have

$$c_\tau^p(\lambda,\mu) = \left\{x : \Delta(\mu)x \in w_\tau^p(\lambda)\right\}.$$

Since $\tau|\lambda| \in \widehat{C_1}$, we obtain from Part (ii) (e) of Theorem 4.9 $w_\tau^p(\lambda) = s_{(\tau|\lambda|)^{\frac{1}{p}}}$. Thus, using the identities

$$\Delta(\mu)^{-1} = C(\mu) = D_{1/\mu}\Sigma$$

we obtain

$$c_\tau^p(\lambda,\mu) = D_{1/\mu}\Sigma s_{|\lambda|^{1/p}},$$

and since $(\tau|\lambda|^{1/p} \in \widehat{C_1}$, then Δ is bijective from $s_{(\tau|\lambda|)^{1/p}}$ into itself, that is, $\Sigma s_{(|\lambda|)^{1/p}} = s_{(|\lambda|)^{1/p}}$ and we conclude $c_\tau^p(\lambda,\mu) = s_{\zeta_p}$. By similar arguments as those used above we obtain $\widetilde{c_\tau^p}(\lambda,\mu) = s^0_{\zeta_p}$.
(ii) Here we obtain

$$c_\tau^{\cdot +p}(\lambda,\mu) = \left\{x : \Delta^+(\mu)x \in w_\tau^p(\lambda)\right\},$$

and since $\tau|\lambda| \in \widehat{C_1}$, we have $w_\tau^p(\lambda) = s_{(\tau|\lambda|)^{1/p}}$. So we obtain

$$c_\tau^{\cdot +p}(\lambda,\mu) = s_{(\tau|\lambda|)^{\frac{1}{p}}}(\Delta^+(\mu)),$$

and from Part (i) (d) in Theorem 4.9, we obtain

$$s_{(\tau|\lambda|)^{\frac{1}{p}}}(\Delta^+(\mu)) = s_{\zeta_p^-} \text{ if } \zeta_p \in \widehat{C_1}. \tag{4.33}$$

We conclude $c_\tau^{\cdot +p}(\lambda,\mu) = s_{\zeta_p^-}$.
As we have seen above we obtain by (ii) (f) of Theorem 4.9

$$\widetilde{c_\tau^{\cdot +p}}(\lambda,\mu) = s^0_{(\tau|\lambda|)^{\frac{1}{p}}}(\Delta^+(\mu)) = s^0_{\zeta_p^-}.$$

This concludes the proof of Part (ii).
(iii)

(a) We have

$$c_\tau^{+\cdot p}(\lambda,\mu)=\left\{x:\Delta(\mu)x\in w_\tau^{+p}(\lambda)\right\}.$$

If $\tau\in\widehat{C_1^+}$, then we have $w_\tau^{+p}(\lambda)=s_{(\tau|\lambda|)^{1/p}}$ and

$$c_\tau^{+\cdot p}(\lambda,\mu)=C(\mu)s_{(\tau|\lambda|)^{\frac{1}{p}}}=D_{\frac{1}{\mu}}\Sigma s_{(\tau|\lambda|)^{\frac{1}{p}}}.$$

We have from Part (i) (a) of Theorem 4.9

$$(\tau|\lambda|)^{1/p}\in\widehat{C_1}\text{ if and only if }\Sigma s_{(\tau|\lambda|)^{1/p}}=s_{(\tau|\lambda|)^{1/p}}$$

and we conclude $c_\tau^{+\cdot p}(\lambda,\mu)=s_{\zeta_p}$. By similar arguments as those used above we obtain $\widetilde{c_\tau^{+\cdot p}}(\lambda,\mu)=s_{\zeta_p}^0$.

(b) Since $w_\tau^{+p}(\lambda)=s_{(\tau|\lambda|)^{1/p}}$ for $\tau\in\widehat{C_1^+}$, we obtain

$$c_\tau^{+p}(\lambda,\mu)=\left\{x:\Delta^+(\mu)x\in w_\tau^{+p}(\lambda)=s_{(\tau|\lambda|)^{\frac{1}{p}}}\right\}=s_{(\tau|\lambda|)^{\frac{1}{p}}}(\Delta^+(\mu)),$$

and we conclude using (4.33). The identity $\widetilde{c_\tau^{+p}}(\lambda,\mu)=s_{\zeta_p^-}$ can be obtained by similar arguments as those used in Part (ii) of the proof. □

Remark 4.12 We have

$$c_\tau^{1}(\lambda,\mu)=[C,\Delta],\ c_\tau^{\cdot+1}(\lambda,\mu)=[C,\Delta^+],\ c_\tau^{+\cdot 1}(\lambda,\mu)=[C^+,\Delta]$$

and $c_\tau^{+1}(\lambda,\mu)=[C^+,\Delta^+]$. For instance, take $p=1$ and $\lambda,\mu\in\mathcal{U}^+$, then we have $\zeta_1=\lambda\tau/\mu$ and

$$\tau|\lambda|\in\widehat{C_1}\text{ implies }c_\tau^{p}(\lambda,\mu)=[C,\Delta]=s_{(\tau\lambda/\mu)}.$$

It can easily be verified that the results stated in Part (ii) of Theorem 4.6 and Parts (ii), (vii) and (viii) of Theorem 4.7 are generalized in Theorem 4.10.

Remark 4.13 We note that the previous sets are BK spaces and we can write for instance that if $\tau|\lambda|\in\widehat{C_1}$ and $(\tau|\lambda|)^{1/p}\in\widehat{C_1}$, then $c_\tau^{+p}(\lambda,\mu)$ is a BK space with respect to the norm $\|\cdot\|_{s_{\xi_p}}$ and $\widetilde{c_\tau^{+p}}(\lambda,\mu)$ has AK.

Corollary 4.6 ([7, Corollary 3, p. 27]) *Let $\lambda,\mu\in\mathcal{U}$ and $\tau\in\mathcal{U}^+$. We assume $1/\tau,\tau|\lambda|\in\Gamma$. Then we have*

(i) $c_\tau^{p}(\lambda,\mu)=c_\tau^{+\cdot p}(\lambda,\mu)=s_{\zeta_p}$.

(ii) $\widetilde{c_\tau^{p}}(\lambda,\mu)=\widetilde{c_\tau^{+\cdot p}}(\lambda,\mu)=s_{\zeta_p}^0$.

Proof Since $\Gamma \subset \widehat{C_1}$, it is enough to note that $\tau|\lambda| \in \Gamma$ if and only if $(\tau|\lambda|)^{1/p} \in \Gamma$ and apply Theorem 4.10. □

To state the next corollary, we need the following elementary lemma.

Lemma 4.10 *Let $q > 0$ be any real and $\tau \in \mathcal{U}^+$ be a nondecreasing sequence. Then we have*

(i) $\tau \in \widehat{C_1}$ *implies* $\tau^q \in \widehat{C_1}$ *for* $q \geq 1$,
(ii) $\tau^q \in \widehat{C_1}$ *implies* $\tau \in \widehat{C_1}$ *for* $0 < q \leq 1$.

Proof Let $q \geq 1$. Since τ is a nondecreasing sequence, we immediately see that for any given integer n we have

$$\sum_{k=1}^{n} \tau_k(\tau_n^{q-1} - \tau_k^{q-1}) = \sum_{k=1}^{n} (\tau_n^{q-1}\tau_k - \tau_k^q) \geq 0$$

and

$$\frac{1}{\tau_n}\left(\sum_{k=1}^{n} \tau_k\right) \geq \frac{1}{\tau_n^q}\left(\sum_{k=1}^{n} \tau_k^q\right). \tag{4.34}$$

Since $\tau \in \widehat{C_1}$ implies $(\sum_{k=1}^{n} \tau_k)/\tau_n = O(1)$ $(n \to \infty)$, we obtain Part (i) using the inequality in (4.34).
Now, writing $\nu = \tau^q \in \widehat{C_1}$ and applying Part (i), we obtain $\tau = \nu^{1/q} \in \widehat{C_1}$ for $0 < q \leq 1$. So we have shown Part (ii). This concludes the proof. □

Corollary 4.7 ([7, Corollary 4, pp. 28–29]) *Let $\tau \in \tau\mathcal{U}^+$, $\lambda \in \mathcal{U}$ and $\tau|\lambda|$ be a nondecreasing sequence.*

(i) If $p \geq 1$, then the condition $(\tau|\lambda|)^{1/p} \in \widehat{C_1}$ implies

$$c_\tau^p(\lambda, \mu) = s_{\zeta_p} \text{ and } \widetilde{c_\tau^p}(\lambda, \mu) = s_{\zeta_p}^0. \tag{4.35}$$

(ii) If $0 < p < 1$, then we have $\tau|\lambda| \in \widehat{C_1}$ if and only if $c_\tau^p(\lambda, \mu) = s_{\zeta_p}$.
(iii) If $\tau|\lambda| \in \Gamma$, then the conditions in (4.35) hold.

Proof (i) If $p \geq 1$ then $0 < 1/p \leq 1$ and by Lemma 4.10, $(\tau|\lambda|)^{1/p} \in \widehat{C_1}$ implies $\tau|\lambda| \in \widehat{C_1}$. So we conclude using Part (i) of Theorem 4.10.
(ii) The necessity follows from Part (i) of Theorem 4.10 and Part (ii) of Lemma 4.10.
To show the sufficiency we first let

$$\widetilde{\tau} = \left((-1)^n \frac{(\tau_n|\lambda_n|)^{\frac{1}{p}}}{|\mu_n|}\right)_{n=1}^{\infty}.$$

We have $\widetilde{\tau} \in c_\tau^p(\lambda, \mu) = s_{\zeta_p}$ and using the convention $\widetilde{\tau}_0 = 0$, we can write

$$|\Delta(\mu)\widetilde{\tau}| = \left(\left|\mu_n(-1)^n \frac{(\tau_n|\lambda_n|)^{\frac{1}{p}}}{\mu_n} - \mu_{n-1}(-1)^{n-1} \frac{(\tau_{n-1}|\lambda_{n-1}|)^{\frac{1}{p}}}{\mu_{n-1}}\right|\right)_{n=1}^{\infty}.$$

So

$$|\Delta(\mu)\widetilde{\tau}|^p = \left(\left((\tau_n|\lambda_n|)^{\frac{1}{p}} + (\tau_{n-1}|\lambda_{n-1}|)^{\frac{1}{p}}\right)^p\right)_{n=1}^{\infty}.$$

Then the condition $\Sigma|\Delta(\mu)\widetilde{\tau}|^p \in s_{\tau|\lambda|}$ implies that there is a real $M > 0$ such that

$$\frac{1}{\tau_n|\lambda_n|}\left(\sum_{k=1}^{n} \tau_k|\lambda_k|\right) \leq \frac{1}{\tau_n|\lambda_n|}\left(\sum_{k=1}^{n} \left((\tau_k|\lambda_k|)^{\frac{1}{p}} + (\tau_{k-1}|\lambda_{k-1}|)^{\frac{1}{p}}\right)^p\right) \leq M$$

for all n. We conclude $\tau|\lambda| \in \widehat{C_1}$. So Part (ii) can be obtained from Theorem 4.10.
(iii) follows from the inclusion $\Gamma \subset \widehat{C_1}$. □

In the following, we denote by $c_\tau^p(\lambda)$ the set $c_\tau^p(\lambda, \lambda)$ and we consider the next identities.

$$c_\tau^p(\lambda) = s_{(\tau^{\frac{1}{p}}|\lambda|^{\frac{1}{p}-1})}, \tag{4.36}$$

and

$$\widetilde{c_\tau^p}(\lambda) = s^0_{(\tau^{\frac{1}{p}}|\lambda|^{\frac{1}{p}-1})}. \tag{4.37}$$

We obtain the following.

Corollary 4.8 ([7, Corollary 5, pp. 29–30]) *Let $\tau \in \mathcal{U}^+$, $\lambda \in \mathcal{U}$ and the sequence $\tau|\lambda|$ be nondecreasing. Then we have*

(i) If $0 < p \leq 1$, then we have

(a) $\tau|\lambda \in \widehat{C_1}$ if and only if the condition in (4.36) holds;
(b) $\tau|\lambda| \in \widehat{C_1}$ implies that the condition in (4.37) holds.

(ii) If $p > 1$, then the condition $(\tau|\lambda|)^{1/p} \in \widehat{C_1}$ implies the conditions in (4.36) and (4.37) hold;
(iii) $\tau|\lambda| \in \Gamma$ implies (4.36) and (4.37).

Proof (i) (a) follows from Part (ii) in Corollary 4.7 with $\lambda = \mu$.
(i) (b) follows from Lemma 4.10 Part (i) of Theorem 4.10.
(ii) follows from Part (i) of Corollary 4.7.
(iii) follows from Corollary 4.6. □

4.11 Computations in W_τ and W_τ^0 and Applications to Statistical Convergence

First, we study some properties of the map $\Delta_\rho^+ \in (w_\infty(\lambda), w_\infty(\lambda))$.

Let $\lambda = (\lambda_n)_{n=1}^\infty \in \mathcal{U}^+$. We have seen that $w_\infty(\lambda)$ and $w^0(\lambda)$ are the sets of sequences that are *strongly* C_1 *bounded and summable to* 0, defined by

$$w_\infty(\lambda) = \left\{ x = (x_n)_{n=1}^\infty \in \omega : \sup_n \frac{1}{\lambda_n} \sum_{k=1}^n |x_k| < \infty \right\}$$

and

$$w^0(\lambda) = \left\{ x = (x_n)_{n=1}^\infty \in \omega : \lim_{n\to\infty} \frac{1}{\lambda_n} \sum_{k=1}^n |x_k| = 0 \right\}.$$

These sets were studied by Malkowsky, with the concept of *exponentially bounded sequences*, see [4]. We recall that Maddox [26, 27] defined and studied the sets $w_\infty(\lambda) = w_\infty$, $w_0(\lambda) = w^0$ and $w(\lambda) = w$ where $\lambda_n = n$ for all n.

We also recall that a nondecreasing sequence $\lambda = (\lambda_n)_{n=1}^\infty \in \mathcal{U}^+$ is said to be *exponentially bounded* if there is an integer $m \geq 2$ such that for all non-negative integers ν there is at least one term $\lambda_n \in I_m^{(\nu)} = [m^\nu, m^{\nu+1} - 1]$. It was shown in [28, Lemma 1] that a nondecreasing sequence $\lambda = (\lambda_n)_{n=1}^\infty$ is exponentially bounded if and only if there are reals $s \leq t$ such that for some subsequence $(\lambda_{n_i})_{i=1}^\infty$, we have

$$0 < s \leq \frac{\lambda_{n_i}}{\lambda_{n_{i+1}}} \leq t < 1 \text{ for all } i = 1, 2, \dots$$

and such a sequence is called an *associated subsequence*. Here we write

$$\|x\|_\lambda = \|x\|_{w_\infty(\lambda)} = \sup_n \left(\frac{1}{\lambda_n} \sum_{k=1}^n |x_k| \right).$$

In [30], it was shown that if $\lambda = (\lambda_n)_{n=1}^\infty \in \mathcal{U}^+$ is exponentially bounded then the class $(w_\infty(\lambda), w_\infty(\lambda))$ is a Banach algebra with the norm

$$\|A\|_{(w_\infty(\lambda), w_\infty(\lambda))} = \sup_{x \neq 0} \left(\frac{\|Ax\|_\lambda}{\|x\|_\lambda} \right). \tag{4.38}$$

We now consider the following matrices for a given sequence $\rho = (\rho_n)_{n=1}^\infty$:

$$\Delta_\rho = \begin{pmatrix} 1 & & & \\ -\rho_1 & 1 & & 0 \\ & \cdot & \cdot & \\ & & -\rho_{n-1} & 1 & \cdot \\ & & & \cdot & \cdot \end{pmatrix} \text{ and } \Delta_\rho^+ = \begin{pmatrix} 1 & -\rho_1 & & \\ & \cdot & \cdot & \\ & & 1 & -\rho_n \\ 0 & & & \cdot & \cdot \end{pmatrix}.$$

It can easily be shown that if $\rho = (\rho_n)_{n=1}^\infty$ and $(\lambda_{n+1}/\lambda_n)_{n=1}^\infty \in \ell_\infty$, then we have $\Delta_\rho^+ \in (w_\infty(\lambda), w_\infty(\lambda))$. We also see that $\Delta_\rho \in (w_\infty(\lambda), w_\infty(\lambda))$ for ρ, $(\lambda_{n-1}/\lambda_n)_{n=2}^\infty \in \ell_\infty$. The following result holds.

Lemma 4.11 ([17, Theorem 3.1, p. 204 and Theorem 3.12, p. 210]) *Let $\lambda \in \mathcal{U}^+$ be an exponentially bounded sequence. Then we have*

(i) If

$$\overline{\lim_{n\to\infty}} \left(\frac{\lambda_{n+1}}{\lambda_n}\right) < \infty \text{ and } \overline{\lim_{n\to\infty}} |\rho_n| < \frac{1}{\overline{\lim_{n\to\infty}} \left(\frac{\lambda_{n+1}}{\lambda_n}\right)}, \tag{4.39}$$

then for given $b \in w_\infty(\lambda)$ the equation $\Delta_\rho^+ x = b$ has a unique solution in $w_\infty(\lambda)$ given by

$$x_n = b_n + \sum_{i=n+1}^{\infty} \left(\prod_{j=n}^{i-1} \rho_j\right) b_i \text{ for all } n. \tag{4.40}$$

(ii) If

$$\overline{\lim_{n\to\infty}} |\rho_n| < \frac{1}{\overline{\lim_{n\to\infty}} \lambda_n^\bullet}, \tag{4.41}$$

then for any given $b \in w_\infty(\lambda)$ the equation $\Delta_\rho x = b$ has a unique solution in $w_\infty(\lambda)$ given by $x_1 = b_1$ and

$$x_n = b_n + \sum_{k=1}^{n-1} \left(\prod_{j=k}^{n-1} \rho_j\right) b_k \text{ for all } n \geq 2. \tag{4.42}$$

Proof (i) We let

$$\Sigma_\rho^{+(N)} = \begin{pmatrix} \left[\Delta_\rho^{+(N)}\right]^{-1} & 0 \\ & 1 \\ 0 & & \cdot \end{pmatrix},$$

where $\Delta_\rho^{+(N)}$ is the finite matrix whose entries are those of the N first rows and columns of Δ_ρ^+. The finite matrix $\Delta_\rho^{+(N)}$ is invertible, since it is an upper triangle. We obtain $\Delta_\rho^+ \Sigma_\rho^{(N)} = (a_{nk})_{n,k=1}^\infty$, with $a_{nn} = 1$ for all n, $a_{n,n+1} = -\rho_n$ for all $n \geq N$ and $a_{nk} = 0$ otherwise. For any given $x \in w_\infty(\lambda)$, we have $(I - \Delta_\rho^+ \Sigma_\rho^{+(N)})x = (\xi_n(x))_{n=1}^\infty$ with $\xi_n(x) = 0$ for all $n \leq N - 1$ and $\xi_n(x) = \rho_n x_{n+1}$ for all $n \geq N$. Now we put

$$K_N = \sup_{n \geq N} \left(\frac{\lambda_{n+1}}{\lambda_n} \right) \sup_{k \geq N} |\rho_k|$$

and show that (4.39) implies $K_N < 1$ for sufficiently large N. For this let $\varepsilon > 0$ be given. We have from (4.39)

$$\overline{\lim_{n \to \infty}} |\rho_n| = l < \infty,$$

since $\overline{\lim}_{n \to \infty} \lambda_{n+1}/\lambda_n \geq 1$, λ being increasing. Then there is an integer N_1 such that

$$\sup_{n \geq N_1} |\rho_n| < l + \varepsilon.$$

Now since $\overline{\lim}_{n \to \infty} \lambda_{n+1}/\lambda_n < \infty$, (4.39) implies $\overline{\lim}_{n \to \infty} \lambda_{n+1}/\lambda_n = L < 1/l$ and as above there is an integer N_2 such that

$$\sup_{n \geq N_2} \left(\frac{\lambda_{n+1}}{\lambda_n} \right) < L + \varepsilon.$$

Since $lL < 1$, for ε small enough taking $N = \max\{N_1, N_2\}$ we then have

$$K_N \leq (l + \varepsilon)(L + \varepsilon) = lL + l\varepsilon + L\varepsilon + \varepsilon^2 < 1.$$

Now

$$\begin{aligned}
\left\| \left(I - \Delta_\rho^+ \Sigma_\rho^{+(N)}\right) x \right\|_{w_\infty(\lambda)} &= \sup_{n \geq N} \left(\frac{1}{\lambda_n} \sum_{k=N}^{n} |\rho_k x_{k+1}| \right) \\
&\leq \sup_{n \geq N} \left[\left(\sup_{k \geq N} |\rho_k| \right) \frac{1}{\lambda_n} \sum_{k=N+1}^{n+1} |x_k| \right] \\
&\leq \sup_{n \geq N} \left(\frac{\lambda_{n+1}}{\lambda_n} \right) \sup_{k \geq N} |\rho_k| \sup_{n \geq N} \left(\frac{1}{\lambda_{n+1}} \sum_{k=N+1}^{n+1} |x_k| \right) \\
&\leq \left[\sup_{n \geq N} \left(\frac{\lambda_{n+1}}{\lambda_n} \right) \sup_{k \geq N} |\rho_k| \right] \|x\|_{w_\infty(\lambda)}.
\end{aligned}$$

Then we have

$$\left\| \left(I - \Delta_\rho^+ \Sigma_\rho^{+(N)}\right) x \right\|_{w_\infty(\lambda)} \leq K_N \|x\|_{w_\infty(\lambda)}$$

and we conclude

$$\left\| I - \Delta_\rho^+ \Sigma_\rho^{+(N)} \right\|_{(w_\infty(\lambda), w_\infty(\lambda))} = \sup_{x \neq 0} \left(\frac{\left\| \left(I - \Delta_\rho^+ \Sigma_\rho^{(N)}\right) x \right\|_{w_\infty(\lambda)}}{\|x\|_{w_\infty(\lambda)}} \right) \leq K_N < 1.$$

Then (4.39) implies

$$\left\|I - \Delta_\rho^+ \Sigma_\rho^{+(N)}\right\|_{(w_\infty(\lambda), w_\infty(\lambda))} \leq K_N < 1$$

and $\Delta_\rho^+ \Sigma_\rho^{+(N)}$ has a unique inverse in the Banach algebra $w_\infty(\lambda)$, $w_\infty(\lambda)$. Since obviously $\Sigma_\rho^{+(N)}$ is bijective from $w_\infty(\lambda)$ into itself the operators defined by $\Delta_\rho^+ \Sigma_\rho^{+(N)}$ and $\Delta_\rho^+ = (\Delta_\rho^+ \Sigma_\rho^{+(N)})(\Sigma_\rho^{+(N)})^{-1}$ are bijective from $w_\infty(\lambda)$ into itself. So for any given $b \in w_\infty(\lambda)$ the equation $\Delta_\rho^+ x = b$ has a unique solution in $w_\infty(\lambda)$. Finally, we obtain

$$(\Delta_\rho^+)^{-1} = \sum_{i=0}^{\infty} (I - \Delta_\rho^+)^i = \left[\left[(\Delta_\rho^+)^T\right]^{-1}\right]^T$$

and an elementary calculation shows

$$(\Delta_\rho^+)^{-1} = \left[\left[(\Delta_\rho^+)^T\right]^{-1}\right]^T = \begin{pmatrix} 1 & \rho_1 & \rho_1\rho_2 & \cdot & & \cdot \\ & 1 & & \rho_2 & \rho_2\rho_3 & \cdot \\ & & & \cdot & \cdot & \\ 0 & & & & 1 & \rho_n \\ & & & & & \cdot \end{pmatrix}.$$

We then obtain (4.40). This completes the proof of Part (i).
(ii) Let

$$\Sigma_\rho^{(N)} = \begin{pmatrix} \left[\Delta_\rho^{(N+1)}\right]^{-1} & 0 \\ & 1 \\ 0 & & \cdot \end{pmatrix},$$

where $\Delta_\rho^{(N+1)}$ is the finite matrix whose entries are those of the $N+1$ first rows and columns of Δ_ρ. The finite matrix $\Delta_\rho^{(N+1)}$ is invertible, since it is a triangle. We obtain $\Sigma_\rho^{(N)}\Delta_\rho = (a_{nk})_{n,k=1}^\infty$, with $a_{nn} = 1$ for all n $a_{n,n-1} = -\rho_{n-1}$ for all $n \geq N+1$ and $a_{nk} = 0$ otherwise. For any given $x \in w_\infty(\lambda)$, we have $(I - \Sigma_\rho^{(N)}\Delta_\rho)x = (\xi_n(x))_{n=1}^\infty$ with $\xi_n(x) = 0$ for all $n \leq N$ and $\xi_n(x) = \rho_{n-1}x_{n-1}$ for all $n \geq N+1$. Now we put

$$K_N' = \sup_{n \geq N+1} (\lambda_n^\bullet) \sup_{k \geq N+1} |\rho_{k-1}|.$$

Then as in Part (i) we have

$$\sup_{n \geq N+1} \left[\left(\sup_{N \geq k} |\rho_k|\right) \lambda_n^\bullet\right] \leq K_N' < 1 \text{ for all } n \geq N. \tag{4.43}$$

Then

$$\begin{aligned}\left\|\left(I-\Sigma_{\rho}^{(N)}\Delta_{\rho}\right)x\right\|_{w_{\infty}(\lambda)} &= \sup_{n\geq N+1}\left(\frac{1}{\lambda_n}\sum_{k=N+1}^{n}|\rho_{k-1}x_{k-1}|\right)\\ &\leq \sup_{n\geq N+1}\left[\left(\sup_{N+1\leq k}|\rho_{k-1}|\right)\lambda_n^{\bullet}\right]\sup_{n\geq N+1}\left(\frac{1}{\lambda_{n-1}}\sum_{k=N}^{n-1}|x_k|\right)\\ &\leq K_N'\|x\|_{w_{\infty}(\lambda)}.\end{aligned}$$

Then using (4.43) we obtain

$$\left\|I-\Sigma_{\rho}^{(N)}\Delta_{\rho}\right\|_{(w_{\infty}(\lambda),w_{\infty}(\lambda))}\leq K_N'<1$$

and we conclude using similar arguments as those used above. □

Remark 4.14 Note that since λ is a nondecreasing sequence we do not need to assume $\overline{\lim}_{n\to\infty}\lambda_n^{\bullet}<\infty$ in Lemma 4.11.

We immediately obtain the following.

Corollary 4.9 *Under (4.41) we have* $w_{\infty}(\lambda)(\Delta_{\rho})=w_{\infty}(\lambda)$.

Now we study the sets $W_{\tau}(A)$, where A is any of the sets $\Delta(\lambda)$, $C(\lambda)$ or $C^{+}(\lambda)$. We put $W_{\tau}=D_{\tau}w_{\infty}$ and $W_{\tau}^{0}=D_{\tau}w_0$ for $\tau\in\mathcal{U}^{+}$, that is,

$$W_{\tau}=\left\{x\in\omega:\|x\|_{W_{\tau}}=\sup_n\left(\frac{1}{n}\sum_{k=1}^{n}\frac{|x_k|}{\tau_k}\right)<\infty\right\}\tag{4.44}$$

and

$$W_{\tau}^{0}=\left\{x\in\omega:\lim_{n\to\infty}\left(\frac{1}{n}\sum_{k=1}^{n}\frac{|x_k|}{\tau_k}\right)=0\right\}.\tag{4.45}$$

It was shown in [15, Proposition 3.1, p. 54] that the sets W_{τ} and W_{τ}^{0} are BK spaces normed by $\|\cdot\|_{W_{\tau}}$ and W_{τ}^{0} has AK. So $W_e=w_{\infty}$ and $W_e^{0}=w_0$. The classes (W_{τ},W_{τ}) and $(W_{\tau}^{0},W_{\tau}^{0})$ are Banach algebras with the norm

$$\|A\|_{(W_{\tau},W_{\tau})}^{*}=\sup_{y\neq 0}\left(\frac{\|Ay\|_{W_{\tau}}}{\|y\|_{W_{\tau}}}\right).$$

In all that follows we use the convention that the terms with subscripts strictly less than 1 are equal to zero. Then we are interested in the study of the following sets where $\lambda,\tau\in\mathcal{U}^{+}$:

$$W_{\tau}(\Delta(\lambda))=\left\{x\in\omega:\sup_n\left(\frac{1}{n}\sum_{k=1}^{n}\frac{1}{\tau_k}|\lambda_k x_k-\lambda_{k-1}x_{k-1}|\right)<\infty\right\},$$

$$W_\tau(C(\lambda)) = \left\{ x \in \omega : \sup_n \frac{1}{n} \sum_{m=1}^{n} \left(\frac{1}{\lambda_m \tau_m} \left| \sum_{k=1}^{m} x_k \right| \right) < \infty \right\},$$

$$W_\tau(C^+(\lambda)) = \left\{ x \in \omega : \sup_n \frac{1}{n} \sum_{m=1}^{n} \left(\frac{1}{\tau_m} \left| \sum_{k=1}^{m} x_k \right| \right) < \infty \right\}.$$

Note that for $\lambda_n = n$ and $\tau = e$, $W_\tau(\Delta(\lambda))$ is the well-known set of all strongly bounded sequences c_∞. We obtain the following result which is a direct consequence of Lemma 4.11.

Proposition 4.6 ([20, Proposition 3.1 pp. 122–123]) *Let $\lambda, \tau \in \mathcal{U}^+$. Then we have*

(i) If $\tau \in \Gamma$, then the operators Δ and Σ are bijective from W_τ into itself and

$$W_\tau(\Delta) = W_\tau,\ W_\tau(\Sigma) = W_\tau.$$

(ii) (a) If $\lambda\tau \in \Gamma$, then

$$W_\tau(C(\lambda)) = W_{\lambda\tau}.$$

(b) If $\tau \in \Gamma$, then

$$W_\tau(\Delta(\lambda)) = W_{\tau/\lambda}.$$

(iii) Let $\tau \in \Gamma^+$. Then

(a) the operators Δ^+ and Σ^+ are bijective from W_τ into itself and

$$W_\tau(\Sigma^+) = W_\tau;$$

(b) the operator $C^+(\lambda)$ is bijective from $W_{\lambda\tau}$ into W_τ and

$$W_\tau(C^+(\lambda)) = W_{\lambda\tau}.$$

Proof (i) By Lemma 4.11, where $\rho_n = \tau_n^\bullet$ and $\lambda_n = n$ for all n, we easily see that if

$$\overline{\lim_{n\to\infty}} \tau_n^\bullet < \frac{1}{\lim_{n\to\infty}\left(\frac{n-1}{n}\right)} = 1,$$

that is, $\tau \in \Gamma$, then $D_{1/\tau}\Delta D_\tau$ is bijective from w_∞ to itself. This means that Δ is bijective from $D_\tau w_\infty$ to itself. Since Σ is also bijective from $D_\tau w_\infty$ to itself, this shows $W_\tau(\Delta) = W_\tau$ and $W_\tau(\Sigma) = W_\tau$.

(ii) We have $x \in W_\tau(C(\lambda))$ if and only if $\Sigma x \in D_{\lambda\tau} w_\infty = W_{\lambda\tau}$. This means $x \in W_{\lambda\tau}(\Sigma)$ and by Part (i), the condition $\lambda\tau \in \Gamma$ implies $W_{\lambda\tau}(\Sigma) = W_{\lambda\tau}$. Then we have $W_\tau(C(\lambda)) = W_{\lambda\tau}$ and $C(\lambda)$ is bijective from $W_{\lambda\tau}$ to W_τ. Since $\Delta(\lambda) = C(\lambda)^{-1}$, we conclude that $\Delta(\lambda)$ is bijective from W_τ to $W_{\lambda\tau}$ and $W_{\lambda\tau}(\Delta(\lambda)) = W_\tau$. We deduce $W_\tau(\Delta(\lambda)) = W_{\tau/\lambda}$ for $\tau \in \Gamma$.

(iii)

(a) By Lemma 4.11 with $\rho_n = \tau_{n+1}/\tau_n$ and $\lambda_n = n$, we have $\Delta_\rho^+ = D_{1/\tau}\Delta^+ D_\tau$ and Δ^+ is bijective from $D_\tau w_\infty = W_\tau$ into itself for $\tau \in \Gamma^+$ and the same holds true for Σ^+. Now the equation $\Sigma^+ x = y$ for $y \in W_\tau$ is equivalent to

$$\sum_{k=n}^{\infty} x_k = y_n \text{ for all } n. \tag{4.46}$$

We deduce that the equation in (4.46) has a unique solution $x = (y_n - y_{n+1})_{n=1}^{\infty} = \Delta^+ y \in W_\tau$ and $W_\tau(\Sigma^+) = W_\tau$.

(iii) (b) We have

$$W_\tau(C^+(\lambda)) = \left\{x \in \omega : \Sigma^+ D_{1/\lambda} x \in W_\tau\right\} = D_\lambda W_\tau(\Sigma^+).$$

Now as we have seen above since $\tau \in \Gamma^+$, we obtain $W_\tau(\Sigma^+) = W_\tau$ and

$$W_\tau(C^+(\lambda)) = D_\lambda W_\tau(\Sigma^+) = D_\lambda W_\tau = W_{\lambda\tau}.$$

This completes the proof. □

4.12 Calculations in New Sequence Spaces

In this section, we study the sets $[C, \Delta]_{W_\tau}$, $[C, C]_{W_\tau}$, $[C^+, \Delta]_{W_\tau}$, $[C^+, C]_{W_\tau}$ and $[C^+, C^+]_{W_\tau}$.

In Sect. 4.8, we defined and studied the sets

$$[A_1, A_2] = [A_1(\lambda), A_2(\mu)] = \{x \in \omega : A_1(\lambda)(|A_2(\mu)x|) \in s_\tau\},$$

where $|x| = (|x_n|)_{n=1}^{\infty}$, A_1 and A_2 of the form $C(\xi)$, $C^+(\xi)$, $\Delta(\xi)$ or $\Delta^+(\xi)$ for $\xi \in \mathcal{U}^+$, and necessary conditions to obtain $[A_1(\lambda), A_2(\mu)]$ in the form s_γ have been given. In the same way, we consider

$$[A_1, A_2]_{W_\tau} = [A_1(\lambda), A_2(\mu)]_{W_\tau} = \{x \in \omega : A_1(\lambda)(|A_2(\mu)x|) \in W_\tau\}$$

for $\lambda, \mu, \tau \in \mathcal{U}^+$. We can explicitly write the sets $[A_1, A_2]_{W_\tau}$ as follows:

$$[C, \Delta]_{W_\tau} = \left\{x \in \omega : \sup_n \left[\frac{1}{n}\sum_{m=1}^{n}\left(\frac{1}{\lambda_m \tau_m}\sum_{k=1}^{m}|\mu_k x_k - \mu_{k-1}x_{k-1}|\right)\right] < \infty\right\},$$

$$[C, C]_{W_\tau} = \left\{x \in \omega : \sup_n \left[\frac{1}{n}\sum_{m=1}^{n}\left(\frac{1}{\lambda_m \tau_m}\sum_{k=1}^{m}\frac{1}{\mu_k}\left|\sum_{i=1}^{k} x_i\right|\right)\right] < \infty\right\},$$

$$[C^+, \Delta]_{W_\tau} = \left\{ x \in \omega : \sup_n \left[\frac{1}{n} \sum_{m=1}^{n} \left(\frac{1}{\tau_m} \sum_{k=m}^{\infty} \frac{1}{\lambda_k} |\mu_k x_k - \mu_{k-1} x_{k-1}| \right) \right] < \infty \right\},$$

$$[C^+, C]_{W_\tau} = \left\{ x \in \omega : \sup_n \left[\frac{1}{n} \sum_{m=1}^{n} \left(\frac{1}{\tau_m} \sum_{k=m}^{\infty} \frac{1}{\lambda_k} \frac{1}{\mu_k} \left| \sum_{i=1}^{k} x_i \right| \right) \right] < \infty \right\},$$

$$[C^+, C^+]_{W_\tau} = \left\{ x \in \omega : \sup_n \left[\frac{1}{n} \sum_{m=1}^{n} \left(\frac{1}{\tau_m} \sum_{k=m}^{\infty} \frac{1}{\lambda_k} \left| \sum_{i=k}^{\infty} \frac{x_i}{\mu_i} \right| \right) \right] < \infty \right\}.$$

We note that if $\lambda_n = \mu_n$ for all n then we obtain the well-known set of sequences that are strongly bounded $[C, \Delta]_{W_e} = c_\infty(\lambda)$. We state the following result.

Theorem 4.11 ([20, Theorem 4.1, pp. 124–125]) *Let $\lambda, \mu\tau \in \mathcal{U}^+$.*

(i) If $\lambda\tau \in \Gamma$, then

$$[C, \Delta]_{W_\tau} = W_{\lambda\tau/\mu}.$$

(ii) If $\lambda\tau, \lambda\mu\tau \in \Gamma$, then

$$[C, C]_{W_\tau} = W_{\lambda\mu\tau}.$$

(iii) If $\tau \in \Gamma^+$ and $\lambda\tau \in \Gamma$, then

$$[C^+, \Delta]_{W_\tau} = W_{\lambda\tau/\mu}.$$

(iv) If $\tau \in \Gamma^+$ and $\lambda\mu\tau \in \Gamma$, then

$$[C^+, C]_{W_\tau} = W_{\lambda\mu\tau}.$$

(v) If $\tau, \lambda\tau \in \Gamma^+$, then

$$[C^+, C^+]_{W_\tau} = W_{\lambda\mu\tau}.$$

Proof In the following we use the fact that for any $\xi \in \mathcal{U}^+$, we have $|x| \in W_\xi$ if and only if $x \in W_\xi$.

(i) We have $C(\lambda)(|\Delta(\mu)x|) \in W_\tau$ if and only if $|\Delta(\mu)x| \in W_\tau(C(\lambda))$ and since $\lambda\tau \in \Gamma$, we obtain by Proposition 4.6 $W_\tau(C(\lambda)) = W_{\lambda\tau}$. Then we have by Part (ii) of Proposition 4.6 $W_{\lambda\tau}(\Delta(\mu)) = W_{\lambda\tau/\mu}$ and we conclude $\Delta(\mu)x \in W_{\lambda\tau}$ if and only if $x \in W_{\lambda\tau}(\Delta(\mu)) = W_{\lambda\tau/\mu}$, that is, $[C, \Delta]_{W_\tau} = W_{\lambda\tau/\mu}$.
(ii) Here we have $C(\lambda)(|C(\mu)x|) \in W_\tau$ if and only if $|C(\mu)x| \in W_\tau(C(\lambda))$, and since $\lambda\tau \in \Gamma$ by Proposition 4.6, we have $W_\tau(C(\lambda)) = W_{\lambda\tau}$. So we have $x \in [C, C]_{W_\tau}$ if and only if $C(\mu)x \in W_{\lambda\tau}$, that is, $x \in W_{\lambda\tau}(C(\mu))$. Then by Part (ii) (a) of Proposition 4.6 $\lambda\mu\tau \in \Gamma$ implies $W_{\lambda\tau}(C(\mu)) = W_{\lambda\mu\tau}$ and we have shown Part (ii).
(iii) For any given $x \in [C^+, \Delta]_{W_\tau}$, we have $\Delta(\mu)x \in W_\tau(C^+(\lambda))$ and for $\tau \in \Gamma^+$, we have $W_\tau(C^+(\lambda)) = W_{\lambda\tau}$. Now the condition $\lambda\tau \in \Gamma$ implies $x \in [C^+, \Delta]_{W_\tau}$ if and only if $x \in W_{\lambda\tau}(\Delta(\mu)) = W_{\lambda\tau/\mu}$ and we have shown Part (iii).

(iv) Let $x \in [C^+, C]_{W_\tau}$. We have that $\tau \in \Gamma^+$ implies $W_\tau(C^+(\lambda)) = W_{\lambda\tau}$ and so $x \in [C^+, C]_{W_\tau}$ if and only if $C(\mu)x \in W_{\lambda\tau}$. Now since $\lambda\mu\tau \in \Gamma$, we have $W_{\lambda\tau}(C(\mu)) = W_{\lambda\mu\tau}$ and we conclude $[C^+, C]_{W_\tau} = W_{\lambda\mu\tau}$.
(v) As above $x \in [C^+, C^+]_{W_\tau}$ if and only if $C^+(\mu)x \in W_\tau(C^+(\lambda))$ and the condition $\tau \in \Gamma^+$ implies $W_\tau(C^+(\lambda)) = W_{\lambda\tau}$. Since $\lambda\tau \in \Gamma^+$, we conclude $W_{\lambda\tau}(C^+(\mu)) = W_{\lambda\mu\tau}$, that is, $[C^+, C^+]_{W_\tau} = W_{\lambda\mu\tau}$. □

Now we are led to study sets of the form $[\Delta, A_2]_{W_\tau}$ for $A_2 \in \{\Delta, C, C^+\}$.
First, we consider the sets $[\Delta, \Delta]_{W_\tau}$, $[\Delta, C]_{W_\tau}$ and $[\Delta, C^+]_{W_\tau}$.
Using the convention $\mu_0 = 0$, and the notation $[\Delta(\mu)x]_k = \mu_k x_k - \mu_{k-1}x_{k-1}$ for $k \geq 1$, we explicitly have

$$[\Delta, \Delta]_{W_\tau} = \left\{ x \in \omega : \sup_n \left(\frac{1}{n} \sum_{k=1}^{n} \frac{1}{\tau_k} |\lambda_k |[\Delta(\mu)x]_k| - \lambda_{k-1}|[\Delta(\mu)x]_{k-1}|| \right) < \infty \right\},$$

$$[\Delta, C]_{W_\tau} =$$
$$\left\{ x \in \omega : \sup_n \left(\frac{1}{n} \sum_{m=1}^{n} \frac{1}{\tau_m} \left| \lambda_m \left| \frac{1}{\mu_m} \sum_{k=1}^{m} x_k \right| - \lambda_{m-1} \left| \frac{1}{\mu_{m-1}} \sum_{k=1}^{m-1} x_k \right| \right| \right) < \infty \right\}$$

and

$$[\Delta, C^+]_{W_\tau} = \left\{ x \in \omega : \sup_n \left(\frac{1}{n} \sum_{m=1}^{n} \frac{1}{\tau_m} \left| \lambda_m \left| \sum_{k=m}^{\infty} \frac{x_k}{\mu_k} \right| - \lambda_{m-1} \left| \sum_{k=m-1}^{\infty} \frac{x_k}{\mu_k} \right| \right| \right) < \infty \right\}.$$

As a direct consequence of Proposition 4.6, we also obtain the following results.

Theorem 4.12 ([20, Proposition 4.2, p. 126]) *Let λ, μ, $\tau \in \mathcal{U}^+$. Then we have.*

(i) If $\tau, \tau/\lambda \in \Gamma$, then
$$[\Delta, \Delta]_{W_\tau} = W_{\tau/\lambda\mu}.$$

(ii) If $\tau, \tau\mu/\lambda \in \Gamma$, then
$$[\Delta, C]_{W_\tau} = W_{\tau\mu/\lambda}.$$

(iii) If $\tau, \tau/\lambda \in \Gamma^+$, then
$$[\Delta, C^+]_{W_\tau} = W_{\tau\mu/\lambda}.$$

Proof (i) Let $x \in [\Delta, \Delta]_{W_\tau}$. Since $\tau \in \Gamma$, we have $W_\tau(\Delta(\lambda)) = W_{\tau/\lambda}$ and $\Delta(\lambda)|\Delta(\mu)x| \in W_\tau$ means $\Delta(\mu)x \in W_{\tau/\lambda}$. We conclude, since $W_{\tau/\lambda}(\Delta(\mu)) = W_{\tau/\lambda\mu}$ for $\tau/\lambda \in \Gamma$.
(ii) By similar arguments as those used above, since $\tau \in \Gamma$, we have $x \in [\Delta, C]_{W_\tau}$ if and only if $C(\mu)x \in W_{\tau/\lambda}$. We conclude, since the condition $\tau\mu/\lambda \in \Gamma$ implies $W_{\tau/\lambda}(C(\mu)) = W_{\tau\mu/\lambda}$.

(iii) Here, under the conditions $\tau, \tau/\lambda \in \Gamma^+$, we have $x \in [\Delta, C^+]_{W_\tau}$ if and only if $x \in W_{\tau/\lambda}(C^+(\mu)) = W_{\tau\mu/\lambda}$. □

The previous results can be applied to the case when w_∞ is replaced by w^0.

Using the Banach algebra $(w^0(\lambda), w^0(\lambda))$ we obtain similar results as those above by replacing $w_\infty(\lambda)$ by $w^0(\lambda)$ and W_τ by $W^0_\tau = D_\tau w^0$. We note that $x \in W^0_\tau$ if and only if

$$\frac{1}{n}\sum_{k=1}^{n}\frac{|x_k|}{\tau_k} \to 0\ (n \to \infty).$$

We can state the following result.

Proposition 4.7 ([20, Proposition 4.3, pp. 126–127]) *Let $\lambda, \mu \in \mathcal{U}^+$. Then we have*

(i) If $\lambda\tau \in \Gamma$, then $[C, \Delta]_{W^0_\tau} = W^0_{\lambda\tau/\mu}$.
(ii) If $\lambda\tau, \lambda\mu\tau \in \Gamma$, then $[C, C]_{W^0_\tau} = W^0_{\lambda\mu\tau}$.
(iii) If $\tau \in \Gamma^+$ and $\lambda\tau \in \Gamma$, then $[C^+, \Delta]_{W^0_\tau} = W^0_{\lambda\tau/\mu}$.
(iv) If $\tau \in \Gamma^+$ and $\lambda\mu\tau \in \Gamma$, then $[C^+, C]_{W^0_\tau} = W^0_{\lambda\mu\tau}$.
(v) If $\tau, \lambda\tau \in \Gamma^+$, then $[C^+, C^+]_{W^0_\tau} = W^0_{\lambda\mu\tau}$.
(vi) If $\tau, \tau/\lambda \in \Gamma$, then $[\Delta, \Delta]_{W^0_\tau} = W^0_{\tau/\lambda\mu}$.
(vii) If $\tau, \tau\mu/\lambda \in \Gamma$, then $[\Delta, C]_{W^0_\tau} = W^0_{\tau\mu/\lambda}$.
(viii) If $\tau, \tau/\lambda \in \Gamma^+$, then $[\Delta, C^+]_{W^0_\tau} = W^0_{\tau\mu/\lambda}$.

We immediately obtain the next remark.

Remark 4.15 It is easy to see that in Proposition 4.7 each of the sets $[A_1, A_2]_{W^0_\tau}$ is equal to $W^0_\tau(A_1A_2)$. This result is a direct consequence of the previous proofs and of the fact that W^0_τ is of absolute type, that is, $|x| \in W^0_\tau$ if and only if $x \in W^0_\tau$.

4.13 Application to A−Statistical Convergence

The results of the previous section can be applied to statistical convergence.

We give conditions for $x_k \to L(S(A))$, where A is one of the infinite matrices

$$D_{1/\tau}C(\lambda)C(\mu),\ D_{1/\tau}\Delta(\lambda)\Delta(\mu),\ \text{or}\ D_{1/\tau}\Delta(\lambda)C(\mu).$$

Then we give conditions for $x_k \to 0(S(A))$, where A is any of the operators

$$D_{1/\tau}C^+(\lambda)\Delta(\mu),\ D_{1/\tau}C^+(\lambda)C(\mu),\ D_{1/\tau}C^+(\lambda)C^+(\mu),\ \text{or}\ D_{1/\tau}\Delta(\lambda)C^+(\mu).$$

The sequence $x = (x_n)_{n=1}^\infty$ is said to be *statistically convergent to the number L*, if

$$\lim_{n\to\infty}\frac{1}{n}|\{k \le n : |x_k - L| \ge \varepsilon\}| = 0 \text{ for all } \varepsilon > 0,$$

where the vertical bars indicate the number of elements in the enclosed set. In this case, we write $x_k \to L(S)$ or $st - \lim x = L$.

Let $A \in (E, F)$ for given $L \in \mathbb{C}$ and we use the notation for every $\varepsilon > 0$

$$I_\varepsilon(A) = \{k \leq n : |[Ax]_k - L| \geq \varepsilon\},$$

(where we assume that every series $[Ax]_k = A_k x = \sum_{m=1}^{\infty} a_{km} x_m$ is convergent for each $k \geq 1$). We say that the sequence $x = (x_n)_{n=1}^{\infty}$ is *$A-$statistically convergent to L* if for every $\varepsilon > 0$

$$\lim_{n \to \infty} \frac{1}{n} |I_\varepsilon(A)| = 0.$$

Then we write $x_k \to L(S(A))$ and for $A = I$, $x_k \to L(S(I))$ means $st - \lim x = L$.

Now we need a lemma where we put $T^{-1}e = \widetilde{l} = (l_n)_{n=1}^{\infty}$ for any given triangle T.

We state the following result.

Lemma 4.12 *If $x - L\widetilde{l} \in w^0(T)$, then x_k is $T-$statistically convergent to L.*

Proof The condition $x - L\widetilde{l} \in w^0(T)$ means $T(x - L\widetilde{l}) \in w^0$. Since

$$Tx - Le = T(x - LT^{-1}e) = T(x - L\widetilde{l})$$

for any $\varepsilon > 0$, we have

$$\begin{aligned} y_n &= \frac{1}{n} \sum_{k=1}^{n} |[Tx]_k - L| = \frac{1}{n} \sum_{k=1}^{n} \left|[T(x - L\widetilde{l})]_k\right| \\ &\geq \frac{1}{n} \sum_{k \in I_\varepsilon(T)} \left|[T(x - L\widetilde{l})]_k\right| \geq \frac{1}{n} \sum_{k \in I_\varepsilon(T)} \varepsilon \\ &\geq \frac{\varepsilon}{n} |\{k \leq n : |[Tx]_k - L| \geq \varepsilon\}|. \end{aligned}$$

We conclude that $x - L\widetilde{l} \in w^0(T)$ implies $y_n \to 0$ $(n \to \infty)$ and $x_k \to L(S(T))$. □

We are led to state the next results.

Theorem 4.13 ([20, Theorem 5.2, pp. 128–129]) *Let $\lambda, \tau, \mu \in \mathcal{U}^+$. Then we have*

(i) *Let $\lambda\tau, \lambda\tau\mu \in \Gamma$. If*

$$\lim_{n \to \infty} \frac{1}{n} \sum_{k=1}^{n} \frac{|x_k - L[\lambda_k \mu_k \tau_k + (\mu_{k-1} + \mu_k)\lambda_{k-1}\tau_{k-1} - \lambda_{k-2}\mu_{k-2}\tau_{k-2}]|}{\lambda_k \mu_k \tau_k} = 0 \tag{4.47}$$

then $x_k \to L(S(D_{1/\tau} C(\lambda) C(\mu)))$, that is, for every $\varepsilon > 0$

$$\lim_{n\to\infty}\frac{1}{n}\left|\left\{k\le n:\left|\frac{1}{\lambda_k\tau_k}\sum_{i=1}^{k}\frac{1}{\mu_i}\left(\sum_{j=1}^{i}x_j\right)-L\right|\ge\varepsilon\right\}\right|=0.$$

(ii) *Let* $\tau, \tau/\lambda \in \Gamma$. *If*

$$\lim_{n\to\infty}\frac{1}{n}\sum_{k=1}^{n}\frac{\lambda_k\mu_k}{\tau_k}\left|x_k-L\left(\frac{1}{\mu_k}\sum_{i=1}^{k}\frac{1}{\lambda_i}\sum_{j=1}^{i}\tau_j\right)\right|=0$$

then $x_k \to L(S(D_{1/\tau}\Delta(\lambda)\Delta(\mu)))$, *that is, for every* $\varepsilon > 0$

$$\lim_{n\to\infty}\frac{1}{n}\left|\left\{k\le n:\left|\frac{1}{\tau_k}\left[\lambda_k[\Delta(\mu)x]_k-\lambda_{k-1}[\Delta(\mu)x]_{k-1}\right]-L\right|\ge\varepsilon\right\}\right|=0.$$

(iii) *Let* $\tau, \tau\mu/\lambda \in \Gamma$. *If*

$$\lim_{n\to\infty}\frac{1}{n}\sum_{k=1}^{n}\frac{\lambda_k}{\mu_k\tau_k}\left|x_k-L\left(\left(\frac{\mu_k}{\lambda_k}-\frac{\mu_{k-1}}{\lambda_{k-1}}\right)\sum_{i=1}^{k-1}\tau_i+\frac{\mu_k}{\lambda_k}\tau_k\right)\right|=0$$

then $x_k \to L(S(D_{1/\tau}\Delta(\lambda)C(\mu)))$, *that is, for every* $\varepsilon > 0$

$$\lim_{n\to\infty}\frac{1}{n}\left|\left\{k\le n:\left|\frac{1}{\tau_k}\left[\left(\frac{\lambda_k}{\mu_k}-\frac{\lambda_{k-1}}{\mu_{k-1}}\right)\sum_{i=1}^{k-1}x_i+\frac{\lambda_k}{\mu_k}x_k\right]-L\right|\ge\varepsilon\right\}\right|=0.$$

Proof (i) First by Part (ii) of Proposition 4.7 and Remark 4.15, we easily see that, for $\lambda\tau, \lambda\tau\mu \in \Gamma$, we have $W_\tau^0(C(\lambda)C(\mu)) = W^0_{\lambda\mu\tau}$. Putting $T = D_{1/\tau}C(\lambda)C(\mu)$, we obtain

$$w^0(T) = W_\tau^0(C(\lambda)C(\mu)) = W^0_{\lambda\mu\tau}. \tag{4.48}$$

Then

$$\tilde{l} = T^{-1}e = \Delta(\mu)\Delta(\lambda)D_\tau e$$

with

$$\begin{aligned} l_n &= [\Delta(\mu)\Delta(\lambda)D_\tau e]_n \\ &= \lambda_n\mu_n\tau_n + (\mu_{n-1}+\mu_n)\lambda_{n-1}\tau_{n-1} - \lambda_{n-2}\mu_{n-2}\tau_{n-2} \text{ for all } n. \end{aligned} \tag{4.49}$$

Using (4.48) and (4.49) we see that the condition in (4.47) is equivalent to $x - L\tilde{l} \in w^0(T)$. By Lemma 4.12, we conclude $x_k \to L(S(T))$.
This completes the proof of Part (i).
(ii) By Part (vi) of Proposition 4.7 and Remark 4.15, we have $W_\tau^0(\Delta(\lambda)\Delta(\mu)) = W^0_{\tau/\lambda\mu}$, since $\tau, \tau/\lambda \in \Gamma$. Then putting $T' = D_{1/\tau}\Delta(\lambda)\Delta(\mu)$, we obtain

$$w^0(T') = W_\tau^0(\Delta(\lambda)\Delta(\mu)) = W_{\tau/\lambda\mu}^0.$$

Since

$$\widetilde{l'} = T'^{-1}e = C(\mu)C(\lambda)D_\tau e,$$

we have

$$l'_n = [C(\mu)C(\lambda)D_\tau e]_n = \frac{1}{\mu_n}\sum_{i=1}^n \frac{1}{\lambda_i}\left(\sum_{j=1}^i \tau_j\right) \text{ for all } n.$$

By Lemma 4.12 we conclude $x_k \to L(S(D_{1/\tau}\Delta(\lambda)\Delta(\mu)))$ for all x with

$$\lim_{n\to\infty}\frac{1}{n}\sum_{k=1}^n |x_k - Ll'_k|\frac{\lambda_k\mu_k}{\tau_k} = 0.$$

This shows Part (ii).

(iii) Since $\tau, \tau\mu/\lambda \in \Gamma$, again we have, by Part (vii) of Proposition 4.7 and Remark 4.15 $W_\tau^0(\Delta(\lambda)C(\mu)) = W_{\tau\mu/\lambda}^0$. Then we let $T'' = D_{1/\tau}\Delta(\lambda)C(\mu)$ and obtain

$$w^0(T'') = W_\tau^0(\Delta(\lambda)C(\mu)) = W_{\tau\mu/\lambda}^0.$$

Writing $\widetilde{l''} = T''^{-1}e = \Delta(\mu)C(\lambda)D_\tau e$ we successively obtain

$$D_\tau e = (\tau_n)_{n=1}^\infty,\ C(\lambda)D_\tau e = \left(\frac{1}{\lambda_n}\left(\sum_{i=1}^n \tau_i\right)\right)_{n=1}^\infty$$

and

$$\Delta(\mu)C(\lambda)D_\tau e = \left(\frac{\mu_n}{\lambda_n}\sum_{i=1}^n \tau_i - \frac{\mu_{n-1}}{\lambda_{n-1}}\sum_{i=1}^{n-1}\tau_i\right)_{n=1}^\infty.$$

So we have for each n

$$l''_n = [\Delta(\mu)C(\lambda)D_\tau e]_n = \left(\frac{\mu_n}{\lambda_n} - \frac{\mu_{n-1}}{\lambda_{n-1}}\right)\sum_{i=1}^{n-1}\tau_i + \frac{\mu_n}{\lambda_n}x_k.$$

We conclude that for every x with

$$\lim_{n\to\infty}\frac{1}{n}\sum_{k=1}^n |x_k - Ll''_k|\frac{\lambda_k}{\mu_k\tau_k} = 0,$$

we have $x_k \to L(S(T''))$. Finally, we easily obtain

$$[T''x]_n = \frac{1}{\tau_n}\left(\frac{\lambda_n}{\mu_n}\sum_{i=1}^{n} x_i - \frac{\lambda_{n-1}}{\mu_{n-1}}\sum_{i=1}^{n-1} x_i\right)$$
$$= \frac{1}{\tau_n}\left[\left(\frac{\lambda_n}{\mu_n} - \frac{\lambda_{n-1}}{\mu_{n-1}}\right)\sum_{i=1}^{n-1} x_i + \frac{\lambda_n}{\mu_n}x_n\right].$$

This shows Part (iii). □

We are led to illustrate the previous results with some examples.

Example 4.2 The condition

$$\lim_{n\to\infty}\frac{1}{n}\sum_{k=1}^{n}\left|\frac{x_k}{2^k} - \frac{7}{4}L\right| = 0$$

for given $L \in \mathbb{C}$ implies for each $\varepsilon > 0$

$$\lim_{n\to\infty}\frac{1}{n}\left|\left\{k \le n : \left|\frac{1}{2^k}\sum_{i=1}^{k}\sum_{j=1}^{i} x_j - L\right| \ge \varepsilon\right\}\right| = 0.$$

Indeed, it is enough to apply Part (i) Theorem 4.13 with $\lambda_k = k$, $\tau_k = 2^k/k$ and $\mu_k = 1$ for all k.

We can also state the next application.

Example 4.3 If $\lim_{n\to\infty}(1/n)\sum_{k=1}^{n}|x_k|/k2^k = 0$, then for each $\varepsilon > 0$

$$\lim_{n\to\infty}\frac{1}{n}\left|\left\{k \le n : \left|\frac{1}{2^k}\left(\frac{1}{k} - \frac{1}{k-1}\right)\sum_{i=1}^{k-1} x_i + \frac{1}{k}x_k\right| \ge \varepsilon\right\}\right| = 0.$$

This result is a direct consequence of Part (iii) of Theorem 4.13 with $\lambda_k = 1$, $\tau_k = 2^k$ and $\mu_k = k$ for all k.

In the following, we use Proposition 4.7 and the expressions of $W_\tau^0(C^+(\lambda)\Delta(\mu)) = [C^+, \Delta]_{W_\tau^0}$, $W_\tau^0(C^+(\lambda)C(\mu)) = [C^+, C]_{W_\tau^0}$, $W_\tau^0(C^+(\lambda)C^+(\mu)) = [C^+, C^+]_{W_\tau^0}$ and $W_\tau^0(\Delta(\lambda)C^+(\mu)) = [\Delta, C^+]_{W_\tau^0}$ obtained above. We now require a lemma which is a direct consequence of Lemma 4.12.

Lemma 4.13 *Let A be an infinite matrix. If $x \in w^0(A)$, then we have*

$$x_k \to 0(S(A)).$$

We deduce the next results.

Theorem 4.14 ([20, Theorem 5.6, p. 131]) *Let $\lambda, \tau, \mu \in \mathcal{U}^+$. Then we have*

(i) *Let* $\tau \in \Gamma^+$ *and* $\lambda\tau \in \Gamma$. *If*

$$\lim_{n\to\infty} \frac{1}{n} \sum_{k=1}^{n} \frac{|x_k|}{\lambda_k \tau_k} \mu_k = 0 \tag{4.50}$$

then $x_k \to 0(S(D_{1/\tau}C^+(\lambda)\Delta(\mu)))$, *that is, for every* $\varepsilon > 0$, *we have*

$$\lim_{n\to\infty} \frac{1}{n} \left| \left\{ k \le n : \left| \frac{1}{\tau_k} \sum_{i=k}^{\infty} \frac{\mu_i x_i - \mu_{i-1} x_{i-1}}{\lambda_i} \right| \ge \varepsilon \right\} \right| = 0. \tag{4.51}$$

(ii) *Let* $\tau \in \Gamma^+$ *and* $\lambda\mu\tau \in \Gamma$. *If*

$$\lim_{n\to\infty} \frac{1}{n} \sum_{k=1}^{n} \frac{|x_k|}{\lambda_k \mu_k \tau_k} = 0, \tag{4.52}$$

then $x_k \to 0(S(D_{1/\tau}C^+(\lambda)C(\mu)))$, *that is, for every* $\varepsilon > 0$

$$\lim_{n\to\infty} \frac{1}{n} \left| \left\{ k \le n : \left| \frac{1}{\tau_k} \sum_{i=k}^{\infty} \frac{1}{\lambda_i} \left(\frac{1}{\mu_i} \sum_{j=1}^{i} x_j \right) \right| \ge \varepsilon \right\} \right| = 0. \tag{4.53}$$

(iii) *Let* $\tau, \lambda\tau \in \Gamma^+$. *If*

$$\lim_{n\to\infty} \frac{1}{n} \sum_{k=1}^{n} \frac{|x_k|}{\lambda_k \mu_k \tau_k} = 0, \tag{4.54}$$

then $x_k \to 0(S(D_{1/\tau}C^+(\lambda)C^+(\mu)))$, *that is, for every* $\varepsilon > 0$

$$\lim_{n\to\infty} \frac{1}{n} \left| \left\{ k \le n : \left| \frac{1}{\tau_k} \sum_{i=k}^{\infty} \frac{1}{\lambda_i} \left(\sum_{j=i}^{\infty} \frac{x_j}{\mu_j} \right) \right| \ge \varepsilon \right\} \right| = 0. \tag{4.55}$$

(iv) *Let* $\tau, \tau/\lambda \in \Gamma^+$. *If*

$$\lim_{n\to\infty} \frac{1}{n} \sum_{k=1}^{n} \frac{\lambda_k |x_k|}{\mu_k \tau_k} = 0,$$

then $x_k \to 0(SD_{1/\tau}(\lambda)C^+(\mu))$, *that is, for every* $\varepsilon > 0$

$$\lim_{n\to\infty} \frac{1}{n} \left| \left\{ k \le n : \frac{1}{\tau_k} \left| (\lambda_k - \lambda_{k-1}) \sum_{i=k-1}^{\infty} \frac{x_i}{\mu_i} + \frac{\lambda_k}{\mu_k} x_k \right| \ge \varepsilon \right\} \right| = 0. \tag{4.56}$$

Proof (i) The condition in (4.50) implies $x \in W^0_{\lambda\tau/\mu}$ and by Proposition 4.7 and Remark 4.15 we have, since $\tau \in \Gamma^+$ and $\lambda\tau \in \Gamma$

$$W^0_{\lambda\tau/\mu} = W^0_\tau(C^+(\lambda)\Delta(\mu))$$

and $x \in W^0_\tau(C^+(\lambda)\Delta(\mu))$. Now it can easily be seen that

$$\left[D_{1/\tau}C^+(\lambda)\Delta(\mu)x\right]_n = \frac{1}{\tau_n}\sum_{i=n}^{\infty}\frac{\mu_i x_i - \mu_{i-1}x_{i-1}}{\lambda_i},$$

so we conclude by Lemma 4.12 with $A = D_{1/\tau}C^+(\lambda)\Delta(\mu)$

$$x_k \to 0(S(D_{1/\tau}C^+(\lambda)(\mu))).$$

This shows Part (i).
(ii) Here the condition in (4.52) means $x \in W^0_{\lambda\mu\tau}$ and we have by Proposition 4.7 and Remark 4.15, since $\tau \in \Gamma^+$ and $\lambda\mu\tau \in \Gamma$, $W^0_{\lambda\mu\tau} = W^0_\tau(C^+(\lambda)C(\mu))$ and $x \in W^0_\tau(C^+(\lambda)C(\mu))$. Now since

$$[D_{1/\tau}C^+(\lambda)C(\mu)x]_n = \frac{1}{\tau_n}\sum_{i=n}^{\infty}\frac{1}{\lambda_i}\left(\frac{1}{\mu_i}\sum_{j=1}^{i}x_j\right),$$

by Lemma 4.12, where $A' = D_{1/\tau}C^+(\lambda)C(\mu)$, we conclude

$$x_k \to 0\left(S(D_{1/\tau}C^+(\lambda)C(\mu))\right).$$

So we have shown Part (ii).
(iii) can be obtained using similar arguments as above with $A'' = D_{1/\tau}C^+(\lambda)C^+(\mu)$, and so

$$x_k \to 0\left(S(D_{1/\tau}C^+(\lambda)C^+(\mu))\right).$$

(iv) can also be obtained similarly. It is enough to put $A''' = D_{1/\tau}\Delta(\lambda)C^+(\mu)$. An elementary computation yields

$$[A'''x]_k = \frac{1}{\tau_k}\left[(\lambda_k - \lambda_{k-1})\sum_{i=k-1}^{\infty}\frac{x_i}{\mu_i} + \frac{\lambda_k}{\mu_k}x_k\right]$$

and we conclude

$$x_k \to 0\left(S(D_{1/\tau}\Delta(\lambda)C^+(\mu))\right),$$

that is, (4.56) holds. □

We state the next example.

Example 4.4 We have for each $\varepsilon > 0$

$$\lim_{n\to\infty}\frac{1}{n}\left|\left\{k\le n : \left|2^k\sum_{i=1}^{\infty}\frac{1}{3^i}\left(\sum_{j=1}^{i}x_j\right)\right|\ge\varepsilon\right\}\right| = 0 \text{ for all } x\in W^0_{3/2}.$$

It is enough to apply Part (ii) of Theorem 4.14 with $\tau_k = 2^{-k}$, $\mu_k = 3^k$ and $\lambda_k = 1$ for all k.

We also have the next example.

Example 4.5 By Part (iii) of Theorem 4.14 with $\lambda_k = \mu_k = k$ and $\tau_k = 2^{-k}$, the condition

$$\lim_{n\to\infty}\frac{1}{n}\sum_{k=1}^{n}2^k\frac{|x_k|}{k^2} = 0$$

implies that for each $\varepsilon > 0$

$$\lim_{n\to\infty}\frac{1}{n}\left|\left\{k\le n : \left|2^k\sum_{i=k}^{\infty}\frac{1}{i}\left(\sum_{j=i}^{\infty}\frac{x_j}{j}\right)\right|\ge\varepsilon\right\}\right| = 0.$$

4.14 Tauberian Theorems for Weighted Means Operators

We start from results on Hardy's Tauberian theorem for Cesàro means. This was formulated as follows, if the sequence $x = (x_n)_{n=1}^{\infty}$ satisfies $\lim_{n\to\infty} C_1 x = L$ and $\Delta x_n = O(1/n)$, then $\lim_{n\to\infty} x = L$. It was shown by Fridy and Khan [22] that the hypothesis $\lim_{n\to\infty} C_1 x = L$ can be replaced by the weaker assumption of the existence of the statistical limit $st - \lim C_1 x = L$, that is, for every $\varepsilon > 0$

$$\lim_{n\to\infty}\frac{1}{n}\left|\{k\le n : |[C_1 x]_k - L|\ge\varepsilon\}\right| = 0.$$

Here it is our aim to show that Hardy's Tauberian theorem for Cesàro means can be extended to the cases when C_1 is successively replaced by the operator of weighted means $\overline{N}_q$ defined in Definition 2.2 and by $C(\lambda)$.

In this way we show in Theorem 4.15 that, under some conditions, if $x = (x_n)_{n=1}^{\infty}$ satisfies $\lim_{n\to\infty}\overline{N}_q x = L_1$ and $\lim_{n\to\infty} Q_n\Delta q_n x_n = L_2$, then

$$\lim_{n\to\infty} x = L_1. \tag{4.57}$$

Similarly in Theorem 4.16, we show that, under some other conditions, (4.57) holds for sequences x that satisfy the conditions

$$\lim_{n\to\infty} \overline{N}_q x = L_1 \text{ and } \lim_{n\to\infty} \frac{Q_n}{q_n} \Delta x_n = L_2.$$

Then in Proposition 4.8, we give an extension of Hardy's Tauberian theorem, where C_1 is replaced by $C(\lambda)$ and we determine sequences μ for which x is convergent when the sequences $C(\lambda)x$ and $(\mu_n \Delta x_n)_{n=1}^{\infty}$ are convergent.

We recall the characterization of (c, c), which we use in all that follows.

Lemma 4.14 (Part **13.** of Theorem 1.23) *We have* $A = (a_{nk})_{n,k=1}^{\infty} \in (c, c)$ *if and only if*

(i) $A \in S_1$*;*
(ii) $\lim_{n\to\infty} \sum_{k=1}^{\infty} a_{nk} = l$ *for some* $l \in \mathbb{C}$*;*
(iii) $\lim_{n\to\infty} a_{nk} = l_k$ *for some* $l_k \in \mathbb{C}$ *and for all* $k \geq 1$.

We recall that a matrix $A = (a_{nk})_{n,k=1}^{\infty} \in (c, c)$ is said to be regular if $x_n \to l$ $(n \to \infty)$ implies $A_n x = \sum_{k=1}^{\infty} a_{nk} x_k$ is convergent for all n and converges to the same limit. We write $x_n \to l$ implies $A_n x \to l$ $(n \to \infty)$. We recall that A is regular if and only if A satisfies the condition in (i) of Lemma 4.14 and $\lim_{n\to\infty} A_n e = 1$ and $\lim_{n\to\infty} a_{nk} = 0$ for all $k \geq 1$.

Let $q = (q_n)_{n=1}^{\infty}$ be a positive sequence, Q be the sequence defined by $Q_n = \sum_{k=1}^{n} q_k$ for all $n \geq 1$. The operator of weighted means $\overline{N}_q$ is defined by the matrix $\overline{N}_q$ of Definition 2.2. It can easily be seen that $\overline{N}_q = D_{1/Q} \Sigma D_q$. In all that follows, we write $x_n = 0$ for any term of sequence with negative subscript.

Now we will give two versions of Tauberian theorems concerning the operator of weighted means $\overline{N}_q$. Then we deal with the operator $C(\lambda)$.

Now we give a first version of a Tauberian theorem for $\overline{N}_q$.

Theorem 4.15 ([12, Theorem 2.1, pp. 3–4])

(i) The following statements are equivalent:

(a) $Q/q \in \ell_{\infty}$,
(b) for any given sequence $(x_n)_{n=1}^{\infty}$, *we have*

$$\lim_{n\to\infty} \frac{q_1 x_1 + \cdots + q_n x_n}{Q_n} = L_1 \text{ if and only if } \lim_{n\to\infty} x_n = L_1$$

for some $L_1 \in \mathbb{C}$.

(ii) We assume

$$\lim_{n\to\infty} \frac{1}{n q_n} \sum_{k=1}^{n} \frac{k}{Q_k} = L \tag{4.58}$$

and

$$\lim_{n\to\infty} \frac{Q_n}{n q_n} - L' \neq 0 \tag{4.59}$$

for some scalars L and L'. Then, for any given sequence $(x_n)_{n=1}^{\infty}$, *the conditions*

$$\lim_{n\to\infty} \frac{q_1x_1 + \cdots + q_nx_n}{Q_n} = L_1 \text{ and } \lim_{n\to\infty} Q_n(q_nx_n - q_{n-1}x_{n-1}) = L_2$$

for some $L_1, L_2 \in \mathbb{C}$ *imply together* $\lim_{n\to\infty} x_n = L_1$.

Proof (i) We have $\overline{N}_q^{-1} = (D_{1/Q}\Sigma D_q)^{-1} = D_{1/q}\Delta D_Q$, that is, $[\overline{N}_q^{-1}]_{n,n-1} = -Q_{n-1}/q_n$, $[\overline{N}_q^{-1}]_{nn} = Q_n/q_n$ for all $n \geq 1$ (with the convention $Q_0 = 0$) and $[\overline{N}_q^{-1}]_{nn} = 0$ otherwise. Since Q is increasing and $Q/q \in \ell_\infty$, we have

$$\left\|\overline{N}_q^{-1}\right\|_{S_1} = \sup_n \left(\frac{Q_n + Q_{n-1}}{q_n}\right) \leq 2\sup_n \frac{Q_n}{q_n} < \infty.$$

Then $\lim_{n\to\infty}(Q_n - Q_{n-1})/q_n = 1$ and we conclude $\overline{N}_q^{-1}$ is regular. This shows the condition in (a) holds if and only if $\overline{N}_q^{-1}$ is regular. So (a) means that, for any $y = (y_n)_{n=1}^\infty$, the condition $y_n = [\overline{N}_qx]_n \to L_1$ implies

$$x_n = [\overline{N}_q^{-1}y]_n \to L_1 \ (n \to \infty).$$

Since $\overline{N}_q$ is regular, we conclude

$$x_n \to L_1 \text{ implies } y_n \to L_1 \ (n \to \infty).$$

So we have shown Part (i).
(ii) Let $x = (x_n)_{n=1}^\infty \in \omega$ and $y = (y_n)_{n=1}^\infty = \overline{N}_qx$. Writing $z = (z_n)_{n=1}^\infty = (Q_n(q_nx_n - q_{n-1}x_{n-1}))_{n=1}^\infty$, we easily see that

$$z = D_Q\Delta D_qx. \tag{4.60}$$

We have $(D_Q\Delta D_q)^{-1} = D_{1/q}\Sigma D_{1/Q}$ and by (4.60) we obtain

$$x = (D_Q\Delta D_q)^{-1}z = D_{1/q}\Sigma D_{1/Q}z.$$

Then

$$\begin{aligned} y &= \overline{N}_qx = \overline{N}_qD_{1/q}\Sigma D_{1/Q}z \\ &= D_{1/Q}\Sigma D_qD_{1/q}\Sigma D_{1/Q}z = D_{1/Q}\Sigma^2D_{1/Q}z \end{aligned}$$

and the infinite matrix Σ^2 is the triangle defined by $[\Sigma^2]_{nk} = n + 1 - k$ for $k \leq n$ and $[\Sigma^2]_{nk} = 0$ otherwise. So we easily obtain

$$y_n = \frac{1}{Q_n}\sum_{k=1}^n \frac{n+1-k}{Q_k}z_k = \frac{n+1}{Q_n}\sum_{k=1}^n \frac{z_k}{Q_k} - \frac{1}{Q_n}\sum_{k=1}^n \frac{k}{Q_kz_k}.$$

Since

$$x_n = \left[D_{1/q}\Sigma D_{1/Q}z\right]_n = \frac{1}{q_n}\sum_{k=1}^{n}\frac{z_k}{Q_k},$$

we obtain

$$y_n = \frac{n+1}{Q_n}q_n x_n - \frac{1}{Q_n}\sum_{k=1}^{n}\frac{k}{Q_k}z_k.$$

Now we consider the triangle

$$\widehat{Q} = \begin{pmatrix} \cdot & & & \\ \cdot & \cdot & & 0 \\ \cdot & \dfrac{1}{(n+1)q_n} & \dfrac{k}{Q_k} & \cdot \\ \cdot & \cdot & & \cdot\ \cdot \end{pmatrix}.$$

The condition in (4.59) implies $1/nq_n \sim L'/Q_n$ $(n \to \infty)$ and since Q_n is increasing we have

$$0 < \lim_{n\to\infty} Q_n \leq \infty$$

and $(1/nq_n)_{n=1}^{\infty} \in c$. So, for each fixed k, the sequence $[\widehat{Q}]_{nk}$ tends to a limit as n tends to infinity. This and the condition in (4.58) imply $\widehat{Q} \in (c, c)$ and, since $z \in c$, we have

$$\frac{1}{(n+1)q_n}\sum_{k=1}^{n}\frac{k}{Q_k}z_k \to l \ (n \to \infty) \text{ for some } l \in \mathbb{C}.$$

Using (4.59) we deduce that if $y_n \to L_1$ and $z_n \to L_2$ $(n \to \infty)$, then we have

$$x_n = \frac{Q_n}{(n+1)q_n}y_n + \frac{1}{(n+1)q_n}\sum_{k=1}^{n}\frac{k}{Q_k}z_k \to L'L_1 + l$$

and $x \in c$. Now since $\overline{N}_q$ is regular and $y_n = [\overline{N}_q x]_n$, we have

$$y_n \to L_1 = L'L_1 + l \ (n \to \infty).$$

We conclude $x_n \to L_1$ $(n \to \infty)$. □

As a consequence of Theorem 4.15 (i), we obtain the next result.

Corollary 4.10 ([12, Corollary 2.2, p. 5]) *Let $x = (x_n)_{n=1}^{\infty}$ be any given sequence. The condition*

$$[\overline{N}_q x]_n \to L \text{ implies } x_n \to L \ (n \to \infty) \tag{4.61}$$

for some $L \in \mathbb{C}$ implies there are $\gamma > 1$ and $K > 0$ such that

$$q_n \geq K\gamma^n \text{ for all } n.$$

Proof The condition in (4.61) implies that $\overline{N}_q^{-1}$ is regular, that is,

$$\frac{Q_n + Q_{n-1}}{q_n} = O(1) \ (n \to \infty).$$

and $Q/q \in \ell_\infty$. Then $q \in \widehat{C}_1$, where $\widehat{C}_1 = \{x : ((\sum_{k=1}^n x_k)/x_n)_{n=1}^\infty \in \ell_\infty\}$ was defined in (4.9), (cf. [6]). We conclude by Part (iii) of Proposition 4.3 (see also [7, Proposition 2.1, p. 1786]). □

As a direct consequence of Part (ii) of Theorem 4.15, we obtain the next corollary.

Corollary 4.11 ([12, Corollary 2.3, p. 6]) *Let $\alpha \geq 0$ and let $(x_n)_{n=1}^\infty$ be a sequence that satisfy*

$$\frac{x_1 + 2^\alpha x_2 + \cdot + n^\alpha x_n}{\sum\limits_{k=1}^n k^\alpha} \to L_1 \text{ and } \left(\sum_{k=1}^n k^\alpha\right)(n^\alpha x_n - (n-1)^\alpha x_{n-1}) \to L_2$$

for some $L_1, L_2 \in \mathbb{C}$. Then $x_n \to L_1$ $(n \to \infty)$.

Proof If $\alpha = 0$ then the conditions in (4.58) and (4.59) are trivially satisfied. Now we let $q_n = n^\alpha$ with $\alpha > 0$ and $\alpha \neq 1$. We obtain

$$\frac{1}{n^{\alpha+1}} \int_0^n x^\alpha \, dx \leq \frac{Q_n}{nq_n} = \frac{\sum\limits_{k=1}^n k^\alpha}{n^{\alpha+1}} \leq \frac{1}{n^{\alpha+1}} \int_1^{n+1} x^\alpha \, dx,$$

and since the sequences $(1/n^{\alpha+1}) \int_0^n x^\alpha \, dx$ and $(1/n^{\alpha+1}) \int_1^{n+1} x^\alpha \, dx$ tend to the same limit $1/(\alpha+1)$ as n tends to infinity, we conclude $\lim_{n\to\infty} Q_n/nq_n = 1/(\alpha+1)$ and (4.59) holds.

Now we deal with the condition in (4.58). For this, we note that for every $k \geq 2$

$$\frac{k}{Q_k} \leq \frac{k}{\int_0^k x^\alpha \, dx} = \frac{\alpha+1}{k^\alpha}.$$

Then

$$\sum_{k=1}^{n} \frac{k}{Q_k} \leq 1 + (\alpha + 1) \sum_{k=2}^{n} \frac{1}{k^{\alpha}}$$
$$\leq 1 + (\alpha + 1) \int_{1}^{n} \frac{dx}{x^{\alpha}}$$
$$\leq 1 + \frac{\alpha + 1}{1 - \alpha} \left(n^{1-\alpha} - 1\right).$$

Thus

$$\frac{1}{n^{\alpha+1}} \sum_{k=1}^{n} \frac{k}{Q_k} \leq \frac{1+\alpha}{1-\alpha} \left(\frac{1}{n^{2\alpha}} - \frac{1}{n^{\alpha+1}}\right) + \frac{1}{n^{\alpha+1}}$$

and

$$\frac{1}{n^{\alpha+1}} \sum_{k=1}^{n} \frac{k}{Q_k} \to 0 \ (n \to \infty).$$

We conclude applying Theorem 4.15.

For $\alpha = 1$, we obtain

$$\frac{1}{nq_n} \sum_{k=1}^{n} \frac{k}{Q_k} = \frac{1}{n^2} \sum_{k=1}^{n} \frac{2}{(k+1)} \leq \frac{2}{n^2} \int_{0}^{n} \frac{dx}{x+1} = \frac{2}{n^2} \log(n+1)$$

and

$$\frac{1}{nq_n} \sum_{k=1}^{n} \frac{k}{Q_k} \to 0 \ (n \to \infty).$$

Since (4.59) trivially holds with $L' = 1/2$, we can apply Theorem 4.15 and conclude $x_n \to L_1$ $(n \to \infty)$.

This completes the proof. □

We immediately deduce the following corollary from the previous proof.

Corollary 4.12 *Let* $(x_n)_{n=1}^{\infty}$ *be any sequence. If*

$$\frac{x_1 + 2x_2 + \cdots + nx_n}{n^2} \to L_1 \text{ and } n^2[nx_n - (n-1)x_{n-1}] \to L_2,$$

then $x_n \to 2L_1$ $(n \to \infty)$.

Now we consider another statement of a Tauberian theorem, where the conditions in (4.58) and (4.59) in Theorem 4.15 are replaced by the convergence of

$$\frac{1}{Q_n} \sum_{k=2}^{n} \frac{q_k Q_{k-1}}{Q_k}$$

and the condition on $Q_n(q_n x_n - q_{n-1}x_{n-1})$ is replaced by a similar condition on another sequence defined by $Q_n(x_n - x_{n-1})/q_n$.

Theorem 4.16 ([12, Theorem 2.5, p. 7]) *We assume*

$$\lim_{n\to\infty} \frac{1}{Q_n} \sum_{k=2}^{n} q_k \frac{Q_{k-1}}{Q_k} = L \tag{4.62}$$

for some scalar L. For any given sequence $(x_n)_{n=1}^{\infty}$, the conditions

$$\lim_{n\to\infty} \frac{q_1 x_1 + \cdots + q_n x_n}{Q_n} = L_1 \text{ and } \lim_{n\to\infty} \frac{Q_n}{q_n}(x_n - x_{n-1}) = L_2 \tag{4.63}$$

for some $L_1, L_2 \in \mathbb{C}$ imply $\lim_{n\to\infty} x_n = L_1$.

Proof We let

$$y = (y_n)_{n=1}^{\infty} = \overline{N}_q x \tag{4.64}$$

and $z = D_{Q/q}\Delta x$. Then we have

$$x = \Sigma D_{q/Q} z \tag{4.65}$$

and

$$y = \overline{N}_q \Sigma D_{q/Q} z = D_{1/Q} \Sigma D_q \Sigma D_{q/Q} z.$$

Since

$$[\Sigma D_q \Sigma]_{nk} = \begin{cases} \sum_{i=k}^{n} q_i & \text{for } k \le n \\ 0 & \text{otherwise,} \end{cases}$$

we obtain

$$\begin{aligned} y_n &= \frac{1}{Q_n} \sum_{k=1}^{n} \left(\sum_{i=k}^{n} q_i \right) \frac{q_k}{Q_k} z_k \\ &= \frac{1}{Q_n} \sum_{k=1}^{n} (Q_n - Q_{k-1}) \frac{q_k}{Q_k} z_k \\ &= \sum_{k=1}^{n} \frac{q_k}{Q_k} z_k - \frac{1}{Q_n} \sum_{k=1}^{n} \frac{Q_{k-1}}{Q_k} q_k z_k. \end{aligned}$$

Using (4.65) we deduce

$$y_n = x_n - \frac{1}{Q_n} \sum_{k=1}^{n} \frac{Q_{k-1}}{Q_k} q_k z_k$$

and

$$x_n = y_n + \frac{1}{Q_n} \sum_{k=1}^{n} \frac{Q_{k-1}}{Q_k} q_k z_k.$$

Now consider the matrix $\widetilde{Q}$ with $(\widetilde{Q})_{nk} = q_k Q_{k-1}/Q_n Q_k$ for $2 \leq k \leq n$ and $(\widetilde{Q})_{nk} = 0$ otherwise. Since Q is increasing, we have $1/Q \in c$ and for each fixed k, $(\widetilde{Q})_{nk}$ tends to a limit as n tends to infinity. This result and the conditions in (4.62) imply $\widetilde{Q} \in (c, c)$. Now we consider the sequence w defined by

$$w_n = \frac{1}{Q_n} \sum_{k=1}^{n} \frac{Q_{k-1}}{Q_k} q_k z_k.$$

The conditions given in (4.63) mean that $y_n \to L_1$ and $z_n \to L_2$ $(n \to \infty)$ and since $\widetilde{Q} \in (c, c)$, we have

$$x_n = y_n + w_n = y_n + [\widetilde{Q}]_n \to L_1 + l \text{ for some } l \in \mathbb{C}.$$

To complete the proof we need to show $l = 0$. For this it is enough to notice that, since $\overline{N}_q$ is regular, the condition $x_n \to L_1 + l$ implies

$$y_n = [\overline{N}_q x]_n \to L_1 + l = L_1 \ (n \to \infty)$$

and $x_n \to L_1$ $(n \to \infty)$.

This concludes the proof. □

This result leads to the next corollary.

Corollary 4.13 *Let* $(x_n)_{n=1}^{\infty}$ *be a sequence with*

$$\lim_{n\to\infty} \frac{1}{\log n} \left(x_1 + \frac{1}{2} x_2 + \cdots + \frac{1}{n} x_n \right) = L_1 \text{ and } \lim_{n\to\infty} n \log n (x_n - x_{n-1}) = L_2.$$

Then $\lim_{n\to\infty} x_n = L_1$.

Proof We have $q_n = 1/n$ for all n, $Q_n = \sum_{k=1}^{n} (1/k)$ and

$$u_n = \frac{1}{Q_n} \sum_{k=2}^{n} q_k \frac{Q_{k-1}}{Q_k} = \frac{1}{Q_n} \sum_{k=2}^{n} \frac{1}{k} - \sigma_n$$

with

$$\sigma_n = \frac{1}{Q_n} \sum_{k=2}^{n} \frac{1}{k^2 Q_k}.$$

Since Q_n tends to infinity as n tends to infinity and $\sigma_n \leq (1/Q_n)\sum_{k=2}^{n}(1/k^2)$, we have $\lim_{n\to\infty} \sigma_n = 0$. Then u_n tends to 1 as n tends to infinity and the condition in (4.62) of Theorem 4.16 is satisfied. Finally, since $Q_n \sim \log n$, we have $Q_n/q_n \sim n \log n$ $(n \to \infty)$ and we conclude by Theorem 4.16. □

Remark 4.16 We note that neither of Theorems 4.15 and 4.16 implies the other. Indeed, we consider the case when $q_n = n$ for all n. Then the sequence $x = e$ satisfies Theorem 4.16, since (4.62) is satisfied with $L = 1$ and $(Q_n/q_n)(x_n - x_{n-1}) = 0$ for all n, but

$$Q_n(q_n x_n - q_{n-1}x_{n-1}) \sim \frac{n^2}{2} \ (n \to \infty),$$

so $Q_n(q_n x_n - q_{n-1}x_{n-1}) \to \infty$ $(n \to \infty)$ and Theorem 4.15 cannot hold.

Furthermore, in the case when $q_n = 1/n$, we have seen in Corollary 4.13 that the condition (4.62) in Theorem 4.16 is satisfied, but (4.59) in Theorem 4.15 is not satisfied.

4.15 The Operator $C(\lambda)$

Now we consider the case when $\overline{N}_q$ is replaced by $C(\lambda)$ and we obtain results that extend some of those given in the previous sections.

By similar arguments as those used in Theorems 4.15 and 4.16, we can state the next proposition, where $\overline{N}_q$ is replaced by $C(\lambda)$. We will see that in the case of $\lambda = \mu$ the sequence λ plays the role of Q with $q = e$.

Proposition 4.8 ([12, Proposition 2.8, p. 9]) *Let $\lambda, \mu \in \mathcal{U}^+$ and*

$$\lim_{n\to\infty} \frac{1}{n}\sum_{k=1}^{n}\frac{k}{\mu_k} = L \tag{4.66}$$

and

$$\lim_{n\to\infty} \frac{\lambda_n}{n} = L'$$

for some scalars L and L′.

Then for any given sequence $(x_n)_{n=1}^{\infty}$ the conditions

$$\frac{x_1 + \cdots + x_n}{\lambda_n} \to l \text{ and } \mu_n(x_n - x_{n-1}) \to l' \ (n \to \infty) \tag{4.67}$$

for some $l, l' \in \mathbb{C}$ together imply that $(x_n)_{n=1}^{\infty}$ is convergent and

$$x_n \to L'l \ (n \to \infty).$$

Proof We let $y_n = (x_1 + \cdots + x_n)/\lambda_n$ and $z_n = \mu_n(x_n - x_{n-1})$. Then we have $y = C(\lambda)x$, $z = D_\mu \Delta x$ and $y = C(\lambda)\Sigma D_{1/\mu} z$. Since $C(\lambda) = D_{1/\lambda}\Sigma$ and $(D_\mu \Delta)^{-1} = \Sigma D_{1/\mu}$, we obtain

$$y = D_{1/\lambda}\Sigma^2 D_{1/\mu} z \text{ and } x = \Sigma D_{1/\mu} z.$$

As we have seen in the proof of Theorem 4.16, we obtain by the calculation of $D_{1/\lambda}\Sigma^2 D_{1/\mu}$

$$y_n = \frac{1}{\lambda_n}\sum_{k=1}^{n} \frac{n-k+1}{\mu_k} z_k.$$

Similarly, we have

$$x_n = \left[\Sigma D_{1/\mu} z\right]_n = \sum_{k=1}^{n} \frac{z_k}{\mu_k}.$$

So, we successively have

$$y_n = \frac{n+1}{\lambda_n} x_n - \frac{1}{\lambda_n}\sum_{k=1}^{n} \frac{k}{\mu_k} z_k \tag{4.68}$$

and

$$x_n = \frac{\lambda_n}{n+1} y_n + \frac{1}{n+1}\sum_{k=1}^{n} \frac{k}{\mu_k} z_k. \tag{4.69}$$

Now let Λ be the triangle defined by $\Lambda_{nk} = k/((n+1)\mu_k)$ for $2 \le k \le n$ and $\Lambda_{nk} = 0$ for $k = 1$ or $k > n$ and for all n and k. By (4.66), we have $\Lambda \in (c, c)$. Now the conditions in (4.67) mean that $y_n \to l$ and $z_n \to l'$ $(n \to \infty)$ and since $\Lambda \in (c, c)$, we deduce

$$x_n = \frac{\lambda_n}{n+1} y_n + [\Lambda z]_n \to L'l + \chi \ (n \to \infty) \text{ for some } \chi \in \mathbb{C}.$$

Now we are led to deal with the cases $L' = 0$ and $L' \neq 0$.

1- We assume $L' = 0$.
Here $x_n \to L'l + \chi = \chi$ and

$$\frac{x_1 + \cdots + x_n}{n} \to \chi.$$

Since $n/\lambda_n \to \infty$ and

$$\frac{x_1 + \cdots + x_n}{\lambda_n} = \frac{x_1 + \cdots + x_n}{n}\frac{n}{\lambda_n} \to l \ (n \to \infty),$$

we conclude $\chi = 0$ and $x_n \to 0 = L'l$ $(n \to \infty)$.

2- We assume $L' \neq 0$.
Here, since $x_n \to L'l + \chi\ (n \to \infty)$, we also have

$$\frac{x_1 + \cdots + x_n}{n} \to L'l + \chi\ (n \to \infty).$$

Then

$$\frac{x_1 + \cdots + x_n}{\lambda_n} = \frac{x_1 + \cdots + x_n}{n} \frac{n}{\lambda_n} \to \frac{L'l + \chi}{L'} = l\ (n \to \infty).$$

So $\chi = 0$ and $x_n \to L'l\ (n \to \infty)$.
This concludes the proof. □

We deduce the following result.

Corollary 4.14 *Let $x = (x_n)_{n=1}^{\infty}$ be a sequence with*

$$(x_1 + \cdots + x_n)/n \to l \text{ and } n(x_n - x_{n-1}) \to l'\ (n \to \infty)$$

for some $l, l' \in \mathbb{C}$. Then we have

(*i*) $x_n \to l\ (n \to \infty)$,
(*ii*) $l' = 0$.

Proof The condition in (4.66) is trivially satisfied and $L = L' = 1$. Since $\Lambda/L = \Lambda$ defined in the proof of Proposition 4.8 is regular, we have here $l' = \chi$. As we have just seen we successively obtain $x_n \to l + \chi\ (n \to \infty)$, $l + \chi = l + l' = l$ and $l' = 0$. □

Remark 4.17 It can be seen that Proposition 4.8 is an extension of Hardy's Tauberian theorem. For this, we show that there is $\mu \in \mathcal{U}^+$ with $(n/\mu_n)_n \in \mathcal{U}^+ \backslash \ell_\infty$ such that (4.66) holds. We take, for instance, $\mu_n = 2^i/i$ when $n = 2^i$ for $i \geq 1$ and $\mu_n = n^3$ otherwise. Let n be any given integer and $I_n = \{2^i : 2^i \leq n\}$. Using the notation $\overline{I'_n} = \overline{I_n} \bigcap [1, n]$ we successively have

$$\begin{aligned} s_n &= \frac{1}{n} \sum_{k=1}^{n} \frac{k}{\mu_k} = \frac{1}{n} \sum_{k \in I_n} \frac{k}{\mu k} + \frac{1}{n} \sum_{k \in \overline{I'_n}} \frac{k}{\mu_k} \\ &\leq \frac{1}{n} \sum_{k \in \{i: 2^i \leq n\}} k + \frac{1}{n} \sum_{k \in \overline{I'_n}} \frac{1}{k^2}. \end{aligned}$$

Putting $N = \max\{i : 2^i \leq n\}$ and $S = \sum_{k=1}^{\infty} 1/k^2$ we deduce

$$s_n \leq \frac{1}{n} \frac{N(N+1)}{2} + \frac{S}{n}.$$

Since $2^N \leq n$, we obtain

$$s_n \leq \frac{1}{2^{N+1}} N(N+1) + \frac{S}{n}.$$

Finally, from the definition of N and the inequality $N \leq \log n / \log 2$, we obtain $N = E(\log n / \log 2)$. So N tends to infinity as n tends to infinity and s_n tends to zero.

Since we have

$$n(x_n - x_{n-1}) = \frac{n}{\mu_n} \mu_n (x_n - x_{n-1})$$

and $(n/\mu_n)_{n=1}^{\infty} \notin \ell_\infty$, the condition $\mu_n(x_n - x_{n-1}) \to l'$ $(n \to \infty)$ does not imply $n(x_n - x_{n-1}) = O(1)$ $(n \to \infty)$ and we have shown that Proposition 4.8 is an extension of Hardy's Tauberian theorem.

More precisely we can state the following result when $C(\lambda)x \in c_0$, where we use the same notations as in the proof of Proposition 4.8.

Proposition 4.9 ([12, Proposition 2.12, p. 11]) *Let* $\lambda, \mu \in \mathcal{U}^+$ *and*

$$\sup_n \left(\frac{1}{n} \sum_{k=1}^{n} \frac{k}{\mu_k} \right) < \infty \tag{4.70}$$

and

$$\sup_n \frac{\lambda_n}{n} < \infty. \tag{4.71}$$

For any given sequence $(x_n)_{n=1}^{\infty}$, *the conditions*

$$\frac{x_1 + \cdots + x_n}{\lambda_n} \to 0 \text{ and } \mu_n(x_n - x_{n-1}) \to 0 \text{ imply } x_n \to 0 \ (n \to \infty).$$

Proof Here the condition in (4.70) implies $\Lambda \in S_1$ and trivially $\lim_{n\to\infty} \Lambda_{nk} = 0$ for all $k \geq 1$. Then we have $\Lambda \in (c_0, c_0)$ and

$$\frac{1}{n+1} \sum_{k=1}^{n} \frac{k}{\mu_k} z_k \to 0 \ (n \to \infty) \text{ for all } z \in c_0.$$

By (4.71), we also have

$$\frac{\lambda_n}{n+1} y_n = \frac{\lambda_n}{n} \frac{n}{n+1} y_n = O(1)o(1) = o(1) \ (n \to \infty)$$

and from the identity in (4.69) in the previous proof, we conclude $x_n \to 0$ $(n \to \infty)$. □

References

1. Connor, J.: On strong matrix summability with respect to a modulus and statistical convergence. Canad. Math. Bull. **32**(2), 194–198 (1989)
2. de Malafosse, B.: Systèmes linéaires infinis admettant une infinité de solutions. Atti. Accad. Peloritana Pericolanti, Cl. I Fis. Mat. Nat. **65**, 49–59 (1988)
3. de Malafosse, B.: On the spectrum of the Cesàro operator in the space s_r. Comm. Fac. Sci. Univ. Ankara Ser. A1 Math. Stat. **48**, 53–71 (1999)
4. de Malafosse, B.: Bases in sequence spaces and expansion of a function in a series of power series. Mat. Vesnik **52**(3–4), 99–112 (2000)
5. de Malafosse, B.: Properties of some sets of sequences and application to the spaces of bounded difference sequences of order μ. Hokkaido Math. J. **31**, 283–299 (2002)
6. de Malafosse, B.: On some BK space. Int. J. Math. Math. Sci. **58**, 1783–1801 (2003)
7. de Malafosse, B.: On the set of sequences that are strongly α-bounded and α-convergent to naught with index p. Rend. Sem. Mat. Univ. Pol. Torino **61**, 13–32 (2003)
8. de Malafosse, B.: Calculations on some sequence spaces. Int. J. Math. Math. Sci. **31**, 1653–1670 (2004)
9. de Malafosse, B.: Sum and product of certain BK spaces and matrix transformations between these spaces. Acta Math. Hung. **104**(3), 241–263 (2004)
10. de Malafosse, B.: The Banach algebra $\mathcal{B}(X)$, where X is a BK space and applications. Mat. Vesnik **57**, 41–60 (2005)
11. de Malafosse, B.: Sum of sequence spaces and matrix transformations. Acta Math. Hung. **113**(3), 289–313 (2006)
12. de Malafosse, B.: Tauberian theorems for the operator of weighted means. Commun. Math. Anal. **5**(2), 1–12 (2008)
13. de Malafosse, B., Malkowsky, E.: Sequence spaces and the inverse of an infinite matrix. Rend. Circ. Mat. Palermo, Serie II **51**, 277–294 (2002)
14. de Malafosse, B., Malkowsky, E.: Matrix transformations in the sets $\chi(\overline{N}_p\overline{N}_q)$ where χ is in the form s_ξ, or $\overset{\circ}{s}_\xi$, or $s_\xi^{(c)}$. Filomat **17**, 85–106 (2003)
15. de Malafosse, B., Malkowsky, E.: Matrix transformations between sets of the form W_ξ and operator generators of analytic semigroups. Jordan J. Math. Stat. **1**(1), 51–67 (2008)
16. de Malafosse, B., Malkowsky, E.: On sequence spaces equations using spaces of strongly bounded and summable sequences by the Cesàro method. Antarctica J. Math. **10**(6), 589–609 (2013)
17. de Malafosse, B., Malkowsky, E.: On the Banach algebra $(w_\infty(\lambda), w_\infty(\lambda))$ and applications to the solvability of matrix equations in $w_\infty(\lambda)$. Pub. Math. Debrecen **85**(1–2), 197–217 (2014)
18. de Malafosse, B., Malkowsky, E.: On the solvability of certain (SSIE) with operators of the form $B(r, s)$. Math. J. Okayama. Univ. **56**, 179–198 (2014)
19. de Malafosse, B., Rakočević, V.: Matrix transformation and statistical convergence. Linear Algebra Appl. **420**, 377–387 (2007)
20. de Malafosse, B., Rakočević, V.: Calculations in new sequence spaces and application to statistical convergence. Cubo A **12**(3), 117–132 (2010)
21. Fast, H.: Sur la convergence statistique. Colloq. Math. **2**, 241–244 (1951)
22. Fridy, J.A., Khan, M.K.: Statistical extensions of some classical Tauberian theorems. Proc. Amer. Math. Soc. **128**, 2347–2355 (2000)
23. Fridy, J.A., Orhan, C.: Lacunary statistical convergence. Pacific J. Math. **160**, 43–51 (1993)
24. Fridy, J.A.: On statistical convergence. Analysis **5**, 301–313 (1985)
25. Hardy, G.H.: A theorem concerning summable series. Proc. Cambridge Philos. Soc. **20**, 304–307 (1920)
26. Maddox, I.J.: Continuous and Köthe-Toeplitz duals of certain sequence spaces. Proc. Cambridge Philos. Soc. **65**, 431–435 (1967)
27. Maddox, I.J.: On Kuttner's theorem. London Math. Soc. **43**, 285–290 (1968)
28. Malkowsky, E.: The continuous duals of the spaces $c_0(\lambda)$ and $c(\lambda)$ for exponentially bounded sequences λ. Acta Sci. Math (Szeged) **61**, 241–250 (1995)

29. Malkowsky, E.: Linear operators between some matrix domains. Rend. Circ. Mat. Palermo, Serie II, Suppl. **68**, 641–655 (2002)
30. Malkowsky, E.: Banach algebras of matrix transformations between spaces of strongly bounded and summable sequences. Adv. Dyn. Syst. Appl. **6**(1), 241–250 (2011)
31. Malkowsky, E., Rakočević, V.: An introduction into the theory of sequence spaces and measures of noncompactness, volume 9(17) of Zbornik radova, Matematčki institut SANU, pp. 143–234. Mathematical Institute of SANU, Belgrade (2000)
32. Mursaleen, M., Edely, O.H.H.: Statistical convergence of double sequences. J. Math. Anal. Appl. **281**(1), 223–231 (2003)
33. Mursaleen, M., Mohiuddine, S.A.: Convergence Methods for Double Sequences and Applications. Springer, New York (2013)
34. Patterson, R.F., Savaş, E.: Lacunary statististical convergence of double sequences. Math. Commun. **10**(1), 55–61 (2005)
35. Steinhaus, H.: Sur la convergence ordinaire et la convergence asymptotique. Colloq. Math. **2**, 73–71 (1951)
36. Wilansky, A.: Summability Through Functional Analysis. Mathematical Studies, vol. 85. North–Holland, Amsterdam, (1984)

Chapter 5
Sequence Spaces Inclusion Equations

In this chapter, we consider new problems and results of *perturbations* on sequence spaces. If E is a set of complex sequences and u is a given sequence with nonzero terms then we write $E_u = \{x = (x_k) : (x_k/u_k) \in E\}$. Starting with the linear spaces F and F' of sequences, we solve the inclusion $F_b \subset F'_x$ for a given positive sequence b, where $x = (x_n)_{n=1}^{\infty}$ is the unknown positive sequence. Then, the question is: what are the solutions for the new inclusion obtained from the previous one where we add a new linear space E of sequences to the set F'_x in the second member of the above inclusion? So, the new problem consists in solving the new inclusion $F_b \subset E + F'_x$. Then, we consider the case when E is the matrix domain of an operator. Until now, there is no general theory to solve these problems in the general case. Nevertheless, if we know the multipliers of E in F and of F' in F, then the inclusion $E + F'_x \subset F_b$, where x is the unknown, is completely solved.

In [2], the sets s_a, s_a^0 and $s_a^{(c)}$ of Example 2.2 were defined for sequences $a \in \mathcal{U}^+$ by $(1/a)^{-1} * E$ and $E = \ell_\infty, c_0, c$, respectively. In [3], the sum $E_a + F_b$ and the product $E_a * F_b$ (cf. (4.6) in Sect. 4.1) were defined where E and F are any of the symbols s, s^0, or $s^{(c)}$. Then in [6] the solvability was determined of sequences spaces inclusion equations $G_b \subset E_a + F_x$ where $E, F, G \in \{s^0, s^{(c)}, s\}$ and some applications were given to sequence spaces inclusions with operators. We recall that the spaces w_∞ and w_0 of strongly bounded and summable sequences are the sets of all sequences $y = (y_n)_{n=1}^{\infty}$ such that $(n^{-1}\sum_{k=1}^{n}|y_k|)_n$ is bounded and tends to zero, respectively (cf. Sect. 4.1). These spaces were studied by Maddox [19] and Malkowsky, Rakočević [22]. In [8, 13, 16] some properties were given of well-known operators defined by the sets $W_a = (1/a)^{-1} * w_\infty$ and $W_a^0 = (1/a)^{-1} * w_0$. In this section, we deal with special *sequence spaces inclusion equations (SSIE), (resp. sequence spaces equations (SSE))*, which are determined by an inclusion, (resp. identity), for which each term is a *sum* or a *sum of products of sets of the form* $(E_a)_T$ and $(E_{f(x)})_T$, where f maps U^+ to itself, E is any linear space of sequences and T is a triangle. Some results on (SSE) and (SSIE) were stated in [1, 4, 5, 8, 9, 14, 15, 18]. In [9], we dealt with the (SSIE) with operators $E_a + (F_x)_\Delta \subset s_x^{(c)}$, where E and F are any of the sets c_0, c, or s_1. Then we gave a solution of the next inclusion equations with operator $s_x^{(c)} + (s_b^0)_\Delta \subset s_b$ and

B. de Malafosse et al., *Operators Between Sequence Spaces and Applications*,
https://doi.org/10.1007/978-981-15-9742-8_5

$s_x^0 + (s_b^0)_\Delta \subset s_b^{(c)}$. We note that the (SSIE) $s_x^{(c)} + (s_b^0)_\Delta \subset s_b$ means $y_n/x_n \to l$ and $(z_n - z_{n-1})/b_n \to 0$ $(n \to \infty)$ together imply $|y_n + z_n| \le K b_n$ for all $y, z \in \omega$ and for some scalars l and K, with $K > 0$. In [15], we determined the set of all positive sequences x for which the *(SSIE)* $(s_x^{(c)})_{B(r,s)} \subset (s_x^{(c)})_{B(r',s')}$ holds, where r, r', s' and s are real numbers, and $B(r, s)$ is the generalized operator of the first difference defined by $(B(r, s)y)_n = r y_n + s y_{n-1}$ for all $n \ge 2$ and $(B(r, s)y)_1 = r y_1$. In this way we determined the set of all positive sequences x for which $(r y_n + s y_{n-1})/x_n \to l$ implies $(r' y_n + s' y_{n-1})/x_n \to l$ $(n \to \infty)$ for all sequences y and some scalar l.

In this chapter, we recall some results stated [8–12, 14]. So we study the sequences spaces inclusions (SSIE) of the form $F \subset E_a + F'_x$ in each of the cases $e \in F$ and $e \notin F$.

5.1 Introduction

In this section, we deal with the (SSIE)

$$F \subset E_a + F'_x, \text{ where } E, F \text{ and } F' \text{ are sequence spaces with } e \in F,$$

and a is a positive sequence.

We obtain the solvability of these (SSIE)'s for $a = (r^n)_{n=1}^\infty$. We consider the (SSIE) $F \subset E_a + F'_x$ as a perturbed inclusion equation of the elementary inclusion equation $F \subset F'_x$. In this way it is interesting to determine the set of all positive sequences a for which the elementary and the perturbed inclusions equations have the same solutions. Then writing, as usual, D_r for the diagonal matrix with $(D_r)_{nn} = r^n$, we study the solvability of the (SSIE) $c \subset D_r * E_\Delta + c_x$ with $E = c_0$ or s_1, where Δ is operator of the first difference. Then we consider the (SSIE) $c \subset D_r * E_{C_1} + s_x^{(c)}$ with $E = c_0$, c or s_1, and $s_1 \subset D_r * (s_1)_{C_1} + s_x$ with $E = c$ or s_1, where C_1 is the Cesàro operator of order 1. For instance, the (SSIE) $s_1 \subset D_r * c_{C_1} + s_x$ is associated with the next statement:

> *The condition* $\sup_n |y_n| < \infty$ *implies that there are sequences* $u, v \in \omega$ *such that* $y = u + v$ *for which* $n^{-1}\sum_{k=1}^n u_k r^{-k} \to l$ *and* $\sup_n(|v_n|/x_n) < \infty$ *for some scalar* l *and for all* y.

For the reader's convenience, we list the characterizations of the classes (c_0, c_0), (c_0, c), (c, c_0), (c, c), (s_1, c) and (ℓ_p, F), where $F = c_0, c$ or ℓ_∞.

Lemma 5.1 *We have*

(i) (Theorem 1.23 7) $A \in (c_0, c_0)$ *if and only if*

$$\|A\|_{(\ell_\infty, \ell_\infty)} = \sup_n \sum_{k=1}^\infty |a_{nk}| < \infty \tag{5.1}$$

holds and

$$\lim_{n\to\infty} a_{nk} = 0 \text{ for all } k; \tag{5.2}$$

(ii) (Theorem 1.23 **12**) $A \in (c_0, c)$ *if and only if (5.1) holds and*

$$\lim_{n\to\infty} a_{nk} = l_k \text{ for all } k \text{ and some scalar } l_k; \tag{5.3}$$

(iii) (Theorem 1.23 **8**) $A \in (c, c_0)$ *if and only if (5.1) and (5.2) hold and*

$$\lim_{n\to\infty} \sum_{k=1}^{\infty} a_{nk} = 0;$$

(iv) (Theorem 1.23 **13**) $A \in (c, c)$ *if and only if (5.1) and (5.3) hold, and*

$$\lim_{n\to\infty} \sum_{k=1}^{\infty} a_{nk} = l \text{ for some scalar } l; \tag{5.4}$$

(v) (Part (iii) of Remark 1.6) $A \in (s_1, c)$ *if and only if (5.3) holds and*

$$\lim_{n\to\infty} \sum_{k=1}^{\infty} |a_{nk}| = \sum_{k=1}^{\infty} |l_k|. \tag{5.5}$$

Lemma 5.2 *Let $p \geq 1$. We write*

$$\|A\|_{(\ell_p,\ell_\infty)} = \begin{cases} \sup_{n,k} |a_{nk}| & \text{for } p = 1 \\ \sup_n \left(\sum_{k=1}^{\infty} |a_{nk}|^q \right)^{1/q} & \text{for } 1 < p < \infty \text{ and } q = p/(p-1). \end{cases}$$

Then we have

(i) (Theorem 1.23 **4** and **5**) $A \in (\ell_p, \ell_\infty)$ *if and only if*

$$\|A\|_{(\ell_p,\ell_\infty)} < \infty; \tag{5.6}$$

(ii) (Theorem 1.23 **9** and **10**) $A \in (\ell_p, c_0)$ *if and only if the conditions in (5.6) and (5.2) are satisfied;*

(iii) (Theorem1.23 **14** and **15**) $A \in (\ell_p, c)$ *if and only if the conditions in (5.6) and (5.3) are satisfied.*

We also use the well-known properties, stated as follows.

Lemma 5.3 (Proposition 2.1) *Let $a, b \in \mathcal{U}^+$ and let $E, F \subset \omega$ be any linear spaces. We have $A \in (E_a, F_b)$ if and only if $D_{1/b} A D_a \in (E, F)$.*

Lemma 5.4 ([4, Lemma 9, p. 45]) *Let T' and T'' be any given triangles and let $E, F \subset \omega$. Then, for any given operator T represented by a triangle, we have $T \in (E_{T'}, F_{T''})$ if and only if $T''TT'^{-1} \in (E, F)$.*

Now we recall that for $\lambda \in \mathcal{U}$, the infinite matrices $C(\lambda)$ and $\Delta(\lambda)$ defined in (4.7) and (4.8) are triangles, and $\Delta(\lambda)$ is the inverse of $C(\lambda)$, that is, $C(\lambda)(\Delta(\lambda)) = \Delta(\lambda)(C(\lambda)y) = y$ for all $y \in \omega$ (for the use of these matrices, see, for instance, [17, 22]). If $\lambda = e$, then we obtain the well-known operator of the first difference represented by $\Delta(e) = \Delta$. We then have $\Delta_n y = y_n - y_{n-1}$ for all $n \geq 1$, with the convention $y_0 = 0$. Usually the notation $\Sigma = C(e)$ is used and then we may write $C(\lambda) = D_{1/\lambda}\Sigma$. We note that $\Delta = \Sigma^{-1}$. The Cesàro operator is defined by $C_1 = C((n)_{n=1}^{\infty})$. We also use the sets W_a and W_a^0 of sequences that are *a-strongly bounded* and *a-strongly convergent to zero* defined for $a \in \mathcal{U}^+$ in (4.44) and (4.45) (cf. [8, 16]). It can easily be seen that $W_a = \{y \in \omega : C_1 D_{1/a}|y| \in s_1\}$. If $a = (r^n)_{n=1}^{\infty}$ the sets W_a and W_a^0 are denoted by W_r and W_r^0. For $r = 1$ we obtain the well-known sets

$$w_\infty = \left\{ y \in \omega : \|y\|_{w_\infty} = \sup_n \left(\frac{1}{n} \sum_{k=1}^{n} |y_k| \right) < \infty \right\}$$

and

$$w_0 = \left\{ y \in \omega : \lim_{n \to \infty} \left(\frac{1}{n} \sum_{k=1}^{n} |y_k| \right) = 0 \right\}$$

called the *spaces of sequences that are strongly bounded* and *strongly summable to zero sequences by the Cesàro method of order* 1.

We will use Lemmas 4.1 and 4.2 on multipliers. The α- and β-duals of a set of sequences E are defined as $E^\alpha = M(E, \ell_1)$ and $E^\beta = M(E, cs)$, respectively, where $cs = c_\Sigma$ is the set of all convergent series.

Now we recall some results that are direct consequence of [20, Theorem 2.4], where

$$\left\| (a_n)_{n=1}^{\infty} \right\|_{\mathcal{M}} = \sum_{\nu=1}^{\infty} 2^\nu \max_{2^\nu \leq k \leq 2^{\nu+1}-1} |a_k|. \tag{5.7}$$

Writing $A_n = (a_{nk})_{k=1}^{\infty}$ for the sequence in the n-th row of the matrix $A = (a_{nk})_{n.k=1}^{\infty}$ obtain the following result:

Lemma 5.5 ([20])

(i) We have $(w_0, \ell_\infty) = (w_\infty, \ell_\infty)$ and $A \in (w_\infty, \ell_\infty)$ if and only if

$$\sup_n (\|A_n\|_{\mathcal{M}}) = \sup_n \left(\sum_{\nu=1}^{\infty} 2^\nu \max_{2^\nu \leq k \leq 2^{\nu+1}-1} |a_{nk}| \right) < \infty; \tag{5.8}$$

(ii) $A \in (w_\infty, c_0)$ *if and only if*

$$\lim_{n\to\infty} \|A_n\|_{\mathcal{M}} = \lim_{n\to\infty} \left(\sum_{\nu=1}^{\infty} 2^\nu \max_{2^\nu \le k \le 2^{\nu+1}-1} |a_{nk}| \right) = 0;$$

(iii) $A \in (w_0, c_0)$ *if and only if (5.8) holds and*

$$\lim_{n\to\infty} a_{nk} = 0 \textit{ for all } k.$$

In the following we use the results stated below.

Lemma 5.6 *Let $p \ge 1$. We have*

(i) $M(c, c_0) = c_0$;
(ii) $M(c, \ell_\infty) = \ell_\infty$;
(iii) $M(c_0, \ell_p) = M(c, \ell_p) = M(\ell_\infty, \ell_p) = \ell_p$;
(iv) $M(\ell_p, F) = \ell_\infty$ *for* $F \in \{c_0, c, s_1, \ell_p\}$.

Proof (i), (ii), (iv) with $F \in \{c_0, c, \ell_\infty\}$ follow from [21, Lemma 3.1, p. 648], Lemma 5.2 and [22, Example 1.28, p. 157]. In Part (iv), it remains to show $M(\ell_p, \ell_p) = \ell_\infty$. We have $\ell_\infty \subset M(\ell_p, \ell_p)$, since for any given sequence $a \in \ell_\infty$, we have

$$\sum_{k=1}^{\infty} |a_k y_k|^p \le \sup_k |a_k|^p \sum_{k=1}^{\infty} |y_k|^p < \infty \text{ for all } y \in \ell_p.$$

Then we have by Lemma 4.2 $M(\ell_p, \ell_p) \subset M(\ell_p, \ell_\infty) = \ell_\infty$ and we have shown $M(\ell_p, \ell_p) = \ell_\infty$.
(iii) Since $a \in M(E, F)$ if and only if $D_a \in (E, F)$, we apply Theorem 1.23 to obtain $M(E, \ell_p) = \ell_p$ for $E \in \{c_0, c, \ell_\infty\}$ immediately from **16, 17, 18** for $p = 1$ and **21, 22, 23** for $1 < p < \infty$.

□

Now we determine the multiplier $M(E, F)$ involving the sets w_0 and w_∞.

Lemma 5.7 *We have*

(i) $M(w_0, F) = M(w_\infty, \ell_\infty) = s_{(1/n)_{n=1}^\infty}$ *for* $F = c_0, c, \ell_\infty$;
(ii) $M(w_\infty, c_0) = M(w_\infty, c) = s^0_{(1/n)_{n=1}^\infty}$;
(iii) $M(\ell_1, w_\infty) = s_{(n)_{n=1}^\infty}$ *and* $M(\ell_1, w_0) = s^0_{(n)_{n=1}^\infty}$;
(iv) $M(E, w_0) = w_0$ *for* $E = s_1, c$;
(v) $M(E, w_\infty) = w_\infty$ *for* $E = c_0, s_1, c$.

Proof (i) First we show $M(w_\infty, \ell_\infty) = s_{(1/n)_{n=1}^\infty}$.
We have $a \in M(w_\infty, \ell_\infty)$ if and only if $D_a = (d_{nk})_{n,k=1}^\infty \in (w_\infty, \ell_\infty)$. Now we apply formula (5.8). We have $d_{nk} = a_n$ for $k = n$ and $a_{nk} = 0$ for $k \ne n$, and, for any given integer n, let ν_n denote the uniquely determined integer for which $2^{\nu_n} \le$

$n \le 2^{\nu_n+1} - 1$. Then $(n+1)/2 \le 2^{\nu_n} \le n$, and, by Lemma Part (i) of Lemma 5.5, we have $D_a \in (w_\infty, \ell_\infty)$ if and only if

$$\sigma_n = \sum_{\nu=0}^{\infty} 2^\nu \max_{2^\nu \le m \le 2^{\nu+1}-1} |d_{nk}| = 2^{\nu_n}|a_n| = O(1)\ (n \to \infty).$$

Since

$$\frac{n+1}{2}|a_n| \le 2^{\nu_n}|a_n| \le n|a_n \text{ for all } n,$$

we conclude $\sup_n \sigma_n < \infty$ if and only if $\sup_n(n|a_n|) < \infty$ and $M(w_\infty, \ell_\infty) = s_{(1/n)_{n=1}^\infty}$. Then we have by Part (iii) of Lemma 5.5 $M(w_0, c_0) = s_{(1/n)_{n=1}^\infty}$. Finally we have

$$s_{(1/n)_{n=1}^\infty} = M(w_0, c_0) \subset M(w_0, c) \subset M(w_0, \ell_\infty) \subset M(w_\infty, \ell_\infty) = s_{(1/n)_{n=1}^\infty}.$$

Thus we have shown Part (i).
(ii) can be shown as above by the use of Part (ii) of Lemma 5.5.
(iii) By [23, Theorem 1], $D_a \in (\ell_1, w_\infty)$ if and only if $\sup_n(|a_n|/n) < \infty$ which means that $a \in s_{(n)_{n=1}^\infty}$. Again by [23, Theorem 1], we have $D_a \in (\ell_1, w_0)$ if and only if $\lim_{n\to\infty}(|a_n|/n) = 0$ and $a \in s^0_{(n)_{n=1}^\infty}$.
(iv) We have $M(s_1, w_0) \subset w_0$, since $e \in s_1$. Now we show $w_0 \subset M(s_1, w_0)$. For this let $a \in w_0$. Then we have for every $y \in s_1$

$$\frac{1}{n}\sum_{k=1}^{n} |a_k y_k| \le \sup_k |y_k| \left(\frac{1}{n}\sum_{k=1}^{n}|a_k|\right) \text{ for all } n$$

and $ay \in w_0$. So we have shown

$$w_0 \subset M(s_1, w_0)$$

and

$$M(s_1, w_0) = w_0.$$

Then we have

$$w_0 = M(s_1, w_0) \subset M(c, w_0) \subset w_0$$

and

$$M(c, w_0) = w_0.$$

This completes the proof of Part (iv).
(v) It remains to show $M(c_0, w_\infty) = w_\infty$. By [22, Lemma 3.56, p. 218], the set $\mathcal{M} = w_\infty^\beta$ is a BK space with AK and is β-perfect, that is, $w_\infty^{\beta\beta} = w_\infty$. By Theorem 1.22 with $X = c_0$ and $Z = \mathcal{M}$, we obtain

$$M(c_0, w_\infty) = M(c_0, w_\infty^{\beta\beta}) = M(\mathcal{M}, c_0^\beta).$$

Since $c_0^\beta = \ell_1$, we conclude $M(c_0, w_\infty) = \mathcal{M}^\beta = w_\infty$.
Now we show the identity $M(w_\infty, c) = s^0_{(1/n)_{n=1}^\infty}$. First we note that

$$M(w_\infty, c) \supset M(w_\infty, c_0) = s^0_{(1/n)_{n=1}^\infty}.$$

It remains to show the inclusion $M(w_\infty, c) \subset s^0_{(1/n)_{n=1}^\infty}$. Since $c \subset (c_0)_\Delta$, we have

$$M(w_\infty, c) \subset M(w_\infty, (c_0)_\Delta).$$

So $a \in M(w_\infty, (c_0)_\Delta)$ implies $\Delta D_a \in (w_\infty, c_0)$. The matrix $\Delta D_a = (\alpha_{nk})_{n,k=1}^\infty$ is the triangle whose the nonzero entries are defined by $a_{nn} = -\alpha_{n,n-1} = a_n$ for all n with $\alpha_{10} = 0$. By the characterization of the class (w_∞, c_0), we have that $\Delta a \in (w_\infty, c_0)$ implies $\|(\Delta a)_n\|_{\mathcal{M}} \to 0$ $(n \to \infty)$, where

$$\|(\Delta a)_n\|_{\mathcal{M}} = \sum_{\nu=0}^{\nu_n - 1} 2^\nu \max_{2^\nu \le k \le 2^{\nu+1}-1} |\alpha_{nk}| + 2^{\nu_n} \max_{2^{\nu_n} \le k \le n} |\alpha_{nk}| \ge 2^{\nu_n} |a_n|$$

and ν_n is an integer uniquely defined by $2^{\nu_n} \le n \le 2^{\nu_n+1} - 1$ for all n. Since $2^{\nu_n} \ge (n+1)/2$, the condition $\|(\Delta a)_n\|_{\mathcal{M}} \to 0$ implies $n|a_n| \to 0$ $(n \to \infty)$. So we have shown the inclusion

$$M(w_\infty, c) \subset s^0_{(1/n)_{n=1}^\infty}.$$

This completes the proof. □

We need some results on the equivalence relation $R_\mathcal{E}$ which is defined using the multiplier of sequence spaces in the following way.

Definition 5.1 For $b \in \mathcal{U}^+$ and for any subset $\mathcal{E}$ of ω, we denote by $cl^\mathcal{E}(b)$ the equivalence class for the equivalence relation $R_\mathcal{E}$ defined by

$$x R_\mathcal{E} y \text{ if } \mathcal{E}_x = \mathcal{E}_y \text{ for } x, y \in \mathcal{U}^+.$$

It can easily be seen that $cl^\mathcal{E}(b)$ is the set of all $x \in \mathcal{U}^+$ such that

$$x/b \in M(\mathcal{E}, \mathcal{E}) \text{ and } b/x \in M(\mathcal{E}, \mathcal{E}),$$

(cf. [18]). Then we have

$$cl^\mathcal{E}(b) = cl^{M(\mathcal{E},\mathcal{E})}(b).$$

For instance, $cl^c(b)$ is the set of all $x \in \mathcal{U}^+$ such that $s_x^{(c)} = s_b^{(c)}$. This is the set of all sequences $x \in \mathcal{U}^+$ such that $x_n \sim Cb_n$ $(n \to \infty)$ for some $C > 0$. In [18] we denoted by $cl^\infty(b)$ the class $cl^{\ell_\infty}(b)$. We recall that $cl^\infty(b)$ is the set of all $x \in \mathcal{U}^+$, such that $K_1 \leq x_n/b_n \leq K_2$ for all n and for some $K_1, K_2 > 0$. Since $M(c_0, c_0) = \ell_\infty$, we have $cl^{c_0}(b) = cl^\infty(b)$.

5.2 The (SSIE) $F \subset E_a + F'_x$ with $e \in F$ and $F' \subset M(F, F')$

Here we are interested in the study of the set of all positive sequences x that satisfy the inclusion $F \subset E_a + F'_x$, where E, F and F' are linear spaces of sequences and a is a positive sequence. For instance, a positive sequence x satisfies the inclusion $c \subset s_a^0 + s_x^{(c)}$ if and only if the next statement holds

The condition $y_n \to l$ implies that there are sequences $u, v \in \omega$ with $y = u + v$ such that $u_n/a_n \to 0$ and $v_n/x_n \to l'$ $(n \to \infty)$ for some scalars l and l' and all sequences y.

We may consider this problem as a *perturbation problem*. If we know the set $M(F, F')$, then the solutions of the *elementary inclusion*$F'_x \supset F$ are determined by $1/x \in M(F, F')$. Now the question is: If $\mathcal{E}$ is a linear space of sequences, what are the solutions of the perturbed inclusion $F'_x + \mathcal{E} \supset F$? An additional question may be the following one: What are the conditions on $\mathcal{E}$ under which the solutions of the elementary and the perturbed inclusions are the same? The solutions of the perturbed inclusion $F \subset E_a + F'_x$ where E, F and F' are linear spaces of sequences cannot be obtained in the general case. So are led to deal with the case when $a = (r^n)_{n=1}^\infty$ for $r > 0$, for which most of these (SSIE) can be totally solved. In the following we write

$$\mathcal{I}_a(E, F, F') = \left\{x \in \mathcal{U}^+ : F \subset E_a + F'_x\right\},$$

where E, F and F' are linear spaces of sequences and $a \in \mathcal{U}^+$. For any set χ of sequences, we put

$$\overline{\chi} = \left\{x \in \mathcal{U}^+ : 1/x \in \chi\right\}.$$

In the following we use the set $\Phi = \{c_0, c, s_1, \ell_p, w_0, w_\infty\}$ with $p \geq 1$. We recall that $c(1)$ is the set of all sequences $\alpha \in \mathcal{U}^+$ that satisfy $\lim_{n\to\infty} \alpha_n = 1$ and consider the condition

$$G \subset G_{1/\alpha} \text{ for all } \alpha \in c(1) \text{ (cf. (4.3))} \tag{5.9}$$

for any given linear space G of sequences. We note that condition (5.9) is satisfied for all $G \in \Phi$.

Theorem 5.1 ([12, Theorem 1, p. 1045]) *Let $a \in \mathcal{U}^+$ and let E, F and F' be linear spaces of sequences. We assume*

$$(a)\ e \in F, \quad (b)\ F' \subset M(F, F'), \quad (c)\ F' \text{ satisfies } (5.9).$$

Then we have

(i) $a \in M(E, c_0)$ *implies* $\mathcal{I}_a(E, F, F') = \overline{F'}$;
(ii) $1/a \in M(F, E)$ *implies* $\mathcal{I}_a(E, F, F') = \mathcal{U}^+$.

Proof (i) Let $x \in \mathcal{I}_a(E, F, F')$. Then there are $\xi \in E$ and $f' \in F'$ such that $1 = a_n\xi_n + x_n f'_n$, hence

$$\frac{1 - a_n\xi_n}{x_n} = f'_n \text{ for all } n.$$

Since $a \in M(E, c_0)$, we have $1 - a_n\xi_n \to 1\ (n \to \infty)$ and

$$\frac{1}{x_n} = \frac{1}{1 - a_n\xi_n} f'_n \text{ for all } n.$$

By the condition in (c), we conclude $x \in \overline{F'}$.
Conversely, the condition $x \in \overline{F'}$ implies $1/x \in F'$, and the condition in (b) implies $1/x \in M(F, F')$. We conclude $F \subset F'_x$ and $x \in \mathcal{I}_a(E, F, F')$. So we have shown Part (i).
(ii) follows from the equivalence of $1/a \in M(F, E)$ and $F \subset E_a$.
This concludes the proof. □

We immediately deduce the following.

Corollary 5.1 ([12, Corollary 1, p. 1045]) *Let E, F and F′ be linear spaces of sequences. We assume*

$$(a)\ e \in F, \quad (b)\ F' \subset M(F, F'), \quad (c)\ E \subset c_0.$$

Then the next statements are equivalent

(i) $F \subset E + F'_x$,
(ii) $F \subset F'_x$,
(iii) $x \in \overline{F'}$.

In some cases where $E = cs$ or ℓ_1, and $F' = \ell_1$, we obtain the next results using the α- and β-duals.

Corollary 5.2 ([12, Corollary 2, p. 1045]) *Let* $a \in \mathcal{U}^+$ *and let F and F′ be linear spaces of sequences. We assume that the conditions in (a), (b), (c) in* Theorem *5.1 are satisfied. Then the set* $\mathcal{I}_a(cs, F, F')$ *of all positive sequences x such that* $F \subset cs_a + F'_x$ *satisfies the next properties.*

(i) $a \in s_1$ *implies* $\mathcal{I}_a(cs, F, F') = \overline{F'}$.
(ii) $1/a \in F^\beta$ *implies* $\mathcal{I}_a(cs, F, F') = \mathcal{U}^+$.

Proof It is enough to show $M(cs, c_0) = s_1$. We have $a \in M(cs, c_0)$ if and only if $D_a\Delta \in (c, c_0)$, and $D_a\Delta$ is the infinite matrix whose nonzero entries are $[D_a\Delta]_{nn} = -[D_a\Delta]_{n,n-1} = a_n$ for all $n \geq 2$, with the convention $[D_a\Delta]_{1,1} = a_1$. By the characterization of (c, c_0) in Lemma 5.1 we conclude $M(cs, c_0) = s_1$.
This completes the proof. □

Corollary 5.3 ([12, Corollary 3, p. 1046]) *Let $a \in \mathcal{U}^+$ and let F and F' be linear spaces of sequences. We assume that the conditions in (a), (b), (c) in* Theorem *5.1 are satisfied. Then the set $\mathcal{I}_a(\ell_1, F, F')$ of all positive sequences x such that $F \subset (\ell_1)_a + F'_x$ satisfies the next properties.*

(i) $a \in s_1$ implies $\mathcal{I}_a(\ell_1, F, F') = \overline{F'}$;
(ii) $1/a \in F^\alpha$ implies $\mathcal{I}_a(\ell_1, F, F') = \mathcal{U}^+$.

Proof This result follows from the identity $M(\ell_1, c_0) = s_1$. □

Corollary 5.4 ([12, Corollary 4, p. 1046]) *Let $a \in \mathcal{U}^+$ and let E and F be linear spaces of sequences. We assume $e \in F$ and $\ell_1 \subset F^\alpha$. Then the set $\mathcal{I}_a(E, F, \ell_1)$ of all positive sequences x such that $F \subset E_a + (\ell_1)_x$ satisfies the implications in (i) and (ii) of* Theorem *5.1 with $F' = \ell_1$.*

Remark 5.1 We obtain similar results to those stated above for the (SSIE) defined by $F_b \subset E_a + F'_x$ with $a, b \in \mathcal{U}^+$. Let $\mathcal{I}_{a,b}(E, F, F')$ be the set of all positive sequences that satisfy the previous (SSIE). Under the conditions (a), (b), (c) in Theorem 5.1 we obtain that $a/b \in M(E, c_0)$ implies $\mathcal{I}_{a,b}(E, F') = \overline{F'_b}$, and $b/a \in M(F, E)$ implies $\mathcal{I}_{a,b}(E, F, F') = \mathcal{U}^+$. This result is a direct consequence of the equivalence of $F_b \subset E_a + F'_x$ and $F \subset E_{a/b} + F'_{x/b}$. For instance, we consider the (SSIE) defined by $c_R \subset c_0 + c_x$ with $R > 0$. A positive sequence x satisfies the (SSIE) $c_R \subset c_0 + c_x$ if and only if the next statement holds

> *The condition $y_n/R^n \to l$ implies there are sequences $u, v \in \omega$ such that $y = u + v$ and $u_n \to 0$ and $v_n/x_n \to l'$ $(n \to \infty)$ for some scalars l and l' and all sequences y.*

We have that $c_R \subset c_0 + c_x$ is equivalent to $c \subset s^0_{1/R} + s^{(c)}_{(x_n/R^n)_n}$. Since $M(c_0, c_0) = s_1$, we have $(R^{-n})_{n=1}^\infty \in s_1$ if and only if $R \geq 1$, and we obtain $\mathcal{I}_{e,(R^{-n})_{n=1}^\infty}(c_0, c) = \overline{c_R}$ for $R \geq 1$, and since $M(c, c_0) = c_0$, we obtain $\mathcal{I}_{e,(R^{-n})_{n=1}^\infty}(c_0, c) = \mathcal{U}^+$ for all $R < 1$.

5.3 The (SSIE) $F \subset E_a + F'_x$ with $E, F, F' \in \{c_0, c, s_1, \ell_p, w_0, w_\infty\}$

In this section, we use the set $\Omega = (\{s_1\} \times (\Phi \setminus \{c\})) \cup (\{c\} \times \Phi)$ with $p \geq 1$ and deal with the perturbed inclusions of the form $F \subset E_a + F'_x$ where $E = c_0$, s_1, ℓ_p, w_0, or w_∞ and $(F, F') \in \Omega$. As a direct consequence of Lemmas 5.6 and 5.7 we obtain.

Lemma 5.8 *We have that $(F, F') \in \Omega$ implies $F' \subset M(F, F')$.*

By Corollary 5.1 and Lemma 5.8 we obtain the following result.

Proposition 5.1 ([12, Proposition 1, p. 1046]) *Let $E \subset c_0$ be a linear space of sequences and let $(F, F') \in \Omega$. Then the next statements are equivalent*

$$(i)\ F \subset E + F'_x, \quad (ii)\ F \subset F'_x, \quad (iii)\ x \in \overline{F'}.$$

Example 5.1 We consider the next statement, where $\widetilde{r} = (r_n)_{n=1}^\infty \in c \cap \mathcal{U}$ with $\lim_{n\to\infty} r_n \neq 0$ and $\widetilde{s} = (s_n)_{n=1}^\infty \in c$. By $\mathcal{I}_{\widetilde{r},\widetilde{s}}$ we denote the set of all positive sequences x that satisfy the next statement.

For every $y \in \omega$, $y_n \to L$ implies that there are sequences $u, v \in \omega$ with $y = u + v$ such that $r_n u_n + s_{n-1} u_{n-1} \to 0$ and $x_n v_n \to L'$ $(n \to \infty)$ for some scalars L and L'.

This statement is equivalent to the (SSIE) defined by $c \subset (c_0)_{B(\widetilde{r},\widetilde{s})} + c_{1/x}$, where $B(\widetilde{r},\widetilde{s})$ is the bidiagonal infinite matrix defined by $[B(\widetilde{r},\widetilde{s})]_{nn} = r_n$ for all n and $[B(\widetilde{r},\widetilde{s})]_{n,n-1} = s_{n-1}$ for all $n \geq 2$, the other entries being equal to zero. By [7, Corollary 5.2.1, p. 15], the operator $B(\widetilde{r},\widetilde{s}) \in (c_0, c_0)$ is bijective if and only if

$$|\lim_{n\to\infty} s_n| < |\lim_{n\to\infty} r_n|. \tag{5.10}$$

So if the condition in (5.10) holds then we obtain $(c_0)_{B(\widetilde{r},\widetilde{s})} = c_0$ and we conclude $\mathcal{I}_{\widetilde{r},\widetilde{s}} = c^+$.

Proposition 5.2 ([12, Proposition 2, p. 1046]) *Let $a \in \mathcal{U}^+$ and let $(F, F') \in \Omega$. We have*

(i) $\mathcal{I}_a(c_0, F, F') = \overline{F'}$ *if* $a \in s_1$, *and* $\mathcal{I}_a(c_0, F, F') = \mathcal{U}^+$ *if* $1/a \in c_0$;
(ii) $\mathcal{I}_a(s_1, F, F') = \overline{F'}$ *if* $a \in c_0$, *and* $\mathcal{I}_a(s_1, F, F') = \mathcal{U}^+$ *if* $1/a \in s_1$;
(iii) $\mathcal{I}_a(\ell_p, F, F') = \overline{F'}$ *if* $a \in s_1$, *and* $\mathcal{I}_a(\ell_p, F, F') = \mathcal{U}^+$ *if* $1/a \in \ell_p$ *for* $p \geq 1$;
(iv) $\mathcal{I}_a(w_0, F, F') = \overline{F'}$ *if* $a \in s_{(1/n)_{n=1}^\infty}$, *and* $\mathcal{I}_a(w_0, F, F') = \mathcal{U}^+$ *if* $1/a \in w_0$.
(v) $\mathcal{I}_a(w_\infty, F, F') = \overline{F'}$ *if* $a \in s^0_{(1/n)_{n=1}^\infty}$, *and* $\mathcal{I}_a(w_\infty, F, F') = \mathcal{U}^+$ *if* $1/a \in w_\infty$.

Proof The proof is a direct consequence of Theorem 5.1 and Lemmas 5.6 and 5.7. Indeed, we successively have $M(E, c_0) = s_1$, for $E = c_0$, or ℓ_p; $M(E, c_0) = c_0$, for $E = c$, or s_1; $M(w_0, c_0) = s_{(1/n)_{n=1}^\infty}$ and $M(w_\infty, c_0) = s^0_{(1/n)_{n=1}^\infty}$. Then we have $M(F, E) = M(s_1, E) = M(c, E)$ for $E \in \Phi \setminus \{c\}$, and $M(s_1, c_0) = c_0$, $M(s_1, s_1) = s_1$, $M(s_1, \ell_p) = \ell_p$, $M(s_1, w_0) = w_0$ and $M(s_1, w_\infty) = w_\infty$. □

In all that follows we write $G^+ = G \cap \mathcal{U}^+$ for any set G of sequences. We may illustrate these results by the next examples.

Example 5.2 Let $p \geq 1$. We consider the system

$$(S) \begin{cases} s_x \subset (\ell_p)_x + w_\infty, \\ w_\infty \subset W_x. \end{cases}$$

We have $s_x \subset (\ell_p)_x + w_\infty$ if and only if $s_1 \subset \ell_p + W_{1/x}$. Then by Part (iii) of Proposition 5.2, the solutions of the (SSIE) $s_1 \subset \ell_p + W_{1/x}$ are determined by $x \in w_\infty$. Then by Lemma 5.7, we have $w_\infty = M(s_1, w_\infty)$ and $x \in w_\infty$ if and only if $s_x \subset w_\infty$. Furthermore, we have $M^+(w_\infty, w_\infty) = \ell_\infty^+$ (cf. [14, Remark 3.4, p. 597]), and $w_\infty \subset W_x$ if and only if $s_1 \subset s_x$. So $x \in \mathcal{U}^+$ satisfies system (S) if and only if $s_1 \subset s_x \subset w_\infty$. This means that $x_n \geq K_1$ and $n^{-1}\sum_{k=1}^n x_k \leq K_2$ for all n and for some K_1 and $K_2 > 0$.

Example 5.3 Let $p \geq 1$, and consider the next statement.

> *The condition $y_n \to l$ implies that there are two sequences $u, v \in \omega$ with $y = u + v$ such that $\sum_{k=1}^\infty |u_k|^p < \infty$ and $n^{-1}\sum_{k=1}^n |v_k|/x_k \leq K$ for all n, for some scalars l and K, with $K > 0$, and for all y.*

This statement is associated with the (SSIE) $c \subset \ell_p + W_x$ and with the set $\mathcal{I}_e(\ell_p, c, w_\infty)$. By Part (iii) of Proposition 5.2, we conclude $\mathcal{I}_e(\ell_p, c, w_\infty) = \overline{w_\infty}$.

Example 5.4 We consider the next statement.

> *The condition $y_n \to l$ implies that there are two sequences $u, v \in \omega$, with $y = u + v$ such that $n^{-1}\sum_{k=1}^n |ku_k| \to 0$ $(n \to \infty)$ and $\sum_{k=1}^\infty (|v_k|/x_k)^p < \infty$ for some scalar $K > 0$ and all y.*

This statement corresponds to the (SSIE) $c \subset W^0_{(1/n)_{n=1}^\infty} + (\ell_p)_x$, and by Part (iv) of Proposition 5.2, the set of all positive sequences that satisfy this (SSIE) is equal to $\mathcal{I}_{(n)_n}(w_0, c, \ell_p) = \overline{\ell_p}$, since $(1/n)_{n=1}^\infty \in s_{(1/n)_{n=1}^\infty}$.

Remark 5.2 Let $a \in \mathcal{U}^+$ and $(F, F') \in \Omega$. Then the conditions (a), (b), (c) of Theorem 5.1 hold. The set $\mathcal{I}_a(cs, F, F')$ of all positive sequences x such that $F \subset cs_a + F'_x$ satisfies the next properties.

(i) $a \in s_1$ implies $\mathcal{I}_a(cs, F, F') = \overline{F'}$.
(ii) $1/a \in F^\beta = \ell_1$ implies $\mathcal{I}_a(cs, F, F') = \mathcal{U}^+$.

When $E = c$, we obtain the following result.

Proposition 5.3 ([12, Proposition 3, p. 1046]) *Let $a \in \mathcal{U}^+$ and $F' \in \Phi$. Then we have*

(i) $\mathcal{I}_a(c, c, F') = \overline{F'}$ if $a \in c_0$, and $\mathcal{I}_a(c, c, F') = \mathcal{U}^+$ if $1/a \in c$.
(ii) $\mathcal{I}_a(c, s_1, F') = \overline{F'}$ if $a \in c_0$, and $\mathcal{I}_a(c, s_1, F') = \mathcal{U}^+$ if $1/a \in c_0$.

Proof The proof follows from Theorem 5.1 and Lemma 5.6. Here we have $M(E, c_0) = M(c, c_0) = c_0$ and $M(F, E) = M(F, c) = c_0$ for $F = s_1$ and $M(F, c) = c$ for $F = c$. $\square$

Now we study the solvability of the

$$F \subset E_r + F'_x \text{ with } E, F' \in \{c_0, c, s_1, \ell_p, w_0, w_\infty\}. \qquad \text{(SSIE)}$$

For $a = (r^n)_{n=1}^{\infty}$, we write $\mathcal{I}_r(E, F, F')$ for the set $\mathcal{I}_a(E, F, F')$. Then we solve the perturbed inclusions $F \subset E_r + F'_x$ where F is either c or s_1 and $E \in \Phi \setminus \{w_0\}$ and $F' \in \Phi$. It can easily be seen that in most of the cases the set $\mathcal{I}_r(E, F, F')$ may be determined by

$$\mathcal{I}_r(E, F, F') = \begin{cases} \overline{F'} & \text{if } r < 1, \\ U^+ & \text{if } r \geq 1 \end{cases} \tag{5.11}$$

or by

$$\mathcal{I}_r(E, F, F') = \begin{cases} \overline{F'} & \text{if } r \leq 1 \\ U^+ & \text{if } r > 1. \end{cases} \tag{5.12}$$

We see that in the first case the elementary inclusion $F \subset F'_x$ and the perturbed inclusion $F \subset E_r + F'_x$ have the same solutions if and only if $r < 1$. Similarly, in the second case the equivalence of the elementary and the perturbed inclusions is satisfied for $r \leq 1$. As a direct consequence of Propositions 5.2 and 5.3 we obtain the following result.

Proposition 5.4 ([12, Proposition 4, p. 1047]) *Let $a \in \mathcal{U}^+$ and $(F, F') \in \Omega$. Then we have*

(i) The sets $\mathcal{I}_r(s_1, F, F')$, $\mathcal{I}_r(c, c, F')$ and $\mathcal{I}_r(w_\infty, F, F')$ are determined by (5.11).
(ii) The sets $\mathcal{I}_r(c_0, F, F')$ and $\mathcal{I}_r(\ell_p, F, F')$ for $p \geq 1$ are determined by (5.12).

Remark 5.3 From Part (ii) in Proposition 5.4, we deduce the equivalence of the (SSIE) $s_1 \subset F'_x$ and the perturbed inclusion $s_1 \subset c_0 + F'_x$, where F' is any of the sets c_0, s_1, ℓ_p for $p \geq 1$, w_0, or w_∞. This is also the case for the (SSIE) $c \subset F'_x$ and the perturbed inclusion $c \subset \ell_p + F'_x$, where F' is any of the sets c_0, c, s_1, ℓ_p for $p \geq 1$, w_0, or w_∞.

Rewriting Proposition 5.4 we obtain.

Corollary 5.5 *Let $r > 0$. Then we have*

(i) Let $F' \in \Phi$. Then

  (a) the solutions of the (SSIE) $c \subset E_r + F'_x$ with $E = c, s_1$, or w_∞ are determined by (5.11);
  (b) the solutions of the (SSIE) $c \subset E_r + F'_x$ with $E = c_0$, or ℓ_p for $p \geq 1$ are determined by (5.12);

(ii) Let $F' \in \Phi \setminus \{c\}$. Then

  (a) the solutions of the (SSIE) $s_1 \subset E_r + F'_x$ with $E = s_1$, or w_∞ are determined by (5.11);
  (b) the solutions of the (SSIE) $s_1 \subset E_r + F'_x$ with $E = c_0$, or ℓ_p for $p \geq 1$ are determined by (5.12).

Remark 5.4 The set $\mathcal{I}_r(w_0, c, F')$ of all the solutions of the (SSIE) $c \subset W_r^0 + F'_x$ where $F' \in \Phi$ is determined for all $r \neq 1$. We obtain

$$\mathcal{I}_r(w_0, c, F') = \begin{cases} \overline{F'} & \text{for } r < 1 \\ U^+ & \text{for } r > 1. \end{cases}$$

Example 5.5 By Part (i) (a) of Corollary 5.5, the solutions of the (SSIE) defined by $c \subset s_r + s_x^0$ are determined by the set of all positive sequences x that satisfy $x_n \to \infty$ $(n \to \infty)$ if $r < 1$. Then the (SSIE) holds for all positive sequences x for $r \geq 1$. Similarly, by Part (i) (b) of Corollary 5.5, the solutions of the (SSIE) defined by $c \subset (\ell_p)_r + W_x$ for $r > 0$ and $p \geq 1$ are determined by $n^{-1} \sum_{k=1}^n x_k^{-1} \leq K$ for all n and for some $K > 0$, for $r \leq 1$, and if $r > 1$ the (SSIE) holds for all positive sequences x.

Example 5.6 By Part (ii) (a) of Corollary 5.5, the solutions of the (SSIE) defined by $s_1 \subset W_r + s_x$ are determined by $1/x \in s_1$ if $r < 1$, and if $r \geq 1$ the (SSIE) holds for all positive sequences x. By Part (ii) (b) of Corollary 5.5 with $E = c_0$ and $F' = w_0$, the solutions of the (SSIE) defined by $s_1 \subset s_r^0 + W_x^0$ are determined by $n^{-1} \sum_{k=1}^n x_k^{-1} \to 0$ $(n \to \infty)$, for $r \leq 1$, and if $r > 1$ the (SSIE) holds for all positive sequences x.

From Theorem 5.1, we obtain the next corollary.

Corollary 5.6 *Let $a \in \mathcal{U}^+$ and E and F be two linear spaces of sequences. We assume*

(a) $e \in F$ (b) $F \subset M(F, F)$ and (c) F satisfies (5.9).

Then we have

(i) $a \in M(E, c_0)$ implies $\mathcal{I}_a(E, F) = \overline{F}$;
(ii) $1/a \in M(F, E)$ implies $\mathcal{I}_a(E, F) = \mathcal{U}^+$.

Now we deal with the (SSIE) $F \subset E_a + F_x$ where F is either c, or s_1 and $E \in \Phi$. We obtain the following by Corollary 5.6 and Lemmas 5.6 and 5.7.

Corollary 5.7 *Let $a \in \mathcal{U}^+$. Then we have*

(i)

(a) $$\mathcal{I}_a(c_0, c) = \begin{cases} \overline{c} & \text{if } a \in s_1 \\ \mathcal{U}^+ & \text{if } 1/a \in c_0; \end{cases}$$

(b) $$\mathcal{I}_a(c, c) = \begin{cases} \overline{c} & \text{if } a \in c_0 \\ \mathcal{U}^+ & \text{if } 1/a \in c; \end{cases}$$

(c) $$\mathcal{I}_a(s_1, c) = \begin{cases} \overline{c} & \text{if } a \in c_0 \\ \mathcal{U}^+ & \text{if } 1/a \in s_1; \end{cases}$$

(d) $$\mathcal{I}_a(\ell_p, c) = \begin{cases} \overline{c} & \text{if } a \in s_1 \\ \mathcal{U}^+ & \text{if } 1/a \in \ell_p \text{ for } p \geq 1; \end{cases}$$

(e) $$\mathcal{I}_a(w_0, c) = \begin{cases} \overline{c} & \text{if } a \in s_{(1/n)_{n=1}^\infty} \\ \mathcal{U}^+ & \text{if } 1/a \in w_0; \end{cases}$$

(f) $$\mathcal{I}_a(w_\infty, c) = \begin{cases} \overline{c} & \text{if } a \in s^0_{(1/n)_{n=1}^\infty} \\ \mathcal{U}^+ & \text{if } 1/a \in w_\infty. \end{cases}$$

(ii)

(a) $$\mathcal{I}_a(c_0, s_1) = \begin{cases} \overline{s_1} & \text{if } a \in s_1 \\ \mathcal{U}^+ & \text{if } 1/a \in c_0; \end{cases}$$

(b) $$\mathcal{I}_a(c, s_1) = \begin{cases} \overline{s_1} & \text{if } a \in c_0 \\ \mathcal{U}^+ & \text{if } 1/a \in c_0; \end{cases}$$

(c) $$\mathcal{I}_a(s_1, s_1) = \begin{cases} \overline{s_1} & \text{if } a \in c_0 \\ \mathcal{U}^+ & \text{if } 1/a \in s_1; \end{cases}$$

(d) $$\mathcal{I}_a(\ell_p, s_1) = \begin{cases} \overline{s_1} & \text{if } a \in s_1 \\ \mathcal{U}^+ & \text{if } 1/a \in \ell_p \text{ for } p \geq 1; \end{cases}$$

(e) $$\mathcal{I}_a(w_0, s_1) = \begin{cases} \overline{s_1} & \text{if } a \in s_{(1/n)_{n=1}^\infty} \\ \mathcal{U}^+ & \text{if } 1/a \in w_0; \end{cases}$$

(f) $$\mathcal{I}_a(w_\infty, s_1) = \begin{cases} \overline{s_1} & \text{if } a \in s^0_{(1/n)_{n=1}}\infty \\ \mathcal{U}^+ & \text{if } 1/a \in w_\infty. \end{cases}$$

Example 5.7 Let $r > 0$. Then by Part (ii) (c) of Corollary 5.7 of the (SSIE) defined by $s_1 \subset s_r + s_x$, are determined by (5.11) with $F' = F = s_1$. The solutions of the (SSIE) defined by $s_1 \subset E_r + s_x$ where $E \in \{c_0, \ell_p\}$ with $p \geq 1$ are determined by (5.12) with $F' = F = s_1$. The solutions of the (SSIE) defined by $c \subset E_r + c_x$, with $E \in \{c, s_1\}$ are determined by (5.11) with $F = c$. The solutions of the (SSIE) $c \subset E_r + c_x$ where $E \in \{c_0, \ell_p\}$ with $p \geq 1$ are determined by (5.12) with $F' = F = c$.

Now we consider the case when $a = (r^n)_{n=1}^\infty$ for $r > 0$. We write E_r for $E_{(r^n)_{n=1}^\infty}$, and the set of all positive sequences x that satisfy

$$F \subset E_r + F_x \tag{5.13}$$

is denoted by $\mathcal{I}_r(E, F)$. Here we explicitly calculate the solutions of the (SSIE) defined by (5.13), where F is either c, or s_1 and $E \in \Phi$. For instance, a positive sequence x satisfies the (SSIE) $c \subset s_r + c_x$ if the next statement holds.

The condition $y_n \to l$ implies that there are sequences $u, v \in \omega$ such that $y = u + v$ for which $\sup_n(|u_n|/r^n) < \infty$ and $v_n/x_n \to l'$ $(n \to \infty)$ for some scalars l and l' and for all y.

We consider the conditions

$$(R^n)_{n=1}^{\infty} \in M(E, c_0) \text{ for all } 0 < R < 1, \tag{5.14}$$

$$(R^n)_{n=1}^{\infty} \in M(E, c_0) \text{ for all } 0 < R \leq 1, \tag{5.15}$$

$$(R^{-n})_{n=1}^{\infty} \in M(F, E) \text{ for all } R \geq 1, \tag{5.16}$$

$$(R^{-n})_{n=1}^{\infty} \in M(F, E) \text{ for all } R > 1. \tag{5.17}$$

We will use the next result which is a direct consequence of Corollary 5.6.

Corollary 5.8 *Let $r > 0$ and let E and F be linear spaces of sequences. We assume that F satisfies the conditions (a), (b), (c) of* Corollary *5.6. Then we have*

(i) If the conditions in (5.14) and (5.16) hold, then the solutions of the (SSIE) defined by (5.13) are determined by (5.11) with $F' = F$.
(ii) If the conditions in (5.15) and (5.17) hold, then the solutions of the (SSIE) defined by (5.13) are determined by (5.12) with $F' = F$.

We obtain the following corollary as a direct consequence of the preceding results.

Corollary 5.9 *Let $r > 0$. Then we have*

(i) (a) The solutions of the (SSIE) $c \subset E_r + c_x$ for $E = c, s_1$, or w_∞ are determined by (5.11) with $F' = c$.
(b) The solutions of the (SSIE) $c \subset E_r + c_x$ with $E = c_0$ or ℓ_p for $p \geq 1$ are determined by (5.12) with $F' = c$.
(ii) (a) The solutions of the (SSIE) $s_1 \subset E_r + s_x$ with $E = s_1$ or w_∞ are determined by (5.11) with $F' = s_1$.
(b) The solutions of the (SSIE) $s_1 \subset E_r + s_x$ with $E = c_0$ or ℓ_p for $p \geq 1$ are determined by (5.12) with $F' = s_1$.

It is interesting to state the next remarks on the elementary (SSIE) successively defined by $c \subset c_0 + c_x$, $c \subset \ell_p + c_x$, $s_1 \subset c_0 + s_x$ and $s_1 \subset \ell_p + s_x$.

Remark 5.5 Obviously we have $c \not\subset c_0$, but we may determine the set I_{c,c_0} of all positive sequences x for which $c \subset c_0 + c_x$. So by Corollary 5.9, we have $x \in I_{c,c_0}$ if and only if $x_n \to L$ $(n \to \infty)$ for some scalar L with $0 < L \leq \infty$. This means that the (SSIE) $c \subset c_x$ and the perturbed equation $c \subset c_0 + c_x$ are equivalent. We obtain a similar result concerning the perturbed (SSIE) defined by $c \subset \ell_p + c_x$ for $p \geq 1$.

Remark 5.6 By Part (ii) (b) of Corollary 5.9, each of the (SSIE) $s_1 \subset c_0 + s_x$ and $s_1 \subset \ell_p + s_x$, for $p \geq 1$ is equivalent to $1/x \in s_1$. This means that the solutions of each of the previous (SSIE) are determined by $x_n \geq K$ for all n and for some $K > 0$. It is interesting to notice that the (SSIE) $s_x \supset s_1$ and each of the perturbed (SSIE) determined by $c_0 + s_x \supset s_1$ and $\ell_p + s_x \supset s_1$ have the same set of solutions.

5.4 Some (SSIE) and (SSE) with Operators

In this section, we study the (SSIE) of the form

$$F \subset D_r * E_\Delta + F_x \tag{5.18}$$

associated with the operator Δ, where the inclusion in (5.18) is considered for any of the cases

$$c \subset D_r * (c_0)_\Delta + c_x,$$
$$c \subset D_r * c_\Delta + c_x,$$
$$c \subset D_r * (s_1)_\Delta + c_x,$$
$$s_1 \subset s_r + s_x$$

and

$$s_1 \subset D_r * (s_1)_\Delta + s_x \text{ for } r > 0.$$

Then we consider the (SSIE) $c \subset D_r * E_{C_1} + s_x^{(c)}$ with $E \in \{c, s_1\}$ and $s_1 \subset D_r * (s_1)_{C_1} + s_x$, where C_1 is the Cesàro operator. Then we solve the (SSE) $D_r * E_{C_1} + s_x^{(c)} = c$ with $E \in \{c_0, c, s_1\}$, and $D_r * (s_1)_{C_1} + s_x = s_1$. We note that since $D_a * E_T = E_{TD_{1/a}}$, where $TD_{1/a}$ is a triangle, for any linear space E of sequences and any triangle T, the previous inclusions and identities can be considered as (SSIE) and (SSE). More precisely, the previous (SSE) can be considered as the perturbed equations of the equations $F_x = F$ with $F = c$, or s_1.

Now we study the (SSIE) of the form

$$F \subset D_r * E_\Delta + F_x.$$

In the next result, among other things, we deal with the (SSIE) $c \subset D_r * (c_0)_\Delta + c_x$, which is associated with the next statement.

> *The condition $y_n \to l$ $(n \to \infty)$ implies that there are sequences $u, v \in \omega$ such that $y = u + v$ and $u_n r^{-n} - u_{n-1} r^{-(n-1)} \to 0$ and $v_n / x_n \to l'$ $(n \to \infty)$ for some scalars l and l' and for all y.*

The corresponding set of sequences is denoted by $\mathcal{I}_r((c_0)_\Delta, c)$. In a similar way, $\mathcal{I}_r((s_1)_\Delta, c)$ is the set of all sequences that satisfy the (SSIE) $c \subset D_r * (s_1)_\Delta + c_x$. We obtain the following result.

Corollary 5.10 *Let $r > 0$. We have*

(i) $\mathcal{I}_r(E, c) = \mathcal{I}_r(s_1, c)$ for $E = c$, $(c_0)_\Delta$, c_Δ or $(s_1)_\Delta$, and $\mathcal{I}_r(s_1, c)$ is determined by (5.11) with $F' = c$.

(ii) $\mathcal{I}_r((s_1)_\Delta, s_1) = \mathcal{I}_r(s_1, s_1)$ is determined by (5.11) with $F' = s_1$.

Proof (i) The identity $\mathcal{I}_r(s_1, c) = \mathcal{I}_r(c, c)$ is a direct consequence of Corollary 5.8, and $\mathcal{I}_r(s_1, c)$ is determined by (5.11) with $F = c$. Indeed, we have $M(E, c_0) = M(c, c_0) = M(s_1, c_0) = c_0$. Then we obtain $M(E, c) = c$ for $E = c$, and $M(E, s_1) = s_1$ for $E = s_1$. Now we deal with $\mathcal{I}((c_0)_\Delta, c)$. We have $(R^n)_{n=1}^\infty \in M((c_0)_\Delta, c_0)$ if and only if $D_R\Sigma \in (c_0, c_0)$. The operator $D_R\Sigma$ is the triangle defined by $(D_R\Sigma)_{nk} = R^n$ for $k \leq n$ and for all n. Then from the characterization of (c_0, c_0), the condition $D_R\Sigma \in (c_0, c_0)$ is equivalent to $nR^n = O(1)$ $(n \to \infty)$, and $R < 1$. Then the condition $(R^{-n})_{n=1}^\infty \in M(c, (c_0)_\Delta)$ implies $\Delta D_{1/R} \in (c, c_0)$. The nonzero entries of $\Delta D_{1/R}$ are given by $[\Delta D_{1/R}]_{nn} = R^{-n}$ and $[\Delta D_{1/R}]_{n,n-1} = -R^{-n+1}$ for all $n \geq 1$, and from the characterization of (c, c_0), we conclude $R \geq 1$. By similar arguments as those used above we obtain $\mathcal{I}_r((s_1)_\Delta, c) = \mathcal{I}_r(c_\Delta, c) = \mathcal{I}_r(s_1, c)$.
(ii) The proof of (ii) is similar and left to the reader. □

We may state an elementary application which can be considered as an exercise and is based on the notion of perturbed sequence spaces inclusions.

Example 5.8 Let $r, \rho > 0$. We consider the system of perturbed (SSIE) defined by

$$(S_1)\begin{cases} c \subset D_r * (c_0)_\Delta + s_x^{(c)} \\ \ell_\infty \subset D_\rho * (s_1)_\Delta + s_{1/x}. \end{cases}$$

By Corollary 5.10, the system (S_1) is equivalent to $c \subset s_x^{(c)}$ and $\ell_\infty \subset s_{1/x}$, that is, $1/x \in c$ and $x \in \ell_\infty$ for r and $\rho < 1$. Since we have $1/x \in c$ if and only if $x_n \to L$ $(n \to \infty)$ for some scalar L with $0 < L \leq \infty$, we conclude that the set $\mathfrak{S}$ of all the positive sequences that satisfy system (S_1) is determined by

$$\mathfrak{S} = \begin{cases} cl^c(e) & \text{if } r, \rho < 1 \\ \varnothing & \text{otherwise.} \end{cases}$$

Now we consider the (SSIE) of the form

$$F \subset D_r * E_{C_1} + F_x, \text{ where } C_1 \text{ is the Cesàro operator and } E, F \in \{c, s_1\}.$$

For instance, the (SSIE) $c \subset D_r * c_{C_1} + s_x^{(c)}$ is associated with the next statement.

The condition $y_n \to l$ $(n \to \infty)$ implies that there are sequences $u, v \in \omega$ such that $y = u + v$ and

$$\frac{1}{n}\sum_{k=1}^{n}\frac{u_k}{r^k} \to l' \text{ and } \frac{v_n}{x_n} \to l'' \ (n \to \infty)$$

for some scalars l, l' and l'' and for all y.

We obtain the following result.

Proposition 5.5 ([12, Proposition 5, p. 1049]) *Let $r > 0$. Then we have*

(i) The solutions of the (SSIE) defined by $c \subset D_r * E_{C_1} + s_x^{(c)}$ with $E \in \{c, s_1\}$ are determined by (5.11) with $F' = c$.

(ii) The solutions of the (SSIE) $s_1 \subset D_r * (s_1)_{C_1} + s_x$ are determined by (5.11) with $F' = s_1$.

Proof (i) Case $E = c$.
Let $R > 0$. Then we have $(R^n)_{n=1}^{\infty} \in M(c_{C_1}, c_0)$ if and only if $D_R C_1^{-1} \in (c, c_0)$. It can easily be seen that the entries of the matrix C_1^{-1} are defined by $[C_1^{-1}]_{nn} = n$, $[C_1^{-1}]_{n,n-1} = -(n-1)$ for all $n \geq 2$ and $[C_1^{-1}]_{1,1} = 1$. Then $D_R C_1^{-1}$ is the triangle whose nonzero entries are given by $[D_R C_1^{-1}]_{nn} = nR^n$, $[D_R C_1^{-1}]_{n,n-1} = -(n-1)R^n$ for all $n \geq 2$ and $[D_R C_1^{-1}]_{1,1} = R$. By the characterization of (c, c_0), this means $\lim_{n\to\infty}(R^n[n-(n-1)]) = \lim_{n\to\infty} R^n = 0$, and $(2n-1)R^n \leq K$ for some $K > 0$ and all n. We conclude $(R^n)_{n=1}^{\infty} \in M(c_{C_1}, c_0)$ if and only if $R < 1$. Then we have $(R^{-n})_{n=1}^{\infty} \in M(c, c_{C_1})$ if and only if $C_1 D_{1/R} \in (c, c)$. But $C_1 D_{1/R}$ is the triangle defined by $[C_1 D_{1/R}]_{nk} = n^{-1}R^{-k}$ for $k \leq n$ and for all n. So the condition $C_1 D_{1/R} \in (c, c)$ is equivalent to $n^{-1}\sum_{k=1}^{n} R^{-k} \to L$ $(n \to \infty)$ for some scalar L and $R \geq 1$. We conclude by Corollary 5.8.
Case $E = s_1$.
We have $(R^n)_{n=1}^{\infty} \in M((s_1)_{C_1}, c_0)$ if and only if $D_R C_1^{-1} \in (s_1, c_0)$, and from the characterization of (s_1, c_0), that is, $\lim_{n\to\infty}[(2n-1)R^n] = 0$ and $R < 1$. Then we have $(R^{-n})_{n=1}^{\infty} \in M(c, (s_1)_{C_1})$ if and only if $C_1 D_{1/R} \in (c, s_1)$, that is, $\sup_n(n^{-1}\sum_{k=1}^{n} R^{-k}) < \infty$ and $R \geq 1$. Again we conclude by Corollary 5.8.
(ii) As we have just seen above, we have $(R^n)_{n=1}^{\infty} \in M((s_1)_{C_1}, c_0)$ if and only if $R < 1$. Then we obtain $(R^{-n})_{n=1}^{\infty} \in M(s_1, (s_1)_{C_1})$ if and only if $C_1 D_{1/R} \in (s_1, s_1)$, that is, $\sup_n(n^{-1}\sum_{k=1}^{n} R^{-k}) < \infty$ and we conclude $R \geq 1$.

□

Remark 5.7 We may deal with the set $\mathcal{I}_a(c_{C_1}, c)$ of all solutions of the (SSIE) $c \subset D_a * c_{C_1} + s_x^{(c)}$. By similar arguments as those used above, we obtain $\mathcal{I}_a(c_{C_1}, c) = \overline{c}$ if $a \in c_0$ and $\mathcal{I}_a(c_{C_1}, c) = \mathcal{U}^+$ if $(n^{-1}\sum_{k=1}^{n} 1/a_k)_{n=1}^{\infty} \in c$. For $a = (r^n)_{n=1}^{\infty}$ we obtain Part (i) of Proposition 5.5. In the case of $a = (n^{\alpha})_{n=1}^{\infty}$ with $\alpha \in \mathbb{R}$, it can easily be shown that

$$\mathcal{I}_a(c_{C_1}, c) = \begin{cases} \overline{c} & \text{if } \alpha < 0 \\ U^+ & \text{if } \alpha \geq 0. \end{cases}$$

We obtain the following result concerning the (SSIE) $c \subset D_r * (c_0)_{C_1} + s_x^{(c)}$.

Corollary 5.11 *The solutions of the (SSIE) $c \subset D_r * (c_0)_{C_1} + s_x^{(c)}$, are determined by*

$$\mathcal{I}_r((c_0)_{C_1}, c) = \begin{cases} \overline{c} & \textit{if } r < 1 \\ U^+ & \textit{if } r > 1. \end{cases}$$

Proof We apply Theorem 5.1. We have $(R^n)_{n=1}^{\infty} \in M((c_0)_{C_1}, c_0)$ if and only if $D_R C_1^{-1} \in (c_0, c_0)$. We conclude $D_R C_1^{-1} \in (c_0, c_0)$ if and only if $((2n-1)R^n)_{n=1}^{\infty} \in$

ℓ_∞ and $R < 1$. Then we have $(R^{-n})_{n=1}^{\infty} \in M(c, (c_0)_{C_1})$ if and only if $C_1 D_{1/R} \in (c, c_0)$, that is, $n^{-1} \sum_{k=1}^{n} R^{-k} \to 0$ $(n \to \infty)$, and $R > 1$. This completes the proof. □

Now we consider an application to the solvability of the following (SSE's) with an operator

$$D_r * E_{C_1} + s_x^{(c)} = c \text{ for } E \in \{c_0, c, s_1\}$$

and

$$D_r * E_{C_1} + s_x = s_1 \text{ for } E \in \{c, s_1\}.$$

For instance, the (SSE) defined by $D_r * (s_1)_{C_1} + c_x = c$ is associated with the next statement.

The condition $y_n \to l$ $(n \to \infty)$ holds if and only if there are sequences $u, v \in \omega$ such that $y = u + v$ and

$$\sup_n \left(\frac{1}{n} \left| \sum_{k=1}^{n} \frac{u_k}{r^k} \right| \right) < \infty \text{ and } \frac{v_n}{x_n} \to l'(n \to \infty)$$

for some scalars l and l' and for all y. Here we also use the (SSIE) defined by $D_r * (c_0)_{C_1} + c_x \subset c$ which is associated with the next statement.

The conditions $n^{-1}(\sum_{k=1}^{n} u_k/r^k) \to 0$ and $v_n/x_n \to l$ together imply $u_n + v_n \to l'$ $(n \to \infty)$ for all sequences u and $v \in \omega$ and for some scalars l and l'.

We write $\mathcal{I}'_a(E, F) = \{x \in \mathcal{U}^+ : E_a + F_x \subset F\}$. We note that since E and F are linear spaces of sequences, we have $x \in \mathcal{I}'_a(E, F)$ if and only if and $E_a \subset F$ and $F_x \subset F$. This means that $x \in \mathcal{I}'_a(E, F)$ if and only if $a \in M(E, F)$ and $x \in M(F, F)$. Then we have

$$\mathbb{S}(E, F) = \mathcal{I}_a(E, F) \cap \mathcal{I}'_a(E, F) = \{x \in \mathcal{U}^+ : E_a + F_x = F\},$$

(see [11, pp. 222–223]).

From Proposition 5.5 and Corollary 5.11 we obtain the next results on the (SSE) $D_r * E_{C_1} + s_x^{(c)} = c$ with $E \in \{c_0, c, s_1\}$, and $D_r * E_{C_1} + s_x = s_1$ with $E \in \{c, s_1\}$.

Proposition 5.6 ([12, Proposition 6, pp. 1050–1051]) *Let $r > 0$. Then we have*

(i) The solutions of the (SSIE) $D_r * E_{C_1} + s_x^{(c)} \subset c$ with $E \in \{c_0, c, s_1\}$ and $D_r * E_{C_1} + s_x \subset s_1$ with $E \in \{c, s_1\}$ are determined by

$$\mathcal{I}'_r\left(E_{C_1}, c\right) = \begin{cases} c^+ & \text{if } r < 1 \\ \varnothing & \text{if } r \geq 1 \end{cases} \text{ for } E \in \{c_0, c, s_1\}, \tag{5.19}$$

and

$$\mathcal{I}'_r(E_{C_1}, s_1) = \begin{cases} s_1^+ & if\, r < 1 \\ \varnothing & if\, r \geq 1 \end{cases} \quad for\ E \in \{c, s_1\}.$$

(ii) The solutions of the perturbed equations $D_r * E_{C_1} + s_x^{(c)} = c$ *with* $E \in \{c_0, c, s_1\}$ *and* $D_r * E_{C_1} + s_x = s_1$ *with* $E \in \{c, s_1\}$ *are determined by*

$$\mathbb{S}_r\left(E_{C_1}, c\right) = \begin{cases} cl^c(e) & if\, r < 1 \\ \varnothing & if\, r \geq 1 \end{cases} \quad for\ E \in \{c_0, c, s_1\},$$

and

$$\mathbb{S}_r(E_{C_1}, s_1) = \begin{cases} cl^\infty(e) & if\, r < 1 \\ \varnothing & if\, r \geq 1 \end{cases} \quad for\ E \in \{c, s_1\}.$$

Proof (i) The inclusion $D_r * (c_0)_{C_1} \subset c$ is equivalent to $D_r C_1^{-1} \in (c_0, c)$, that is, $D_r C_1^{-1} \in S_1$ and $(n + n - 1)r^n \leq K$ for all n and some $K > 0$. So we have $D_r * (c_0)_{C_1} \subset c$ if and only if $r < 1$. Then the inclusion $s_x^{(c)} \subset c$ is equivalent to $x \in c$. So the (SSIE) $D_r * (c_0)_{C_1} + s_x^{(c)} \subset c$ is equivalent to $r < 1$ and $x \in c$ and the identity in (5.19) holds for $E = c_0$.

Case $E = c$.
The inclusion $D_r * c_{C_1} \subset c$ is equivalent to $D_r C_1^{-1} \in (c, c)$, that is, $[n - (n - 1)]r^n = r^n \to L$ $(n \to \infty)$ for some scalar L, and $[n + (n - 1)]r^n = O(1)$ $(n \to \infty)$. This implies $r < 1$. Using similar arguments as those above we conclude that (5.19) holds for $E = c$.
The proof of the case $E = s_1$ is similar and left to the reader.

(ii) is obtained from Part (i) and Part (i) of Proposition 5.5 for $\mathbb{S}_r(E_{C_1}, c)$ with $E = c$ or s_1 and is obtained from Part (i) and Corollary 5.11 for $\mathbb{S}_r((c_0)_{C_1}, c)$. Then the determination of the set $\mathbb{S}_r((s_1)_{C_1}, s_1)$ is obtained from Part (i) and Part (ii) of Proposition 5.5.
It remains to determine the set $\mathbb{S}_r(c_{C_1}, s_1)$. For this we deal with the solvability of the (SSIE) $s_1 \subset D_r * c_{C_1} + s_x$. As we have seen in the proof of Proposition 5.5, we have $(r^n)_{n=1}^\infty \in M(c_{C_1}, c_0)$ if and only if $D_r C_1^{-1} \in (c, c_0)$, that is, $r < 1$. Then we have $(r^{-n})_{n=1}^\infty \in M(s_1, c_{C_1})$ if and only if $C_1 D_{1/r} \in (s_1, c)$. Since $\lim_{n\to\infty}[C_1 D_{1/r}]_{nk} = 0$ for all $k \geq 1$, we have by Part (v) of Lemma 5.1 $C_1 D_{1/r} \in (s_1, c)$ if and only if $n^{-1} \sum_{k=1}^n r^{-k} \to 0$ $(n \to \infty)$ and $r > 1$. So we have shown $\mathcal{I}_r(c_{C_1}, s_1) = \overline{s_1}$ for $r < 1$ and $\mathcal{I}_r(c_{C_1}, s_1) = \overline{s_1}$ for $r > 1$. We conclude for the set $\mathbb{S}_r(c_{C_1}, s_1)$ using the determination of $\mathcal{I}'_r(c_{C_1}, s_1)$ given in Part (i).
This completes the proof. ☐

Example 5.9 For instance, we have $x \in \mathbb{S}_r(c_{C_1}, s_1)$ if and only if the next statement holds.

The condition $\sup_n |y_n| < \infty$ *holds if and only if there are sequences* $u, v \in \omega$ *such that* $y = u + v$ *for which* $n^{-1} \sum_{k=1}^{n} u_k/r^k \to l \ (n \to \infty)$ *and* $\sup_n (|v_n|/x_n) < \infty$ *for some scalar* l *and all sequences* y.

We have $\mathbb{S}_r(c_{C_1}, s_1) \neq \varnothing$ if and only if $r < 1$, and then $\mathbb{S}_r(c_{C_1}, s_1) = cl^{\infty}(e)$, that is, $x \in \mathbb{S}_r(c_{C_1}, s_1)$ if and only if there are $K_1, K_2 > 0$ such that $K_1 \leq x_n \leq K_2$ for all n.

Remark 5.8 By similar arguments as those used above, the set $\mathcal{I}'_a(c_{C_1}, c)$ of all sequences $x \in \mathcal{U}^+$ for which $D_a * c_{C_1} + s_x^{(c)} \subset c$ is determined by $\mathcal{I}'_a(c_{C_1}, c) = c^+$ if $a \in s_{(1/n)_{n=1}^{\infty}}$ and $\mathcal{I}'_a(c_{C_1}, c) = \varnothing$ if $a \notin s_{(1/n)_{n=1}^{\infty}}$. Indeed, we have $a \in M(c_{C_1}, c)$ if and only if $D_a C_1^{-1} \in (c, c)$. But the matrix $D_a C_1^{-1}$ is the triangle whose the nonzero entries are given by $[D_a C_1^{-1}]_{nn} = na_n$, $[D_a C_1^{-1}]_{n,n-1} = -(n-1)a_n$ for all $n \geq 2$ and $[D_a C_1^{-1}]_{1,1} = a_1$. From the characterization of (c, c), the condition $D_a C_1^{-1} \in (c, c)$ is equivalent to

$$\lim_{n\to\infty} (a_n[n - (n-1)]) = \lim_{n\to\infty} a_n = L$$

for some scalar L, and $(2n-1)a_n \leq K$ for some $K > 0$ and all n. We conclude $a \in M(c_{C_1}, c)$ if and only if $a \in s_{(1/n)_{n=1}^{\infty}}$. Since $s_{(1/n)_{n=1}^{\infty}} \subset c_0$, this result and Remark 5.7 together imply $\mathcal{S}_a(c_{C_1}, c) = cl^c(e)$ if $a \in s_{(1/n)_{n=1}^{\infty}}$, and $\mathcal{S}_a(c_{C_1}, c) = \varnothing$ if $a \notin s_{(1/n)_{n=1}^{\infty}}$. So we obtain the result of Part (i) in Proposition 5.6, where $a = (r^n)_{n=1}^{\infty}$. In this way, we also obtain $\mathbb{S}_{(n^{-\alpha})_{n=1}^{\infty}}(c_{C_1}, c) = cl^c(e)$ if $\alpha \geq 1$ and $\mathbb{S}_{(n^{-\alpha})_{n=1}^{\infty}}(c_{C_1}, c) = \varnothing$ if $\alpha < 1$.

Conclusion. In this section, we studied the solvability of the (SSIE) of the form $F \subset (E_a)_T + F'_x$ *for some triangle* T *and* $e \in F$. *In future, we will be led to deal with the (SSIE) with operators of the form* $F \subset (E_a)_A + F'_x$ *in each of the cases* $e \in F$ *and* $e \notin F$, *where* A *is an infinite matrix which can be a triangle, or a band matrix, or the inverse of a band matrix.*

5.5 The (SSIE) $F \subset E_a + F'_x$ for $e \notin F$

In this section, we consider a class of sequences spaces inclusions equations that are direct applications of Theorem 5.2 and involve the sets c_0, c, ℓ_p for $1 \leq p \leq \infty$, w_0 and w_∞. As we have seen above, the solutions of the perturbed inclusion $F \subset E_a + F'_x$, where E, F and F' are linear spaces of sequences, cannot be obtained in the general case. So we are led to deal with the case when $a = (r^n)_{n=1}^{\infty}$ with $r > 0$ for which most of these (SSIE) can be totally solved. Here also we write $\mathcal{I}_a(E, F, F') = \{x \in \mathcal{U}^+ : F \subset E_a + F'_x\}$, where E, F and F' are linear spaces of sequences and $a \in \mathcal{U}^+$. For any set χ of sequences, we let $\overline{\chi} = \{x \in \mathcal{U}^+ : 1/x \in \chi\}$. So we have $F'_x \supset F$ if and only if $x \in \overline{M(F, F')}$. For instance, $\overline{s_{(1/n)_{n=1}^{\infty}}}$ is the set whose elements are determined by $x_n \geq Cn$ for all n and some $C > 0$. We note that for any given $b \in \mathcal{U}^+$, we also have $F'_x \supset F_b$ if and only if $b/x \in M(F, F')$, that is, $x \in [\overline{M(F, F')}]_b$ or $x \in \overline{M(F, F')_{1/b}}$. We recall that $\Phi = \{c_0, c, \ell_\infty, \ell_p, w_0, w_\infty\}$ with $p \geq 1$.

Now we state a general result for the study of the (SSIE) of the form $F \subset E_a + F'_x$ where e does not necessarily belong to F. Then among other things, we deal with the solvability of the next (SSIE)

$$c_0 \subset E_a + F'_x \text{ for } E, F' \in \{c_0, c, s_1\}, \tag{1-}$$

$$c_0 \subset (\ell_p)_a + F'_x \text{ for } F' \in \{c_0, c, s_1\} \text{ and } p \geq 1, \tag{2-}$$

$$c_0 \subset E_a + W_x \text{ for } E \in \Phi, \tag{3-}$$

$$\ell_1 \subset E_a + W_x \text{ for } E \in \Phi, \tag{4-}$$

$$\ell_p \subset E_a + F'_x \text{ with } E, F' \in \{c_0, c, s_1, \ell_p\}, \tag{5-}$$

$$w_0 \subset E_a + F'_x \text{ with } E, F' \in \{c_0, c, s_1\}, \tag{6-}$$

$$w_0 \subset E_a + F'_x \text{ with } E \in \{c_0, c, s_1, w_0, w_\infty\},\ F' \in \{w_0, w_\infty\}, \tag{7-}$$

$$w_\infty \subset E_a + s_x \text{ with } E \in \{c_0, s_1\}, \tag{8-}$$

$$w_\infty \subset E_a + W_x \text{ with } E \in \{c_0, s_1, w_\infty\}, \tag{9-}$$

$$c_0 \subset W_a + F'_x \text{ for } F' \in \{c_0, c, s_1\}. \tag{10-}$$

We denote by $\mathcal{U}_1^+$ the set of all sequences α with $0 < \alpha_n \leq 1$ for all n and consider the condition

$$G \subset G_{1/\alpha} \text{ for all } \alpha \in \mathcal{U}_1^+ \tag{5.20}$$

for any given linear space G of sequences. Then we introduce the linear space of sequences H which contains the spaces E and F'. The proof of the next theorem is based on the fact that if H satisfies the condition in (5.20) we then have

$$H_\alpha + H_\beta = H_{\alpha+\beta} \text{ for all } \alpha, \beta \in \mathcal{U}^+$$

(cf. [14, Proposition 5.1, pp. 599–600]). We note that c does not satisfy this condition, but each of the sets c_0, ℓ_∞, ℓ_p, $(p \geq 1)$, w_0 and w_∞ satisfies the condition in (5.20).

In the following, we write $M(F, F') = \chi$. The next result is used to determine a new class of (SSIE).

Theorem 5.2 ([11, Theorem 9, p. 216]) *Let $a \in \mathcal{U}^+$ and let E, F and F' be linear subspaces of ω. We assume that*

(a) χ satisfies condition (5.9).
(b) There is a linear space of sequences H that satisfies the condition in (5.20) and the conditions

$$(\alpha)\ E, F' \subset H \quad \text{and} \quad (\beta)\ M(F, H) = \chi.$$

Then we have

(i) $a \in M(\chi, c_0)$ implies $\mathcal{I}_a(E, F, F') = \overline{\chi}$;
(ii) $a \in \overline{M(F, E)}$ implies $\mathcal{I}_a(E, F, F') = \mathcal{U}^+$.

Proof (i) Let $a \in M(\chi, c_0)$ and let $x \in \mathcal{I}_a(E, F, F')$. As we have just seen, we have

$$F \subset H_a + H_x \subset H_{a+x},$$

which implies

$$\xi = \frac{1}{a+x} \in \chi.$$

Then it can easily be shown that

$$\frac{1}{x} = \frac{\xi}{1 - a\xi}.$$

Then the conditions $\xi \in \chi^+$ and $a \in M(\chi, c_0)$ imply $1 - a_n\xi_n \to 1$ $(n \to \infty)$. By the condition in (a), we conclude $x \in \overline{\chi}$.
Conversely, let $x \in \overline{\chi}$. Then we successively obtain $1/x \in \chi = M(F, F')$, $F \subset F'_x$, $F \subset E_a + F'_x$ and $x \in \mathcal{I}_a(E, F, F')$. We conclude $\mathcal{I}_a(E, F, F') = \overline{\chi}$.
(ii) follows from the equivalence of $a \in \overline{M(F, E)}$ and $F \subset E_a$.
This completes the proof.

□

Corollary 5.12 *Let $a \in \mathcal{U}^+$, and E, F and F' be linear subspaces of ω. We assume χ satisfies condition (5.9) and $E \subset F'$, where F' satisfies the condition in (5.20). Then we have*

(i) $a \in M(\chi, c_0)$ implies $\mathcal{I}_a(E, F, F') = \overline{\chi}$;
(ii) $a \in \overline{M(F, E)}$ implies $\mathcal{I}_a(E, F, F') = \mathcal{U}^+$.

In the next remarks we compare Theorems 5.2 and 5.1 when $e \in F$.

Remark 5.9 Under some conditions we may compare Theorems 5.2 and 5.1. Let $a \in \mathcal{U}^+$ and let E and F be linear subspaces of ω. We denote by $\mathcal{I}_a(E, F)$ the set of all positive sequences x that satisfy $F \subset E_a + E_x$. We assume that the conditions $e \in F$ and $E \subset M(F, E)$ hold and that E satisfies the condition in (5.9). Then we have $\chi = M(F, E) = E$ and trivially $M(\chi, c_0) = M(E, c_0)$, so the condition (i) in Theorem 5.2 is equivalent to the condition (i) in Theorem 5.1.

Remark 5.10 Let $a \in \mathcal{U}^+$ and let E, F and F' be linear subspaces of ω. We assume

$$\text{(a) } e \in F, \quad \text{(b) } E \subset F', \quad \text{(c) } F' \subset M(F, F') \quad \text{and} \quad \text{(d) } F' \text{ satisfies (5.9).}$$

Then we have $\chi = M(F, F') = F'$ and $M(\chi, c_0) \subset M(E, c_0)$. So it can easily be seen that Theorem 5.1 implies Theorem 5.2. This result can be illustrated by the next example. We consider the (SSIE) $c \subset s_a^0 + s_x$. By Theorem 5.1, we obtain that $a \in \ell_\infty$ implies $\mathcal{I}_a(c_0, c, s_1) = \overline{s_1}$, and since $\chi = M(c, s_1) = s_1$, we obtain that $a \in c_0$ implies $\mathcal{I}_a(c_0, c, s_1) = \overline{\chi} = \overline{s_1}$ in Theorem 5.2. We obtain similar results for the (SSIE) $c \subset s_a^0 + F'_x$ where $F' \in \{\ell_\infty, w_0, w_\infty\}$.

5.6 Some Applications

Now we apply the results of the previous sections.

We deal with the inclusion

$$F \subset E_a + F'_x, \text{ where either } e \notin F \text{ or } F' \not\subset M(F, F'),$$

since the case of $e \in F$ and $F' \subset M(F, F')$ was studied in [12].

The (SSIE) $c_0 \subset E_a + F'_x$.

Now we deal with the set $\mathcal{I}_a(E, c_0, F')$ of all solutions of the (SSIE) $c_0 \subset E_a + F'_x$ for $E, F' \in \{c_0, c, s_1\}$. Then we deal with the (SSIE) $c_0 \subset (\ell_p)_a + F'_x$ for $F' \in \{c_0, c, s_1\}$ and $p \geq 1$. Then we study the (SSIE) $c_0 \subset E_a + W_x$ for $E \in \Phi$. We obtain the next results as a direct consequence of Theorem 4.16 and Lemmas 5.6 and 5.7.

Proposition 5.7 *Let $a \in \mathcal{U}^+$ and let $p \geq 1$. Then we have*

(i) *Let $E, F' \in \{c_0, c, s_1\}$. Then*

  (a) *the condition $a \in c_0$ implies $\mathcal{I}_a(E, c_0, F') = \overline{s_1}$;*
  (b) *the condition $a \in \overline{s_1}$ implies $\mathcal{I}_a(E, c_0, F') = \mathcal{U}^+$.*

(ii) *Let $F' \in \{c_0, c, s_1\}$. Then*

  (a) *the condition $a \in c_0$ implies $\mathcal{I}_a(\ell_p, c_0, F') = \overline{s_1}$;*
  (b) *the condition $a \in \overline{\ell_p}$ implies $\mathcal{I}_a(\ell_p, c_0, F') = \mathcal{U}^+$.*

(iii) *Let $E \in \Phi$. Then*

(a) the condition $a \in s^0_{(1/n)_{n=1}^\infty}$ *implies* $\mathcal{I}_a(E, c_0, w_\infty) = \overline{w_\infty}$*;*
(b) $\mathcal{I}_a(E, c_0, w_\infty) = \mathcal{U}^+$ *in the next cases*

$$(\alpha)\ a \in \overline{s_1} \text{ for } E \in \{c_0, c, s_1\},$$
$$(\beta)\ a \in \overline{E} \text{ for } E \in \{\ell_p, w_0, w_\infty\}.$$

Proof (i) We take $H = s_1$ in Theorem 5.2. Then we have $\chi = M(F, F') = M(F, H) = s_1$ for $F' \in \{c_0, c, s_1\}$ and we obtain $\mathcal{I}_a(E, c_0, F') = \overline{s_1}$ if $a \in c_0$. Then we have $M(F, E) = M(c_0, E) = s_1$ for $E \in \{c_0, c, s_1\}$ which shows Part (i).
(ii) Again, with $H = s_1$, we obtain $\chi = M(c_0, F') = M(c_0, H) = s_1$ for $F' \in \{c_0, c, s_1\}$. Since $M(\chi, c_0) = c_0$, we obtain $\mathcal{I}_a(\ell_p, c_0, F') = \overline{s_1}$ if $a \in c_0$ for $F' \in \{c_0, c, s_1\}$. Then by Lemma 5.6, we obtain $M(F, E) = M(c_0, \ell_p) = \ell_p$ and $\mathcal{I}_a(\ell_p, c_0, F') = \mathcal{U}^+$ if $a \in \overline{\ell_p}$.
(iii)

(iii) (a) Here we take $H = F' = w_\infty$. So we obtain $\chi = M(c_0, F') = w_\infty$, and since $M(w_\infty, c_0) = s^0_{(1/n)_{n=1}^\infty}$, we obtain the statement in (iii) (a).
(iii) (b) The cases (α) and (β) follow from Lemma 5.6, where $M(c_0, E) = s_1$ for $E \in \{c_0, c, s_1\}$ and $M(c_0, E) = E$ for $E \in \{\ell_p, w_0, w_\infty\}$.

□

The (SSIE) $\ell_1 \subset E_a + W_x$ *and* $\ell_p \subset E_a + F'_x$*.*

Now we deal with the set $\mathcal{I}_a(E, \ell_1, w_\infty)$ of all solutions of the (SSIE) $\ell_1 \subset E_a + W_x$ for $E \in \Phi$. Then we consider the set $\mathcal{I}_a(E, \ell_p, F')$ of all solutions of the (SSIE) $\ell_p \subset E_a + F'_x$ with $E, F' \in \{c_0, c, s_1, \ell_p\}$. We obtain the following result.

Proposition 5.8 *Let* $a \in \mathcal{U}^+$ *and let* $p \geq 1$*. Then we have*

(i) Let $E \in \Phi = \{c_0, c, s_1, \ell_p, w_0, w_\infty\}$*. Then*

(a) the condition $a \in s_{(1/n)_{n=1}^\infty}$ *implies* $\mathcal{I}_a(E, \ell_1, w_\infty) = \overline{s_{(n)_{n=1}^\infty}}$*;*
(b) $\mathcal{I}_a(E, \ell_1, w_\infty) = \mathcal{U}^+$ *in the next cases*

$$(\alpha)\, a \in \overline{s_1} \text{ and } E \in \{c_0, c, s_1, \ell_p\}, \quad (\beta)\, a \in \overline{s^0_{(n)_{n=1}^\infty}} \text{ and } E = w_0$$

and

$$(\gamma)\, a \in \overline{s_{(n)_{n=1}^\infty}} \text{ and } E = w_\infty.$$

(ii) Let $E, F' \in \{c_0, c, s_1, \ell_p\}$*. Then*

$$a \in c_0 \text{ implies } \mathcal{I}_a(E, \ell_p, F') = \overline{s_1} \tag{5.21}$$

and

$$a \in \overline{s_1} \text{ implies } \mathcal{I}_a(E, \ell_p, F') = \mathcal{U}^+. \tag{5.22}$$

Proof (i) (a) We take $F' = H = w_\infty$ in Theorem 5.2, so we have $\chi = M(\ell_1, H) = s_{(n)_{n=1}^\infty}$ by Lemma 5.7. Then we have $M(\chi, c_0) = s^0_{(1/n)_{n=1}^\infty}$. This shows (i) (a).
(b) Now we show (i) (b)

Case (α). By Lemma 5.7, we have $M(\ell_1, E) = s_1$ for $E \in \{c_0, c, s_1\}$. Then we consider the case $E = \ell_p$. By Theorem 1.22 with $X = \ell_1$ and $Z = \ell_q$ we have $M(\ell_1, \ell_p) = M(\ell_1, \ell_q^\beta) = M(\ell_q, s_1) = s_1$. This concludes the proof of (i) (b) (α).
Cases (β) *and* (γ).
We obtain these cases from Lemma 5.7, where $M(\ell_1, w_0) = w_0$ and $M(\ell_1, w_\infty) = w_\infty$.

Thus we have shown Part (i).
(ii) We take $H = s_1$. So we obtain $\chi = M(\ell_1, F') = M(\ell_1, H) = s_1$ for $F' \in \{c_0, c, s_1, \ell_p\}$, and since $M(\chi, c_0) = c_0$ we obtain (5.21). Then by Lemma 5.6, we have $M(\ell_p, E) = s_1$ for $E \in \{c_0, c, s_1, \ell_p\}$, which shows (5.22). This completes the proof.

□

The (SSIE) of $w_0 \subset E_a + F'_x$.

Now we study the set $\mathcal{I}_a(E, w_0, F')$ of all solutions of the (SSIE) $w_0 \subset E_a + F'_x$ in each of the cases $E, F' \in \{c_0, c, s_1\}$ and $E \in \{c_0, c, s_1, w_0, w_\infty\}$ and $F' \in \{w_0, w_\infty\}$.

Proposition 5.9 *Let* $a \in \mathcal{U}^+$. *Then we have*

(i) Let $E, F' \in \{c_0, c, s_1\}$. *Then*

(a) the condition $a \in s^0_{(n)_{n=1}^\infty}$ *implies* $\mathcal{I}_a(E, w_0, F') = \overline{s_{(1/n)_{n=1}^\infty}}$;
(b) the condition $a \in \overline{s_{(1/n)_{n=1}^\infty}}$ *implies* $\mathcal{I}_a(E, w_0, F') = \mathcal{U}^+$.

(ii) Let $E \in \{c_0, c, s_1, w_0, w_\infty\}$ *and* $F' \in \{w_0, w_\infty\}$. *Then*

(a) the condition $a \in c_0$ *implies* $\mathcal{I}_a(E, w_0, F') = \overline{s_1}$;
(b) the identity $\mathcal{I}_a(E, w_0, F') = \mathcal{U}^+$ *holds in the next cases*
(α) $a \in \overline{s_{(1/n)_{n=1}^\infty}}$ *for* $E \in \{c_0, c, s_1\}$;
(β) $a \in \overline{s_1}$ *for* $E \in \{w_0, w_\infty\}$.

Proof (i) follows from Lemma 5.7, where we have

$$M(w_0, Y) = s_{(1/n)_{n=1}^\infty} \text{ for } Y \in \{c_0, c, s_1\}. \tag{5.23}$$

Then it is enough to apply Theorem 5.2 with $H = s_1$.
(ii) Here we take $H = w_\infty$. Then we have $\chi = M(w_0, F') = M(w_0, H) = s_1$ for $F' \in \{w_0, w_\infty\}$. The result is obtained from Lemma 5.7, where (5.23) holds and $M(w_0, Y) = s_1$ for $Y \in \{w_0, w_\infty\}$.

□

The (SSIE) $w_\infty \subset E_a + F'_x$.

Now we study the set $\mathcal{I}_a(E, w_\infty, s_1)$ of all solutions of the (SSIE) $w_\infty \subset E_a + s_x$ with $E \in \{c_0, s_1\}$. Then we consider the (SSIE) $w_\infty \subset E_a + W_x$ with $E \in \{c_0, s_1, w_\infty\}$. We note that although $e \in w_\infty$, Theorem 5.1 cannot be applied, since we have $w_\infty \not\subset M(w_\infty, F')$, where F' is either of the sets s_1 or w_∞, and we have $M(w_\infty, s_1) = s_{(1/n)_{n=1}^\infty}$ and $M^+(w_\infty, w_\infty) = s_1^+$. We obtain the following result.

Proposition 5.10 *Let $a \in \mathcal{U}^+$. Then we have*

(i) Let $E \in \{c_0, s_1\}$. Then

(a) the condition $a \in s^0_{(n)_{n=1}^\infty}$ implies $\mathcal{I}_a(E, w_\infty, s_1) = \overline{s_{(1/n)_{n=1}^\infty}}$;
(b) the identity $\mathcal{I}_a(E, w_\infty, s_1) = U^+$ holds in the next cases
(α) $a \in \overline{s^0_{(1/n)_{n=1}^\infty}}$ for $E = c_0$;
(β) $a \in \overline{s_{(1/n)_{n=1}^\infty}}$ for $E = s_1$.

(ii) Let $E \in \{c_0, s_1, w_\infty\}$. Then

(a) the condition $a \in c_0$ implies $\mathcal{I}_a(E, w_\infty, w_\infty) = \overline{s_1}$;
(b) the identity $\mathcal{I}_a(E, w_\infty, w_\infty) = U^+$ holds in the next cases
(α) $a \in \overline{s^0_{(1/n)_{n=1}^\infty}}$ for $E = c_0$;
(β) $a \in \overline{s_{(1/n)_{n=1}^\infty}}$ for $E = s_1$;
(γ) $a \in \overline{s_1}$ for $E = w_\infty$.

Proof (i) We take $F' = H = s_1$ and $\chi = M(w_\infty, H) = s_{(1/n)_{n=1}^\infty}$ by Lemma 5.7. Then we have $M(\chi, c_0) = s^0_{(n)_{n=1}^\infty}$. This shows (i) (a). The statements in (i) (a) and (i) (b) are direct consequences of Lemma 5.7, where we have $M(w_\infty, c_0) = s^0_{(1/n)_{n=1}^\infty}$ and $M(w_\infty, s_1) = s_{(1/n)_{n=1}^\infty}$.
(ii) Here we take $H = w_\infty$, then by [14, Remark 3.4, p. 597], we have $\chi^+ = M^+(w_\infty, H) = s_1^+$ and $\mathcal{I}_a(E, w_\infty, w_\infty) = \overline{s_1}$ if $a \in M(\chi^+, c_0) = c_0$ for $E \in \{c_0, s_1, w_\infty, w_\infty\}$. Then we have $M(w_\infty, c_0) = s^0_{(1/n)_{n=1}^\infty}$, $M(w_\infty, s_1) = s_{(1/n)_{n=1}^\infty}$ and $M^+(w_\infty, w_\infty) = s_1^+$ and we conclude by Theorem 5.2.
This completes the proof. □

The (SSIE) defined by $c_0 \subset W_a + F'_x$.

We may add to the previous classes the (SSIE) $c_0 \subset W_a + F'_x$ for which the previous theorems cannot be applied. We state the next result.

Proposition 5.11 *Let $a \in \mathcal{U}^+$ and $\mathcal{I}_a(w_\infty, c_0, F')$ be the set of all solutions of the (SSIE) $c_0 \subset W_a + F'_x$. We assume $F' \subset s_1$ and $M(c_0, F') = \ell_\infty$. Then we have*

$$na_n \to 0\ (n \to \infty) \text{ implies } \mathcal{I}_a(w_\infty, c_0, F') = \overline{s_1} \tag{5.24}$$

and

$$a \in \overline{w_\infty} \text{ implies } \mathcal{I}_a(w_\infty, c_0, F') = \mathcal{U}^+. \tag{5.25}$$

Proof Let $x \in \mathcal{I}_a(w_\infty, c_0, F')$ and assume $na_n \to 0$ $(n \to \infty)$. Then, since $w_\infty \subset s_{(n)_{n=1}^\infty}$ and $F' \subset s_1$, we deduce $c_0 \subset s_{(na_n)_{n=1}^\infty} + s_x = s_{(na_n+x_n)_{n=1}^\infty}$, which implies $((na_n + x_n)^{-1})_{n=1}^\infty \in M(c_0, s_1) = s_1$. So there is $K > 0$ such that $x_n \geq K - na_n$, and since $na_n \to 0$ $(n \to \infty)$, there is $C > 0$ such that $x_n \geq C$ for all n. So we obtain the statement in (5.24).
Then $a \in \overline{w_\infty}$ implies $c_0 \subset W_a$ and (5.25) holds. □

Corollary 5.13 *Let $a \in \mathcal{U}^+$ and $F' \in \{c_0, c, s_1\}$. Then (5.24) and (5.25) hold.*

From the results of Propositions 5.7, 5.8, 5.9 and 5.10, we obtain the next corollaries where r is a positive real.

Corollary 5.14 *(i) Let $E, F' \in \{c_0, c, s_1\}$. Then the solutions of the (SSIE) $c_0 \subset E_r + F'_x$ are determined by (5.11) with $\chi = s_1$.*
(ii) Let $E \in \{c_0, c, s_1, w_\infty\}$. Then the solutions of the (SSIE) $c_0 \subset E_r + W_x$ are determined by (5.11) with $\chi = w_\infty$.

Remark 5.11 We may illustrate Part (i) of Corollary 5.14 as follows. Let $E, F' \in \{c_0, c, s_1\}$. Then each of the next statements is equivalent to $r < 1$, where

(a) $\mathcal{I}_r(E, c_0, F') = \overline{s_1}$.
(b) For every $x \in \mathcal{U}^+$, we have $c_0 \subset E_r + F'_x$ if and only if $c_0 \subset F'_x$.
(c) For every $x \in \mathcal{U}^+$, we have $c_0 \subset E_r + F'_x$ if and only if $x_n \geq K$ for all n.

Corollary 5.15 *Let $p \geq 1$. Then we have*

(i) Let $E \in \Phi = \{c_0, c, s_1, \ell_p, w_0, w_\infty\}$. Then the solutions of the (SSIE) $\ell_1 \subset E_r + W_x$ are determined by (5.11) with $\chi = s_{(n)_{n=1}^\infty}$.
(ii) Let $E, F' \in \{c_0, c, s_1, \ell_p\}$. Then the solutions of the (SSIE) $\ell_p \subset E_r + F'_x$ are determined by (5.11) with $\chi = s_1$.

Corollary 5.16 *(i) Let $E, F' \in \{c_0, c, s_1\}$. Then the solutions of the (SSIE) $w_0 \subset E_r + F'_x$ are determined by (5.12) with $\chi = s_{(1/n)_{n=1}^\infty}$.*
(ii) Let $E, F' \in \{w_0, w_\infty\}$. Then the solutions of the (SSIE) $w_0 \subset E_r + F'_x$ are determined by (5.11) with $\chi = s_1$.

Corollary 5.17 *(i) Let $E \in \{c_0, s_1\}$. Then the solutions of the (SSIE) $w_\infty \subset E_r + s_x$ are determined by (5.12) with $\chi = s_{(1/n)_{n=1}^\infty}$.*
(ii) The solutions of the (SSIE) $w_\infty \subset W_r + W_x$ are determined by (5.11) with $\chi = s_1$.

Remark 5.12 By Corollaries 5.16 and 5.17, we easily see that the solutions of each of the (SSIE) $w_0 \subset c_0 + F'_x$, $w_0 \subset c + F'_x$, $w_0 \subset \ell_\infty + F'_x$, with $F' \in \{c_0, c, s_1\}$, and $w_\infty \subset c_0 + s_x$, $w_\infty \subset \ell_\infty + s_x$ are determined by $x_n \geq Kn$ for all n and some $K > 0$.

Example 5.10 Let $\mathcal{I}_{1/2}(c, \ell_1, w_\infty)$ be the set of all positive sequences x that satisfy the next statement.

The condition $\sum_{k=1}^{\infty}|y_k| < \infty$ *implies that there are sequences* $u, v \in \omega$ *with* $y = u + v$ *such that* $2^n u_n \to L$ $(n \to \infty)$ *and* $n^{-1}\sum_{k=1}^{n} x_k|v_k| \leq K$ *for all* n *and for all sequences* y *and for some scalars* L *and* K *with* $K > 0$.

The set $\mathcal{I}_{1/2}^{;}(c, \ell_1, w_\infty)$ is associated with the (SSIE) $\ell_1 \subset c_{1/2} + W_{1/x}$, which is equivalent to $x \in s_{(n)_{n=1}^{\infty}}$. This means that $\mathcal{I}_{1/2}^{;}(c, \ell_1, w_\infty)$ is the set of all positive sequences x that satisfy $x_n \leq K'n$ for all n and some $K' > 0$.

By Propositions 5.7, 5.9 and 5.10 and Corollary 5.13, we obtain the next corollary concerning the solvability of some (SSIE) of the form $F \subset E_r + F'_x$ for $r \neq 1$.

Corollary 5.18 *Let* $p \geq 1$. *Then we have*

(i) *The solutions of each of the (SSIE) defined by* $c_0 \subset (\ell_p)_r + F'_x$ *with* $F' \in \{c_0, c, s_1\}$ *and by* $w_\infty \subset E_r + W_x$ *with* $E \in \{c_0, s_1\}$ *are determined by the set*

$$\mathcal{I}^1 = \begin{cases} \overline{s_1} & \text{if } r < 1 \\ U^+ & \text{if } r > 1. \end{cases}$$

(ii) *The solutions of each of the (SSIE) defined by* $c_0 \subset E_r + W_x$ *with* $E \in \{\ell_p, w_0, w_\infty\}$ *are determined by*

$$\mathcal{I}^2 = \begin{cases} \overline{w_\infty} & \text{if } r < 1 \\ U^+ & \text{if } r > 1. \end{cases}$$

(iii) *The solutions of each of the (SSIE) defined by* $w_0 \subset E_r + F'_x$ *with* $E \in \{c_0, c, s_1\}$, $F' \in \{w_0, w_\infty\}$ *are determined by*

$$\mathcal{I}^3 = \begin{cases} \overline{s_1} & \text{if } r < 1 \\ U^+ & \text{if } r > 1. \end{cases}$$

Example 5.11 We consider the system

$$\begin{cases} w_0 \subset c + s_{1/x}^0 \\ \ell_1 \subset W_{1/2} + W_x. \end{cases}$$

Let $\mathbb{S}$ be the set of all positive sequences that satisfy the system. By Part (i) of Corollary 5.16 and Part (i) of Corollary 5.15, the solutions of the (SSIE) $w_0 \subset c + s_{1/x}^0$ are determined by $x \in s_{(1/n)_{n=1}^{\infty}}$ and those of the (SSIE) $\ell_1 \subset W_{1/2} + W_x$ are determined by $x \in \overline{s_{(n)_{n=1}^{\infty}}}$. Since we have $x \in \overline{s_{(n)_{n=1}^{\infty}}}$ if and only if $1/x \in s_{(n)_{n=1}^{\infty}}$ and $(1/nx_n)_{n=1}^{\infty} \in s_1$, we conclude

$$\mathbb{S} = s_{(1/n)_{n=1}^{\infty}} \cap \overline{s_{(n)_{n=1}^{\infty}}} = cl^{\infty}\left(\left(\frac{1}{n}\right)_{n=1}^{\infty}\right)$$

$$= \left\{ x \in \mathcal{U}^+ : \frac{K_1}{n} \le x_n \le \frac{K_2}{n} \text{ for all } n \text{ and some } K_1, K_2 > 0 \right\}.$$

Now we deal with the next (SSIE's)

$$c_0 \subset E_a + s_x^0 \text{ and } \ell_p \subset E_a + (\ell_p)_x \text{ for } E \in \{c_0, c, s_1, \ell_p\},$$

$$w_0 \subset E_a + W_x^0 \text{ for } E \in \{w_0, w_\infty\}$$

and

$$w_\infty \subset W_a + W_x \text{ for } E \in \{c_0, c, s_1, \ell_p\}.$$

Then we determine the solvability of the next (SSIE's)

$$c_0 \subset E_r + s_x^0 \text{ for } E \in \{c_0, c, s_1, w_\infty\}$$

and

$$\ell_p \subset E_r + (\ell_p)_x \text{ for } E \in \{c_0, c, s_1, \ell_p\}.$$

For instance, the (SSIE) $\ell_p \subset s_r + (\ell_p)_x$ is equivalent to the next statement.

The condition $\sum_{k=1}^\infty |y_k|^p < \infty$ implies that there are sequences $u, v \in \omega$ with $y = u + v$ such that $\sup_n(|u_n|/r^n) < \infty$ *and $\sum_{k=1}^\infty |v_k|/x_k)^p < \infty$ for all sequences y.*

In the case of $F' = F$ and $M(F, F) = s_1$ in Theorem 5.2, we immediately obtain the next corollary.

Corollary 5.19 *Let $a \in \mathcal{U}^+$, and let E, F be linear spaces of sequences. We assume the next statements hold*

(a) There is a linear space $H \subset \omega$ that satisfies the condition in (5.20) such that $E, F \subset H$;
(b) $M(F, F) = M(F, H) = s_1$.

Then we have

(i) $a \in c_0$ implies $\mathcal{I}_a(E, F) = \overline{s_1}$;
(ii) $a \in \overline{M(F, E)}$ implies $\mathcal{I}_a(E, F) = \mathcal{U}^+$.

Example 5.12 Here we use the set $cs = c_\Sigma$ of all convergent series. Let $a \in c_0$ and denote by $\mathcal{I}_a(cs, c_0)$ the set of all positive sequences x that satisfy the next statement.

The condition $y_n \to 0$ $(n \to \infty)$ implies that there are sequences $u, v \in \omega$ such that $y = u + v$ for which the series $\sum_{k=1}^\infty u_k/a_k$ is convergent and $v_n/x_n \to 0$ $(n \to \infty)$ for all sequences y.

This statement is associated with the (SSIE) $c_0 \subset cs_a + s_x^0$. Corollary 5.19 can be applied with $H = c_0$, $\chi = s_1$ and $\mathcal{I}_a(cs, c_0) = \overline{s_1}$. We have $1/a \in M(c_0, cs)$ if and only if $\sum_{k=1}^\infty 1/a_k < \infty$. So if $1/a \in cs$, then the (SSIE) $c_0 \subset cs_a + s_x^0$ is satisfied for all positive sequences x.

When $F = H$, by Corollary 5.19 we obtain the next results.

Corollary 5.20 *Let $a \in \mathcal{U}^+$, and let E, F be linear spaces of sequences with $E \subset F$. We assume F satisfies the condition in (5.20), and $M(F, F) = s_1$. Then we have*

(i) $a \in c_0$ implies $\mathcal{I}_a(E, F) = \overline{s_1}$;
(ii) $a \in \overline{M(F, E)}$ implies $\mathcal{I}_a(E, F) = \mathcal{U}^+$.

Corollary 5.21 *Let $a \in \mathcal{U}^+$. Then*

(i) (a) Let $E \in \{c_0, c, s_1\}$. Then the solutions of the (SSIE) $c_0 \subset E_a + s_x^0$ are determined by

$$\mathcal{I}_a(E, c_0) = \begin{cases} \overline{s_1} & \text{if } a \in c_0 \\ U^+ & \text{if } a \in \overline{s_1}. \end{cases} \tag{5.26}$$

(b) The solutions of the (SSIE) $c_0 \subset (\ell_p)_a + s_x^0$ for $p \geq 1$ are determined by

$$\mathcal{I}_a(\ell_p, c_0) = \begin{cases} \overline{s_1} & \text{if } a \in c_0 \\ U^+ & \text{if } a \in \overline{\ell_p}. \end{cases}$$

(ii) Let $E \in \{c_0, c, s_1, \ell_p\}$. Then the solutions of the next (SSIE) $\ell_p \subset E_a + (\ell_p)_x$ are determined by

$$\mathcal{I}_a(E, \ell_p) = \begin{cases} \overline{s_1} & \text{if } a \in c_0 \\ U^+ & \text{if } a \in \overline{s_1}. \end{cases}$$

(iii) Let $E \in \{w_0, w_\infty\}$. Then the sets $\mathcal{I}_a(E, w_0)$ and $\mathcal{I}_a(w_\infty, w_\infty)$ of all solutions of each of the (SSIE) $w_0 \subset E_a + W_x^0$ and $w_\infty \subset W_a + W_x$ satisfy

$$\mathcal{I}_a(E, w_0) = \mathcal{I}_a(w_\infty, w_\infty) = \mathcal{I}_a(E, c_0)$$

determined in (5.26).

Proof (i) (a) It is enough to take $H = s_1$, in Corollary 5.19. So we have $M(c_0, c_0) = M(c_0, H) = H$.

(b) It is enough to take $H = c_0$.

(ii) We take $H = s_1$ in Corollary 5.19. Again we obtain $M(\ell_p, \ell_p) = M(\ell_p, H) = H$. Since by Lemma 5.6, we have $M(\ell_p, E) = s_1$ for $E \in \{c_0, c, s_1, \ell_p\}$ we conclude for Part (ii).

(iii) Here we take $H = w_\infty$ and apply Corollary 5.19 using the identities $M(w_0, w_0) = M(w_0, w_\infty) = M(w_\infty, w_\infty) = s_1$ (cf. [14, Remark 3.4, p. 597]).

This concludes the proof. □

As a direct consequence of Corollary 5.21 and Proposition 5.11, we obtain the following.

Corollary 5.22 *Let $a \in c_0^+$ and let $p \geq 1$. Then each of the next (SSIE)*

$$c_0 \subset E_a + s_x^0 \text{ with } E \in \{c_0, c, s_1, \ell_p\}, \tag{i}$$

$$\ell_p \subset E_a + (\ell_p)_x \text{ with } E \in \{c_0, c, s_1, \ell_p\}, \tag{ii}$$

$$w_0 \subset E_a + W_x^0 \text{ with } E \in \{w_0, w_\infty\} \tag{iii}$$

and

$$w_\infty \subset W_a + W_x \tag{iv}$$

has the same set of solutions which is determined by $\mathcal{I}_a = \overline{s_1}$.

Remark 5.13 We note that by Corollary 5.13, the solutions of the (SSIE) $c_0 \subset W_a + s_x^0$ are determined by $\mathcal{I}_a = \overline{s_1}$ if $a \in s^0_{(1/n)_{n=1}^\infty}$.

We immediately deduce the following result from Corollaries 5.15, 5.16, 5.17, 5.13 and 5.14.

Corollary 5.23 *Let $r > 0$ and $p \geq 1$. Then the solutions of the next (SSIE)*

(i) $c_0 \subset E_r + s_x^0$ *for* $E \in \{c_0, c, s_1\}$,
(ii) $c_0 \subset W_r + F'_x$ *for* $F' \in \{c_0, c, s_1\}$,
(iii) $\ell_p \subset E_r + (\ell_p)_x$ *for* $E \in \{c_0, c, s_1, \ell_p\}$,
(iv) $w_0 \subset E_r + W_x^0$ *for* $E \in \{w_0, w_\infty\}$,
(v) $w_\infty \subset W_r + W_x$

are determined by (5.11) with $\chi = s_1$.

Remark 5.14 Let $r > 0$ and assume that F satisfies the condition in (5.20), $E \subset F$ and $M(F, F) = s_1$. Then we have

(i) the condition $r < 1$ implies $\mathcal{I}_r(E, F) = \overline{s_1}$;
(ii) the condition $(r^{-n})_{n=1}^\infty \in M(F, E)$ implies $\mathcal{I}_r(E, F) = \mathcal{U}^+$.

Remark 5.15 We easily deduce that under the hypotheses given in Remark 5.14 and the condition $M(F, E) = M(F, F) = s_1$, the (SSIE) defined by $F \subset E_r + F_x$ is totally solved and $\mathcal{I}_r(E, F) = \mathcal{I}_r(E, s_1)$ is defined by (5.11) with $\chi = s_1$. It is the case of the (SSIE) defined by $E \subset E_r + E_x$ for $r > 0$, $E \in \{c_0, s_1, \ell_p, w_0, w_\infty\}$ with $p \geq 1$. We also note that, by Lemma 5.1, where $E = c$, the solutions of the (SSIE) $c \subset c_r + c_x$ are defined by (5.11) with $\chi = c$.

Example 5.13 In this example, we use the set *bv of sequences of bounded variation* defined by

$$bv = (\ell_1)_\Delta = \left\{ y \in \omega : \sum_{k=1}^\infty |y_k - y_{k-1}| < \infty \right\}$$

with the convention $y_0 = 0$. For $a \in c_0$, we consider the (SSIE) $c_0 \subset bv_a + s_x^0$, which is associated with the next statement.

The condition $y_n \to 0$ $(n \to \infty)$ implies that there are sequences $u, v \in \omega$ such that $y = u + v$ for which

$$\sum_{k=1}^{\infty} \left| \frac{u_k}{a_k} - \frac{u_{k-1}}{a_{k-1}} \right| < \infty \text{ and } \frac{v_n}{x_n} \to 0 \ (n \to \infty)$$

for all sequences y.

Since $\Sigma \in (\ell_1, s_1)$, we deduce $bv \subset s_1$ and we can apply Corollary 5.19 with $H = s_1$. Then we have $\mathcal{I}_a(bv, c_0) = \overline{s_1}$.

Finally we state a result on the solvability of the (SSE)

$$E_r + (\ell_p)_x = \ell_p \text{ with } E \in \{c_0, c, s_1, \ell_p\}.$$

We use the set $\mathcal{I}'_r(E, F)$ of all positive sequences x that satisfy $E_r + F_x \subset F$. We recall that $x \in \mathcal{I}'_r(E, F)$ if and only if $(r^n)_{n=1}^{\infty} \in M(E, F)$ and $x \in M(F, F)$.

Proposition 5.12 *Let $r > 0$. Then the set $\mathbb{S}_r(E, \ell_p)$ of all solutions of the (SSE) defined by $E_r + (\ell_p)_x = \ell_p$ with $E = c_0$, c, s_1 or ℓ_p for $p \geq 1$ is determined by*

$$\mathbb{S}_r(E, \ell_p) = \begin{cases} cl^{\infty}(e) & \text{if } r < 1 \\ \varnothing & \text{if } r \geq 1 \end{cases} \text{ for } E = c_0, c \text{ or } s_1 \tag{5.27}$$

and

$$\mathbb{S}_r(\ell_p, \ell_p) = \begin{cases} cl^{\infty}(e) & \text{if } r < 1 \\ s_1^+ & \text{if } r = 1 \\ \varnothing & \text{if } r > 1. \end{cases} \tag{5.28}$$

Proof We have $E_r + (\ell_p)_x \subset \ell_p$ if and only if $(r^n)_{n=1}^{\infty} \in M(E, \ell_p)$ and $x \in M(\ell_p, \ell_p) = s_1$. Now it can easily be seen that $M(E, \ell_p) = \ell_p$ for $E = c_0$, c, s_1, and $M(\ell_p, \ell_p) = s_1$. Then we have

$$\mathcal{I}'_r(E, \ell_p) = \begin{cases} s_1^+ & \text{if } r < 1 \\ \varnothing & \text{if } r \geq 1 \end{cases} \text{ for } E \in \{c_0, c, s_1\}.$$

Then we have

$$\mathcal{I}'_r(\ell_p, \ell_p) = \begin{cases} s_1^+ & \text{if } r \leq 1 \\ \varnothing & \text{if } r > 1. \end{cases}$$

Since $\mathbb{S}_r(E, \ell_p) = \mathcal{I}_r(E, \ell_p) \cap \mathcal{I}'_r(E, \ell_p)$ and using Corollary 5.23, we obtain the statements in (5.27) and (5.28). This concludes the proof. □

References

1. Farés, A., de Malafosse, B.: Sequence spaces equations and application to matrix transformations. Int. Math. Forum **3**(19), 911–927 (2008)
2. de Malafosse, B.: On some BK space. Int. J. Math. Math. Sci. **58**, 1783–1801 (2003)
3. de Malafosse, B.: Sum of sequence spaces and matrix transformations. Acta Math. Hung. **113**(3), 289–313 (2006)
4. de Malafosse, B.: Application of the infinite matrix theory to the solvability of certain sequence spaces equations with operators. Mat. Vesn. **54**(1), 39–52 (2012)
5. de Malafosse, B.: Applications of the summability theory to the solvability of certain sequence spaces equations with operators of the form $B(r, s)$. Commun. Math. Anal. **13**(1), 35–53 (2012)
6. de Malafosse, B.: Solvability of certain sequence spaces equations with operators. Demonstr. Math. **46**(2), 299–314 (2013)
7. de Malafosse, B.: On some Banach algebras and applications to the spectra of the operator $B(\widetilde{r}, \widetilde{s})$ mapping in new sequence spaces. Appl. Math. Lett. **2**, 7–18 (2014)
8. de Malafosse, B.: On the spectra of the operator of the first difference on the spaces W_τ and W_τ^0 and application to matrix transformations. Gen. Math. Notes **22**(2), 7–21 (2014)
9. de Malafosse, B.: Solvability of sequence spaces equations using entire and analytic sequences and applications. J. Ind. Math. Soc. **81**(1–2), 97–114 (2014)
10. de Malafosse, B.: Solvability of sequence spaces equations of the form $(E_a)_\Delta + Fx = F_b$. Fasc. Math. **55**, 109–131 (2015)
11. de Malafosse, B.: On new classes of sequence spaces inclusion equations involving the sets c_0, c, ℓ_p $(1 \leq p \leq \infty)$, w_0 and w_∞. J. Ind. Math. Soc. **84**(3–4), 211–224 (2017)
12. de Malafosse, B.: Application of the infinite matrix theory to the solvability of sequence spaces inclusion equations with operators. Ukr. Math. Zh. **71**(8), 1040–1052 (2019)
13. de Malafosse, B., Malkowsky, E.: Matrix transformations between sets of the form W_ξ and operator generators of analytic semigroups. Jordanian J. Math. Stat. **1**(1), 51–67 (2008)
14. de Malafosse, B., Malkowsky, E.: On sequence spaces equations using spaces of strongly bounded and summable sequences by the Cesàro method. Antarct. J. Math. **10**(6), 589–609 (2013)
15. de Malafosse, B., Malkowsky, E.: On the solvability of certain (SSIE) with operators of the form $B(r, s)$. Math. J. Okayama Univ. **56**, 179–198 (2014)
16. de Malafosse, B., Rakočević, V.: Calculations in new sequence spaces and application to statistical convergence. Cubo A **12**(3), 117–132 (2010)
17. de Malafosse, B., Rakočević, V.: Series summable (C, λ, μ) and applications. Linear Algebra Appl. **436**(11), 4089–4100 (2012)
18. de Malafosse, B., Rakočević, V.: Matrix transformations and sequence spaces equations. Banach J. Math. Anal. **7**(2), 1–14 (2013)
19. Maddox, I.J.: On Kuttner's theorem. J. Lond. Math. Soc. **43**, 285–290 (1968)
20. Malkowsky, E.: The continuous duals of the spaces $c_0(\lambda)$ and $c(\lambda)$ for exponentially bounded sequences λ. Acta Sci. Math. (Szeged) **61**, 241–250 (1995)
21. Malkowsky, E.: Linear operators between some matrix domains. Rend. Circ. Mat. Palermo, Serie II, Suppl. **68**, 641–655 (2002)
22. Malkowsky, E., Rakočević, V.: An introduction into the theory of sequence spaces and measures of noncompactness. Zbornik radova (Matematčki institut SANU) **9**(17), 143–234 (2000). Mathematical Institute of SANU, Belgrade
23. Malkowsky, E., Rakočević, V.: The measure of noncompactness of linear operators between spaces of strongly C_1 summable and bounded sequences. Acta Math. Hung. **89**(1–2), 29–45 (2000)

Chapter 6
Sequence Space Equations

In this chapter, we extend the results of Chap. 5, and first consider the solvability of the equation $F_x = F_b$ for a given positive sequence b, where $x = (x_n)_{n=1}^{\infty}$ is the unknown positive sequence. Now, the question is what are the solutions of the new equation obtained from the previous one where we add a new linear space of sequences E to the set F_x in the first member of the above equation? So, the new problem consists in solving the equation $E + F_x = F_b$ and determining whether its solutions are the same as above, or not. Then, we will consider the case when E is the matrix domain of an operator A in $\mathcal{E}$. We refer to [5–7, 9] for more related results.

We also solve the sequence spaces equations (SSE) $E_a + F_x = F_b$ where E and F are linear spaces of sequences of the form $c_0, c, \ell_\infty, \ell_p$ $(p \geq 1)$, w_0, w_∞, $\mathbf{\Gamma}$ or $\mathbf{\Lambda}$. We recall that

$$\mathbf{\Lambda} = \left\{ x = (x_n)_{n=1}^{\infty} \in \omega : \sup_n \left(|x_n|^{1/n} \right) < \infty \right\} \tag{6.1}$$

and

$$\mathbf{\Gamma} = \left\{ x = (x_n)_{n=1}^{\infty} \in \omega : \lim_{n \to \infty} \left(|x_n|^{1/n} \right) = 0 \right\} \tag{6.2}$$

are the *sets of analytic* and *entire sequences*, respectively. Then we deal with a class of (SSE) with operators of the form $E_T + F_x = F_b$, where T is either Δ or Σ and E is any of the sets $c_0, c, \ell_\infty, \ell_p$ $(p \geq 1)$, w_0, Γ, or Λ and $F = c, \ell_\infty$ or Λ. Then we solve the (SSE) defined by $(E_a)_\Delta + s_x^{(c)} = s_b^{(c)}$, where E is either c_0 or ℓ_p, and the (SSE) $(E_a)_\Delta + s_x^0 = s_b^0$, where E is either c or ℓ_∞.

B. de Malafosse et al., *Operators Between Sequence Spaces and Applications*,
https://doi.org/10.1007/978-981-15-9742-8_6

6.1 Introduction

It is well known that the sets $\mathbf{\Gamma}$ and $\mathbf{\Lambda}$ of *analytic* and *entire sequences*, respectively, defined by (6.1) and (6.2) are complete metric spaces with respect to the metric d, where

$$d(y, z) = \sup_n |y_n - z_n|^{1/n} \text{ for all } y \text{ and } z;$$

furthermore $\mathbf{\Lambda}$ is an FK space (Example 1.4(c)), while $\mathbf{\Gamma}$ is not a linear metric space (Example 1.1(b)).
For a detailed study of the sets $\mathbf{\Lambda}$ and $\mathbf{\Gamma}$ we refer the reader to [13]. For $a \in \mathcal{U}^+$ we write $\mathbf{\Lambda}_a = D_a = D_a\mathbf{\Lambda}$. So $y \in \mathbf{\Lambda}_a$ if and only if

$$\sup_n \left(\frac{|y_n|}{a_n}\right)^{1/n} < \infty$$

and $\mathbf{\Lambda}_a$ is called the set of all *a-analytic sequences*. For $a = e$, we write $\mathbf{\Lambda}_e = \mathbf{\Lambda}$ Similarly we put $\mathbf{\Gamma}_a = D_a\mathbf{\Gamma}$, and $y \in \mathbf{\Gamma}_a$ if and only if

$$\lim_{n\to\infty} \left(\frac{|y_n|}{a_n}\right)^{1/n} = 0;$$

we write $\mathbf{\Gamma}_e = \mathbf{\Gamma}$ and $\mathbf{\Gamma}_a$ is the set of all *a-entire sequences.*

First we prove an elementary lemma.

Lemma 6.1 *(i) For any α and $K > 0$, there is $L > 0$ such that $\alpha K^n \leq L^n$ for all n.*
(ii) For any α, $K, K' > 0$, there is $L' > 0$ such that $K^n + \alpha K'^n \leq L'^n$ for all n.

Proof (i) It is enough to take $L = \alpha K$ if $\alpha > 1$, and $L = K$ if $\alpha \leq 1$.
(ii) is a consequence of Part (i), since there is $L > 0$ such that

$$K^n + \alpha K'^n \leq K^n + L^n \leq (K + L)^n \text{ for all } n.$$

□

We have for any given $a \in \mathcal{U}^+$

$$[C(a)a]_n = \frac{a_1 + \cdots + a_n}{a_n} \text{ for all } n.$$

We recall that

$$\widehat{C_1} = \left\{a \in \mathcal{U}^+ : \sup_n [C(a)a]_n < \infty\right\},$$
$$\widehat{C} = \{a \in \mathcal{U}^+ : C(a)a \in c\}$$

and also use the set

$$\widehat{C}_\Lambda = \{a \in \mathcal{U}^+ : [C(a)a]_n \leq k^n \text{ for all } n \text{ and some } k > 0\}.$$

We note that if $a, b \in \widehat{C_1}$, then $a + b$ and $ab \in \widehat{C_1}$. For any given $R > 0$, it can easily be seen that $(R^n)_{n=1}^\infty \in \widehat{C_1}$ if and only $R > 1$. By (4.11 in Proposition 4.3), the condition $a \in \widehat{C_1}$ implies that there are $K > 0$ and $\gamma > 1$ such that

$$a_n \geq K\gamma^n \text{ for all } n.$$

It is known that $\widehat{C}$ is equal to the set $\widehat{\Gamma}$ of all $x \in \mathcal{U}^+$ with $\lim_{n\to\infty}(x_{n-1}/x_n) < 1$. Here we state some results that are consequences of [1, Proposition 2.1, p. 1786], [3, Proposition 9, p. 300], and of the fact that $(s_a)_\Delta \subset s_a$ is equivalent to $D_{1/a}\Sigma D_a \in S_1$ and $a \in \widehat{C_1}$.

We also have the next elementary result.

Lemma 6.2 *Let $a, b \in \mathcal{U}^+$ and assume $E_a = E_b$, where $E = c_0$ or ℓ_∞. Then $a \in \widehat{C_1}$ if and only if $b \in \widehat{C_1}$.*

Now we state the next result.

Proposition 6.1 *Let $a, b \in \mathcal{U}^+$. Then*

(i) $E_{a+b} = E_a + E_b$, where E is any of the sets c_0, ℓ_∞, $\mathbf{\Gamma}$ or $\mathbf{\Lambda}$.
(ii) The following statements are equivalent where $E = c_0$ or ℓ_∞.

(a) $a \in \widehat{C_1}$,
(b) $(E_a)_\Delta \subset E_a$,
(c) $(E_a)_\Delta = E_a$.

(iii) $(s_a^{(c)})_\Delta = s_a^{(c)}$ if and only if $a \in \widehat{\Gamma}$.
(iv)

(a) We have $\mathbf{\Lambda}_a = \mathbf{\Lambda}_b$ if and only if $\mathbf{\Gamma}_a = \mathbf{\Gamma}_b$, and the equality $\mathbf{\Lambda}_a = \mathbf{\Lambda}_b$ is equivalent to the statement

$$k_1^n \leq \frac{a_n}{b_n} \leq k_2^n \text{ for all } n \text{ and some } k_1, k_2 > 0, \tag{6.3}$$

(b) $\mathbf{\Lambda}_a(\Delta) = \mathbf{\Lambda}_b$ if and only if $\mathbf{\Lambda}_a = \mathbf{\Lambda}_b$ and $a \in \widehat{C}_\Lambda$.

Proof (i) As we have seen above in [9, Proposition 5.1, pp. 599–600], it was stated that if E satisfies the condition in (5.20), then $E_a + E_b = E_{a+b}$ for all $a, b \in \mathcal{U}^+$.
(ii) and (iii) follow from Theorem 4.2.
(iv)

(a) We will see in Proposition 6.2 that $M(\mathbf{\Gamma},\mathbf{\Gamma}) = \mathbf{\Lambda}$. So the condition $\mathbf{\Gamma}_a = \mathbf{\Gamma}_b$ is equivalent to $a/b, b/a \in M(\mathbf{\Gamma},\mathbf{\Gamma}) = \mathbf{\Lambda}$. This shows $\mathbf{\Lambda}_a = \mathbf{\Lambda}_b$ if and only if $\mathbf{\Gamma}_a = \mathbf{\Gamma}_b$.

Now we show $\mathbf{\Lambda}_a = \mathbf{\Lambda}_b$ if and only if (6.3) holds. We have that $\mathbf{\Lambda}_a \subset \mathbf{\Lambda}_b$ implies $a \in \mathbf{\Lambda}_b$ and $a/b \in \mathbf{\Lambda}$.
Similarly $\mathbf{\Lambda}_b \subset \mathbf{\Lambda}_a$ implies $b/a \in \mathbf{\Lambda}$.
So we have shown that $\mathbf{\Lambda}_a = \mathbf{\Lambda}_b$ implies (6.3).
Now we show that (6.3) implies $\mathbf{\Lambda}_a = \mathbf{\Lambda}_b$.
First (6.3) implies $a/b \in \mathbf{\Lambda}$. Then, for any $y \in \mathbf{\Lambda}_a$, there is $\xi \in \mathbf{\Lambda}$ such that $y = a\xi = b[(a/b)\xi] \in \mathbf{\Lambda}_b$ and we have shown that $a/b \in \mathbf{\Lambda}$ implies $\mathbf{\Lambda}_a \subset \mathbf{\Lambda}_b$.
Similarly we obtain that $b/a \in \mathbf{\Lambda}$ implies $\mathbf{\Lambda}_b \subset \mathbf{\Lambda}_a$. We conclude that (6.3) is equivalent to $\mathbf{\Lambda}_a = \mathbf{\Lambda}_b$.
(b) *Necessity.*
First, the identity $(\mathbf{\Lambda}_a)_\Delta = \mathbf{\Lambda}_b$ is equivalent to $\Delta^{-1}\mathbf{\Lambda}_a = \mathbf{\Lambda}_b$, where $\Delta^{-1} = \Sigma$. We deduce $((-1)^n b_n)_{n=1}^\infty \in \mathbf{\Lambda}_b = (\mathbf{\Lambda}_a)_\Delta$ and $\Sigma a \in \mathbf{\Lambda}_b$. So there are k_1' and $k_2 > 0$ such that

$$\frac{b_{n-1} + b_n}{a_n} \le k_1'^n, \ \frac{a_1 + \cdots + a_n}{b_n} \le k_2^n \text{ for all } n \tag{6.4}$$

with $b_0 = 0$, and (6.3) holds with $k_1 = 1/k_1'$.
So we have shown that $(\mathbf{\Lambda}_a)_\Delta = \mathbf{\Lambda}_b$ implies $\mathbf{\Lambda}_a = \mathbf{\Lambda}_b$.
Now we show that $(\mathbf{\Lambda}_a)_\Delta = \mathbf{\Lambda}_b$ implies $a \in \widehat{C}_\Lambda$.
We have seen that $(\mathbf{\Lambda}_a)_\Delta = \mathbf{\Lambda}_b$ implies (6.4). So we have

$$\frac{1}{k_1^n}\frac{a_1 + \cdots + a_n}{a_n} \le \frac{a_1 + \cdots + a_n}{b_n} \le k_2^n \text{ for all } n$$

and $a \in \widehat{C}_\Lambda$. This shows the necessity of Part (b).
Sufficiency. We assume $a \in \widehat{C}_\Lambda$ and $\mathbf{\Lambda}_a = \mathbf{\Lambda}_b$.
Then let $y \in (\mathbf{\Lambda}_a)_\Delta$. We have $y = \Sigma z$ for some $z \in \mathbf{\Lambda}_a$. So there is $h \in \ell_\infty$ such that $z_n = a_n h_n^n$ and

$$\frac{|y_n|}{a_n} = \frac{\left|\sum_{k=1}^n z_k\right|}{a_n} \le \frac{a_1|h_1| + \cdots + a_n|h_n^n|}{a_n} \le \frac{a_1 K + \cdots + a_n K^n}{a_n}$$
$$\le \frac{a_1 + \cdots + a_n}{a_n} \max\{K, K^n\} \text{ for all } n \text{ and some } K > 0.$$

Now by Lemma 6.1 and since $a \in \widehat{C}_\Lambda$, we obtain

$$\frac{|y_n|}{a_n} \le k^n \max\{K, K^n\} \le k'^n \text{ for some } k \text{ and } k' \text{ and all } n,$$

and $y \in \mathbf{\Lambda}_a$.
Conversely, let $y \in \mathbf{\Lambda}_a$. Then $\Delta y \in \mathbf{\Lambda}_a$. Again, by Lemma 6.1, we obtain

$$\begin{aligned}\frac{|y_n - y_{n-1}|}{a_n} &\leq \frac{a_n|h_n^n| + a_{n-1}|h_{n-1}^{n-1}|}{a_n}\\ &\leq K^n + k^n K^{n-1} = K^n + \frac{1}{k}(kK)^n\\ &\leq K'^n \text{ for all } n \text{ and some } K' > 0.\end{aligned}$$

This shows the sufficiency and concludes the proof. □

Now we recall the characterizations of the classes

$$(c_0(p), c_0(q)) \text{ and } (c_0(p), \ell_\infty(q))$$

and determine the sets

$$M(\mathbf{\Gamma}, F),\ M(\mathbf{\Lambda}, F),\ M(E, \mathbf{\Lambda}) \text{ and } M(E, \mathbf{\Gamma}) \text{ for } E, F \in \{c_0, c, \ell_\infty, \Gamma, \Lambda\}.$$

Let $p = (p_n)_{n=1}^\infty \in \mathcal{U}^+ \bigcap \ell_\infty$ be a sequence and put . In Example 1.1, we introduced the sets

$$\ell_\infty(p) = \left\{y = (y_n)_{n=1}^\infty : \sup_n |y_n|^{p_n} < \infty\right\},$$
$$c_0(p) = \left\{y = (y_n)_{n=1}^\infty : \lim_{n\to\infty} |y_n|^{p_n} = 0\right\}.$$

It was shown there that set $c_0(p)$ is a complete paranormed space with $g(y) = \sup_n |y_n|^{p_n/L}$, where $L = \max\{1, \sup_n p_n\}$ ([12, Theorem 6]) and $\ell_\infty(p)$ is a paranormed space with g only if $\inf_n p_n > 0$ in which case $\ell_\infty(p) = \ell_\infty$ ([14, Theorem 9]). So we can state the next lemma, where for any given integer k, we denote by $\mathbb{N}_k$ the set of all integers $n \geq k$.

Lemma 6.3 ([11, Theorem 5.1.13]) *Let $p, q \in \mathcal{U}^+ \bigcap \ell_\infty$. Then*

(i) $A \in (c_0(p), c_0(q))$ if and only if for all $N \in \mathbb{N}_1$ there is $M \in \mathbb{N}_2$ such that

$$\sup_n \left(N^{1/q_n} \sum_{k=1}^\infty |a_{nk}| M^{-1/p_k}\right) < \infty \textit{ and } \lim_{n\to\infty} |a_{nk}|^{p_n} = 0 \textit{ for all } k.$$

(ii) $A \in (c_0(p), \ell_\infty(q))$ if and only if there is $M \in \mathbb{N}_2$ such that

$$\sup_n \left(\sum_{k=1}^\infty |a_{nk}| M^{-1/p_k}\right)^{q_n} < \infty.$$

Example 6.1 In this way we have $A \in (\mathbf{\Gamma},\mathbf{\Lambda})$ if and only if there is an integer $M \geq 2$ such that $\sup_n(\sum_{k=1}^{\infty} |a_{nk}|M^{-k})^{1/n} < \infty$, since $\Gamma = c_0(p)$ and $\Lambda = \ell_\infty(p)$ with $p_n = 1/n$ for all n.

To show the next proposition we need the following lemma.

Lemma 6.4 ([5, Proposition 4.1, p. 104]) *We have* $\mathbf{\Lambda}= \bigcup_{r>0} s_r$ *and* $\mathbf{\Gamma}= \bigcap_{r>0} s_r$.

Proof The proof of $\mathbf{\Lambda}= \bigcup_{r>0} s_r$ was given in [4].
We show $\mathbf{\Gamma}\subset \bigcap_{r>0} s_r$. For this let $a \in\mathbf{\Gamma}$. Then there is a sequence $\varepsilon = (\varepsilon_n)_{n=1}^{\infty} \in c_0$ such that $a_n = \varepsilon_n^n$ for all n and

$$\frac{|a_n|}{r^n} = \left(\frac{\varepsilon_n}{r}\right)^n = o(1)\ (n \to \infty) \text{ for all } r > 0,$$

so $a \in \bigcap_{r>0} s_r$ and $\mathbf{\Gamma}\subset \bigcap_{r>0} s_r$.
Conversely, let $a \in \bigcap_{r>0} s_r$. Then $a \in s_r$ and $|a_n| \leq M_r r^n$ for all n, for all $r > 0$ and for some $M_r > 0$. Since for every $r > 0$ we have $M_r^{1/n} \to 1\ (n \to \infty)$, there is an integer N_r such that

$$M_r^{1/n} \leq 2 \text{ and } |a_n|^{1/n} \leq 2r \text{ for all } n \geq N_r.$$

We conclude $|a_n|^{1/n} \to 0\ (n \to \infty)$ and $a \in\mathbf{\Gamma}$. □

Now we state the next result, where we have $\Gamma_1 = \Gamma$ and $\Lambda_1 = \Lambda$.

Proposition 6.2 ([5, Proposition 4.1, p. 104]) *We have*

(i) $M(\mathbf{\Gamma}, F) =\mathbf{\Lambda}$ *for* $F \in \{c_0, c, \ell_\infty,\mathbf{\Gamma},\mathbf{\Lambda}\}$;
(ii) $M(\mathbf{\Lambda}, F) =\mathbf{\Gamma}$ *for* $F \in \{c_0, c, \ell_\infty,\mathbf{\Gamma}\}$;
(iii) $M(E,\mathbf{\Lambda}) =\mathbf{\Lambda}$ *for* $E \in \{c_0, c, \ell_\infty,\mathbf{\Gamma},\mathbf{\Lambda}\}$;
(iv) $M(E,\mathbf{\Gamma}) =\mathbf{\Gamma}$ *for* $E \in \{c_0, c, \ell_\infty,\mathbf{\Gamma}\}$.

Proof (i) We show $M(\Gamma, \Gamma) = \Lambda$.
By Part (i) of Lemma 6.3, we have $a \in M(\mathbf{\Gamma},\mathbf{\Gamma})$, that is, $D_a \in(\mathbf{\Gamma},\mathbf{\Gamma})$ if and only if for all integers $N \geq 1$, there is an integer $M \geq 2$ such that

$$|a_n|^{1/n} \leq K\frac{M}{N} \text{ for all } n \text{ and some } K > 0.$$

This means that $a \in\mathbf{\Gamma}$ and we have shown $M(\mathbf{\Gamma},\mathbf{\Gamma}) =\mathbf{\Lambda}$.
Now we show $M(\mathbf{\Gamma},\mathbf{\Lambda}) =\mathbf{\Lambda}$.
By Part (i) of Lemma 6.3, we have $D_a \in (\mathbf{\Gamma},\mathbf{\Lambda})$ if and only if there is an integer $M \geq 2$ such that

$$\sup_n \left(|a_n|M^{-n})\right)^{1/n} < \infty.$$

This means $a \in \Lambda$. So we have $M(\mathbf{\Gamma},\mathbf{\Lambda}) =\mathbf{\Lambda}$. Since $\mathbf{\Gamma}\subset c_0 \subset c \subset s_1 \subset \mathbf{\Gamma}$, we conclude

$$\Lambda = M(\Gamma, \Gamma) \subset M(\Gamma, F) \subset M(\Gamma, \Lambda) = \Lambda,$$

where F is any of the sets c_0, c or ℓ_∞.
Thus we have shown Part (i)
(ii) We show $\Gamma \subset M(\Lambda,\Lambda)$.
Let $a \in\Gamma$.Then, for each $y \in\Lambda$, we have $|a_n y_n|^{1/n} \leq K|a_n|^{1/n}$ for all n and some $K > 0$. Since $|a_n|^{1/n} \to 0$ $(n \to \infty)$, we conclude $ay \in\Gamma$.
Now we show $M(\Lambda, \ell_\infty) \subset\Gamma$.
By Lemma 6.4, we have $\Lambda= \bigcup_{r>0} s_r$ and $M(\Lambda, \ell_\infty) \subset M(s_r, \ell_\infty)$ for all $r > 0$. Hence we obtain that $a \in M(\Lambda, \ell_\infty)$ implies $|a_n|r^n \leq K$ for all $n \geq 1, r > 0$ and some K. This shows $a \in \bigcap_{r>0} s_r = \Gamma$. Then we have

$$\Gamma \subset M(\Lambda, \Gamma) \subset M(\Lambda, F) \subset M(\Lambda, \ell_\infty) \subset \Gamma,$$

where F is any of the sets c_0 or ℓ_∞.
(iii) We have $\Lambda\subset M(\Lambda,\Lambda)$.
Let $a\in\Lambda$. Then we have $ay\in\Lambda$ for each $y\in\Lambda$, since

$$|a_n y_n|^{1/n} = |a_n|^{1/n} |y_n|^{1/n} \leq KK' \text{ for all } n.$$

Now we obtain

$$\Lambda \subset M(\Lambda, \Lambda) \subset M(E, \Lambda) \subset M(\Gamma, \Lambda) = \Lambda,$$

where E is any of the sets c_0, c or ℓ_∞.
This concludes the proof of Part (iii).
(iv) First we show $M(c_0,\Gamma) \subset\Gamma$.
Let $a \in M(c_0,\Gamma)$. Since $\Gamma= \bigcap_{r>0} s_r$ by Lemma 6.4, we have $a \in M(c_0, s_r)$ and $(a_n r^{-n})_{n=1}^\infty \in M(c_0, s_1) = s_1$ for all $r > 0$. This shows $a \in \bigcap_{r>0} s_r = \Gamma$ and $M(c_0,\Gamma) \subset\Gamma$. Now we can write

$$\Gamma \subset M(\Lambda, \Gamma) \subset M(E, \Gamma) \subset M(c_0, \Gamma) \subset \Gamma \text{ for } E = c_0, c \text{ or } \ell_\infty.$$

This concludes the proof of Part (iv).

□

6.2 The (SSE) $E_a + F_x = F_b$ with $e \in F$

In this section, we apply the previous results to the solvability of the (SSE) $E_a + F_x = F_b$ with $\mathbf{e} \in F$ and we state the next result, where we do not use the homomorphism in $\mathcal{U}^+$ defined by $a \mapsto E_a$. We also need the following conditions in the next result.

$$e \in F, \tag{6.5}$$

and

$$F \subset M(F, F). \tag{6.6}$$

Theorem 6.1 ([5, Theorem 5.1, pp. 106–107]) *Let $a, b \in \mathcal{U}^+$ and E and F be two linear subspaces of ω. We assume F satisfies the conditions in (5.9), (6.5) and (6.6), and*

$$M(E, F) \subset M(E, c_0). \tag{6.7}$$

Then

$$\mathbb{S}(E, F) = \begin{cases} cl^F(b) & \text{if } a/b \in M(E, F) \\ \varnothing & \text{otherwise.} \end{cases} \tag{6.8}$$

Proof First, trivially the equation

$$E_a + F_x = F_b \tag{6.9}$$

is equivalent to

$$E_a + F_x \subset F_b \tag{6.10}$$

and

$$F_b \subset E_a + F_x. \tag{6.11}$$

The identity in (6.9) implies (6.10), which is equivalent to $a/b \in M(E, F)$ and $x/b \in M(F, F)$. We show that if $a/b \in M(E, F)$, then the inclusion in (6.11) is equivalent to $F_b \subset F_x$.

Necessity. By the condition in (6.5), there are sequences $\xi \in E$ and $\xi' \in F$ such that

$$b = a\xi + x\xi',$$
$$\frac{x}{b}\xi' = e - \frac{a}{b}\xi$$

and

$$\frac{b}{x} = \frac{\xi'}{e - \frac{a}{b}\xi}.$$

By (6.7), we have $a/b \in M(E, F) \subseteq M(E, c_0)$ and $(a/b)\xi \in c_0$. Then we obtain

$$\frac{b_n}{x_n} = \frac{\xi'_n}{1 - \frac{a_n}{b_n}\xi_n} \sim \xi'_n \ (n \to \infty)$$

and by (5.9) we have $b/x \in F$, but (6.6) successively implies $b/x \in M(F, F)$ and $F_b \subset F_x$.

Sufficiency. The converse is trivial since $F_b \subset F_x$ implies $F_b \subset E_a + F_x$. So we

have shown that for $a/b \in M(E, F)$ the inclusion $F_b \subset E_a + F_x$ is equivalent to $F_b \subset F_x$. We conclude that the identity in Eq. (6.9) is equivalent to $a/b \in M(E_1, F)$ and $F_x = F_b$.
Finally we have $\mathbb{S}(E, F) = \varnothing$ if $a/b \notin M(E_1, F)$, since $E_a \subset E_a + F_x = F_b$ implies $E_a \subset F_b$ and $a/b \in M(E, F)$.
This completes the proof. □

We immediately deduce the following.

Corollary 6.1 *Let E and F be any of the sets c_0, c, ℓ_∞, $\mathbf{\Gamma}$ or $\mathbf{\Lambda}$ and assume $e \in F$, and $M(E, F) \subseteq M(E, c_0)$. Then $\mathbb{S}(E, F)$ is defined by (6.8).*

Proof The proof follows the same lines as above. Here we have $F \subset M(F, F)$, where F is any of the sets c_0, c, ℓ_∞, $\mathbf{\Gamma}$ or $\mathbf{\Lambda}$. □

Now we give an application.

Let $a, b \in \mathcal{U}^+$. We consider the sets $S(E, c)$, where E is any of the sets c_0, c, ℓ_∞, Γ or Λ. We obtain the next results as a direct consequence of Theorem 6.1 and Corollary 6.1.

Proposition 6.3 ([5, Proposition 5.1, p. 108]) *Let $a, b \in \mathcal{U}^+$. Then*

$$\mathbb{S}(c_0, c) = \begin{cases} cl^c(b) & \text{if } a/b \in \ell_\infty \\ \varnothing & \text{otherwise;} \end{cases}$$

$$\mathbb{S}(\ell_\infty, c) = \begin{cases} cl^c(b) & \text{if } a/b \in c_0 \\ \varnothing & \text{otherwise;} \end{cases}$$

$$\mathbb{S}(\mathbf{\Gamma}, c) = \begin{cases} cl^c(b) & \text{if } a/b \in \mathbf{\Lambda} \\ \varnothing & \text{otherwise;} \end{cases}$$

$$\mathbb{S}(\mathbf{\Lambda}, c) = \begin{cases} cl^c(b) & \text{if } a/b \in \mathbf{\Gamma} \\ \varnothing & \text{otherwise.} \end{cases}$$

Now we give an example that illustrates the previous proposition.

Example 6.2 We consider the set of all $x \in \mathcal{U}^+$ such that for every sequence y we have $y_n \to l_1$ $(n \to \infty)$ if and only if there are sequences $u, v \in \omega$ for which $y = u + v$ and

$$|u_n|^{1/n} \to 0 \text{ and } x_n v_n \to l_2 \ (n \to \infty) \text{ for some } l_1 \text{ and } l_2.$$

Since this set corresponds to the equation $\Gamma + s_{1/x}^{(c)} = c$, by *Proposition* 6.3 it is equal to the set of all sequences that tend to a positive limit.

In the case when $E = F = c$, we obtain the next theorem.

Theorem 6.2 ([5, Theorem 5.2, p. 108]) *Let $a, b \in \mathcal{U}^+$. Then*

$$\mathbb{S}(c, c) = \begin{cases} cl^{\mathbf{c}}(b) & \text{if } a/b \in c_0 \\ s_b^{(c)} \cap \mathcal{U}^+ & \text{if } a/b \in c \setminus c_0 \\ \varnothing & \text{if } a/b \notin c. \end{cases}$$

Proof First let $a/b \in c_0$. Then $x \in \mathbb{S}(c, c)$ implies $s_x^{(c)} \subset s_b^{(c)}$ and $x \in s_b^{(c)}$. Then since $b \in s_b^{(c)}$, the condition $x \in \mathbb{S}(c, c)$ also implies that there are sequences $\varphi, \psi \in c$ such that $b = a\varphi + x\psi$. So we have

$$\frac{b}{x} = \frac{\psi}{e - \frac{a}{b}\varphi}.$$

But since $a/b \in c_0$, it follows that $(a/b)\varphi \in c_0$ and $b/x \in c$. Since $x \in s_b^{(c)}$, we conclude $x \in cl^c(b)$.
Conversely, if $s_x^{(c)} = s_b^{(c)}$ we deduce $s_a^{(c)} + s_x^{c)} = s_a^{(c)} + s_b^{(c)} = s_b^{(c)}$. So we have shown $\mathbb{S}(s^{(c)}, s^{(c)}) = cl^{\mathbf{c}}(b)$ if $a/b \in c_0$.
Now let $a/b \in c \setminus c_0$. Then $a_n/b_n \to L \neq 0\,(n \to \infty)$ and $s_a^{(c)} = s_b^{(c)}$. So $x \in \mathbb{S}(c, c)$ implies

$$s_a^{(c)} + s_x^{(c)} = s_b^{(c)} + s_x^{(c)} = s_b^{(c)}. \tag{6.12}$$

Then we have $s_x^{(c)} \subset s_b^{(c)}$ and $x \in s_b^{(c)}$, hence $\mathbb{S}(c, c) \subset s_b^{(c)}$.
Conversely, if $x \in s_b^{(c)}$ then $s_x^{(c)} \subset s_b^{(c)}$ and as above we obtain (6.12) and $x \in \mathbb{S}(c, c)$. Thus we have shown $\mathbb{S}(c, c) = s_b^{(c)}$. Now since $s_a^{(c)} \subset s_a^{(c)} + s_x^{(c)} = s_b^{(c)}$, we deduce $a/b \in c$. So if $a/b \notin c$ then $\mathbb{S}(c, c) = \varnothing$.
This concludes the proof. □

We also have the next proposition, where we write $cl^{\ell_\infty}(b) = cl^\infty(b)$.

Proposition 6.4 ([5, Proposition 5.2, p. 109]) *Let $a, b \in \mathcal{U}^+$. Then*

$$\mathbb{S}(c_0, \ell_\infty) = \begin{cases} cl^\infty(b) & \text{if } a/b \in \ell_\infty \\ \varnothing & \text{otherwise;} \end{cases}$$

$$\mathbb{S}(\mathbf{\Gamma}, \ell_\infty) = \begin{cases} cl^\infty(b) & \text{if } a/b \in \mathbf{\Lambda} \\ \varnothing & \text{otherwise;} \end{cases}$$

$$\mathbb{S}(\mathbf{\Lambda}, \ell_\infty) = \begin{cases} cl^\infty(b) & \text{if } a/b \in \mathbf{\Gamma} \\ \varnothing & \text{otherwise.} \end{cases}$$

We conclude with the next proposition.

Proposition 6.5 ([5, Proposition 5.3, p. 109]) *Let $a, b \in \mathcal{U}^+$. Then we have*

$$\mathbb{S}(\mathbf{\Gamma}, \mathbf{\Lambda}) = \begin{cases} cl^{\mathbf{\Lambda}}(b) & \text{if } a/b \in \mathbf{\Lambda} \\ \varnothing & \text{otherwise.} \end{cases}$$

We obtain the following result.

Theorem 6.3 *Let $a, b \in \mathcal{U}^+$ and E and F be linear subspaces of ω. We assume*

(i) $M(F, F) = \chi$ satisfies condition (5.9),
(ii) $M(E, F) \subset M(\chi, c_0)$,
(iii) There is a linear subspace H of ω that satisfies the condition in (5.20) and the conditions (a) and (b), where

(a) $E, F \subset H$,
(b) $M(F, F) = M(F, H)$.

Then $\mathbb{S}(E, F)$ is defined by (6.8).

Proof Let $x \in \mathbb{S}(E, F)$. Then we have $E_a + F_x \subset F_b$ and $a/b \in M(E, F)$ and

$$x/b \in M(F, F). \tag{6.13}$$

Then we have $F_b \subset E_a + F_x$ and by Part (iii) (a), we obtain

$$F_b \subset H_a + H_x \subset H_{a+x}.$$

This implies

$$\xi = \frac{b}{a + x} \in \chi.$$

Then it can easily be shown that

$$\frac{b}{x} = \frac{\xi}{e - \frac{a}{b}\xi}.$$

So the conditions $\xi \in M(F, F)$ and $a/b \in M(E, F) \subset M(\chi, c_0)$ imply

$$\lim_{n\to\infty}\left(1 - \frac{a_n \xi_n}{b_n}\right) = 1$$

and $b/x \in M(F, F)$. Using the condition in (6.13), we conclude $a/b \in M(E, F)$ and $x \in cl^{M(F,F)}(b)$.
Conversely, we assume $a/b \in M(E, F)$ and $x \in cl^{M(F,F)}(b)$. Then we have $E_a + F_x = E_a + F_b = F_b$ and $x \in \mathbb{S}(E, F)$.
This concludes the proof. □

We also have the following corollary.

Corollary 6.2 *Let $a, b \in \mathcal{U}^+$ and E and F be linear subspaces of ω with $E \subset F$. We assume*

(i) F satisfies the condition in (5.20),
(ii) $M(F, F) = \chi$ satisfies the condition in (5.9),
(iii) $M(E, F) \subset M(\chi, c_0)$.

Then the set of all the solutions of the (SSE) $E_a + F_x = F_b$ is determined by (6.8).

6.3 Some Applications

In this section, we provide some more applications of the results in the previous section.

First we need the next lemma.

Lemma 6.5 *Let $p \geq 1$. Then we have*

(i) $M(w_\infty,\mathbf{\Gamma}) = M(\ell_p, \mathbf{\Gamma}) = \mathbf{\Gamma}$
(ii) $M(\mathbf{\Gamma},w_\infty) = M(\mathbf{\Gamma}, \ell_p) = \mathbf{\Lambda}$.

Proof (i) First we show $M(w_\infty,\mathbf{\Gamma}) = \mathbf{\Gamma}$.
Since $\ell_\infty \subset w_\infty \subset \mathbf{\Lambda}$, we have by Lemma 6.2

$$\mathbf{\Gamma} = M(\mathbf{\Lambda}, \mathbf{\Gamma}) \subset M(w_\infty, \mathbf{\Gamma}) \subset M(\ell_\infty, \mathbf{\Gamma}) = \mathbf{\Gamma}$$

and $M(w_\infty,\mathbf{\Gamma}) = \mathbf{\Gamma}$.
Now we show $M(\ell_p,\mathbf{\Gamma}) = \mathbf{\Gamma}$.
First we have

$$\mathbf{\Gamma} = M(c_0, \mathbf{\Gamma}) \subset M(\ell_p, \mathbf{\Gamma}),$$

since $\ell_p \subset c_0$.
Now we show $M(\ell_p,\mathbf{\Gamma}) \subset \mathbf{\Gamma}$.
For this let $r > 0$. We have $a \in M(\ell_p, s_r)$ if and only if $D_{(a_n/r^n)_{n=1}^\infty} \in M(\ell_p, s_1)$, that is, $(a_n/r^n)_{n=1}^\infty \in s_1$ and $a \in s_r$. So we have $M(\ell_p, s_r) = s_r$. Since $\mathbf{\Gamma} = \cap_{r>0} s_r$, we conclude $M(\ell_p,\mathbf{\Gamma}) \subset M(\ell_p, s_r)$ for all $r > 0$ and $M(\ell_p,\mathbf{\Gamma}) \subset \mathbf{\Gamma}$. We conclude $M(\ell_p,\mathbf{\Gamma}) = \mathbf{\Gamma}$.
(ii) Since $\ell_\infty \subset w_\infty \subset \mathbf{\Lambda}$, we obtain

$$\mathbf{\Lambda} = M(\mathbf{\Gamma}, \ell_\infty) \subset M(\mathbf{\Gamma}, w_\infty) \subset M(\mathbf{\Gamma}, \mathbf{\Lambda}) = \mathbf{\Lambda}.$$

This shows $M(\mathbf{\Gamma}, w_\infty) = \mathbf{\Lambda}$.
Now we show $M(\mathbf{\Gamma}, \ell_p) = \mathbf{\Lambda}$. By Lemma 6.2, we have $M(\mathbf{\Gamma},\mathbf{\Gamma}) = M(\mathbf{\Gamma},\mathbf{\Lambda}) = \mathbf{\Lambda}$. Then we have

$$\mathbf{\Lambda} = M(\mathbf{\Gamma}, \mathbf{\Gamma}) \subset M(\mathbf{\Gamma}, \ell_p) \subset M(\mathbf{\Gamma}, \mathbf{\Lambda}) = \mathbf{\Lambda}.$$

This completes the proof of Part (ii)
This completes the proof. □

As a direct consequence of Theorem 6.3 we obtain the next corollary.

Corollary 6.3 *Let $a, b \in \mathcal{U}^+$ and $p \geq 1$. Then we have*

(i) The sets of all solutions of the (SSE) $E_a + \mathbf{\Gamma}_x = \mathbf{\Gamma}_b$, where $c_0 \subset E \subset \mathbf{\Lambda}$ are determined by

$$\mathbb{S}(E, \mathbf{\Gamma}) = \begin{cases} cl^{\mathbf{\Lambda}}(b) & \text{if } a/b \in \mathbf{\Gamma} \\ \varnothing & \text{otherwise.} \end{cases}$$

(ii) The sets of all solutions of the (SSE) defined by $s_a^{(c)} + s_x^0 = s_b^0$ and $s_a + s_x^0 = s_b^0$ satisfy $\mathbb{S}(c, c_0) = \mathbb{S}(s_1, c_0)$ and are determined by

$$\mathbb{S}(s_1, c_0) = \begin{cases} cl^{\infty}(b) & \text{if } a/b \in c_0 \\ \varnothing & \text{otherwise.} \end{cases}$$

(iii) Let $p > 1$ and let $q = p/(p-1)$. Then the solutions of the (SSE) $(\ell_p)_a + (\ell_1)_x = (\ell_1)_b$ are determined by

$$\mathbb{S}(\ell_p, \ell_1) = \begin{cases} cl^{\infty}(b) & \text{if } a/b \in \ell_q \\ \varnothing & \text{otherwise.} \end{cases}$$

Proof (i) We have $\chi = M(\mathbf{\Gamma}, \mathbf{\Gamma}) = \mathbf{\Lambda}$, $\mathbf{\Lambda}$ satisfies the condition in (5.9). Since $c_0 \subset E$, we have

$$M(E, \mathbf{\Gamma}) \subset M(c_0, \mathbf{\Gamma}) = \mathbf{\Gamma} = M(\chi, c_0).$$

For $H = \mathbf{\Lambda}$, we have $\chi = M(F, H) = M(\mathbf{\Gamma}, \mathbf{\Lambda}) = \mathbf{\Lambda}$ and Theorem 6.3 can be applied. So we obtain Part (i).
(ii) We limit our study to the case of the (SSE) $s_a^{(c)} + s_x^0 = s_b^0$.
Then we have

$$\chi = M(F, F) = M(c_0, c_0) = \ell_\infty$$

and

$$M(E, F) = M(c, c_0) = c_0 = M(\chi, c_0).$$

Taking $H = \ell_\infty$, we obtain $M(F, H) = M(c_0, \ell_\infty) = M(F, F) = c_0$ and Theorem 6.3 can be applied.
The case of the (SSE) $s_a + s_x^0 = s_b^0$ can be obtained in a similar way.
So we have shown Part (ii).
(iii) We apply Corollary 6.2 with $\chi = M(\ell_1, \ell_1) = \ell_\infty$ and

$$M(E, F) = M(\ell_p, \ell_1) = \ell_q.$$

Then we obtain $M(E, F) \subset c_0 = M(\chi, c_0)$. If we take $H = \ell_\infty$, then we have

$$M(F, H) = M(\ell_p, \ell_\infty) = \ell_\infty = M(F, F).$$

This completes the proof. □

Example 6.3 We consider the statement $n^{-1}|y_n|^{1/n} \to 0$ $(n \to \infty)$ if and only if there are sequences u and v for which $y = u + v$

$$\frac{1}{n}\sum_{k=1}^{n} |u_k| \to 0 \text{ and } \left(\frac{|v_n|}{x_n}\right)^{1/n} \to 0 \ (n \to \infty) \text{ for all } y.$$

By Stirling's formula, we have $(n!)^{1/n} \sim ne^{-1}$ $(n \to \infty)$ and $n^{-1}|y_n|^{1/n} \to 0$ if and only if $(|y_n|/n!)^{1/n} \to 0$, that is, $y \in \mathbf{\Gamma}_{(n!)_{n=1}^{\infty}}$. This statement is equivalent to

$$w_0 + \mathbf{\Gamma}_x = \mathbf{\Gamma}_{(n!)_{n=1}^{\infty}} \tag{6.14}$$

and to $x \in \mathbb{S}(w_0, \mathbf{\Gamma})$. By Part (ii) of Corollary 6.3,the solutions of equation (6.14) are determined by $K_1^n n! \leq x_n \leq K_2^n n!$ for all n and some $K_1, K_2 > 0$.

Again as a direct consequence of Theorem 6.3 we obtain the next proposition.

Proposition 6.6 *Let $a, b \in \mathcal{U}^+$ and $p \geq 1$. Each of the (SSE) $E_a + (\ell_p)_x = (\ell_p)_b$ where $E = c_0$, c or ℓ_∞ is regular and*

$$\mathbb{S}(E, \ell_p) = \begin{cases} cl^\infty(b) & \text{if } a/b \in \ell_p \\ \varnothing & \text{otherwise.} \end{cases}$$

Proof We only consider the case $E = \ell_\infty$, since the proofs of the other cases are similar. By Lemma 5.6, we have $M(\ell_p, \ell_p) = \ell_\infty$ and by Lemma 5.7, we have

$$M(E, F) = M(\ell_\infty, \ell_p) = \ell_p.$$

and then

$$M(E, F) \subset M(M(F, F), c_0) = c_0.$$

We take $H = \ell_\infty$ and obtain

$$M(F, H) = M(\ell_p, \ell_\infty) = \ell_\infty = M(\ell_p, \ell_p).$$

This completes the proof. □

Now we consider the (SSE)

$$E_a + W_x^0 = W_b^0 \text{ with } E \in \{s_1, c\}.$$

Here for given $a, b \in \mathcal{U}^+$, we determine the set of all $x \in \mathcal{U}^+$ that satisfy the statement

$y \in W_b^0$ if and only if there are sequences $u, v \in \omega$ for which $y = u + v$,

$$\frac{u_n}{a_n} \to l \text{ and } \frac{1}{n}\sum_{k=1}^{n} \frac{|v_k|}{x_k} \to 0 \ (n \to \infty)$$

for some scalar l and all sequences y.

Let ϖ denote the set of all sequences $\tau \in \mathcal{U}^+$ such that $\tau \in w_0$ implies $\tau \in c_0$. We note that $(r^n)_{n=1}^{\infty} \in \varpi$ for all $r > 0$. Now we state the next theorem.

Theorem 6.4 ([9, Theorem 7.1, pp. 604–605]) *Let $E = s_1$ or c and $a/b \in \varpi$. Then the set $\mathbb{S}(E, w_0)$ of all positive sequences such that $E_a + W_x^0 = W_b^0$ is determined by*

$$\mathbb{S}(E, w_0) = \begin{cases} cl^{\infty}(b) & \text{if } a/b \in c_0 \\ \varnothing & \text{if } a/b \notin c_0. \end{cases}$$

Proof Let $x \in \mathbb{S}(E, w_0)$. Then we have

$$E_a + W_x^0 \subset W_b^0 \tag{6.15}$$

and

$$W_b^0 \subset E_a + W_x^0. \tag{6.16}$$

The inclusion in (6.15) implies

$$\frac{a}{b} \in M(E_1, w_0) = w_0 \tag{6.17}$$

and

$$W_x^0 \subset W_b^0. \tag{6.18}$$

Since $a/b \in \varpi$, the condition in (6.17) implies $a/b \in c_0$. Then we have by [9, Remark 3.4, p. 597] $M(w_0, w_0) = \ell_\infty$, and the condition in (6.18) implies $s_x \subset s_b$ and $x \in s_b$. Now we consider the inclusion in (6.16). We have

$$W_b^0 \subset E_a + W_x^0 \subset W_a + W_x = W_{a+x}. \tag{6.19}$$

Thus the condition in (6.19) implies $W_b^0 \subset W_{a+x}$ and

$$\left(\frac{b_n}{u_n + x_n}\right)_{n=1}^{\infty} \in M(w_0, w_\infty) = \ell_\infty.$$

So there is $K > 0$ such that

$$\frac{x_n}{b_n} \geq K - \frac{a_n}{b_n}.$$

Since $a/b \in c_0$, there is $K_1 > 0$ such that $x_n/b_n \geq K_1 > 0$ for all n. We conclude that $x \in \mathbb{S}(E, w_0)$ implies $x \in cl^\infty(b)$ and $a/b \in c_0$.
Conversely, we assume $x \in cl^\infty(b)$ and $a/b \in c_0$. Then $s_x = s_b$ which implies $W_x^0 = W_b^0$ and $E_a + W_x^0 = E_a + W_b^0$. Now since $M(E_1, w_0) = w_0$ for $E = s_1$ or c by Lemma 5.7, it follows that $a/b \in c_0$ implies $a/b \in M(E_1, w_0)$ for $E = s_1$ or c. Then $E_a \subset W_b^0$ and $E_a + W_b^0 = W_b^0$.
This concludes the proof. □

Example 6.4 Since $(1/n)_{n=1}^\infty \in \varpi$, the solutions of the (SSE) $c + W_x^0 = W_{(n)_{n=1}^\infty}$ are determined by $K_1 n \leq x_n \leq K_2 n$ for all n and some $K_1, K_2 > 0$.

Now we consider the (SSE)

$$E_r + W_x^0 = W_u^0.$$

For any real $r > 0$ and for any linear space E_1, let E_r denote the set $E_{(r^n)_{n=1}^\infty} = D_{(r^n)_{n=1}^\infty} E_e$: for instance, we have $W_r^0 = W^0_{(r^n)_{n=1}^\infty}$. Similarly we use the notation $cl^\infty(u) = cl^\infty((u^n)_{n=1}^\infty)$.

Theorem 6.5 ([9, Theorem 7.3, pp. 606–607]) *Let $r, u > 0$ and E be a linear subset of ω. Also let $\widetilde{\mathbb{S}}(E, w_0)$ be the set of the solutions of the (SSE)*

$$E_r + W_x^0 = W_u^0.$$

We assume

(i) $E_1 \subset w_\infty$.
(ii) For any real $\rho > 0$, the condition

$$(\rho^n)_{n=1}^\infty \in M(E_1, w_0)$$

holds if and only if $\rho < 1$.

Then

$$\widetilde{\mathbb{S}}(E, w_0) = \begin{cases} cl^\infty(u) & \text{if } r < u \\ \varnothing & \text{if } r \geq u. \end{cases}$$

Proof Let $x \in \widetilde{\mathbb{S}}(E_1, w_0)$. Then we have

$$E_r + W_x^0 \subset W_u^0 \tag{6.20}$$

and

$$W_u^0 \subset E_r + W_x^0. \tag{6.21}$$

The inclusion in (6.20) implies

$$\left((r/u)^n\right)_{n=1}^{\infty} \in M(E_1, w_0) \tag{6.22}$$

and

$$W_x^0 \subset W_u^0. \tag{6.23}$$

By (ii), the condition in (6.22) implies $\rho = r/u < 1$ and $r < u$, and as we have just seen, since $M(w_0, w_0) = \ell_\infty$, we obtain $s_x \subset s_u$ and $x \in s_u$.
Now we consider the inclusion in (6.21). Then

$$W_u^0 \subset E_r + W_x^0 \subset W_r + W_x.$$

Since

$$W_r + W_x = W_{(r^n + x_n)_{n=1}^{\infty}},$$

we obtain

$$W_u^0 \subset W_{(r^n + x_n)_{n=1}^{\infty}}$$

and

$$\left(\frac{u^n}{r^n + x_n}\right)_{n=1}^{\infty} \in M(w_0, w_\infty) = \ell_\infty.$$

So there is $K > 0$ such that $x_n/u^n \geq K - \rho^n$. Since $\rho < 1$, there is $K_1 > 0$ such that $x_n/u^n \geq K_1 > 0$ for all n. We conclude that $x \in \widetilde{\mathbb{S}}(E_1, W^0)$ implies $x \in cl^\infty(u)$ and $r < u$.
Conversely, we assume $x \in cl^\infty(u)$ and $r < u$. Then $s_x = s_u$ which implies $W_x^0 = W_u^0$ and $E_r + W_x^0 = E_r + W_u^0$. Now by (ii), since $\rho = r/u < 1$, we have $(\rho^n)_{n=1}^{\infty} \in M(E_1, w_0)$, $E_r \subset W_u^0$ and $E_r + W_u^0 = W_u^0$.
This concludes the proof. □

Corollary 6.4 ([9, Corollary 7.4, pp. 607–608]) *Let $r, u > 0$. Then the sets $\widetilde{\mathbb{S}}(E, w_0)$ of solutions of the (SSE)*

$$E_r + W_x^0 = W_u^0 \text{ with } E = \ell_\infty, c \text{ or } w_\infty$$

are given by

$$\widetilde{\mathbb{S}}(E, w_0) = \begin{cases} cl^\infty(u) & \text{if } r < u \\ \varnothing & \text{if } r \geq u \end{cases} \tag{6.24}$$

and for $E = w_0$, we have

$$\widetilde{\mathbb{S}}(w_0, w_0) = \begin{cases} cl^\infty(u) & \text{if } r < u \\ s_u \cap U^+ & \text{if } r = u \\ \varnothing & \text{if } r > u. \end{cases}$$

Proof 1- Case $E = \ell_\infty$ or c.
It is enough to apply Theorem 6.5. First we have $E_1 \subset w_\infty$ and the condition in (i) of Theorem 6.5 is satisfied. Then we have $a/b = (\rho^n)_{n=1}^\infty \in \varpi$, since $n^{-1}\sum_{k=1}^n \rho^k \to 0$ implies $0 < \rho < 1$.

2- Case $E = w_\infty$.
It is enough to show Theorem 6.5 with $E = w_\infty$. We have $(\rho^n)_{n=1}^\infty \in M(w_\infty, w_0)$. Since $M(w_\infty, w_0) \subset M(\ell_\infty, (c_0)_{C_1})$, we have $C_1 D_\rho \in (\ell_\infty, c_0)$. The matrix $C_1 D_\rho$ is the triangle defined by $[C_1 D_\rho]_{nk} = \rho^k/n$, for $k \leq n$, and the condition $C_1 D_\rho \in (\ell_\infty, c_0)$ implies

$$\frac{1}{n}\sum_{k=1}^n \rho^k \to 0\ (n \to \infty)$$

and $\rho < 1$.
Conversely, if $0 < \rho < 1$, then we have $n\rho^n \to 0\ (n \to \infty)$ and

$$(\rho^n)_{n=1}^\infty \in \left(s^0_{(1/n)_{n=1}^\infty}\right)^+ = M^+(w_\infty, w_0) \subset M^+(w_\infty, w_\infty).$$

We conclude the condition in (ii) of Theorem 6.5 is satisfied for $E = w_\infty$.

3- Case $E = w_0$. The (SSE) $W_r^0 + W_x^0 = W_u^0$ is equivalent to $W_u^0 = W^0_{(r^n + x_n)_{n=1}^\infty}$ and to $s_u = s_{(r^n+x_n)_{n=1}^\infty} = s_r + s_x$. We conclude by [10, Corollary 12, p. 918]. □

Example 6.5 It can easily be shown that the solutions of the (SSE) $w_\infty + W_x^0 = W_2^0$ are determined by $K_1 2^n \leq x_n \leq K_2 2^n$ for all n and for some K_1 and $K_2 > 0$.

Example 6.6 For $u > 1$, the solutions of the (SSE) $c + W_x^0 = W_u^0$ are determined by $K_1 u^n \leq x_n \leq K_2 u^n$ for all n and for some K_1 and $K_2 > 0$. If $u \leq 1$, then the (SSE) has no solution.

Example 6.7 The statement

$y \in w_0$ if and only if there are sequences u and $v \in \omega$ for which $y = u + v$,

$$2^n u_n \to l \text{ and } \frac{1}{n}\sum_{k=1}^n \frac{|v_k|}{x_k} \to 0\ (n \to \infty)$$

for some scalar l and for all sequences y

is equivalent to the (SSE) $s_{1/2}^{(c)} + W_x^0 = w_0$. The solutions of the (SSE) are determined by $K_1 \leq x_n \leq K_2$ for all n and for some K_1 and $K_2 > 0$.

We have the next result.

Lemma 6.6 *Let $a, b \in \mathcal{U}^+$ and E and F be any of the sets c_0, c, ℓ_∞, Γ or Λ. Then the set $\mathbb{S}$ of all $x \in \mathcal{U}^+$, which satisfy the system $E_a + F_x = F_b$ and $F_b \subset F_x$, is determined by $\mathbb{S} = \mathbb{S}(E, F)$ defined in (6.8).*

Proof First the (SSE) $E_a + F_x = F_b$ implies $E_a \subset F_b$ and $F_x \subset F_b$, that is, $a/b \in M(E, F)$ and $x/b \in M(F, F)$. Since $F_b \subset F_x$, we conclude that $x \in \mathbb{S}$ implies $a/b \in M(E, F)$ and $x \in cl^F(b)$.
Conversely, we assume $a/b \in M(E, F)$ and $x \in cl^F(b)$. Then we have

$$E_a + F_x = E_a + F_b = F_b,$$

and $x \in \mathbb{S}$.
This completes the proof. □

We close this section with the solvability of the (SSE)

$$E_a + (F_x)_\Delta = F_b, \text{ where } F_b \subset F_x \text{ and } E, F \in \{c_0, c, \ell_\infty, \Gamma, \Lambda\}.$$

For $a, b \in \mathcal{U}^+$, let

$$\mathbb{S}(E, F) = \left\{x \in \mathcal{U}^+ : E_a + (F_x)_\Delta = F_b \text{ and } F_b \subset F_x\right\},$$

where E and F are any of the sets c_0, c, ℓ_∞, $\mathbf{\Gamma}$ or $\mathbf{\Lambda}$. For instance, if $E = F = c$, then the condition $x \in \mathbb{S}(c, c)$ means that the statements in (a) and (b) hold, where

(a) $b_n/x_n \to l_1 \ (n \to \infty)$, for some $l_1 \in \mathbb{C}$,
(b) $y_n/b_n \to l \ (n \to \infty)$ *if and only if there are sequences* u *and* $v \in \omega$ *such that* $y = u + v$ *and*

$$\frac{u_n}{a_n} \to l' \text{ and } \frac{\Delta v_n}{x_n} \to l'' \ (n \to \infty)$$

for some scalars l, l' *and* l'' *and for all sequences* $y \in \omega$.

We obtain the next result.

Theorem 6.6 *Let* E *and* F *be any of the sets* c_0, c, ℓ_∞, $\mathbf{\Gamma}$ *or* $\mathbf{\Lambda}$.

(i) We assume $M(F, F) = \ell_\infty$.

(a) If $b \notin \widehat{C}_1$, *then* $\mathbb{S}(E, F) = \varnothing$.
(b) If $b \in \widehat{C}_1$, *then*

$$\mathbb{S}(E, F) = \begin{cases} cl^\infty(b) & \text{if } a/b \in M(E, F) \\ \varnothing & \text{otherwise.} \end{cases}$$

(ii) We assume $F = c$.

(a) If $b \notin \widehat{\Gamma}$, *then* $\mathbb{S}(E, F) = \varnothing$.
(b) If $b \in \widehat{\Gamma}$, *then*

$$\mathbb{S}(E,F)=\begin{cases} cl^c(b) & \text{if } a/b\in M(E,c)\\ \varnothing & \text{otherwise.}\end{cases}$$

(iii) *We assume* $F=\mathbf{\Lambda}$.

(a) *If* $b\notin\widehat{C}_\Lambda$, *then* $\mathbb{S}(E,F)=\varnothing$.
(b) *If* $b\in\widehat{C}_\Lambda$, *then*

$$\mathbb{S}(E,F)=\begin{cases} cl^\Lambda(b) & \text{if } a/b\in M(E,\mathbf{\Lambda})=\mathbf{\Lambda}\\ \varnothing & \text{otherwise.}\end{cases}$$

Proof (i) First we note that the condition $M(F,F)=\ell_\infty$ implies $F=s_1$ or c_0. Let $x\in\mathbb{S}(E,F)$. We successively obtain $(F_x)_\Delta\subset F_x, D_{1/x}\Sigma D_x\in S_1$ and $x\in\widehat{C}_1$. By Part (ii) of Proposition 6.1, we have $(F_x)_\Delta=F_x$, $E_a+F_x=F_b$ and $F_b\subset F_x$. By Lemma 6.6, we have $a/b\in M(E,F)$ and $x\in cl^\infty(b)$. We conclude $x\in\widehat{C}_1, a/b\in M(E,F)$ and $F_x=F_b$. Since $F=c_0$ or ℓ_∞, the identity $F_x=F_b$ is equivalent to $s_x=s_b$ and by Lemma 6.2, $x\in\widehat{C}_1$ implies $b\in\widehat{C}_1$. So we have shown that $x\in\mathbb{S}(E,F)$ implies

$$b\in\widehat{C}_1,\ \frac{a}{b}\in M(E,F)\text{ and } F_x=F_b. \tag{6.25}$$

Now we show that (6.25) implies $x\in\mathbb{S}(E,F)$. The conditions $b\in\widehat{C}_1$ and $F_x=F_b$ together imply $x\in\widehat{C}_1$. So we have

$$E_a+(F_x)_\Delta=E_a+F_x=E_a+F_b,$$

and since $a/b\in M(E,F)$, we successively obtain $E_a\subset F_b$, $E_a+F_b=F_b$ and $E_a+(F_x)_\Delta=F_b$. We conclude $x\in\mathbb{S}(E,F)$ if and only if the conditions in (6.25) hold. So we have shown Part (i).
(ii), (iii) Parts (ii) and (iii) can be shown similarly. We note that for Part (ii), we have $F=c$ and $M(F,F)=c$.

□

6.4 The (SSE) with Operators

In this section, we deal with a class of (SSE) with operators of the form $E_T+F_x=F_b$, where T is either Δ or Σ and E is any of the sets c_0, c, ℓ_∞, ℓ_p for $p\geq 1$, w_0, $\mathbf{\Gamma}$ or $\mathbf{\Lambda}$ and $F=c$, ℓ_∞ or $\mathbf{\Lambda}$. For instance, the solvability of the (SSE) with operator defined by the equation $\mathbf{\Gamma}_\Sigma+\mathbf{\Lambda}_x=\mathbf{\Lambda}_b$ consists in determining the set of all positive sequences $x=(x_n)_{n=1}^\infty$ that satisfy the statement:

$\sup_n((|y_n|/b_n)^{1/n} < \infty$ if and only if there are sequences u and $v \in \omega$ with $y = u + v$ such that

$$\lim_{n\to\infty}\left|\sum_{k=1}^{n} u_k\right|^{1/n} = 0 \text{ and } \sup_n\left[\left(\frac{|v_n|}{x_n}\right)^{1/n}\right] < \infty$$

for all sequences y.

Then we deal with the solvability of the (SSE) with operator $E_T + F_x = F_b$ where a and b are any positive sequences. We also apply the results of Sect. 5.5 to solve the (SSE) of the form $E_\Delta + F_x = F_b$ with the operator Δ of the first difference, where $E = c_0, c, \ell_\infty, \ell_p$ for $p \geq 1$, w_0, $\mathbf{\Gamma}$ or $\mathbf{\Lambda}$ and $F = c$ or ℓ_∞. Then using the operator Σ of the sum, we solve the (SSE) of the form $E_\Sigma + F_x = F_b$, where $E = c_0, c, \ell_\infty, \ell_p$ for $p \geq 1$, w_0 $\mathbf{\Gamma}$ or $\mathbf{\Lambda}$ and $F = c, \ell_\infty$, and the (SSE) $\mathbf{\Gamma}_\Sigma + \mathbf{\Gamma}_x = \mathbf{\Gamma}_b$.

For the reader's convenience, we state the next lemma, which is Part (iii) of Remark 1.6.

Lemma 6.7 *Let $A = (a_{nk})_{nk=1}^\infty$ be an infinite matrix. Then we have*

(i) If $\lim_{n\to\infty} a_{nk} = 0$ for all k, then $A \in (\ell_\infty, c)$ if and only if

$$\lim_{n\to\infty}\sum_{k=1}^{\infty}|a_{nk}| = 0; \tag{6.26}$$

(ii) ((6.1) in Part **6** *of* Theorem *1.23) $A \in (\ell_\infty, c_0)$ if and only if (6.26) holds.*

We recall that $cs_0 = (c_0)_\Sigma$ is the set of all series that converge to zero, that is,

$$cs_0 = \left\{y : \left(\sum_{k=1}^{n} y_k\right)_{n=1}^{\infty} \in c_0\right\}.$$

Now we consider the (SSE) $E + F_x = F_b$. To simplify we still write $\mathbb{S}(E, F)$ for the set of all positive sequences x that satisfy the (SSE) $E + F_x = F_b$.

In the following, we use the fact that the condition in (5.9) is true for any of the spaces $F = c$, s_1 or $\mathbf{\Lambda}$.

Regular sequence spaces equations.

Definition 6.1 We say that $\mathbb{S}(E, F)$ (or the equation $E + F_x = F_b$) is *regular* if $\mathbb{S}(E, F)$ is defined by (6.8) with $a = e$.

We note that $E + F_x = F_b$ is not regular, in general. Indeed for $E = F = \ell_\infty$, we have $M(\ell_\infty, \ell_\infty) = \ell_\infty$ and if $1/b \in \ell_\infty \setminus c_0$ and $s_1 = s_b$, then $\mathbb{S}(\ell_\infty, \ell_\infty) = s_b \cap \mathcal{U}^+ \neq cl^{M(F,F)}(b)$, (cf. [10, Theorem 11, pp. 916–917]). In particular, the solutions of the (SSE) $\ell_\infty + s_x = \ell_\infty$ are determined by $x \in \ell_\infty \cap \mathcal{U}^+$, that is, $0 < x_n \leq M$

for all n and some $M > 0$. We also note that the set $S(c, c)$ is not regular since we have by Theorem 6.2 $\mathbb{S}(c, c) = cl^c(b)$ for $1/b \in c_0$; $\mathbb{S}(c, c) = c_b$ for $1/b \in c \setminus c_0$, and $\mathbb{S}(c, c) = \varnothing$ for $1/b \notin c$.

As a direct consequence of Theorem 6.1 we obtain the following result.

Lemma 6.8 *Let $b \in \mathcal{U}^+$ and E and F be two linear subspaces of ω. We assume that F satisfies the conditions in (5.9), (6.5), (6.6) and that (6.7) holds. Then $\mathbb{S}(E, F)$ is regular, that is,*

$$\mathbb{S}(E, F) = \begin{cases} cl^F(b) & \text{if } 1/b \in M(E, F) \\ \varnothing & \text{if } 1/b \notin M(E, F). \end{cases}$$

As a direct consequence of the preceding results and using Lemma 6.8, we obtain the next results.

Lemma 6.9 *Let $b \in \mathcal{U}^+$ and $p \geq 1$. Then each of the next (SSE) is regular, where*

(i) $\mathbf{\Gamma} + \mathbf{\Lambda}_x = \mathbf{\Lambda}_b$;
(ii) $E + c_x = c_b$ *for* $E = \mathbf{\Gamma}, \mathbf{\Lambda}, c_0, \ell_\infty$ w_0 *and* ℓ_p;
(iii) $E + s_x = s_b$ *for* $E = \mathbf{\Gamma}, \mathbf{\Lambda}, c_0, w_0$ *and* ℓ_p.

Proof (i) follows from Proposition 6.5.
(ii) follows from Proposition 6.3 for $E = \mathbf{\Gamma}, \mathbf{\Lambda}, c_0, \ell_\infty$. The case $E = w_0$ is a direct consequence of Theorem 6.1 and of the equalities $M(w_0, c) = M(w_0, c_0) = s_{(1/n)_{n=1}^\infty}$. The case $E = \ell_p$ follows from Theorem 6.1 and the identity $M(\ell^p, c) = \ell_\infty$, where $p \geq 1$.
(iii) follows from Proposition 6.4 for $E = \mathbf{\Gamma}, \mathbf{\Lambda}, c_0$. The case $E = w_0$ is a direct consequence of the identity $M(w_0, s_1) = s_{(1/n)_{n=1}^\infty}$, and we conclude by Theorem 6.1. The case $E = \ell_p$ follows from Theorem 6.1 and the identity $M(\ell^p, \ell_\infty) = \ell_\infty$ where $p \geq 1$.
This concludes the proof. □

More precisely we obtain the following lemma which is a direct consequence of Lemma 6.9.

Lemma 6.10 *Let $b \in \mathcal{U}^+$. Then we have*

(i)

$$\mathbb{S}(\ell_\infty, c) = \begin{cases} cl^c(b) & \text{if } 1/b \in c_0 \\ \varnothing & \text{otherwise.} \end{cases} \tag{a}$$

(b) *Let F be any of the sets c, s_1 or $\mathbf{\Lambda}$. Then we have*

$$\mathbb{S}(\mathbf{\Gamma}, F) = \begin{cases} cl^F(b) & \text{if } 1/b \in \mathbf{\Lambda} \\ \varnothing & \text{otherwise.} \end{cases}$$

(ii) *We have for $F = c$ or s_1:*

(a) Let $p \geq 1$. We have $\mathbb{S}(\ell_p, F) = \mathbb{S}(c_0, F)$ and

$$\mathbb{S}(c_0, F) = \begin{cases} cl^F(b) & \text{if } 1/b \in s_1 \\ \varnothing & \text{otherwise.} \end{cases}$$

$$\mathbb{S}(w_0, F) = \begin{cases} cl^F(b) & \text{if } 1/b \in s_{(1/n)_{n=1}^{\infty}} \\ \varnothing & \text{otherwise;} \end{cases} \tag{b}$$

$$\mathbb{S}(\Lambda, F) = \begin{cases} cl^F(b) & \text{if } 1/b \in \Gamma \\ \varnothing & \text{otherwise;} \end{cases} \tag{b}$$

Remark 6.1 The results for $\mathbb{S}(\ell_p, c)$ and $\mathbb{S}(\ell_p, \ell_\infty)$ follow from Lemma 5.6, where $M(\ell_p, c) = M(\ell_p, \ell_\infty) = \ell_\infty$.

Example 6.8 We consider the set of all $x \in \mathcal{U}^+$ that satisfy the statement:

for every sequence y, we have $y_n \to l_1$ $(n \to \infty)$ if and only if there are sequences u and $v \in \omega$ for which $y = u + v$ and

$$|u_n|^{1/n} \to 0 \text{ and } x_n v_n \to l_2 \ (n \to \infty)$$

for some l_1 and l_2.

Since this set corresponds to the equation $\Gamma + s_{1/x}^{(c)} = c$, by *Lemma* 6.3 it is equal to the set of all sequences that tend to a positive limit.

Example 6.9 It can easily be shown that the solutions of the (SSE) $w_\infty + W_x^0 = W_2^0$ are determined by $K_1 2^n \leq x_n \leq K_2 2^n$ for all n and for some K_1 and $K_2 > 0$.

Example 6.10 For $u > 1$, the solutions of the (SSE) $c + W_x^0 = W_u^0$ are determined by $K_1 u^n \leq x_n \leq K_2 u^n$ for all n and some K_1 and $K_2 > 0$, and if $u \leq 1$, then the (SSE) has no solution.

Example 6.11 The (SSE) $s_{1/2}^{(c)} + W_x^0 = w_0$ is equivalent to the statement

$y \in w_0$ if and only if there are sequences u and $v \in \omega$ for which $y = u + v$,

$$2^n u_n \to l \text{ and } \frac{1}{n} \sum_{k=1}^{n} \frac{|v_k|}{x_k} \to 0 \ (n \to \infty)$$

for some scalar l and for all sequences y.

The solutions of the (SSE) are determined by $K_1 \leq x_n < K_2$ for all n and some K_1 and $K_2 > 0$.

Now we give an application to the solvability of the (SSE)

$$E_T + F_x = F_b \text{ with } e \in F.$$

Let $b \in \mathcal{U}^+$, and E and F be two subsets of ω. We deal with the (SSE) with operators

$$E_T + F_x = F_b, \tag{6.27}$$

where T is a triangle and $x \in \mathcal{U}^+$ is the unknown. The equation in (6.27) means

for every $y \in \omega$, we have $y/b \in F$ if and only if there are sequences u and $v \in \omega$ such that $y = u + v$ such that

$$Tu \in E \text{ and } v/x \in F.$$

We assume $e \in F$. By $\mathbb{S}(E_T, F)$, we denote the set of all $x \in \mathcal{U}^+$ that satisfy the (SSE) in (6.27). We obtain the next result which is a direct consequence of Lemma 6.8, where we replace E by E_T.

Proposition 6.7 ([6, Proposition 6.1, p. 94–95]) *Let $b \in \mathcal{U}^+$ and let E and F be linear vector spaces of sequences. We assume F satisfies the conditions*

(i) $e \in F$,
(ii) $F \subset M(F, F)$,
(iii) *F satisfies condition (5.9)*

and

$$M(E_T, F) \subset M(E_T, c_0). \tag{6.28}$$

Then the set $\mathbb{S}(E_T, F)$ is regular, that is,

$$\mathbb{S}(E_T, F) = \begin{cases} cl^F(b) & \text{if } 1/b \in M(E_T, F) \\ \varnothing & \text{if } 1/b \notin M(E_T, F). \end{cases}$$

We may adapt the previous result using the notations of matrix transformations instead of the multiplier of sequence spaces. So we obtain the following.

Corollary 6.5 ([6, Corollary 6.1, p. 94–95]) *Let $b \in \mathcal{U}^+$ and let E and F be linear spaces of sequences. We assume F satisfies conditions* (i), (ii) *and* (iii) *in* Proposition 6.7 *and that*

$$D_\alpha T^{-1} \in (E, F) \text{ implies } D_\alpha T^{-1} \in (E, c_0) \text{ for all } \alpha \in \omega. \tag{6.29}$$

Then we have

$$\mathbb{S}(E_T, F) = \begin{cases} cl^F(b) & \text{if } D_{1/b}T^{-1} \in (E, F) \\ \varnothing & \text{if } D_{1/b}T^{-1} \notin (E, F). \end{cases}$$

Proof This result is a direct consequence of Proposition 6.7 and of the fact that the condition $1/b \in M(E_T, F)$ is equivalent to $D_{1/b} \in (E_T, F)$ and to $D_{1/b}T^{-1} \in (E, F)$. □

6.5 Some (SSE's) with the Operators Δ and Σ

In this section, we give some applications to the solvability of the $(SSE's)$

$$E_\Delta + F_x = F_b \text{ and } E_\Sigma + F_x = F_b.$$

We apply Proposition 6.7 and Lemma 6.10 to solve (SSE) of the form $E_T + F_x = F_b$ in each of the cases $T = \Delta$ and $T = \Sigma$. We obtain a class of (SSE) that are regular, that is, for which $\mathbb{S}(E, F)$ is regular.

Now we solve each of the $(SSE's)$

$$(c_0)_\Delta + c_x = c_b \text{ and } (c_0)_\Delta + s_x = s_b.$$

The solvability of the first (SSE) means that

for every $y \in \omega$ we have $y_n/b_n \to l_1$ $(n \to \infty)$ if and only if there are sequences u and $v \in \omega$ such that $y = u + v$ and

$$u_n - u_{n-1} \to 0 \text{ and } \frac{v_n}{x_n} \to l_2 \ (n \to \infty)$$

for some scalars l_1 and l_2.

Proposition 6.8 ([6, Proposition 7.1, p. 95]) *Let $b \in \mathcal{U}^+$ and $F = c$ or ℓ_∞. Then we have*

$$\mathbb{S}((c_0)_\Delta, F) = \begin{cases} cl^F(b) & \text{if } 1/b \in s_{(1/n)_{n=1}^\infty} \\ \varnothing & \text{if } 1/b \notin s_{(1/n)_{n=1}^\infty}. \end{cases}$$

Proof The condition $\alpha \in M((c_0)_\Delta, s_1)$ means $D_\alpha \Sigma \in (c_0, s_1) = S_1$ and is equivalent to $n\alpha_n = O(1)$ $(n \to \infty)$. So $M((c_0)_\Delta, s_1) = s_{(1/n)_{n=1}^\infty}$. In the same way, by the characterization of (c_0, c_0), we obtain $M((c_0)_\Delta, c_0) = s_{(1/n)_{n=1}^\infty}$. Then we have

$$s_{(1/n)_{n=1}^\infty} = M((c_0)_\Delta, c_0) \subset M((c_0)_\Delta, c) \subset M((c_0)_\Delta, s_1) = s_{(1/n)_{n=1}^\infty},$$

and $M((c_0)_\Delta, F) = s_{(1/n)_{n=1}^\infty}$ for $F = s_1, c$ or c_0. We conclude by Proposition 6.7. □

Example 6.12 Let $\alpha \geq 0$. The (SSE) $(c_0)_\Delta + c_x = c_{(n^\alpha)_{n=1}^\infty}$ has solutions if and only if $\alpha > 1$. These solutions are determined by $\lim_{n\to\infty} x_n/n^\alpha > 0$ $(n \to \infty)$. If $0 \leq \alpha < 1$, then the (SSE) has no solution. We note that the (SSE) $(c_0)_\Delta + c_x = c$ has no solution.

Example 6.13 Let $u > 0$. Then the set of all positive sequences x that satisfy the (SSE) $(c_0)_\Delta + s_x = s_u$ is empty if $u \leq 1$, and if $u > 1$ it is equal to the set of all sequences that satisfy $K_1 u^n \leq x_n \leq K_2 u^n$ for all n and some K_1 and $K_2 > 0$.

Now we solve each of the $(SSE's)$

$$bv_p + c_x = c_b \text{ and } bv_p + s_x = s_b,$$

where

$$bv_p = (\ell_p)_\Delta = \left\{ y \in \omega : \sum_{k=1}^{\infty} |y_k - y_{k-1}|^p < \infty \right\} \text{ for } 0 < p < \infty$$

is the set of all sequences of *p-bounded variation.* The solvability of the second (SSE) consists in determining the set of all positive sequences x, such that the next statement holds.

For every $y \in \omega$, we have $\sup_n (|y_n|/b_n) < \infty$ if and only if there are sequences u and $v \in \omega$ with $y = u + v$ such that

$$\sum_{k=1}^{\infty} |u_n - u_{n-1}|^p < \infty \text{ and } \sup_n \left(\frac{|v_n|}{x_n} \right) < \infty.$$

We obtain the next proposition.

Proposition 6.9 ([6, Proposition 7.2, p. 96]) *Let $b \in \mathcal{U}^+$, $p > 1$ and $q = p/(p-1)$. We have for $F = c$ or ℓ_∞*

$$\mathbb{S}(bv_p, F) = \begin{cases} cl^F(b) & \text{if } \left(\frac{n^{1/q}}{b_n} \right)_{n=1}^{\infty} \in s_1 \\ \varnothing & \text{if } \left(\frac{n^{1/q}}{b_n} \right)_{n=1}^{\infty} \notin s_1. \end{cases}$$

Proof We have $\alpha \in M(bv_p, \ell_\infty)$ if and only if $D_\alpha \Sigma \in (\ell_p, \ell_\infty)$. We obtain from the characterization of (ℓ_p, ℓ_∞) in (1.1) of **1** in Theorem 1.23

$$n|\alpha_n|^q = O(1) \ (n \to \infty). \tag{6.30}$$

So we have $M(bv_p, \ell_\infty) = s_{(n^{-1/q})_{n=1}^{\infty}}$. Now we obtain $\alpha \in M(bv_p, c_0)$ if and only if (6.30) holds and

$$\alpha_n \to 0 \ (n \to \infty). \tag{6.31}$$

But trivially the condition in (6.30) implies that in (6.31). So we have

$$s_{(n^{-1/q})_{n=1}^{\infty}} = M(bv_p, c_0) \subset M(bv_p, c) \subset M(bv_p, \ell_\infty) = s_{(n^{-1/q})_{n=1}^{\infty}}$$

and $M(bv_p, F) = s_{(n^{-1/q})_{n=q}^{\infty}}$ for $F = c_0$, c or ℓ_∞. We apply Proposition 6.7, where the condition $1/b \in M(bv_p, \ell_\infty) = s_{(n^{-1/q})_{n=1}^{\infty}}$ means $(n^{1/q}/b_n)_{n=1}^{\infty} \in s_1$.
This concludes the proof. □

Example 6.14 The (SSE) defined by $bv_2 + c_x = c$ has no solution since $q = 2$ and $(\sqrt{n}/b_n)_{n=1}^{\infty} \notin s_1$.

Example 6.15 Let $p > 1$ and $r > 0$. The set $S = \mathbb{S}(bv_p, c)$ of all solutions of the (SSE) $bv_p + c_x = c_{(n^r)_{n=1}^{\infty}}$ is empty if $r < (p-1)/p$. If $r \geq (p-1)/p$, then it is determined by $\lim_{n\to\infty}(x_n/n^r) > 0$. For any given $r \neq 1$, we have $S \neq \varnothing$ if and only if $p \leq 1/(1-r)$.

Solvability of the (SSE) $(w_0)_\Delta + F_x = F_b$.

Here a positive sequence x is a solution of the (SSE) $(w_0)_\Delta + c_x = c_b$ if the next statement holds.

For every $y \in \omega$, we have $y_n/b_n \to l_1$ $(n \to \infty)$ if and only if there are sequences u and $v \in \omega$ with $y = u + v$ such that

$$\frac{1}{n}\sum_{k=1}^{n}|u_k - u_{k-1}| \to 0 \text{ and } \frac{v_n}{x_n} \to l_2 \ (n \to \infty)$$

for some scalars l_1 and l_2.

We obtain a similar statement for the (SSE) $(w_0)_\Delta + s_x = s_b$.
The next proposition holds.

Proposition 6.10 ([6, Proposition 7.3, p. 98]) *Let $b \in \mathcal{U}^+$. Then we have for $F = c$ or ℓ_∞*

$$\mathbb{S}((w_0)_\Delta, F) = \mathbb{S}((c_0)_\Delta, F) = \mathbb{S}(w_0, F) = \begin{cases} cl^F(b) & \text{if } 1/b \in s_{(1/n)_{n=1}^{\infty}} \\ \varnothing & \text{otherwise.} \end{cases}$$

Proof We have $\alpha \in M((w_0)_\Delta, \ell_\infty)$ if and only if

$$D_\alpha \Sigma \in (w_0, \ell_\infty). \tag{6.32}$$

Now we define the integer ν_n by

$$2^{\nu_n} \leq n \leq 2^{\nu_n+1} - 1. \tag{6.33}$$

Then, by the characterization of (w_0, ℓ_∞), the condition in (6.32) means there is $K > 0$ such that

$$\sigma_n = \sum_{\nu=0}^{\infty} 2^\nu \max_{2^\nu \le k \le 2^{\nu+1}-1} |(D_\alpha \Sigma)_{nk}| = |\alpha_n| \sum_{\nu=0}^{\nu_n} 2^\nu = \alpha_n| \left(2^{\nu_n+1} - 1\right) \le K \text{ for all } n. \tag{6.34}$$

Then we have from (6.33) $D_\alpha \Sigma \in (w_0, \ell_\infty)$ if and only if

$$n|\alpha_n| \le \left(2^{\nu_n+1} - 1\right) |\alpha_n| \le K \text{ for all } n \text{ and some } K > 0,$$

and we obtain $M((w_0)_\Delta, \ell_\infty) \subset s_{(1/n)_{n=1}^\infty}$.
Now we show $s_{(1/n)_{n=1}^\infty} \subset M((w_0)_\Delta, \ell_\infty)$.
Let $\alpha \in s_{(1/n)_{n=1}^\infty}$. Then we have $n|\alpha_n| \le K$ for all n, and it follows from by (6.34) and (6.33) that

$$\sigma_n = \left(2^{\nu_n+1} - 1\right) |\alpha_n| \le (2n-1)|\alpha_n| \le 2K \text{ for all } n.$$

This shows $s_{(1/n)_{n=1}^\infty} \subset M((w_0)_\Delta, \ell_\infty)$ and $M((w_0)_\Delta, \ell_\infty) = s_{(1/n)_{n=1}^\infty}$.
We obtain by similar arguments as above $M((w_0)_\Delta, c_0) = s_{(1/n)_{n=1}^\infty}$. Then we have

$$s_{(1/n)_{n=1}^\infty} = M\left((w_0)_\Delta, c_0\right) \subset M\left((w_0)_\Delta, c\right) \subset M\left((w_0)_\Delta, \ell_\infty\right) = s_{(1/n)_{n=1}^\infty}.$$

Finally, we have $M((w_0)_\Delta, F) = s_{(1/n)_{n=1}^\infty}$ for $F = c_0\, c$ or ℓ_∞. We conclude by Proposition 6.7 and Lemma 6.10.
This completes the proof. □

Example 6.16 The (SSE) $(w_0)_\Delta + c_x = c$ has no solution.

Example 6.17 The solutions of the (SSE) $(w_0)_\Delta + s_x = s_{(n)_{n=1}^\infty}$ are determined by $K_1 n \le x_n \le K_2 n$ for all n and some $K_1, K_2 > 0$.

Now we deal with the solvability of the (SSE) with an operator

$$\Lambda_\Delta + F_x = F_b.$$

Proposition 6.11 ([6, Proposition 7.4, p. 99]) *Let $b \in \mathcal{U}^+$. We have for $F = c$ or s_1*

$$\mathbb{S}(\mathbf{\Lambda}_\Delta, F) = \mathbb{S}(\mathbf{\Lambda}, F) = \begin{cases} cl^F(b) & \text{if } 1/b \in \mathbf{\Gamma} \\ \varnothing & \text{otherwise.} \end{cases}$$

Proof By Lemma 6.1, $\Delta \in (\mathbf{\Lambda},\mathbf{\Lambda})$ is bijective, since $e \in \widehat{C}_{\mathbf{\Lambda}}$. Indeed, we have $n \le K^n$ for all n and some $K > 1$, hence $\mathbf{\Lambda}_\Delta = \mathbf{\Lambda}$. Also it follows from Lemma 6.2 that $M(\mathbf{\Lambda}_\Delta, F) = M(\mathbf{\Lambda}, F) = \mathbf{\Gamma}$ for $F = c_0\; c$ or s_1 and we apply Lemma 6.10.
This concludes the proof. □

Now we solve the (SSE's)

$$E_\Sigma + F_x = F_b, \text{ where } E \in \{c, c_0, w_0, \mathbf{\Lambda}, \mathbf{\Gamma}, \ell_p\} \text{ for } p > 1, \text{ and } F \in \{c, \ell_\infty\},$$

$$\boldsymbol{\Gamma}_\Sigma + \boldsymbol{\Lambda}_x = \boldsymbol{\Lambda}_b \text{ and } (\ell_\infty)_\Sigma + c_x = c_b.$$

First we deal with the (SSE's)

$$\chi + F_x = F_b, \text{ where } \chi \in \{cs, bs, cs_0\} \text{ and } (\ell_p)_\Sigma + F_x = F_b, \text{ where } F \in \{c, \ell_\infty\}.$$

For instance, x is a solution of the (SSE) $cs + c_x = c_b$ if the next statement holds.

For every $y \in \omega$, we have $y_n/b_n \to l_1$ $(n \to \infty)$ if and only if there are sequences u and $v \in \omega$ with $y = u + v$ and the series $\sum_{k=1}^{\infty} u_k$ is convergent and $v_n/x_n \to l_2$ $(n \to \infty)$ for some scalars l_1 and l_2.

Proposition 6.12 ([6, Proposition 7.5, p. 99–100]) *Let $b \in \mathcal{U}^+$. Then we have*

$$\mathbb{S}(bs, c) = \mathbb{S}(\ell_\infty, c) = \begin{cases} cl^c(b) & \text{if } 1/b \in c_0 \\ \varnothing & \text{otherwise.} \end{cases} \tag{i}$$

For $F = c$ or ℓ_∞, we have

$$\mathbb{S}(cs, F) = \mathbb{S}(cs_0, F) = \mathbb{S}((\ell_p)_\Sigma, F) = \mathbb{S}(c_0, F)$$

with $p \geq 1$, and

$$\mathbb{S}(c_0, F) = \begin{cases} cl^F(b) & \text{if } 1/b \in s_1 \\ \varnothing & \text{otherwise.} \end{cases}$$

Proof (i) We have $\alpha \in M(bs, c)$ if and only if $D_\alpha \in (\ell_\infty(\Sigma), c)$ and $D_\alpha\Delta \in (\ell_\infty, c)$. The matrix $D_\alpha\Delta$ is the triangle with $(D_\alpha\Delta)_{nn} = -(D_\alpha\Delta)_{n,n-1} = \alpha_n$ for all n, with the convention $(D_\alpha\Delta)_{1,0} = 0$, the other entries being equal to zero. Trivially we have $\lim_{n\to\infty}(D_\alpha\Delta)_{nk} = 0$ for all k and $\lim_{n\to\infty}\sum_{k=1}^{\infty}|(D_\alpha\Delta)_{nk}| = 0$ which implies $M(bs, c) = c_0$. Since $bs = \ell_\infty(\Sigma) \subset \ell_\infty$, we conclude

$$c_0 = M(\ell_\infty, c_0) \subset M(bs, c_0) \subset M(bs, c) = c_0,$$

and Proposition 6.7 and Lemma 6.10 can be applied.
(ii) Case of $\mathbb{S}(cs_0, F)$.
Since $c_0 \subset c \subset \ell_\infty$ and $cs_0 = (c_0)_\Sigma \subset c_0$, we obtain

$$s_1 = M(c_0, c_0) \subset M(cs_0, c_0) \subset M(cs_0, c) \subset M(cs_0, \ell_\infty). \tag{6.35}$$

Now $\alpha \in M(cs_0, \ell_\infty)$ if and only if $D_\alpha \in (cs_0, \ell_\infty)$ and $D_\alpha\Delta \in (c_0, \ell_\infty)$. Since $(c_0, \ell_\infty) = S_1$, we have $|\alpha_n| + |\alpha_{n-1}| \leq K$ for all n and some $K > 0$ and $\alpha \in s_1$. So $M(cs_0, \ell_\infty) = s_1$. Using (6.35), we conclude $M(cs_0, F) = s_1$ for $F = c_0, c$ or ℓ_∞, and Proposition 6.7 and Lemma 6.10 can be applied.
Case of $\mathbb{S}(cs, F)$.

By similar arguments as those used above and noting that $cs = c_\Sigma$, we obtain

$$s_1 = M(cs, c_0) \subset M(cs, c) \subset M(cs, \ell_\infty) = s_1. \tag{6.36}$$

Case of $\mathbb{S}((\ell_p)_\Sigma, F)$.
Let $p > 1$. First $\alpha \in M((\ell_p)_\Sigma, \ell_\infty)$ implies $D_\alpha \Delta \in (\ell_p, \ell_\infty)$. By the characterization of (ℓ_p, ℓ_∞) in (5.1) of **5** in Theorem 1.23, we obtain $|\alpha_n|^q = O(1)$ $(n \to \infty)$ and $\alpha \in s_1$. This means $M((\ell_p)_\Sigma, \ell_\infty) \subset s_1$. We have $(\ell_p)_\Sigma \subset \ell_p$, since $\Delta \in (\ell_p, \ell_p)$ and

$$s_1 = M(\ell_p, c_0) \subset M\left((\ell_p)_\Sigma, c_0\right) \subset M\left((\ell_p)_\Sigma, c\right) \subset M\left((\ell_p)_\Sigma, \ell_\infty\right) \subset s_1.$$

So Proposition 6.7 and Lemma 6.10 can be applied.
In the case $p = 1$, using similar arguments as those above and by the characterizations of (ℓ_1, ℓ_∞) in (4.1) of **4** in Theorem 1.23 and (ℓ_1, c_0) in (4.1) and (7.1) of **9** in Theorem 1.23, we obtain $M((\ell_1)_\Sigma, F) = s_1$, where $F = c_0\ c$ or ℓ_∞.
This concludes the proof of Part (ii).

□

Now we solve the following (SSE's) with an operator

$$(w_0)_\Sigma + s_x = s_b \text{ and } (w_0)_\Sigma + c_x = c_b.$$

We note that x is a solution of the second (SSE) if

for every $y \in \omega$, we have $y_n/b_n \to l_1$ $(n \to \infty)$ if and only if there are sequences u and $v \in \omega$ such that $y = u + v$ and

$$\frac{1}{n}\sum_{k=1}^{n}\left|\sum_{i=1}^{k} u_i\right| \to 0 \textit{ and } \frac{v_n}{x_n} \to l_2\ (n \to \infty)$$

for some scalars l_1 and l_2.

First we state a lemma.

Lemma 6.11 *We have*

$$M((w_0)_\Sigma, \ell_\infty) = M((w_0)_\Sigma, c_0) = M((w_\infty)_\Sigma, \ell_\infty) = s_{(1/n)_{n=1}^\infty}.$$

Proof We have $M((w_0)_\Sigma, c_0) = M((w_0)_\Sigma, \ell_\infty)$. Indeed,

$$\alpha \in M\left((w_0)_\Sigma, c_0\right) \text{ if and only if } D_\alpha \Delta \in (w_0, c_0).$$

But, by Lemma 5.5, $D_\alpha \Delta \in (w_0, c_0)$ if and only if $D_\alpha \Delta \in (w_0, \ell_\infty)$. So $\alpha \in M((w_0)_\Sigma, c_0)$ if and only if $\alpha \in M((w_0)_\Sigma, \ell_\infty)$ and $M((w_0)_\Sigma, c_0) = M((w_0)_\Sigma, \ell_\infty)$.
Now we show $M((w_\infty)_\Sigma, \ell_\infty) = s_{(1/n)_{n=1}^\infty}$. For this let $\alpha \in M((w_\infty)_\Sigma, \ell_\infty)$. Then we

have $D_\alpha \Delta \in (w_\infty, \ell_\infty)$. Let ν_n for each non-negative integer n be the uniquely defined integer with $2^{\nu_n} \le n \le 2^{\nu_n+1} - 1$. Then we obtain

$$\sigma_n = \sum_{\nu=0}^{\infty} 2^\nu \max_{2^\nu \le k \le 2^{\nu+1}-1} |(D_\alpha \Delta)_{nk}| \ge |\alpha_n| 2^{\nu_n} \ge \frac{n+1}{2} |\alpha_n|.$$

But by Lemma 5.5, $D_\alpha \Delta \in (w_\infty, \ell_\infty)$ implies $\sigma \in \ell_\infty$ and $\alpha \in s_{(1/n)_{n=1}^\infty}$. So we have shown $M((w_\infty)_\Sigma, \ell_\infty) \subset s_{(1/n)_{n=1}^\infty}$.
Conversely, we show $s_{(1/n)_{n=1}^\infty} \subset M((w_\infty)_\Sigma, \ell_\infty)$. We have $w_\infty \subset s_{(1/n)_{n=1}^\infty}$. Since $\Delta \in (s_{(1/n)_{n=1}^\infty}, s_{(1/n)_{n=1}^\infty})$, we obtain $(s_{(1/n)_{n=1}^\infty})_\Sigma \subset s_{(1/n)_{n=1}^\infty}$ and

$$M\left((w_\infty)_\Sigma, \ell_\infty\right) \supset M\left(\left(s_{(1/n)_{n=1}^\infty}\right)_\Sigma, \ell_\infty\right) \supset M\left(s_{(1/n)_{n=1}^\infty}, \ell_\infty\right) = s_{(1/n)_{n=1}^\infty}.$$

We conclude $M((w_\infty)_\Sigma, \ell_\infty) = s_{(1/n)_{n=1}^\infty}$ and obtain $M((w_0)_\Sigma, c_0) = s_{(1/n)_{n=1}^\infty}$ using a similar arguments as those above.
This concludes the proof. □

Proposition 6.13 ([6, Proposition 7.6, p. 102–103]) *Let $b \in \mathcal{U}^+$ and $F = c$ or ℓ_∞. Then we have*

$$S\left((w_0)_\Sigma, F\right) = \mathbb{S}(w_0, F) = \begin{cases} cl^F(b) & \text{if } 1/b \in s_{(1/n)_{n=1}^\infty} \\ \varnothing & \text{otherwise.} \end{cases}$$

Proof First $(w_0)_\Sigma \subset w_0$ implies $M(w_0, c_0) \subset M((w_0)_\Sigma, c_0)$ and we obtain by Lemma 6.11

$$\begin{aligned} s_{(1/n)_{n=1}^\infty} &= M\left(w_0, c_0\right) \subset M\left((w_0)_\Sigma, c_0\right) \subset M\left((w_0)_\Sigma, c\right) \\ &\subset M\left((w_0)_\Sigma, \ell_\infty\right) = s_{(1/n)_{n=1}^\infty}. \end{aligned}$$

Then we have $M((w_0)_\Sigma, F) = s_{(1/n)_{n=1}^\infty}$ for $F = c_0, c$ or ℓ_∞, and Proposition 6.7 and Lemma 6.10 can be applied. □

Remark 6.2 From Propositions 6.8, 6.10 and 6.13, we have

$$\mathbb{S}(\chi, F) = \mathbb{S}(w_0, F) \text{ for } \chi \in \{(w_0)_\Sigma, (w_0)_\Delta, (c_0)_\Delta\}.$$

Finally, we deal with the (SSE's)

$$\mathbf{\Gamma}_\Sigma + F_x = F_b, \text{ where } F \in \{c, \ell_\infty, \mathbf{\Lambda}\},$$

and

$$\mathbf{\Lambda}_\Sigma + F_x = F_b \text{ for } \in \{c, \ell_\infty\}.$$

A positive sequence x is a solution of the (SSE) $\Gamma_\Sigma + c_x = c_b$ if the next statement holds

$\lim_{n\to\infty} y_n/b_n = l$ if and only if there are sequences u and $v \in \omega$ with $y = u + v$ such that $\lim_{n\to\infty} |\sum_{k=1}^n u_k|^{1/n} = 0$ and $\lim_{n\to\infty} v_n/x_n = l'$ for some scalars l and l' and for all $y \in \omega$.

We obtain the next result.

Theorem 6.7 ([6, Theorem 7.1, p. 102]) *Let $b \in \mathcal{U}^+$. Then*

(i) for $F = c$, ℓ_∞ or $\mathbf{\Lambda}$, we have

$$\mathbb{S}(\mathbf{\Gamma}_\Sigma, F) = \mathbb{S}(\mathbf{\Gamma}, F) = \begin{cases} cl^F(b) & \text{if } 1/b \in \mathbf{\Lambda} \\ \varnothing & \text{otherwise;} \end{cases}$$

(ii) for $F = c$ or ℓ_∞, we have

$$\mathbb{S}(\mathbf{\Lambda}_\Sigma, F) = \mathbb{S}(\mathbf{\Lambda}, F) = \begin{cases} cl^F(b) & \text{if } 1/b \in \mathbf{\Gamma} \\ \varnothing & \text{otherwise.} \end{cases}$$

Proof (i) First let $\alpha \in M(\mathbf{\Gamma}_\Sigma, \mathbf{\Lambda})$. Then $D_\alpha \Delta \in (\mathbf{\Gamma}, \mathbf{\Lambda})$, and we obtain by Lemma 4.14

$$\left[|\alpha_n|(M^{-n} + M^{-n+1})\right]^{1/n} \le K \text{ for all } n \text{ and some } K > 0 \text{ and } M \ge 2.$$

This implies

$$|\alpha_n|^{1/n} \le \frac{KM}{(1+M)^{1/n}} \le K' \text{ for all } n \text{ and some } K' > 0.$$

We conclude $M(\mathbf{\Gamma}_\Sigma, \mathbf{\Lambda}) \subset \mathbf{\Lambda}$. Then it can easily be seen that $\mathbf{\Gamma}_\Sigma \subset \mathbf{\Gamma}$, since $\Delta \in (\mathbf{\Gamma}, \mathbf{\Gamma})$. Then we have by Lemma 6.2

$$\mathbf{\Lambda} = M(\mathbf{\Gamma}, c_0) \subset M(\mathbf{\Gamma}_\Sigma, c_0) \subset M(\mathbf{\Gamma}_\Sigma, c) \subset M(\mathbf{\Gamma}_\Sigma, \ell_\infty) \subset M(\mathbf{\Gamma}_\Sigma, \mathbf{\Lambda}) \subset \mathbf{\Lambda}.$$

So we obtain $M(\mathbf{\Gamma}_\Sigma, F) = \mathbf{\Lambda}$ for $F = c_0$ c, ℓ_∞ or $\mathbf{\Lambda}$ and Proposition 6.7 and Lemma 6.10 can be applied.

This concludes the proof of Part (i).

(ii) As we have seen in Proposition 6.11, the operator $\Delta \in (\mathbf{\Lambda}, \mathbf{\Lambda})$ is bijective and this is also the case for $\Sigma \in (\mathbf{\Lambda}, \mathbf{\Lambda})$. Then we have $\mathbf{\Lambda}_\Sigma = \mathbf{\Lambda}$ and we conclude by Lemma 6.10 that Part (ii) holds.

□

Example 6.18 The solutions of the (SSE) $\mathbf{\Gamma}_\Sigma + \mathbf{\Lambda}_x = \mathbf{\Lambda}_u$ with $u > 0$ are determined by $k_1^n \le x_n \le k_2^n$ for all n and some $k_1, k_2 > 0$.

Example 6.19 Each of the (SSE) $\mathbf{\Lambda}_\Sigma + F_x = F_u$, where $F = c$ or s_1 has no solution for any given $u > 0$.

6.6 The Multiplier $M((E_a)_\Delta, F)$ and the (SSIE) $F_b \subset (E_a)_\Delta + F_x$

In this section, we determine the multiplier

$$M((E_a)_\Delta, F), \text{ where } E \in \{c_0, \ell_p\} \text{ and } F \in \{c_0, c, s_1\}.$$

Then we deal with the (SSIE)

$$F_b \subset (E_a)_\Delta + F_x,$$

where E and F are sequence spaces with $E \subset s_1$ and $c_0 \subset F \subset s_1$.

In the following we use the factorable matrix $D_\alpha \Sigma D_\beta$ with $\alpha, \beta \in \omega$ defined by $(D_\alpha \Sigma D_\beta)_{nk} = \alpha_n \beta_k$ for $k \leq n$ for all n, the other entries being equal to zero.

Lemma 6.12 ([7, Lemma 6, p. 115]) *Let $a \in \mathcal{U}^+$ and let $p > 1$. Then we have*

(i) The condition $a \notin cs$ implies

$$M\left(\left(s_a^0\right)_\Delta, F\right) = s_{\left(\frac{1}{\sum_{k=1}^n a_k}\right)_{n=1}^\infty} \text{ for } F = c_0, c \text{ or } s_1. \tag{6.37}$$

(ii) The condition $a^q \notin cs$ implies with $q = p/(p-1)$

$$M\left(\left(\ell_a^p\right)_\Delta, F\right) = s_{\left(\left(\sum_{k=1}^n a_k^q\right)^{-1/q}\right)_{n=1}^\infty} \text{ for } F = c_0, c \text{ or } s_1. \tag{6.38}$$

Proof (i) We have $\alpha \in M((s_a^0)_\Delta, c_0)$ if and only if $D_\alpha \Sigma D_a \in (c_0, c_0)$. It follows from the characterization of (c_0, c_0) by (1.1) and (7.1) in **7** of Theorem 1.23 that

$$|\alpha_n| \sum_{k=1}^n a_k \leq K \text{ for all } n \text{ and some } K > 0 \tag{6.39}$$

and

$$\alpha \in c_0. \tag{6.40}$$

But since $a \notin cs$, the condition in (6.39) implies (6.40), and so $\alpha \in M((s_a^0)_\Delta, c_0)$ if and only if (6.39) holds. This shows the identity in (6.37) for $F = c_0$.
In a similar way, the identity (6.37) for $F = s_1$ can easily be shown. From the inclusions $M((s_a^0)_\Delta, c_0) \subset M((s_a^0)_\Delta, c) \subset M((s_a^0)_\Delta, s_1)$, we conclude that the identity in (6.37) holds for $F = c$.
(ii) We have $\alpha \in M((\ell_a^p)_\Delta, c_0)$ if and only if $D_\alpha \Sigma D_a \in (\ell^p, c_0)$. We have by the characterization of (ℓ^p, c_0) in (5.1) and (7.1) of **10** in Theorem 1.23

$$|\alpha_n|^q \sum_{k=1}^n a_k^q \leq K \text{ for all } n \text{ and some } K > 0 \tag{6.41}$$

and (6.40) holds. But since $a^q \notin cs$, the condition in (6.41) implies (6.40), and we so $\alpha \in M((\ell_a^p)_\Delta, c_0)$ if and only if (6.41) holds. So we have shown that the identity in (6.38) holds for $F = c_0$.
In a similar way the identity in (6.38) with $F = s_1$ can easily be shown. We conclude the proof using the inclusions $M((\ell_a^p)_\Delta, c_0) \subset M((\ell_a^p)_\Delta, c) \subset M((\ell_a^p)_\Delta, s_1)$. □

Now we study some properties of the (SSIE)

$$F_b \subset (E_a)_\Delta + F'_x.$$

Let E and F be two linear subspaces of ω. We write $\mathcal{I}((E_a)_\Delta, F, F')$ for the set of all $x \in \mathcal{U}^+$ such that $F_b \subset (E_a)_\Delta + F'_x$. It can easily be seen that the sets $(E_a)_\Delta$ and F_x are linear spaces of sequences, and we have $z \in (E_a)_\Delta + F'_x$ if and only if there are $\xi \in E$ and $f' \in F'$ such that $z_n = \sum_{k=1}^n a_k \xi_k + f'_n x_n$. To simplify we write

$$\mathcal{I}_E^{F,F'} = \mathcal{I}\left((E_a)_\Delta, F, F'\right).$$

We need the next lemma.

Lemma 6.13 ([8]) *Let $a, b \in \mathcal{U}^+$. Then we have $(\chi_a)_\Delta + (\chi_b)_\Delta = (\chi_{a+b})_\Delta$ for $\chi = s_1$ or c_0.*

Proof Since the inclusion $(\chi_{a+b})_\Delta \subset (\chi_a)_\Delta + (\chi_b)_\Delta$ is trivial, it is enough to show $(\chi_a)_\Delta + (\chi_b)_\Delta \subset (\chi_{a+b})_\Delta$. For this, let $y \in (\chi_a)_\Delta + (s_b)_\Delta$. Since $(\chi_\alpha)_\Delta = (\Sigma D_\alpha)\chi$ with $\alpha \in \mathcal{U}^+$, there are sequences u and $v \in \chi$ such that

$$y_n = \sum_{k=1}^n a_k u_k + \sum_{k=1}^n b_k v_k = \sum_{k=1}^n (a_k + b_k) z_k = (\Sigma D_a + \Sigma D_b)_n\, z,$$

where $z_k = (a_k u_k + b_k v_k)/(a_k + b_k)$ for all k. Since $0 < a_k/(a_k + b_k) < 1$ and $0 < b_k/(a_k + b_k) < 1$, we have $|z_k| \leq |u_k| + |v_k|$ for all k, $(|u_k| + |v_k|)_{k=1}^\infty \in \ell_\infty$ for $\chi = s_1$ and $(|u_k| + |v_k|)_{k=1}^\infty \in c_0$ for $\chi = c_0$. This shows $y \in (\Sigma D_{a+b})\chi = (\chi_{a+b})_\Delta$ and $(\chi_a)_\Delta + (\chi_b)_\Delta \subset (\chi_{a+b})_\Delta$.
This completes the proof. □

Remark 6.3 As a direct consequence of the preceding lemma we have $\Sigma D_a \chi + \Sigma D_b \chi = (\Sigma D_{a+b})\chi$ for $\chi = s_1$ or c_0.

In the following, we use the sequence $\sigma = (\sigma_n)_{n=1}^\infty$ defined for $a, b \in \mathcal{U}^+$ by

$$\sigma_n = \frac{1}{b_n} \sum_{k=1}^n a_k \text{ for all } n.$$

We recall that, for any given $b \in \mathcal{U}^+$, $\overline{s}_b = \overline{s_{1/b}}$ is the set of all sequences x such that $x_n \geq Kb_n$ for all n and some $K > 0$. We note that $s_b \cap \overline{s_{1/b}} = cl^\infty(b)$. First we state the next lemma (cf. [8]).

Lemma 6.14 *Let $a, b \in \mathcal{U}^+$ and E, F and F' be linear spaces of sequences that satisfy $F \supset c_0$ and $E, F' \subset \ell_\infty$. Then we have*

(i) *If $\sigma \in c_0^+$, then $I_E^{FF'} \subset \overline{s_{1/b}}$.*
(ii) *If $a \in c_0^+$ and $b = e$, then $I_E^{FF'} \cap c \subset cl^c(e)$.*

Proof (i) Let $x \in I_E^{FF'}$. Then we have $F_b \subset (E_a)_\Delta + F'_x$. Since $E, F' \subset s_1$, we obtain

$$(E_a)_\Delta + F'_x = (\Sigma D_a)E + D_x F' \subset (\Sigma D_a)s_1 + D_x s_1.$$

Now we let $\tau = (\sum_{k=1}^n a_k)_{n=1}^\infty$. We obtain by elementary calculations

$$D_{1/\tau}\Sigma D_a \in (s_1, s_1)$$

and

$$(\Sigma D_a)s_1 \subset s_\tau.$$

Then we have

$$s_b^0 \subset (s_a)_\Delta + s_x \subset s_{\tau+x}$$

and

$$\frac{b}{\tau + x} \in M(c_0, s_1) = s_1.$$

So there is $K > 0$ such that

$$\sum_{k=1}^n a_k + x_n \geq Kb_n \text{ for all } n.$$

Hence we have

$$x_n \geq b_n(K - \sigma_n) \text{ for all } n$$

and since $\sigma \in c_0$, there is $K' > 0$ such that $x_n \geq K'b_n$ for all n and consequently $x \in \overline{s_{1/b}}$.
(ii) First we show $s_x \subset (s_{x+x^-})_\Delta$. This inclusion is equivalent to $D_{1/(x+x^-)}\Delta D_x \in (s_1, s_1)$, where

$$\left[D_{1/(x+x^-)}\Delta D_x\right]_{nn} = \frac{x_n}{x_{n-1} + x_n}$$

and

$$\left[D_{1/(x+x^-)}\Delta D_x\right]_{n,n-1} = -\frac{x_{n-1}}{x_{n-1} + x_n} \text{ for all } n,$$

the other entries being naught. Now let $x \in \mathcal{I}_E^{FF'} \cap c$ with $b = e$. Then we have $x \in c$ and $c_0 \subset (s_a)_\Delta + s_x$. This inclusion implies

$$c_0 \subset (\Sigma D_a)s_1 + (\Sigma D_{x+x^-})s_1$$

and we obtain by Lemma 6.13

$$(\Sigma D_a)s_1 + (\Sigma D_{x+x^-})s_1 = (\Sigma D_{a+x+x^-})s_1 = (s_{a+x+x^-})_\Delta.$$

We deduce

$$c_0 \subset (s_{a+x+x^-})_\Delta.$$

So there is $K > 0$ such that

$$\frac{1}{a_n + x_n + x_{n-1}} \leq K$$

and

$$x_n + x_{n-1} \geq \frac{1}{K} - a_n \text{ for all } n.$$

Since $a \in c_0$, there is $M > 0$ such that

$$x_n + x_{n-1} \geq M \text{ for all } n. \tag{6.42}$$

Then $x \in c$ implies

$$\lim_{n\to\infty}(x_n + x_{n-1}) = 2\lim_{n\to\infty} x_n \geq M$$

and $\lim_{n\to\infty} x_n > 0$, which implies $s_x^{(c)} = c$. So we have shown $\mathcal{I}_E^{FF'} \cap c \subset cl^c(e)$. This concludes the proof.

□

6.7 The (SSE) $(E_a)_\Delta + s_x^{(c)} = s_b^{(c)}$

Here we solve the (SSE)

$$(E_a)_\Delta + s_x^{(c)} = s_b^{(c)} \text{ where } E \in \{c_0, \ell_p\} \text{ for } p > 1.$$

For instance, the (SSE)

$$(s_a^0)_\Delta + s_x^{(c)} = s_b^{(c)}$$

is equivalent to the statement

$$\frac{y_n}{b_n} \to l_1 \ (n \to \infty)$$

if and only if there are two sequences u and v with $y = u + v$ *such that*

$$\frac{\Delta_n u}{a_n} \to 0 \text{ and } \frac{v_n}{x_n} \to l_2 \ (n \to \infty)$$

for all sequences y and for some scalars l_1 *and* l_2.

For any given $a, b \in \mathcal{U}^+$ we write $\mathbb{S}((E_a)_\Delta, F)$ for the set of all solutions of the (SSE) $(E_a)_\Delta + F_x = F_b$, where E and F are linear spaces. These results extend some results stated in Propositions 6.8 and 6.9, since the (SSE) $(s_a^0)_\Delta + s_x^{(c)} = s_b^{(c)}$ reduces to $(c_0)_\Delta + s_x^{(c)} = s_b^{(c)}$, and the (SSE) $(\ell_a^p)_\Delta + s_x^{(c)} = s_b^{(c)}$ reduces to $(\ell^p)_\Delta + s_x^{(c)} = s_b^{(c)}$ for $a = e$. We have the following.

Theorem 6.8 ([7, Theorem 1, pp. 117–118]) *Let* $a, b \in \mathcal{U}^+$. *Then we have*

(i) *The set* $\mathbb{S}((s_a^0)_\Delta, c)$ *of all solutions of the (SSE)* $(s_a^0)_\Delta + s_x^{(c)} = s_b^{(c)}$ *is determined in the following way:*

(a) *If* $a \notin cs$, *(that is,* $\sum_k a_k = \infty$*), then we have*

$$\mathbb{S}((s_a^0)_\Delta, c) = \begin{cases} cl^c(b) & \text{if } \sigma \in s_1 \\ \varnothing & \text{if } \sigma \notin s_1. \end{cases}$$

(b) *If* $a \in cs$, *then we have*

$$\mathbb{S}((s_a^0)_\Delta, c) = \begin{cases} cl^c(b) & \text{if } \frac{1}{b} \in c_0 \\ cl^c(e) & \text{if } \frac{1}{b} \in c \setminus c_0 \\ \varnothing & \text{if } \frac{1}{b} \notin c. \end{cases} \tag{6.43}$$

(ii) *The set* $\mathbb{S}((\ell_a^p)_\Delta, c)$ *with* $p > 1$ *of all solutions of the (SSE)* $(\ell_a^p)_\Delta + s_x^{(c)} = s_b^{(c)}$ *is determined in the following way:*

(a) *If* $a^q \notin cs$, *then*

$$\mathbb{S}((\ell_a^p)_\Delta, c) = \begin{cases} cl^c(b) & \text{if } \left(\frac{a_1^q + \cdots + a_n^q}{b_n^q}\right)_{n=1}^\infty \in s_1 \\ \varnothing & \text{if } \left(\frac{a_1^q + \cdots + a_n^q}{b_n^q}\right)_{n=1}^\infty \notin s_1. \end{cases}$$

(b) *If* $a^q \in cs$, *then* $\mathbb{S}((\ell_a^p)_\Delta, c) = \mathbb{S}((s_a^0)_\Delta, c)$ *is defined by (6.43).*

Proof (i)

(a) First consider the case $a \notin cs$.
By Lemma 6.12, we have $M((s_a^0)_\Delta, c) = M((s_a^0)_\Delta, c_0)$ and we can apply Proposition 6.7, where $1/b \in M((s_a^0)_\Delta, c)$ if and only if $\sigma \in s_1$.
(a) *Case* $a \in cs$.
We deal with the three cases

$$(\alpha)1/b \notin c, \quad (\beta)1/b \in c_0 \quad \text{and} \quad (\gamma)1/b \in c \setminus c_0.$$

(α) We have $\mathbb{S}((s_a^0)_\Delta, c) = \varnothing$. Indeed, if we assume there is $x \in \mathbb{S}((s_a^0)_\Delta, c)$, then $(s_a^0)_\Delta \subset s_b^{(c)}$ and $D_{1/b}\Sigma D_a \in (c_0, c)$. From the characterization of (c_0, c) in (1.1) and (11.2) of **12** in Theorem 1.23, we deduce $1/b \in c$, which is a contradiction. We conclude $\mathbb{S}((s_a^0)_\Delta, c) = \varnothing$.
(β) Let $1/b \in c_0$. Then $x \in \mathbb{S}((s_a^0)_\Delta, c)$ implies

$$x \in s_b^{(c)} \tag{6.44}$$

and $s_b^{(c)} \subset (s_a^0)_\Delta + s_x^{(c)}$. Since $b \in s_b^{(c)}$, there are sequences $\varepsilon \in c_0$ and $\varphi \in c$ such that

$$\frac{b_n}{x_n}\left(1 - \frac{1}{b_n}\sum_{k=1}^{n} a_k \varepsilon_k\right) = \varphi_n \text{ for all } n.$$

We deduce $b/x \in c$, since $\sigma \in c_0$. Using the condition in (6.44) we conclude that $x \in \mathbb{S}((s_a^0)_\Delta, c)$ implies $s_x^{(c)} = s_b^{(c)}$.
Conversely, we assume $s_x^{(c)} = s_b^{(c)}$. Then we have

$$(s_a^0)_\Delta + s_x^{(c)} = (s_a^0)_\Delta + s_b^{(c)} = s_b^{(c)},$$

since $\sigma \in s_1$ and $1/b \in c$. We conclude $\mathbb{S}((s_a^0)_\Delta, c) = cl^c(b)$.
(γ) Here we have $\lim_{n\to\infty} b_n = L > 0$ and $s_b^{(c)} = c$ and we are led to study the (SSE)

$$(s_a^0)_\Delta + s_x^{(c)} = c.$$

We have that $x \in \mathbb{S}((s_a^0)_\Delta, c)$ implies $x \in c$. Then, by Part (ii) of Lemma 6.14 with $E = c_0$ and $F = F' = c$, we obtain $\mathbb{S}((s_a^0)_\Delta, c) \subset cl^c(e)$.
Conversely, $x \in cl^c(e)$ implies $s_x^{(c)} = c$, and since $a \in cs$, we have $(s_a^0)_\Delta \subset c$ and $(s_a^0)_\Delta + s_x^{(c)} = c$. We conclude $\mathbb{S}((s_a^0)_\Delta, c) = cl^c(e)$.

This completes the proof of Part (i).
(ii)

(a) *Case* $a^q \notin cs$.
By Lemma 6.12, we have $M((\ell_a^p)_\Delta, c) = M((\ell_a^p)_\Delta, c_0)$. Then we can apply Proposition 6.7, where $1/b \in M((\ell_a^p)_\Delta, c)$ if and only if

$$\left(\frac{\sum_{k=1}^{n} a_k^q}{b_n^q}\right)_{n=1}^{\infty} \in s_1.$$

(b) *Case* $a^q \in cs$.
As above we deal with the three

$$(\alpha) 1/b \notin c, \quad (\beta) 1/b \in c_0 \quad \text{and} \quad (\gamma) 1/b \in c \setminus c_0.$$

(α) We have that $x \in \mathbb{S}((\ell_a^p)_\Delta, c)$ implies $(\ell_a^p)_\Delta \subset s_b^{(c)}$ and $D_{1/b}\Sigma D_a \in (\ell^p, c)$. From the characterization of (ℓ^p, c) in (5.1) and (11.2) of **15** in Theorem 1.23, we deduce $1/b \in c$. We conclude that if $1/b \notin c$, then $\mathbb{S}((\ell_a^p)_\Delta, c) = \varnothing$.
(β) We have that $x \in \mathbb{S}((\ell_a^p)_\Delta, c)$ implies

$$x \in s_b^{(c)} \tag{6.45}$$

and

$$s_b^{(c)} \subset \left(\ell_a^p\right)_\Delta + s_x^{(c)}. \tag{6.46}$$

Since $b \in s_b^{(c)}$, there are sequences $\lambda \in \ell_p$ and $\varphi \in c$ such that

$$\frac{b_n}{x_n}\left(1 - \frac{1}{b_n}\sum_{k=1}^{n} a_k \lambda_k\right) = \varphi_n \text{ for all } n.$$

From the characterization of (ℓ_p, c_0) in (5.1) and (7.1) of **10** in Theorem 1.23, we have $D_{1/b}\Sigma D_a \in (\ell_p, c_0)$, since $1/b \in c_0$ and $a^q \in cs$ together imply

$$\left(\frac{a_1^q + \cdots + a_n^q}{b_n^q}\right)_{n=1}^{\infty} \in s_1.$$

We deduce

$$\left(D_{1/b}\Sigma D_a\right)_n \lambda = \frac{1}{b_n}\sum_{k=1}^{n} a_k \lambda_k \to 0 \ (n \to \infty),$$

and $b/x \in c$. Using the condition in (6.45), we conclude that $x \in \mathbb{S}((\ell_a^p)_\Delta, c)$ implies $s_x^{(c)} = s_b^{(c)}$.
Conversely, we assume $s_x^{(c)} = s_b^{(c)}$. Since $1/b \in c_0$ and $a^q \in cs$ together imply $D_{1/b}\Sigma D_a \in (\ell_p, c)$, **10** in Theorem 1.23, we successively obtain $(\ell_a^p)_\Delta \subset s_b^{(c)}$,

$$(\ell_a^p)_\Delta + s_x^{(c)} = (\ell_a^p)_\Delta + s_b^{(c)} = s_b^{(c)}$$

and $x \in \mathbb{S}((\ell_a^p)_\Delta, c)$. We conclude $\mathbb{S}((\ell_a^p)_\Delta, c) = cl^c(b)$.
(γ) Here we have $s_b^{(c)} = c$ and we are led to study the (SSE)

$$(\ell_a^p)_\Delta + s_x^{(c)} = c.$$

We have that $x \in \mathbb{S}((\ell_a^p)_\Delta, c)$ implies $x \in c$, that is, $x_n \to l$ $(n \to \infty)$. Then by Part (ii) of Lemma 6.14 with $E = \ell^p$ and $F = F' = c$, we obtain $\mathbb{S}((\ell_a^p)_\Delta, c) \subset cl^c(e)$. Since $a^q \in cs$, we have $\Sigma D_a \in (\ell^p, c)$ and $(\ell_a^p)_\Delta \subset c$. This implies $cl^c(e) \subset \mathcal{S}((\ell_a^p)_\Delta, c)$. We conclude $\mathbb{S}((\ell_a^p)_\Delta, c) = cl^c(e)$.

This completes the proof. □

Now we study the equation

$$s_x^{(c)} = s_b^{(c)}$$

and the perturbed equation

$$(s_a^0)_\Delta + s_x^{(c)} = s_b^{(c)}.$$

In view of perturbed equations we state the following. Let b be a positive sequence. Then the equation

$$s_x^{(c)} = s_b^{(c)} \tag{6.47}$$

is equivalent to $\lim_{n\to\infty} x_n/b_n = l$ $(n \to \infty)$ for some $l > 0$. Then the (SSE)

$$(s_a^0)_\Delta + s_x^{(c)} = s_b^{(c)} \tag{6.48}$$

can be considered as a perturbed equation of (6.47), and the question is what are the conditions on a for which the perturbed equation and the (SSE) defined by (6.47) have the same solutions. As a direct consequence of Theorem 6.8 we obtain the next corollary.

Corollary 6.6 *Let $a, b \in \mathcal{U}^+$. Then we have*

(i) If $1/b \in c$, then the equations in (6.47) and (6.48) are equivalent if and only if $a \in cs \cup (\omega \setminus cs \cap (s_b)_\Sigma)$.
(ii) If $1/b \notin c$, then the perturbed equation in (6.48) has no solutions.

Proof (i) is an immediate consequence of Theorem 6.8.
(ii) Let $a \notin cs$. The condition $\sigma \in s_1$ should imply $1/b_n \le K(\sum_{k=1}^n a_k)^{-1}$ for all n and some $K > 0$ and $1/b \in c_0$, which is contradictory. So the perturbed equation in (6.48) has no solutions. The case $a \in cs$ is a direct consequence of Part (i) (b) of Theorem 6.8. □

Remark 6.4 We may state a similar result for the perturbed equation $(\ell_a^p)_\Delta + s_x^{(c)} = s_b^{(c)}$.

Now we state the next elementary results, where $b, b^q \in \widehat{C_1}$.

Corollary 6.7 ([7, Corollary 4, p. 121]) *Let $a, b \in \mathcal{U}^+$. Then we have*

(i) *Let $b \in \widehat{C_1}$. Then the set $\mathbb{S}((s_a^0)_\Delta, c)$ of all positive $x \in \mathcal{U}^+$ such that $(s_a^0)_\Delta + s_x^{(c)} = s_b^{(c)}$ is determined as follows.*

(a) *Let $a \notin cs$. Then we have*

$$\mathbb{S}((s_a^0)_\Delta, c) = \begin{cases} cl^c(b) & \text{if } a/b \in s_1 \\ \varnothing & \text{if } a/b \notin s_1. \end{cases} \tag{6.49}$$

(b) *Let $a \in cs$. Then we have $\mathbb{S}((s_a^0)_\Delta, c) = cl^c(b)$.*

(ii) *Let $p > 1$ and $b^q \in \widehat{C_1}$ with $q = p/(p-1)$. Then the set $\mathbb{S}((\ell_a^p)_\Delta, c)$ of all $x \in \mathcal{U}^+$ such that $(\ell_a^p)_\Delta + s_x^{(c)} = s_b^{(c)}$ is determined in the following way:*

(a) *Let $a^q \notin cs$. Then $\mathbb{S}((\ell_a^p)_\Delta, c) = \mathbb{S}((s_a^0)_\Delta, c)$ defined by (6.49).*
(b) *Let $a^q \in cs$. Then $\mathbb{S}((\ell_a^p)_\Delta, c) = \mathbb{S}((s_a^0)_\Delta, c) = cl^c(b)$.*

Proof (i)

(a) We have $\sigma \in s_1$ if and only if $a \in (s_b)_\Sigma$. But as we saw in Proposition 6.1, we have $b \in \widehat{C_1}$ if and only if $(s_b)_\Delta = s_b$. This implies that $\Delta \in (s_b, s_b)$ is bijective and so is for $\Sigma = \Delta^{-1}$. So we have $(s_b)_\Sigma = s_b$. We have $\sigma \in s_1$ if and only if $a/b \in s_1$, and we conclude by Theorem 6.8. This completes the proof of Part (i) (a).
(b) follows from the fact that $b \in \widehat{C_1}$ implies $1/b \in c_0$.

(ii)

(a) Here we have

$$\left(\frac{a_1^q + \cdots + a_n^q}{b_n^q}\right)_{n=1}^\infty \in s_1 \text{ if and only if } a^q \in (s_{b^q})_\Sigma,$$

and as we have just seen, we have $(s_{b^q})_\Sigma = s_{b^q}$, since $b^q \in \widehat{C_1}$. So we obtain Part (a).
(b) The condition $b^q \in \widehat{C_1}$ implies that there are $C > 0$ and $\gamma > 1$ such that $b_n^q \geq C\gamma^n$ for all n. So we have $b_n \geq C^{1/q}\gamma^{n/q}$ for all n, and $1/b \in c_0$. We conclude by Theorem 6.8.

This completes the proof. □

Remark 6.5 We note that we have for $b \in \widehat{C_1}$

$$\mathbb{S}((s_a^0)_\Delta, c) \neq \varnothing \text{ if and only if } a \in (cs \cup (\omega \setminus cs \cap s_b)) \cap \mathcal{U}^+.$$

Example 6.20 We consider the (SSE) with operator defined by

$$\left(s^0_{(n^{-\alpha})_{n=1}^\infty}\right)_\Delta + s^{(c)}_x = s^{(c)}_b, \tag{6.50}$$

with $0 < \alpha \le 1$ and $b \in \widehat{C_1}$. We have $a/b = (n^{-\alpha}/b_n)_n$. The condition $b \in \widehat{C_1}$ implies that there are $K > 0$ and $\gamma > 1$ such that $b_n \ge K\gamma^n$ for all n. This implies $a/b \in c_0$. We apply Corollary 6.7 and conclude that the solutions of the (SSE) in (6.50) satisfy the condition $x_n \sim Cb_n$ $(n \to \infty)$ for some $C > 0$.

Example 6.21 Let $b^q \in \widehat{C_1}$. It can easily be shown that the solutions of the (SSE) $(\ell^p_{(n^\alpha)_{n=1}^\infty})_\Delta + s^{(c)}_x = s^{(c)}_b$ are defined by $x_n \sim Cb_n$ $(n \to \infty)$ for some $C > 0$ and all reals α.

Remark 6.6 We note that if $a \in \widehat{C_1}$, the set $\mathbb{S}(s^0_a, c)$ is determined by Part (i) of Corollary 6.7. Indeed, by Part (ii) Theorem 4.2, the condition $a \in \widehat{C_1}$ implies $(s^0_a)_\Delta = s^0_a$, and we conclude from the solvability of the (SSE) $s^0_a + s^{(c)}_x = s^{(c)}_b$ given in Proposition 6.3.

Remark 6.7 If $\overline{\lim}_{n\to\infty}(a_{n-1}/a_n) < 1$, then we have $(\ell^p_a)_\Delta = \ell^p_a$, (cf. [2, Theorem 6.5 p. 3200]). So we obtain $\mathbb{S}((\ell^p_a)_\Delta, c) = cl^c(b)$ if $a/b \in s_1$, and $\mathbb{S}((\ell^p_a)_\Delta, c) = \varnothing$ if $a/b \notin s_1$.

We obtain the next corollary the proof of which is elementary and left to the reader.

Corollary 6.8 ([7, Corollary 5, pp. 122–123]) *Let $R, \overline{R} > 0$, and denote by $\mathbb{S}_{R,\overline{R}}$ the set of all positive sequences x that satisfy the (SSE) $(s^0_R)_\Delta + s^{(c)}_x = s^{(c)}_{\overline{R}}$. Then we obtain*

(i) Case $R < 1$.
We have

$$\mathbb{S}_{R,\overline{R}} = \begin{cases} cl^c\left(\overline{R}\right) & \text{if } \overline{R} \ge 1 \\ \varnothing & \text{if } \overline{R} < 1. \end{cases}$$

(ii) Case $R = 1$.
We have

$$\mathbb{S}_{R,\overline{R}} = \begin{cases} cl^c\left(\overline{R}\right) & \text{if } \overline{R} > 1 \\ \varnothing & \text{if } \overline{R} \le 1. \end{cases}$$

(iii) Case $R > 1$.
We have

$$\mathbb{S}_{R,\overline{R}} = \begin{cases} cl^c\left(\overline{R}\right) & \text{if } R \le \overline{R} \\ \varnothing & \text{if } R > \overline{R}. \end{cases}$$

As a direct consequence of the preceding we can state the next remark.

Remark 6.8 Let $R, \overline{R} > 0$. Then $\mathbb{S}_{R,\overline{R}} \neq \varnothing$ if and only if $R = 1 < \overline{R}$ or $1 < R \leq \overline{R}$ or $R < 1 \leq \overline{R}$. For instance, the set of all positive sequences that satisfy the (SSE) $(s_R^0)_\Delta + s_x^{(c)} = s_2^{(c)}$ is nonempty if and only if $R \leq 2$.

Now we consider the next statement:

the condition $n^\alpha y_n \to l_1$ $(n \to \infty)$ holds if and only if there are two sequences u and v with $y = u + v$ such that

$$r^n(u_n - u_{n-1}) \to 0 \text{ and } x_n v_n \to l_2 \ (n \to \infty)$$

for some scalars l_1 and l_2 and for all $y \in \omega$.

The set of all x that satisfy the previous statement is equivalent to the (SSE)

$$\left(s_{1/r}^0\right)_\Delta + s_{1/x}^{(c)} = s_{(1/n^\alpha)_{n=1}^\infty}^{(c)}. \tag{6.51}$$

6.8 More Applications

In this section, we study some more applications of the results of the previous section.

We obtain the following corollary.

Corollary 6.9 ([7, Corollary 6, pp. 123–124]) *Let $r > 0$, α be a real and $\widetilde{\mathbb{S}}_{r,\alpha}$ denote the set of all positive sequences x that satisfy the (SSE) defined by (6.51). Then we obtain*

(i) If $r < 1$, then $\widetilde{\mathbb{S}}_{r,\alpha} = \varnothing$.
(ii) If $r = 1$, then

$$\widetilde{\mathbb{S}}_{r,\alpha} = \begin{cases} cl^c\left((n^\alpha)_{n=1}^\infty\right) & \text{if } \alpha \leq -1 \\ \varnothing & \text{if } \alpha > -1. \end{cases}$$

(iii) If $r > 1$, then

$$\widetilde{\mathbb{S}}_{r,\alpha} = \begin{cases} cl^c\left((n^\alpha)_{n=1}^\infty\right) & \text{if } \alpha \leq 0 \\ \varnothing & \text{if } \alpha > 0. \end{cases}$$

Proof We note that $r < 1$ implies $a = (r^{-n})_{n=1}^\infty \notin cs$. So the statement in Part (i) follows from the equivalence $\sigma_n \sim (1 - r^{-1}n^\alpha r^{-n})$ $(n \to \infty)$ and $\sigma \notin s_1$ for $r < 1$. Let $r = 1$. Then we have $\sigma_n \sim n^{\alpha+1}$ $(n \to \infty)$ and $\sigma \in s_1$ if and only if $\alpha \leq -1$, and we conclude by Theorem 6.8.

This shows Part (ii)

Finally, for $r > 1$, we have $a \in cs$ and $1/b = (n^\alpha)_{n=1}^\infty \in c$ if and only if $\alpha \leq 0$, and we conclude by Theorem 6.8.

This completes the proof. □

We immediately deduce the next remark.

Remark 6.9 We have $\widetilde{\mathbb{S}}_{r,\alpha} \neq \emptyset$ if and only if $r = 1 \leq \alpha$ or $r > 1$ and $\alpha \leq 0$. We also have $\widetilde{\mathbb{S}}_{r,0} \neq \emptyset$ if and only if $r > 1$.

Example 6.22 We consider the statement

> $y_n/n \to l_1$ $(n \to \infty)$ *holds if and only if there are two sequences u and v, with* $y = u + v$ *such that* $u_n - u_{n-1} \to 0$ *and* $x_n v_n \to l_2$ $(n \to \infty)$ *for some scalars* l_1 *and* l_2 *and for all* $y \in \omega$.

This statement holds if and only if $x \in \widetilde{\mathbb{S}}_{1,-1}$, that is, $x_n/n \to L$ $(n \to \infty)$ with $L > 0$.

For the (SSE) $(s^0_{(1/n^\alpha)})_\Delta + s^{(c)}_{1/x} = s^{(c)}_{1/r}$, we obtain the next result by Theorem 6.8.

Corollary 6.10 ([7, Corollary 7, p. 124]) *Let* $r > 0$, α *be a real and* $\mathbb{S}_{\alpha,r}$ *be the set of all* $x \in \mathcal{U}^+$ *such that* $(s^0_{(1/n^\alpha)_{n=1}^\infty})_\Delta + s^{(c)}_{1/x} = s^{(c)}_{1/r}$. *Then the next statements are equivalent.*

(i) $\mathbb{S}_{\alpha,r} \neq \emptyset$;
(ii) $\mathbb{S}_{\alpha,r} = cl^c\left((r^n)_{n=1}^\infty\right)$;
(iii) $r \leq 1 < \alpha$ *or* $\alpha \leq 1$ *and* $r < 1$.

Proof This result is a direct consequence of the equivalences

$$\int_n = \sum_{k=1}^n k^{-\alpha} \sim \frac{n^{1-\alpha}}{1-\alpha} \ (n \to \infty) \text{ if } \alpha \neq 1; \text{ and } \int_n \sim \log n \ (n \to \infty) \text{ if } \alpha = 1.$$

Then if $\alpha \neq 1$, we have $(r^n n^{1-\alpha})_{n=1}^\infty \in \ell_\infty$ if and only if $r \leq 1 < \alpha$ or α and $r < 1$. If $\alpha = 1$, we have $(r^n \log n)_{n=1}^\infty \in \ell_\infty$ if and only if $r < 1$. This concludes the proof. □

Example 6.23 For $r = 1/2$, we have $\mathbb{S}_{\alpha,1/2} = cl^c((2^{-n})_{n=1}^\infty)$ for all reals α.

Now let $\mathbb{S}_{\alpha,\beta}$ for all reals α and β be the set of all positive sequences $x = (x_n)_{n=1}^\infty$ that satisfy the following statement.

> *For every y, the condition* $n^\beta y_n \to l_1$ $(n \to \infty)$ *holds if and only if there are two sequences u and v with* $y = u + v$ *such that* $n^\alpha(u_n - u_{n-1}) \to 0$ *and* $x_n v_n \to l_2$ $(n \to \infty)$ *for some scalars* l_1 and l_2.

This statement leads to the solvability of the (SSE) $(s^0_{(1/n^\alpha)_{n=1}^\infty})_\Delta + s^{(c)}_{1/x} = s^{(c)}_{(1/n^\beta)_{n=1}^\infty}$. We obtain the next result which can be proved by similar arguments as those above.

Corollary 6.11 ([7, Corollary 8, p. 125]) *Let* α *and* β *be reals. Then we have*

(i) *If* $\alpha < 1$, *then*

$$\mathbb{S}_{\alpha,\beta} = \begin{cases} cl^c\left((n^\beta)_{n=1}^\infty\right) & \text{if } \beta \leq \alpha - 1 \\ \emptyset & \text{if } \beta > \alpha - 1. \end{cases}$$

(ii) *If* $\alpha = 1$, *then*

$$\mathbb{S}_{\alpha,\beta} = \begin{cases} cl^c\left(\left(n^\beta\right)_{n=1}^\infty\right) & \text{if } \beta < 0 \\ \emptyset & \text{if } \beta \geq 0. \end{cases}$$

(iii) *If* $\alpha > 1$, *then*

$$\mathbb{S}_{\alpha,\beta} = \begin{cases} cl^c\left(\left(n^\beta\right)_{n=1}^\infty\right) & \text{if } \beta \leq 0 \\ \emptyset & \text{if } \beta > 0. \end{cases}$$

Corollary 6.12 *We have* $\mathbb{S}_{\alpha,\beta} \neq \emptyset$ *if and only if* $\beta \leq \alpha - 1 < 0$, *or* $\alpha = 1$ *and* $\beta < 0$, *or* $\alpha > 1$ *and* $\beta \leq 0$.

Example 6.24 As a direct consequence of the preceding results, note that the (SSE) $(s^0_{(n^{-\alpha})_{n=1}^\infty})_\Delta + s^{(c)}_{1/x} = c$ is equivalent to $x_n \to L$ $(n \to \infty)$ with $L > 0$ for all $\alpha > 1$.

In the next corollary, we deal with the following statement for reals α and β and $p > 1$:

the condition $n^\beta y_n \to l_1$ *holds if and only if there are sequences* u *and* $v \in \omega$ *with* $y = u + v$ *such that*

$$\sum_{k=1}^\infty (k^\alpha |u_k - u_{k-1}|)^p < \infty \text{ and } x_n v_n \to l_2 \ (n \to \infty)$$

for all $y \in \omega$ *and some scalars* l_1 and l_2.

This is equivalent to the (SSE)

$$\left(\ell^p_{(n^{-\alpha})_{n=1}^\infty}\right)_\Delta + s^{(c)}_{1/x} = s^{(c)}_{(n^{-\beta})_{n=1}^\infty}. \tag{6.52}$$

We obtain the next result.

Corollary 6.13 ([7, Corollary 10, pp. 125–126]) *Let* α *and* β *be reals and* $\mathbb{S}_p(c)$ *be the set of all solutions of the (SSE) (6.52). Then we have*

(i) *If* $\alpha q \geq 1$, *then*

$$\mathbb{S}_p(c) = \begin{cases} cl^c\left(\left(n^\beta\right)_{n=1}^\alpha\right) & \text{if } \beta < 0 \\ \emptyset & \text{if } \beta \geq 0. \end{cases}$$

(ii) *If* $\alpha q < 1$, *then*

$$\mathbb{S}_p(c) = \begin{cases} cl^c\left(\left(n^\beta\right)_{n=1}^\infty\right) & \text{if } \alpha - \beta \geq \frac{1}{q} \\ \emptyset & \text{if } \alpha - \beta < \frac{1}{q}. \end{cases}$$

Proof The proof follows from the fact that

$$\sigma_n \sim \frac{n^{(\beta-\alpha)q+1}}{1-\alpha q}\ (n \to \infty),$$

if $\alpha q \neq 1$; and $\sigma_n \sim n^{\beta q}\log n$ $(n \to \infty)$ if $\alpha q = 1$. Then it can easily be seen that $\sigma \in \ell_\infty$ if and only if $\alpha - \beta \geq 1/q$ for $\alpha q < 1$, or $\beta < 0$ for $\alpha q \geq 1$. We conclude by Theorem 6.8. □

We deal for reals β with the statement:

$n^\beta y_n \to l_1$ if and only if $y = u + v$ with

$$\sum_{k=1}^{\infty}\left(\frac{|u_k - u_{k-1}|}{k}\right)^2 < \infty \textit{ and } x_n v_n \to l_2\ (n \to \infty)$$

for all sequences y and some scalars l_1 and l_2.

This statement is equivalent to the (SSE) $(\ell^2_{(n)_{n=1}^\infty})_\Delta + s_{1/x}^{(c)} = s^{(c)}_{(n^{-\beta})_{n=1}^\infty}$ and this (SSE) has solutions if and only if $\beta \leq -3/2$.

Example 6.25 We note that the set of all solutions of the (SSE) $(\ell^2_{(1/\sqrt{n})_{n=1}^\infty})_\Delta + s_x^{(c)} = s^{(c)}_{(\log n)_{n=1}^\infty}$ are determined by $\lim_{n\to\infty}(x_n/\log n) > 0$. This result comes from the equivalence $\sum_{k=1}^n (1/\sqrt{k})^2 \sim \log n$ $(n \to \infty)$.

Now we solve the (SSE)

$$(E_a)_\Delta + s_x^0 = s_b^0, \text{ where } E \in \{c, \ell_\infty\}.$$

For $E = c$, the solvability of the previous (SSE) consists in determining the set of all positive sequences $x = (x_n)_{n=1}^\infty$ that satisfy the next statement.

For every sequence y the condition $y_n/b_n \to 0$ $(n \to \infty)$ holds if and only if there are two sequences u and v with $y = u + v$ such that

$$\frac{u_n - u_{n-1}}{a_n} \to l \textit{ and } \frac{v_n}{x_n} \to 0\ (n \to \infty)$$

for some scalar l.

We also consider the (SSE)

$$(s_a^{(c)})_\Delta + s_x^0 = s_b^0$$

as a perturbed equation of the equation $s_x^0 = s_b^0$, which is equivalent to $K_1 \leq x_n/b_n \leq K_2$ for all n and some $K_1, K_2 > 0$. We obtain the equivalence of these two equations under some conditions on a and b.

By Lemma 6.14, we obtain the solvability of the (SSE's)

$$(s_a^{(c)})_\Delta + s_x^0 = s_b^0 \text{ and } (s_a)_\Delta + s_x^0 = s_b^0.$$

Theorem 6.9 ([7, Theorem 2, p. 127]) *The set $\mathbb{S}_E^0$ of all solutions of the (SSE) $(E_a)_\Delta + s_x^0 = s_b^0$, where $E = c$ or ℓ_∞ is determined by*

$$\mathbb{S}_E^0 = \begin{cases} cl^\infty(b) & \text{if } \sigma \in c_0 \\ \varnothing & \text{if } \sigma \notin c_0. \end{cases}$$

Proof Let $x \in \mathbb{S}_E^0$. Then the inclusion $(E_a)_\Delta + s_x^0 \subset s_b^0$ holds. This implies $(E_a)_\Delta \subset s_b^0$ and $D_{1/b}\Sigma D_a \in (E, c_0)$, whence

$$D_{1/b}\Sigma D_a \in (c, c_0)$$

since $E \supset c$ and

$$\sigma_n \to 0 \ (n \to \infty). \tag{6.53}$$

Now we have $s_x^0 \subset s_b^0$ and

$$x \in s_b. \tag{6.54}$$

Then we consider the (SSIE)

$$s_b^0 \subset (E_a)_\Delta + s_x^0. \tag{6.55}$$

By Part (i) of Lemma 6.14 with $E \subset s_1$ and $F = F' = c_0$, we obtain $\mathbb{S}_E^0 \subset \overline{s_{1/b}}$. Using the condition in (6.54) we conclude that $x \in \mathbb{S}_E^0$ implies $x \in cl^\infty(b)$. Conversely, we assume $x \in cl^\infty(b)$ and that (6.53) holds. Since $1/b \in c_0$, we have $(E_a)_\Delta \subset s_b^0$ for $E = c_0$ or s_1 and obtain

$$(E_a)_\Delta + s_x^0 = (E_a)_\Delta + s_b^0 = s_b^0$$

and $x \in \mathbb{S}_E^0$. We conclude $\mathbb{S}_E^0 = cl^\infty(b)$. □

For $a = e$ we easily obtain the next result.

Corollary 6.14 *The set $\mathbb{S}(E_\Delta, c_0)$ of all solutions of the (SSE) $E_\Delta + s_x^0 = s_b^0$ where $E = c$ or ℓ_∞ is determined by*

$$\mathbb{S}(E_\Delta, c_0) = \begin{cases} cl^\infty(b) & \text{if } (n/b_n)_{n=1}^\infty \in c_0 \\ \varnothing & \text{if } (n/b_n)_{n=1}^\infty \notin c_0. \end{cases}$$

Example 6.26 The equation

$$E_\Delta + s_x^0 = s^0_{(n^\alpha)_{n=1}^\infty}, \text{ where } E \in \{c, \ell_\infty\}$$

has solutions if and only if $\alpha > 1$. So the equation $E_\Delta + s_x^0 = c_0$ has no solution, and the solutions of the equation $E_\Delta + s_x^0 = s^0_{(n^2)_{n=1}^\infty}$ are determined by $K_1 n^2 \leq x_n \leq K_2 n^2$ for all n and some $K_1, K_2 > 0$.

We close this section with the solvability of the (SSE)

$$(E_a)_\Delta + s_x^0 = s_b^0 \text{ for special sequences } a \text{ and } b,$$

in particular, we consider the $(SSE's)$

$$\left(s^{(c)}_{(n^{-\alpha})_{n=1}^\infty}\right)_\Delta + s^0_{1/x} = s^0_{(n^{-\beta})_{n=1}^\infty} \tag{6.56}$$

$$\left(s^{(c)}_{R}\right)_\Delta + s^0_{x} = s^0_{\overline{R}} \tag{6.57}$$

$$\left(s^{(c)}_{(n^{-\alpha})_{n=1}^\infty}\right)_\Delta + s^0_{1/x} = s^0_{1/R} \tag{6.58}$$

$$\left(s^{(c)}_{1/R}\right)_\Delta + s^0_{1/x} = s^0_{(n^{-\beta})_{n=1}^\infty} \tag{6.59}$$

with reals α and β and $R, \overline{R} > 0$. The following result holds which we state without proof.

Proposition 6.14 ([7, Proposition 1, p. 129])

(i) The (SSE) (6.56) has solutions if and only if $\beta < \alpha - 1 < 0$ or $\alpha \geq 1$ and $\beta < 0$.
(ii) The (SSE) (6.57) has solutions if and only if $R \leq 1 < \overline{R}$ or $1 < R < \overline{R}$.
(iii) The (SSE) (6.58) has solutions if and only if $R < 1 < \alpha$ or $R < \alpha = 1$ or α and $R < 1$.
(iv) The (SSE) (6.59) has solutions if and only if $R = 1$ and $\beta < -1$, or $R > 1$ and $\beta < 0$.

Example 6.27 The (SSE) $(s^{(c)}_{1/2})_\Delta + s^0_x = s^0_{\overline{R}}$ has solutions if and only if $\overline{R} > 1$.

Example 6.28 Let τ and τ' be reals. Then the system of (SSE)

$$\begin{cases} c_\Delta + s_x^0 = s^0_{(n^\tau)_{n=1}^\infty} \\ \left(s^{(c)}_{1/2}\right)_\Delta + s^0_x = s^0_{(n^{\tau'})_{n=1}^\infty} \end{cases},$$

where x is the unknown, has solutions if and only if $\tau = \tau' > 1$. Then x is a solution of the system if and only if $C_1 n^\tau \leq x_n \leq C_2 n^\tau$ for all n and some $C_1, C_2 > 0$. This is a direct consequence of Part (iv) in Proposition 6.14 and of the elementary fact that $s_{(n^\tau)_{n=1}^\infty} = s_{(n^{\tau'})_{n=1}^\infty}$ if and only if $\tau = \tau'$.

Example 6.29 Let $\mathbb{S}_c^0$ be the set of all positive sequences that satisfy the following statement.

For every sequence y, *the condition* $y_n/n \to 0$ $(n \to \infty)$ *holds if and only if there are two sequences* u *and* v *with* $y = u + v$ *such that* $\sqrt{n}(u_n - u_{n-1}) \to L$ *and* $x_n v_n \to 0$ $(n \to \infty)$ *for some scalar* L.

By Part (i) of Proposition 6.14, we have $x \in \mathbb{S}_c^0$ if and only if $K_1/n \le x_n \le K_2/n$ for all n and some $K_1, K_2 > 0$.

References

1. de Malafosse, B.: On some BK space. Int. J. Math. Math. Sci. **58**, 1783–1801 (2003)
2. de Malafosse, B.: On the Banach algebra $\mathcal{B}(\ell_p(\alpha))$. Int. J. Math. Math. Sci. **60**, 3187–3203 (2004)
3. de Malafosse, B.: Sum of sequence spaces and matrix transformations. Acta Math. Hung. **113**(3), 289–313 (2006)
4. de Malafosse, B.: On the sets of ν-analytic and ν-entire sequences and matrix transformations. Int. Math. Forum **2**(36), 1795–1810 (2007)
5. de Malafosse, B.: Solvability of sequence spaces equations using entire and analytic sequences and applications. J. Ind. Math. Soc. **81**(1–2), 97–114 (2014)
6. de Malafosse, B.: On sequence spaces equations of the form $E_T + F_x = F_b$ for some triangle T. Jordan J. Math. Stat. **8**(1), 79–105 (2015)
7. de Malafosse, B.: Solvability of sequence spaces equations of the form $(E_a)_\Delta + Fx = F_b$. Fasc. Math. **55**, 109–131 (2015)
8. de Malafosse, B.: Extension of the results on the $(SSIE)$ and the (SSE) of the form $F \subset \mathcal{E} + F_x'$ and $\mathcal{E} + F_x = F$. Fasc. Math. **59**, 107–123 (2017)
9. de Malafosse, B., Malkowsky, E.: On sequence spaces equations using spaces of strongly bounded and summable sequences by the Cesàro method. Antarctica J. Math. **10**(6), 589–609 (2013)
10. Farés, A., de Malafosse, B.: Sequence spaces equations and application to matrix transformations. Int. Math. Forum **3**(19), 911–927 (2008)
11. Grosse-Erdmann, K.-G.: Matrix transformations between the sequence spaces of Maddox. J. Math. Anal. Appl. **180**, 223–238 (1993)
12. Maddox, I.J.: Some properties of paranormed sequence spaces. London J. Math. Soc. **2**(1), 316–322 (1969)
13. Rao, K.C., Srinivasalu, T.C.: Matrix operators on analytic and entire sequences. Bull. Malays. Math. Sci. Soc. (Second series) **14**, 422–436 (1991)
14. Simons, S.: The sequence spaces $\ell(p_\nu)$ and $m(p_\nu)$. London Math. Soc. **3**(15), 422–436 (1965)

Chapter 7
Solvability of Infinite Linear Systems

Here we apply the theory of matrix transformations to the solvability of infinite systems of linear equations. So, we obtain some results in the spectral theory. Many results were published in this domain by Bilgiç and Furkan in 2008 [7], Furkan, Bilgiç and Başar in 2010 [27], Akhmedov and Başar in 2006 and 2007 [2, 3], Akhmedov and El-Shabrawy in 2011 [4], de Malafosse in 2014 [22], Srivastava and Kumar in 2010, 2012 and 2018 [39–41], Başar and Karaisa in 2013 and 2014 [6, 30] and Das in 2017 [10]. More related results on the spectra of operators on sequence spaces can be found in [31, 42–44].

We also deal with the Hill equation which was studied by Brillouin [8], Ince [29], and more recently by Hochstadt (1963) [28], Magnus and Winkler (1966) [33] and Rossetto [38]. These authors used an *infinite determinant* and showed that the Hill equation has a nonzero solution if the determinant is zero. B. Rossetto provided an algorithm to calculate the *Floquet exponent* from the generalization of the *notion of a characteristic equation* and of a *truncated infinite determinant*. Here, we consider a Banach algebra in which we may obtain the inverse of an infinite matrix and obtain a new method to calculate the *Floquet exponent*. Furthermore, we determine the solutions of the infinite linear system associated with the Hill equation with a second member and give a method to approximate them. Finally, we present a study of the Mathieu equation which can be written as an infinite tridiagonal linear system of equations.

7.1 Banach Algebras of Infinite Matrices

In the following, we consider infinite linear systems given by the matrix equation

$$Ax = b, \tag{7.1}$$

where $A=(a_{nk})_{n,k=1}^{\infty}$ is an infinite matrix and $x = (x_1, x_2, \dots)^T$ and $b = (b_1, b_2 \dots)^T \in \omega$ are *column matrices*, that is,

B. de Malafosse et al., *Operators Between Sequence Spaces and Applications*,
https://doi.org/10.1007/978-981-15-9742-8_7

$$Ax = \begin{pmatrix} a_{11} & . & . & a_{1k} & . \\ . & . & . & . & . \\ a_{n1} & . & . & a_{nk} & . \\ . & . & . & . & . \\ . & . & . & . & . \end{pmatrix} \begin{pmatrix} x_1 \\ . \\ x_k \\ . \\ . \end{pmatrix} = \begin{pmatrix} b_1 \\ . \\ b_n \\ . \\ . \end{pmatrix}$$

whenever the series defined by $A_n x = \sum_{k=1}^{\infty} a_{nk} x_k$ are convergent for all n. When $A \in (X, Y)$ for given $X, Y \subset \omega$, the question is "if $b = (b_1, b_2, \dots)^T \in Y$ is a given sequence, does the equation in (7.1) have a solution in X" ? In this way we are led to determine if this solution is unique in the set X and if it is possible to explicitly compute this solution or give an approximation for it. There is no direct answer to these problems. We can write the equation in (7.1) in the form of the infinite linear system

$$\sum_{k=1}^{\infty} a_{nk} x_k = b_n \text{ for } n = 1, 2, \dots .$$

We note that if X is a BK space with AK, X has a Schauder basis with the elements $e^{(n)}$ for $n = 1, 2, \dots$. We recall that in this case every operator $L \in \mathcal{B}(X) = \mathcal{B}(X, X)$ is given by the infinite $A = (a_{nk})_{n,k=1}^{\infty}$ such that $L(x) = Ax$ for every $x \in X$, where $a_{nk} = L(e^{(k)})_n$ for all n and k.

So we are led to consider the *left or right inverse* of an infinite matrix and we need to determine an adapted *Banach algebra* in which we can define the product of infinite matrices. It is useful to use the next well-known elementary property. For any Banach space X of sequences, the condition

$$\|I - A\|^*_{\mathcal{B}(X)} < 1$$

implies that A is invertible and

$$A^{-1} = \sum_{i=0}^{\infty} (I - A)^i \in \mathcal{B}(X).$$

Let, for instance,

$$\mathcal{A} = \begin{pmatrix} 1 & 1/2 & 1/2 & . & . & 1/2 & . \\ & 1/2 & 1/3 & 1/3 & . & 1/3 & . \\ & & . & . & . & . & . \\ & & & 1/n & 1/(n+1) & 1/(n+1) & . \\ & \mathbf{0} & & & . & . & . \\ & & & & & . & . \\ & & & & & & . \end{pmatrix}.$$

We will see in Sect. 7.2 that using a convenient Banach algebra the system $Ax = b$ for b has a solution in a well-chosen space. Then we will see in Sect. 7.2 that if C_1^T

denotes the transpose of the matrix of the Cesàro operator, the equation $C_1^T x = b$ also has a solution in a well-chosen space.

Now we consider the Banach algebra S_τ.

We recall that the set s_τ is a BK space with the norm $\|x\|_{s_\tau} = \sup_n |x_n|/\tau_n$ (Example 2.2). Also the set

$$S_\tau = \left\{ A = (a_{nk})_{n,k=1}^\infty : \|A\|_{S_\tau} = \sup_n \left(\frac{1}{\tau_n} \sum_{k=1}^\infty |a_{nk}|\tau_k \right) < \infty \right\} \tag{7.2}$$

is a Banach algebra with identity normed by $\|A\|_{S_\tau}$ (cf. [17]). We recall that if $A \in (s_\tau, s_\tau)$, then $\|Ax\|_{s_\tau} \leq \|A\|_{S_\tau} \|x\|_{s_\tau}$ for all $x \in s_\tau$. Thus we obtain the following result where we put $\mathbb{B}(s_\tau) = \mathcal{B}(s_\tau) \bigcap (s_\tau, s_\tau)$.

Lemma 7.1 *For any given $\tau \in \mathcal{U}^+$, we have $\mathbb{B}(s_\tau) = S_\tau = (s_\tau, s_\tau)$.*

We consider the Banach algebra $\mathcal{B}(\chi)$ for $\chi \in \{s_\tau, s_\tau^0, s_\tau^{(c)}, (\ell_p)_\tau\}$ with $1 \leq p < \infty$, and give an explicit expression for the norm $\|A\|^*_{\mathcal{B}((\ell_p)_\tau)}$ for $1 \leq p \leq \infty$, and establish some properties of the equation $Ax = b$ for $A \in \mathcal{B}(\chi)$ and $b \in \chi$ with $\chi \in \{s_\tau, s_\tau^0, s_\tau^{(c)}, (\ell_p)_\tau\}$ and $1 \leq p < \infty$.

We recall the next result where, as before, $\mathcal{U}$ is the set of all sequences $u = (u_n)_{n=1}^\infty$ with $u_n \neq 0$ for all n, and $\mathcal{L}$ denotes the set of *lower triangular infinite matrices*, that is, $A \in \mathcal{L}$ if $a_{nk} = 0$ for $k > n$ and all n.

Lemma 7.2 *(i)* (Theorem 2.3) *Let $T \in \mathcal{L}$ be a triangle, that is, $t_{nn} \neq 0$ for all n and X a BK space. Then $X(T) = X_T$ is a BK space with the norm*

$$\|x\|_{X(T)} = \|Tx\|_X. \tag{7.3}$$

(ii) (Part *(b)* of *Theorem 2.4). Let $T = D_a = (a_n \delta_{nk})_{n,k=1}^\infty$ be a diagonal matrix with $a \in \mathcal{U}$ and X be a BK space with AK. Then $X(T)$ has AK with the norm given by (7.3).*

We summarize; Part (i) of the next lemma is an immediate consequence of Lemma 7.2, and Part (ii) is obvious.

Lemma 7.3 *(i) Let $\tau \in \mathcal{U}^+$. Then s_τ, s_τ^0 and $s_\tau^{(c)}$ are BK spaces with the norm $\|\cdot\|_{s_\tau}$ and s_τ^0 has AK. The set $(\ell_p)_\tau$ for $1 \leq p < \infty$ is a BK space with AK with the norm $\|\cdot\|_{(\ell_p)_\tau}$.*

(ii) Let χ be any of the spaces s_τ, s_τ^0 or $s_\tau^{(c)}$. Then we have

$$|P_n(x)| = |x_n| \leq \tau_n \|x\|_{s_\tau} \text{ for all } n \geq 1 \text{ and all } x \in \chi.$$

Remark 7.1 We recall that if X is a BK space with the norm $\|\cdot\|_X$, then by Part 1.10(a) Theorem 1.10,

$$(X, X) \subset \mathcal{B}(X).$$

We recall some more definitions and results. We write $\bar{B}_X(0,1) = \{x \in X : \|x\|_X \leq 1\}$ for the closed unit ball in a BK space X. Thus we obtain

$$\|A\|^*_{\mathcal{B}(X)} = \sup_{x \neq 0}\left(\frac{\|Ax\|_X}{\|x\|_X}\right) = \sup_{x \in \bar{B}_X(0,1)} \|Ax\|_X \text{ for all } A \in \mathcal{B}(X).$$

Also for every $a = (a_n)_{n=1}^{\infty} \in X$ such that the series $\sum_{n=1}^{\infty} a_n x_n$ is convergent for all $x \in X$, the identity

$$\|a\|^*_X = \sup_{x \in \bar{B}_X(0,1)} \left|\sum_{n=1}^{\infty} a_n x_n\right|$$

is defined and finite (cf. (1.15) and subsequent remark).

Furthermore, the following results hold.

Lemma 7.4 (Theorem 1.9)

Let X be a BK space. Then $A \in (X, \ell_\infty)$ if and only if

$$\sup_{n \geq 1}\left(\|A_n\|^*_X\right) = \sup_{n \geq 1}\left(\sup_{x \in \bar{B}_X(0,1)}\left|\sum_{k=1}^{\infty} a_{nk} x_k\right|\right) < \infty$$

Lemma 7.5 (Example 1.11 or [35, Theorem 1.2.3, p. 155]) *For every $a \in \ell_1$, $\|a\|^*_{c_0} = \|a\|^*_c = \|a\|^*_{\ell_\infty} = \|a\|_{\ell_1} = \sum_{n=1}^{\infty} |a_n|$.*

It is clear that if X is a BK space, then $A \in (X, s_\beta)$ if and only if

$$\sup_n \left\|\frac{1}{\beta_n} A_n\right\|^*_X < \infty.$$

Since there is no characterization of the class $((\ell_p)_\tau, (\ell_p)_\tau)$ for $1 < p < \infty$ and $p \neq 2$, we need to define a subset $\widehat{\mathcal{B}}_p(\tau)$ of $((\ell_p)_\tau, (\ell_p)_\tau)$ permitting us to obtain the inverse of some well-chosen matrix map $A \in ((\ell_p)_\tau, (\ell_p)_\tau)$. Thus we are led to define the number

$$N_{p,\tau}(A) = \left[\sum_{n=1}^{\infty}\left(\sum_{k=1}^{\infty}\left(|a_{nk}|\frac{\tau_k}{\tau_n}\right)^q\right)^{p-1}\right]^{1/p}$$

for $1 < p < \infty$ and $q = p/(p-1)$. Now we can state the following proposition.

Proposition 7.1 ([19, Proposition 7, pp. 45–46]) *Let $\tau \in \mathcal{U}^+$. Then*

(i) for every $A \in S_\tau$

$$\|A\|^*_{\mathcal{B}(s_\tau)} = \|A\|_{S_\tau} = \sup_n\left(\frac{1}{\tau_n}\sum_{k=1}^{\infty}|a_{nk}|\tau_k\right);$$

(ii) (a) $\mathcal{B}((\ell_1)_\tau) = ((\ell_1)_\tau, (\ell_1)_\tau)$ *and* $A \in \mathcal{B}((\ell_1)_\tau)$ *if and only if*

$$A^T \in S_{1/\tau};$$

(b) $\|A\|^*_{\mathcal{B}((\ell_1)_\tau)} = \|A^T\|_{S_{1/\tau}}$ *for all* $A \in \mathcal{B}((\ell_1)_\tau)$.

(iii) For $1 < p < \infty$, *we have* $\widehat{\mathcal{B}}_p(\tau) \subset \mathcal{B}((\ell_p)_\tau)$, *where*

$$\widehat{\mathcal{B}}_p(\tau) = \left\{A = (a_{nk})_{n,k=1}^\infty : N_{p,\tau}(A) < \infty\right\},$$

and for every $A \in \widehat{\mathcal{B}}_p(\tau)$

$$\|A\|^*_{\mathcal{B}((\ell_p)_\tau)} \le N_{p,\tau}(A).$$

Proof (i) First we have

$$\|Ax\|_{s_\tau} = \sup_n\left(\frac{1}{\tau_n}\left|\sum_{k=1}^\infty a_{nk}x_k\right|\right) = \sup_n\left(\frac{1}{\tau_n}|A_n x|\right) \text{ for all } x \in s_\tau, \tag{7.4}$$

then

$$\|A\|^*_{\mathcal{B}(s_\tau)} = \sup_{x\in B_{s_\tau}(0,1)}\left(\sup_n\left(\frac{1}{\tau_n}|A_n x|\right)\right) = \sup_n\left(\frac{1}{\tau_n}\sup_{x\in B_{s_\tau}(0,1)}|A_n x|\right).$$

Writing $x = \tau y$ in (7.4), we obtain

$$\|A\|^*_{\mathcal{B}(s_\tau)} = \sup_n\left(\frac{1}{\tau_n}\sup_{y\in B_{s_1}(0,1)}(|A_n(\tau y)|)\right) = \sup_n\left(\frac{1}{\tau_n}\|A_n D_\tau\|^*_{\ell_\infty}\right).$$

Now we have

$$\|A_n D_\tau\|^*_{\ell_\infty} = \left\|(a_{nk}\tau_k)_{k=1}^\infty\right\|_{\ell_1} = \sum_{k=1}^\infty |a_{nk}|\tau_k \text{ for all } n.$$

We conclude

$$\|A\|^*_{\mathcal{B}(s_\tau)} = \sup_n\left(\sum_{k=1}^\infty |a_{nk}|\frac{\tau_k}{\tau_n}\right).$$

(ii) By Theorem 1.10, we have $\mathcal{B}(\ell_1) = (\ell_1, \ell_1)$ and the condition $A \in ((\ell_1)_\tau, (\ell_1)_\tau)$ is equivalent to $A^T \in S_{1/\tau}$ by Theorem 1.22, and it is clear that the identity in (b) holds.

(iii) Let $A \in \widehat{\mathcal{B}}_p(\tau)$ be a given infinite matrix and take any $x \in \ell_p$. Then

$$\|Ax\|_{\ell_p}^p = \left\| \left(\sum_{k=1}^{\infty} a_{nk}x_k \right)_{n=1}^{\infty} \right\|_{\ell_p}^p = \sum_{n=1}^{\infty} \left| \sum_{k=1}^{\infty} a_{nk}x_k \right|^p \leq \sum_{n=1}^{\infty} \left(\sum_{k=1}^{\infty} |a_{nk}x_k| \right)^p,$$

and from the Hölder inequality ((A.3)), we obtain for all n

$$\sum_{k=1}^{\infty} |a_{nk}x_k| \leq \left(\sum_{k=1}^{\infty} |a_{nk}|^q \right)^{1/q} \left(\sum_{k=1}^{\infty} |x_k|^p \right)^{1/p} = \left(\sum_{k=1}^{\infty} |a_{nk}|^q \right)^{1/q} \cdot \|x\|_{\ell_p},$$

with $q = p/(1-p)$. We deduce

$$\|Ax\|_{\ell_p}^p \leq \sum_{n=1}^{\infty} \left[\left(\sum_{k=1}^{\infty} |a_{nk}|^q \right)^{1/q} \|x\|_{\ell_p} \right]^p = \sum_{n=1}^{\infty} \left(\sum_{k=1}^{\infty} |a_{nk}|^q \right)^{p/q} \|x\|_{\ell_p}^p$$

and since $p/q = p - 1$, we have

$$\|Ax\|_{\ell_p} \leq \left[\sum_{n=1}^{\infty} \left(\sum_{k=1}^{\infty} |a_{nk}|^q \right)^{p-1} \right]^{1/p} \|x\|_{\ell_p}$$

and

$$\|A\|_{\mathcal{B}(\ell_p)}^* = \sup_n \left(\frac{\|Ax\|_{\ell_p}}{\|x\|_{\ell_p}} \right) \leq \left[\sum_{n=1}^{\infty} \left(\sum_{k=1}^{\infty} |a_{nk}|^q \right)^{p-1} \right]^{1/p}.$$

We have shown that if $A \in \widehat{\mathcal{B}}_p(e)$, then $A \in \mathcal{B}(\ell_p)$. So if $A \in \widehat{\mathcal{B}}_p(\tau)$ and $D_{1/\tau}AD_\tau \in \widehat{\mathcal{B}}_p(e)$, then we have $D_{1/\tau}AD_\tau \in \mathcal{B}(\ell_p)$ and $A \in \mathcal{B}((\ell_p)_\tau)$. This concludes the proof.

□

Remark 7.2 In [18], we obtained additional results on the Banach algebra $\mathcal{B}(\ell_p)$ and in [24], we dealt with the Banach algebra $(w_\infty(\Lambda), w_\infty(\Lambda))$ and gave some applications to the solvability of matrix equations in $w_\infty(\Lambda)$.

7.2 Solvability of the Equation $Ax = b$

We use Banach algebra techniques to solve infinite linear systems.

We start with the following elementary but very useful result for $a = (a_n)_{n=1}^{\infty} \in \mathcal{U}$ (cf. [19, Proposition 8, p. 47]).

Proposition 7.2 *Let $X \subset \omega$ be a BK space. We assume $D_{1/a}A \in (X, X)$ and*

$$\|I - D_{1/a}A\|^*_{\mathcal{B}(X)} < 1. \tag{7.5}$$

Then the equation $Ax = b$ with $D_{1/a}b = (b_n/a_n)_{n=1}^{\infty} \in X$ has a unique solution in X given by $x = (D_{1/a}A)^{-1}D_{1/a}b$.

Proof First we see that, since $D_{1/a}A \in (X, X)$ and condition (7.5) holds, $D_{1/a}A$ is invertible in $\mathcal{B}(X)$. Since $\mathcal{B}(X)$ is a Banach algebra of operators with identity,

$$(D_{1/a}A)^{-1}(D_{1/a}Ax) = \left[(D_{1/a}A)^{-1}o(D_{1/a}A)\right]x = Ix = x \text{ for all } x \in X.$$

Thus the equation $Ax = b$ with $D_{1/a}b \in X$ is equivalent to $D_{1/a}Ax = D_{1/a}b$ which in turn is $x = (D_{1/a}A)^{-1}(D_{1/a}b) \in X$. This concludes the proof. □

We can state a similar result in a more general case. As we have seen in Sect. 4, we let

$$\Gamma_\tau = \left\{A = (a_{nk})_{n,k=1}^{\infty} \in S_\tau : \|I - A\|_{S_\tau} < 1\right\},$$

and let $\Gamma_r = \Gamma_{(r^n)_{n=1}^{\infty}}$. Then, for $1 < p < \infty$, we let

$$\Gamma'_{p,\tau} = \left\{A = (a_{nk})_{n,k=1}^{\infty} \in ((\ell_p)_\tau, (\ell_p)_\tau) : N_{p,\tau}(I - A) < 1\right\}.$$

We note that, since S_τ is a Banach algebra, the condition $A \in \Gamma_\tau$ means that A is invertible and $A^{-1} \in S_\tau$.

In the following we let $|a| = (|a_n|)_{n=1}^{\infty}$ for any given sequence $a = (a_n)_{n=1}^{\infty}$.

Corollary 7.1 ([19, Corollary 9, p. 47]) *Let $\tau \in \mathcal{U}^+$ and A be an infinite matrix. We assume $D_{1/a}A \in \Gamma_\tau$ with $a \in \mathcal{U}$. Then we have*

(i) (a) For any given $b \in s_{|a|\tau}$, the equation $Ax = b$ has a unique solution in s_τ given by

$$x^0 = (D_{1/a}A)^{-1}(D_{1/a}b) = A^{-1}b \tag{7.6}$$

with $A^{-1} \in (s_{|a|\tau}, s_\tau)$.

(b) If $\lim_{n\to\infty}(a_{nk}/a_{nn}\tau_n) = 0$ for all $k \geq 1$, then, for any given $b \in s^0_{|a|\tau}$, the equation $Ax = b$ has a unique solution in s^0_τ given by (7.6), and $A^{-1} \in (s^0_{|a|\tau}, s^0_\tau)$.

(c) If $\lim_{n\to\infty}(a_{nk}/a_{nn}\tau_n) = l_k$ for some l_k for all $k \geq 1$, and

$$\lim_{n\to\infty}\left(\frac{1}{\tau_n a_{nn}}\sum_{k=1}^{\infty} a_{nk}\tau_k\right) = l \text{ for some } l$$

then for any given $b \in s^{(c)}_{|a|\tau}$, the equation $Ax = b$ has a unique solution in $s^{(c)}_\tau$ given by (7.6) with $A^{-1} \in (s^{(c)}_{|a|\tau}, s^{(c)}_\tau)$.

(ii) (a) If $A^T D_{1/a} \in \Gamma_{1/\tau}$, then for any given $b \in (\ell_1)_{|a|\tau}$, the equation $Ax = b$ has a unique solution in $(\ell_1)_\tau$ given by (7.6) with $A^{-1} \in (\ell_1(|a|\tau), \ell_1(\tau))$.

(b) *Let* $p > 1$. *If* $D_{1/a}A \in \Gamma'_{p,\tau}$, *then for any given* $b \in (\ell_p)_{|a|\tau}$, *the equation* $Ax = b$ *has a unique solution in* $(\ell_p)_\tau$ *given by (7.6) with* $A^{-1} \in ((\ell_p)_{|a|\tau}, (\ell_p)_\tau)$.

Proof (i)

(a) If $D_{1/a}A \in \Gamma_\tau$, then we have by Part (i) of Proposition 7.1,

$$\|I - D_{1/a}A\|_{S_\tau} = \|I - D_{1/a}A\|^*_{\mathcal{B}(s_\tau)} < 1.$$

Thus $D_{1/a}A$ is invertible in $\mathcal{B}(s_\tau) \bigcap S_\tau$. Then $(D_{1/a}A)^{-1} \in S_\tau$, that is, $A^{-1} \in (s_{|a|\tau}, s_\tau)$ and we conclude by Proposition 7.2.
(ii) Since the set s_τ^0 is a BK space with AK, we have $\mathcal{B}(s_\tau^0) = (s_\tau^0, s_\tau^0)$. Also, since $D_{1/a}A \in \Gamma_\tau$ and $\lim_{n\to\infty}(a_{nk}/a_{nn}\tau_n) = 0$ for all $k \geq 1$, we deduce $D_{1/\tau}(D_{1/a}A)D_\tau \in (c_0, c_0)$. So $D_{1/a}A \in (s_\tau^0, s_\tau^0)$. Then we have

$$\|I - D_{1/a}A\|^*_{\mathcal{B}(s_\tau^0)} = \sup_{x \in B_{s_\tau}} \left(\|(I - D_{1/a}A)x\|_{s_\tau}\right) = \|I - D_{1/a}A\|_{S_\tau} < 1$$

and we conclude by Proposition 7.2.
(c) Here we have

$$\|I - D_{1/a}A\|^*_{\mathcal{B}(s_\tau^{(c)})} = \|I - D_{1/a}A\|_{S_\tau} < 1.$$

Then $(D_{1/a}A)^{-1} \in S_\tau \bigcap \mathcal{B}(s_\tau^{(c)})$, so $(D_{1/a}A)^{-1}$ is an operator represented by an infinite matrix, since $(D_{1/a}A)^{-1} \in S_\tau$ and the condition $(D_{1/a}A)^{-1} \in \mathcal{B}(s_\tau^{(c)})$ implies $(D_{1/a}A)^{-1} \in (s_\tau^{(c)}, s_\tau^{(c)})$. We conclude again by Proposition 7.2.

(ii)

(a) By Part (ii) (b) of Proposition 7.1, the condition $A^T D_{1/a} \in \Gamma_{1/\tau}$ implies

$$\|I - D_{1/a}A\|^*_{\mathcal{B}(\ell_1(\tau))} = \|I - (D_{1/a}A)^t\|_{S_{1/\tau}} < 1$$

and

$$(D_{1/a}A)^{-1} \in \mathcal{B}(\ell_1(\tau)) = ((\ell_1)_\tau, (\ell_1)_\tau).$$

Thus $A^{-1} \in ((\ell_1)_{|a|\tau}, (\ell_1)_\tau)$ and we conclude by Proposition 7.2.
(b) By Part (iii) of Proposition 7.1, we have

$$(D_{1/a}A)^{-1} \in \mathcal{B}\left((\ell_p)_\tau\right) = \left((\ell_p)_\tau, (\ell_p)_\tau\right),$$

and so $A^{-1} \in ((\ell_p)_{|a|\tau}, (\ell_p)_\tau)$. Again we conclude by Proposition 7.2.

This completes the proof. □

As direct consequences of the preceding we consider the next examples.

Example 7.1 We consider the matrix

$$A_1 = \begin{pmatrix} 1 & a & a^2 & . & . \\ c & 1 & a & a^2 & \\ c^2 & c & 1 & a & \\ . & . & . & . & . \end{pmatrix}$$

([32, Example 3, p. 158]). We easily see that if

$$0 < a < 1 \text{ for } 0 < c < 1 \text{ and } 2a + 2 - 3ac < 1, \tag{7.7}$$

then $A_1 \in \Gamma_1$. Indeed, we have

$$\|I - A_1\|_{S_1} \leq \sum_{n=1}^{\infty} a^n + \sum_{n=1}^{\infty} c^n = \frac{a}{1-a} + \frac{c}{1-c} < 1.$$

We conclude the equation $A_1 x = b$ for any $b \in s_1$ has a unique solution in s_1 given by

$$x^0 = \sum_{i=0}^{\infty} (I - A_1)^i b = A_1^{-1} b. \tag{7.8}$$

Now if (7.7) is satisfied, then we also have

$$\|I - A_1^T\|_{S_1} \leq \frac{a}{1-a} + \frac{c}{1-c} < 1,$$

and for any $b \in \ell_1$, the equation $A_1 x = b$ has a unique solution in ℓ_1 given by (7.8).

Example 7.2 For the matrix $\mathcal{A}$ given in Sect. 7.1, it can easily be verified that $\mathcal{A} \in (s_r, s_r) = S_r$ and $\mathcal{A} \notin \Gamma_r$ for all $r < 1$. If we define

$$A' = D_{(n)_{n=1}^{\infty}} \mathcal{A} \Delta^+,$$

then

$$A' = \begin{pmatrix} 1 & -1/2 & & & \mathbf{0} \\ & . & . & & \\ & & 1 & -1/(n+1) & \\ \mathbf{0} & & & & \end{pmatrix}$$

and

$$\|I - A'\|_{S_r} = \frac{r}{2}.$$

Let $r < 1$. Then Δ^+ is bijective from s_r to itself. Since $\mathcal{A}$ and $\Delta^+ \in S_r$, we have

$$\mathcal{A}(\Delta^+ x') = \left(\mathcal{A}\Delta^+\right) x' \text{ for all } x' \in s_r.$$

So the equation $\mathcal{A}x = b$ with $x = \Delta^+ x'$ is equivalent to

$$D_{(n)_{n=1}^\infty}\left[\mathcal{A}(\Delta^+ x')\right] = D_{(n)_{n=1}^\infty}\left[(\mathcal{A}\Delta^+)x'\right] = (D_{(n)_{n=1}^\infty}\mathcal{A}\Delta^+)x' = D_{(n)_{n=1}^\infty} b.$$

We conclude that the equation $\mathcal{A}x = b$ has a solution given by

$$x = A'^{-1} D_{(n)_{n=1}^\infty} b = \Delta^+ \left(D_{(n)_{n=1}^\infty}\mathcal{A}\Delta^+\right)^{-1} D_{(n)_{n=1}^\infty} b$$

in the set

$$X_1 = \bigcup_{r<1} s_r$$

for all b with $D_{(n)_{n=1}^\infty} b = (nb_n)_{n=1}^\infty \in X_1$.

Example 7.3 Now we consider the case of the transpose of the Cesàro operator. We have

$$C_1^T = \begin{pmatrix} 1 & 1/2 & 1/3 & . & . & . & . \\ & 1/2 & 1/3 & . & . & . & . \\ & & . & . & . & . & . \\ & & & 1/n & 1/(n+1) & 1/(n+2) & . \\ 0 & & & & . & . & . \\ & & & & & . & . \\ & & & & & & . \end{pmatrix}.$$

Here we easily see that for every $r > 0$, we have $C_1^T \notin \Gamma_r$, but for every $r < 1$, we obtain

$$\|I - D_{(n)_{n=1}^\infty} C_1^T \Delta^+\|_{S_r} = \sup_n \left[n \sum_{k=n}^{\infty} \left(\frac{1}{k} - \frac{1}{k+1} \right) r^{k-n+1} \right] < 1. \tag{7.9}$$

So the equation

$$C_1^T x = b \tag{7.10}$$

with $x = \Delta^+ x'$ is equivalent to

$$D_{(n)_{n=1}^\infty}\left[C_1^T(\Delta^+ x')\right] = D_{(n)_{n=1}^\infty} b.$$

Now since $C_1^T, \Delta^+ \in S_r$, we have $C_1^T(\Delta^+ x') = (C_1^T \Delta^+)x'$ and

$$D_{(n)_{n=1}^\infty}\left[C_1^T\left(\Delta^+ x'\right)\right] = \left(D_{(n)_{n=1}^\infty} C_1^T \Delta^+\right) x' \text{ for all } x' \in s_r. \tag{7.11}$$

So (7.9) implies $x' = (D_{(n)_{n=1}^\infty} C_1^T \Delta^+)^{-1} D_{(n)_{n=1}^\infty} b$ is the unique solution of (7.11). Using the fact that Δ^+ is bijective from s_r to itself for all $r < 1$, we conclude that equation (7.10) has a unique solution in the set $X_1 = \bigcup_{r<1} s_r$ given by

$$x = \Delta^+ \left(D_{(n)_{n=1}^\infty} C_1^T \Delta^+ \right)^{-1} D_{(n)_{n=1}^\infty} b$$

for all b with $D_{(n)_{n=1}^\infty} b = (nb_n)_{n=1}^\infty \in X_1$.

Concerning operators mapping from $(\ell_p)_\tau$ to itself we obtain the next application.

Corollary 7.2 ([18, Proposition 4.5, p. 3194]) *Let $A = (a_{nk})_{n,k=1}^\infty$ be an infinite matrix, and $\tau > 0$, $0 < \rho < 1$ and $1 < p < \infty$ satisfy the inequality*

$$\frac{\rho^p}{(1-\rho^q)^{p-1}} < \tau p - 1 \text{ with } q = \frac{p}{p-1}; \tag{7.12}$$

and we assume

$$\begin{cases} |a_{nk}| \leq \dfrac{1}{n^\tau} \text{ for } 1 \leq k < n-1 \\ a_{nn} = 1 \text{ for } n \geq 1 \\ a_{nk} = 0 \text{ otherwise.} \end{cases}$$

Then $A \in ((\ell_p)_{1/\rho}, (\ell_p)_{1/\rho})$ and for any $b \in (\ell_p)_{1/\rho}$, the equation $Ax = b$ has a unique solution in $(\ell_p)_{1/\rho}$.

Proof We have

$$\sigma_n = \sum_{\substack{k=1 \\ k \neq n}}^{\infty} |a_{nk}|^q \left(\frac{1}{\rho}\right)^{(k-n)q} \leq \frac{1}{n^{\tau q}} \sum_{k=1}^{n-1} \left(\frac{1}{\rho}\right)^{(k-n)q} \leq \frac{1}{n^{\tau q}} \frac{\rho^q}{1-\rho^q},$$

and so

$$\sum_{n=1}^{\infty} \sigma_n^{p-1} \leq \frac{\rho^p}{(1-\rho^q)^{p-1}} \sum_{n=2}^{\infty} \frac{1}{n^{\tau p}},$$

since $p - 1 = p/q$ and $\tau p - 1 > 0$ in (7.12). Now from the inequality

$$\sum_{n=2}^{\infty} \frac{1}{n^{\tau p}} \leq \int_1^\infty \frac{dx}{x^{\tau p}} = \frac{1}{\tau p - 1},$$

and using (7.12), we obtain

$$\left[N_{p,1/\rho}(I - A)\right]^p = \sum_{n=1}^{\infty} \sigma_n^{p-1} \leq \frac{\rho^p}{(1-\rho^q)^{p-1}} \cdot \frac{1}{\tau p \quad 1} < 1.$$

We conclude applying Part (ii) of Corollary 7.1. □

Remark 7.3 If we put $\rho = 1/2$, $p = q = 2$ and $\tau > 2/3$, then (7.12) holds and A is bijective from $(\ell_2)_2$ to itself.

7.3 Spectra of Operators Represented by Infinite Matrices

In this section, we give some applications of the preceding results to various domains, where the solvability of linear equations is used.

We recall that the *spectrum of an operator* $A \in (X, X)$ is the set $\sigma(A, X)$ of all complex numbers λ such that $A - \lambda I$ is not invertible. We denote by

$$\rho(A, X) = \mathbb{C} \setminus \sigma(A, X)$$

the *resolvent set of* A. There are many results on the spectrum and the fine spectrum of operators such as Δ, the Cesàro operator C_1 and the operator of weighted means mapping in sets of the form s_1, c_0, c, or ℓ_p (cf. ,for instance, Altay, Basar et al [5]). Here we focus on the cases when the operators Δ, $\Delta^T = \Delta^+$, and $\overline{N}_q$ are mapping between sets of the form s_τ. In the following, we need to explicitly compute the inverse of any Toeplitz operator mapping between sets of the form s_r. So we use the isomorphism $\varphi : f \to A$ from the algebra of the power series into the algebra of the corresponding matrices.

A recent paper [25] contains results on the spectra of special operators between generalized spaces of sequences that are strongly bounded or strongly convergent to zero.

First we recall that we can associate with any power series $f(z) = \sum_{k=0}^{\infty} a_k z^k$ defined in the open disc $|z| < R$ the upper triangular infinite matrix $A = \varphi(f) \in \bigcup_{0<r<R} S_r$ defined by

$$\varphi(f) = \begin{pmatrix} a_0 & a_1 & a_2 & . \\ & a_0 & a_1 & . \\ \mathbf{0} & & a_0 & . \\ & & & . \end{pmatrix}$$

(see [14]). Practically we write $\varphi[f(z)]$ instead of $\varphi(f)$. We have the following result.

Lemma 7.6 *(i) The map $\varphi : f \to A$ is an isomorphism from the algebra of the power series defined in $|z| < R$ into the algebra of the corresponding matrices $\bar{A}$.*

(ii) Let $f(z) = \sum_{k=0}^{\infty} a_k z^k$, with $a_0 \neq 0$, and assume that the power series $1/f(z) = \sum_{k=0}^{\infty} a'_k z^k$ has $R' > 0$ as radius of convergence. Then we have

$$\varphi\left(\frac{1}{f}\right) = [\varphi(f)]^{-1} \in \bigcup_{0<r<R'} S_r.$$

Now we give an application to the spectrum of the operator of the first differences mapping s_r to itself.

First we need to recall the definition of Δ^h. For $h \in \mathbb{C} \setminus \mathbb{N}$, we obtain

$$(1-z)^h = 1 + \sum_{i=1}^{\infty} \frac{-h(-h+1)\cdots(-h+i-1)}{i!} z^k \text{ for } |z| < 1. \tag{7.13}$$

If we write

$$\binom{-h+i-1}{i} = \begin{cases} \dfrac{-h(-h+1)\cdots(-h+i-1)}{i!} & \text{if } i > 0, \\ 1 & \text{if } i = 0 \end{cases}$$

for any given $h \in \mathbb{C}$, then we obtain

$$(\Delta^+)^h = \varphi[(1-z)^h] = \varphi\left[\sum_{i=0}^{\infty} \binom{-h+i-1}{i} z^i\right] \text{ for } |z| < 1.$$

We deduce $\Delta^h = (\varkappa_{nk})_{n,k=1}^{\infty}$ with

$$\varkappa_{nk} = \begin{cases} \binom{-h+n-k-1}{n-k} & \text{if } k \leq n \\ 0 & \text{if } k > n. \end{cases}. \tag{7.14}$$

We begin with the next results.

Lemma 7.7 ([14, Theorem 4, pp. 290–292]) *Let $r > 0$. Then we have*

(i) $\sigma(\Delta, s_r) = \overline{D}(1, 1/r)$;
(ii) $\sigma(\Delta^+, s_r) = \overline{D}(1, r)$.

Proof (i) First we let

$$\Lambda_\lambda(A) = \frac{1}{\lambda - 1}(\lambda I - A),$$

where A is a given matrix and $\lambda \neq 1$ is any complex number. We suppose that $\lambda \notin \sigma(\Delta, s_r)$. This means that $\lambda I - \Delta$ for $\lambda \neq 1$ is bijective from s_r into itself. Then $\lambda I - \Delta$ is invertible, since it is a triangle. The infinite matrix $(\lambda I - \Delta)^{-1}$ is also bijective from s_r into itself, and by Lemma 7.1, we have $(s_r, s_r) = S_r$ and $(\lambda I - \Delta)^{-1} \in S_r$. We are led to explicitly compute the inverse of $\lambda I - \Delta$. We have $(\lambda I - \Delta)^T = \varphi(\lambda - 1 + z)$ and

$$\left[(\lambda I - \Delta)^T\right]^{-1} = \varphi\left(\frac{1}{\lambda - 1 + z}\right).$$

Since

$$\frac{1}{\lambda - 1 + z} = \sum_{k=0}^{\infty}(-1)^k \frac{z^k}{(\lambda - 1)^{k+1}} \text{ for } |z| < |\lambda - 1|,$$

we obtain $(\lambda I - \Delta)^{-1} = (\xi_{nk})_{n,k=1}^{\infty}$, where

$$\xi_{nk} = \begin{cases} \dfrac{(-1)^{n-k}}{(\lambda - 1)^{n-k+1}} & \text{if } k \leq n \\ 0 & \text{otherwise.} \end{cases}$$

The condition $(\lambda I - \Delta)^{-1} \in S_r$ is then equivalent to

$$\chi = r \sup_n \left(\sum_{k=1}^{n} \frac{1}{(|\lambda - 1|r)^{n-k+1}} \right) = r \sup_n \left[\frac{\left(\frac{1}{|\lambda-1|r}\right)^{n+1} - \frac{1}{|\lambda-1|r}}{\frac{1}{|\lambda-1|r} - 1} \right] < \infty. \tag{7.15}$$

We see that (7.15) is equivalent to $1/(|\lambda - 1|r) < 1$. This shows that $\lambda \notin \overline{D}(1, 1/r)$. Conversely, we assume $\lambda \notin \overline{D}(1, 1/r)$. Since

$$\Lambda_\lambda(\Delta) = \frac{1}{\lambda - 1}(\lambda I - \Delta) = \begin{pmatrix} 1 & & & \\ 1/(\lambda - 1) & 1 & & \mathbf{0} \\ & 1/(\lambda - 1) & 1 & \\ \mathbf{0} & & & \ddots \end{pmatrix},$$

we obtain

$$\|\Lambda_\lambda(\Delta) - I\|_{S_r} = \frac{1}{|\lambda - 1|r} < 1,$$

and $\lambda I - \Delta$ is bijective from s_r into itself.
(ii) If $\lambda I - \Delta^+$ is bijective from s_r into itself, we have $(\lambda I - \Delta^+)^{-1} = (\xi_{nk}^+)_{n,k=1}^{\infty}$, where

$$\xi_{nk}^+ = \begin{cases} \dfrac{(-1)^{k-n}}{(\lambda - 1)^{k-n+1}} & \text{if } k \geq n \\ 0 & \text{otherwise.} \end{cases}$$

The condition $(\lambda I - \Delta^+)^{-1} \in S_r$ is equivalent to

$$\chi' = \sup_n \left(\sum_{k=n}^{\infty} \frac{1}{|\lambda - 1|^{k-n+1}} r^{k-n} \right) < \infty,$$

and to $r/|\lambda - 1| < 1$, which shows $\lambda \notin \overline{D}(1, r)$.
Conversely, we take $\lambda \notin \overline{D}(1, r)$. By similar arguments as those above, we have $\lambda I - \Delta^+ = \varphi(\lambda - 1 + z)$ and

$$\left\|\Lambda_\lambda(\Delta^+) - I\right\|_{S_r} = \left\|\frac{1}{\lambda - 1}\varphi(-z)\right\|_{S_r} = \frac{r}{|\lambda - 1|} < 1,$$

which shows that $\lambda I - \Delta^+$ is bijective from s_r into itself.
This completes the proof. □

Concerning the triangle $\Sigma = \Delta^{-1}$ we recall that $\Sigma_{nk} = 1$ for $k \leq n$. We can state the following result.

Proposition 7.3 ([14, Theorem 5, p. 292]) *Let $r > 1$. Then*

(i) $1/\lambda \in \overline{D}(1, 1/r)$ *if and only if* $\lambda \in \sigma(\Sigma, s_r)$.
(ii) *For all* $\lambda \notin \sigma(\Sigma, s_r)$, $\lambda I - \Sigma$ *is bijective from* s_r *into itself and*

$$[(\lambda I - \Sigma)^{-1}]_{nk} = \begin{cases} \eta_{nn} = \dfrac{1}{1-\lambda} & \text{for all } n \\ \eta_{nk} = \dfrac{1}{1-\lambda)^2} \cdot \left(\dfrac{-\lambda}{1-\lambda}\right)^{n-k-1} & \text{for } k \leq n \\ \eta_{nk} = 0 & \text{otherwise.} \end{cases} \tag{7.16}$$

Proof (i) follows from Part (i) of Lemma 7.7.
(ii) The operators Σ and Δ are bijective from s_r into itself for $r > 1$. We have $(\Sigma - \lambda I)^T = \varphi(1 - \lambda + z/(1-z))$ and

$$\frac{1}{1 - \lambda + \frac{z}{1-z}} = \frac{1-z}{1 - \lambda + \lambda z} = \frac{1}{1-\lambda} - \frac{1}{(1-\lambda)^2}\sum_{n=1}^{\infty}\left(\frac{-\lambda}{1-\lambda}\right)^{n-1} z^n.$$

So we obtain the entries of $(\lambda I - \Sigma)^{-1}$ in the statement in (7.16). □

Now we are interested in the study of the spectrum of the operator $\overline{N}_q$ considered as operator from s_τ into itself, or from s_τ^0 into itself.
We state the following theorem without proof and use the notation

$$J_q = \left\{\frac{q_n}{Q_n} : n \geq 1\right\}.$$

Theorem 7.1 ([20, Theorem 11, p. 417]) *We assume q is nondecreasing, $q/Q \in c_0$ and $\tau \in \Gamma$. Then we have*

(i) $s_\tau(\overline{N}_q - \lambda I) = s_\tau$ *for all* $\lambda \notin J_q \cup \{0\}$;
(ii) $s_\tau^0(\overline{N}_q - \lambda I) = s_\tau^0$ *for all* $\lambda \notin J_q \cup \{0\}$ *and*

$$\sigma\left(\overline{N}_q, s_\tau\right) = \sigma\left(\overline{N}_q, s_\tau^0\right) = J_q \cup \{0\}.$$

We immediately deduce from the preceding the following result, see also [12, Proposition 1, pp. 57–58].

Corollary 7.3 *Let $r > 1$. Then we have*

$$\sigma(C_1, s_r) = \sigma\left(C_1, s_r^0\right) = \{0\} \cup \{1/n : n \geq 1\}.$$

7.4 Matrix Transformations in $\chi(\Delta^m)$

In this section, we establish some properties of the set $\chi(\Delta)$ for $\chi = s_\tau, s_\tau^0, s_\tau^{(c)}$ or $(\ell_p)_\tau$. The characterization of the class $(\chi(\Delta^m), \chi(\Delta^m))$ given in [14] is complicated. So we deal with the subset $(\chi(\Delta^m), \chi(\Delta^m)_{\mathcal{L}'}$ of infinite Toeplitz triangles that map $\chi(\Delta^m)$ into itself.

First, we consider the sets $X = \chi(\Delta^m)$ for $\chi = s_\tau, s_\tau^0, s_\tau^{(c)}$ or $(\ell_p)_\tau$ with $1 \leq p < \infty$.

We recall that
$$\Gamma = \left\{\tau \in U^+ : \overline{\lim}_{n\to\infty}\left(\frac{\tau_{n-1}}{\tau_n}\right) < 1\right\}.$$

In the next lemma we recall the results stated above.

Lemma 7.8 *We have $\widehat{C} = \widehat{\Gamma} \subset \Gamma \subset \widehat{C_1}$.*

Lemma 7.9 *Let $\tau \in \mathcal{U}^+$ and $m \geq 1$ be an integer. Then*

(i) *$s_\tau^{(c)}(\Delta^m) = s_\tau^{(c)}$ if and only if $\tau \in \widehat{\Gamma}$.*
(ii) *If $\tau \in \Gamma$, then $s_\tau(\Delta^m) = s_\tau$, $s_\tau^0(\Delta^m) = s_\tau^0$ and $(\ell_p)_\tau(\Delta^m) = (\ell_p)_\tau$ for $1 \leq p < \infty$.*

Proof (i) follows from [16, Theorem 2.6, p. 1789].

(ii) By [16, Proposition 2.1, p. 1786], the condition $\tau \in \Gamma$ implies $\tau \in \widehat{C_1}$, so Δ and Δ^m are bijective from s_τ to itself and from s_τ^0 to itself; then $s_\tau(\Delta^m) = s_\tau$ and $s_\tau^0(\Delta^m) = s_\tau^0$.

It remains to show that if $\tau \in \Gamma$, then $(\ell_p)_\tau(\Delta) = (\ell_p)_\tau$ for $1 \leq p < \infty$ (cf. [18, Theorem 6.5, p. 3200]). If we let

$$l = \overline{\lim}_{n\to\infty}\left(\frac{\tau_{n-1}}{\tau_n}\right) < 1,$$

for given ε_0, such that $0 < \varepsilon_0 < 1 - l$, there exists N_0 such that

$$\sup_{n \geq N_0+1}\left(\frac{\tau_{n-1}}{\tau_n}\right) \leq l + \varepsilon_0 < 1.$$

Now we consider the infinite matrix

$$\Sigma_\tau^{(N_0)} = \begin{pmatrix} \left[\Delta_\tau^{(N_0)}\right]^{-1} & & 0 \\ & 1 & \\ 0 & & 1 \\ & & & \cdot \end{pmatrix},$$

where $\Delta_\tau^{(N_0)}$ is the finite matrix whose entries are those of $\Delta_\tau = D_{1/\tau}\Delta D_\tau$ for all $n, k \le N_0$. We obtain

$$Q = \Sigma_\tau^{(N_0)}\Delta_\tau = (q_{nk})_{n,k=1}^{\infty},$$

with

$$q_{nk} = \begin{cases} 1 & \text{for } k = n \\ -\dfrac{\tau_k}{\tau_{k+1}} & \text{for } k = n - 1 \ge N_0 \\ 0 & \text{otherwise.} \end{cases}$$

We obtain for every $x \in \ell_p$

$$(I - Q)x = \left(0, \dots, 0, \frac{\tau_{N_0}}{\tau_{N_0+1}} x_{N_0}, \dots, \frac{\tau_{n-1}}{\tau_n} x_{n-1}, \dots\right)^T,$$

where $\tau_{N_0} x_{N_0} / \tau_{N_0+1}$ is in the $(N_0 + 1)$ position. So we obtain

$$\|(I - Q)x\|_{l_p}^p = \sum_{n=N_0+1}^{\infty} \left(\frac{\tau_{n-1}}{\tau_n}\right)^p |x_{n-1}|^p \le \sup_{n \ge N_0+1} \left[\left(\frac{\tau_{n-1}}{\tau_n}\right)^p\right] \left(\sum_{n=N_0}^{\infty} |x_n|^p\right)$$

and

$$\|I - Q\|_{\mathcal{B}(\ell_p)}^* = \sup_{x \ne 0} \left(\frac{\|(I - Q)x\|_{\ell_p}}{\|x\|_{\ell_p}}\right) \le \left[\sup_{n \ge N_0+1} \left(\frac{\tau_{n-1}}{\tau_n}\right)^p\right]^{1/p}.$$

Since $\tau_{n-1}/\tau_n \le l + \varepsilon_0 < 1$ for all $n \ge N_0 + 1$, we deduce

$$\sup_{n \ge N_0+1} \left(\frac{\tau_{n-1}}{\tau_n}\right)^p < 1.$$

Hence

$$\|I - Q\|_{\mathcal{B}(\ell_p)} \le \left[\sup_{n \ge N_0+1} \left(\frac{\tau_{n-1}}{\tau_n}\right)^p\right]^{1/p} < 1.$$

We have shown that Q is invertible in $\mathcal{B}(\ell_p)$. Now let $b \in \ell_p$. The equations

$$\Delta_\tau x = b \text{ and } Qx = \Sigma^{(N_0)} b$$

are equivalent in ℓ_p. Since $Q^{-1} \in \mathcal{B}(\ell_p)$, using similar arguments as those in Proposition 7.2, we obtain

$$Q^{-1}(Qx) = (Q^{-1}Q)x = x = (\Delta_\tau)^{-1}b \text{ for all } x \in \ell_p.$$

This shows that the map Δ_τ is bijective from ℓ_p to ℓ_p and Δ is bijective from $(\ell_p)_\tau$ to $(\ell_p)_\tau$.

This completes the proof. □

Now we study the equation $Ax = b$ for $b \in \chi(\Delta)$, where $\chi \in \{s_\tau, s_\tau^0, s_\tau^{(c)}\}$.

First we recall the characterizations of the set $(\chi(\Delta), \chi(\Delta)$ for $\chi = s_\tau, s_\tau^0$, or $s_\tau^{(c)}$. For this we consider the following properties

$$\lim_{l\to\infty}\left(\sum_{j=1}^{l}\tau_j\left|\sum_{k=l}^{\infty}\left(\frac{a_{nk}-a_{n-1,k}}{\tau_n}\right)\right|\right) = 0 \text{ for all } n; \tag{7.17}$$

$$\sup_n\left|\sum_{j=1}^{\infty}\tau_j\left(\sum_{k=j}^{\infty}\left(\frac{a_{nk}-a_{n-1,k}}{\tau_n}\right)\right)\right| < \infty; \tag{7.18}$$

$$\lim_{n\to\infty}\left[\tau_j\left(\sum_{k=j}^{\infty}\left(\frac{a_{nk}-a_{n-1,k}}{\tau_n}\right)\right)\right] = 0 \text{ for all } j; \tag{7.19}$$

$$\lim_{l}\left(\sum_{j=1}^{l}\tau_j\left|\sum_{k=l}^{\infty}\left(\frac{a_{nk}-a_{n-1,k}}{\tau_n}\right)\right|\right) < \infty \text{ for all } n; \tag{7.20}$$

$$\lim_{n\to\infty}\left(\tau_j\left(\sum_{k=j}^{\infty}\left(\frac{a_{nk}-a_{n-1,k}}{\tau_n}\right)\right)\right) = l_j \text{ for all } j; \tag{7.21}$$

$$\lim_{n\to\infty}\sum_{k=1}^{\infty}\left[\left(\frac{a_{nk}-a_{n-1,k}}{\tau_n}\right)\left(\sum_{j=1}^{k}\tau_j\right)\right] = l \quad \text{exists.} \tag{7.22}$$

As a direct consequence of [23, Corollary 3.1, p. 669], and [34, Theorem 1], we obtain the following

Proposition 7.4 *Let $\tau \in \mathcal{U}^+$. Then*

(i) $A \in (s_\tau(\Delta), s_\tau(\Delta))$ if and only if (7.17) and (7.18) hold;
(ii) $A \in (s_\tau^0(\Delta), s_\tau^0(\Delta))$ if and only if (7.18), (7.19) and (7.20) hold;
(iii) $A \in (s_\tau^{(c)}(\Delta), s_\tau^{(c)}(\Delta))$ if and only if (7.18), (7.20), (7.21) and (7.22) hold.

We note that $A \in (\chi(\Delta^m), \chi(\Delta^m))$ if and only if

$$D_{1/\tau}\Delta^m A \in (\chi(\Delta^m), Y) \text{ with } Y \in \{\ell_\infty, c_0, c\},$$

(see [34]). Now we give necessary conditions for a matrix map to be bijective from $\chi(\Delta^m)$ into itself when $\chi \in \{s_\tau, s_\tau^0, s_\tau^{(c)}, (\ell_p)_\tau\}$ and $p \geq 1$. We state the next proposition.

Proposition 7.5 ([19, Proposition 13, p. 51]) *Let $\rho \in (0, 1)$ and $m \geq 1$ be an integer.*

(i) Let $b \in \chi(\Delta^m)$ and assume that there is $\rho \in (0, 1)$ such that

$$\left\|\Delta^m[(I - A)x]\right\|_{s_\tau} \leq \rho\|\Delta^m x\|_{s_\tau} \text{ for all } x \in \chi(\Delta^m). \tag{7.23}$$

Then

(a) $A \in (\chi(\Delta^m), \chi(\Delta^m))$ and, for any $b \in \chi(\Delta^m)$, the equation $Ax = b$ has a unique solution in $\chi(\Delta^m)$.

(b) If (7.23) holds and $\tau \in \widehat{\Gamma}$, then $A \in (\chi, \chi)$ and, for any $b \in \chi$ the equation $Ax = b$ has a unique solution in χ.

(ii)

(a) If $\tau \in \Gamma$ and $A^T \in \Gamma_{1/\tau}$, then $A \in (\chi, \chi)$ and, for any $b \in \chi$, the equation $Ax = b$ has a unique solution in $(\ell_1)_\tau(\Delta^m)$.

(b) Let $1 < p < \infty$. If $\tau \in \Gamma$ and $A \in \Gamma'_{p,\tau}$, then $A \in (\chi, \chi)$ and, for any $b \in \chi$, the equation $Ax = b$ has a unique solution in $(\ell_p)_\tau(\Delta^m)$.

Proof (i)

(a) It is enough to show the statement in the case when $X = s_\tau$. Using Lemma 7.2, the condition in (7.23) means that

$$\|I - A\|^*_{\mathcal{B}(s_\tau(\Delta^m))} = \sup_{x \neq 0}\left(\frac{\|\Delta^m[(I - A)x]\|_{s_\tau}}{\|\Delta^m x\|_{s_\tau}}\right) \leq \rho < 1.$$

Hence A is invertible in $\mathcal{B}(s_\tau(\Delta^m))$ and by Proposition 7.2, the equation $Ax = b$ has a unique solution in $s_\tau(\Delta^m)$ for $b \in s_\tau(\Delta^m)$.

(b) Since $\widehat{\Gamma} \subset \Gamma$, we easily deduce by Lemma 7.8 and Part (ii) of Lemma 7.9 that $\tau \in \widehat{\Gamma}$ implies $\chi(\Delta^m) = \chi$, $\|x\|_\chi = \|x\|_{s_\tau}$ for $\chi \in \{s_\tau, s_\tau^0, s_\tau^{(c)}\}$. We conclude, since the condition in (7.20) means that

$$\|I - A\|_{B(s_\tau)} = \|I - A|_{S_\tau} \leq \rho < 1.$$

(ii) By Lemma 7.9, the condition $\tau \in \Gamma$ implies that Δ is bijective from $(\ell_p)_\tau$ to itself for any given p with $1 \leq p < \infty$. Then Δ^m is also bijective and $(\ell_p)_\tau(\Delta^m) = (\ell_p)_\tau$. We conclude applying Part (ii) of Corollary 7.1 with $a = e$.
This completes the proof. □

Now we consider an interesting special case which is useful for many applications, when A is an infinite tridiagonal matrix.

An application

Let $\gamma, \eta \in \mathbb{C}$ and

$$M(\gamma,\eta)=\begin{pmatrix} 1 & \eta & & & \\ \gamma & 1 & \eta & \mathbf{O} & \\ & \gamma & 1 & \eta & \\ & \mathbf{O} & . & . & . \end{pmatrix}. \tag{7.24}$$

We obtain the following result as a direct consequence of Proposition 7.5.

Corollary 7.4 ([19, Example 14, pp. 51–52]) *For $\tau \in \mathcal{U}^+$ and*

$$\xi_\tau = |\gamma| \sup_{n\geq 2}\left(\frac{\tau_{n-1}}{\tau_n}\right) + |\eta| \sup_{n\geq 1}\left(\frac{\tau_{n+1}}{\tau_n}\right) < 1, \tag{7.25}$$

the equation

$$M(\gamma,\eta)x = b \tag{7.26}$$

with $b \in s_\tau(\Delta)$ has a unique solution $\overset{\circ}{x} = [M(\gamma,\eta)]^{-1}b \in s_\tau(\Delta)$. Furthermore if $\tau \in \Gamma$, then the equation (7.26) with $b \in s_\tau$ has a unique solution $\overset{\circ}{x} \in s_\tau$.

Proof First we have $[M(\gamma,\eta) - I]x = (\gamma x_{n-1} + \eta x_{n+1})_{n=1}^\infty$ with $x_0 = 0$. Then we easily obtain $\Delta[(M(\gamma,\eta) - I)x] = (y_n)_{n=1}^\infty$ with

$$y_n = \eta(x_{n+1} - x_n) + \gamma(x_{n-1} - x_{n-2}) = 0 \text{ and } x_{-1} = x_0 = 0.$$

Hence

$$\begin{aligned}\frac{1}{\tau_n}|y_n| &\leq |\eta|\cdot|x_{n+1}-x_n|\frac{1}{\tau_{n+1}}\cdot\frac{\tau_{n+1}}{\tau_n} + |\gamma|\cdot|x_{n-1}-x_{n-2}|\frac{1}{\tau_{n-1}}\cdot\frac{\tau_{n-1}}{\tau_n}\\ &\leq |\eta|\cdot\frac{\tau_{n+1}}{\tau_n}\cdot\|\Delta x\|_{s_\tau} + |\gamma|\cdot\frac{\tau_{n-1}}{\tau_n}\|\Delta x\|_{s_\tau}.\end{aligned}$$

Thus

$$\sup_n\left(\frac{1}{\tau_n}|y_n|\right) = \|\Delta[(I - M(\gamma,\eta))x]\|_{s_\tau} \leq \xi_\tau\|\Delta x\|_{s_\tau} < \|\Delta x\|_{s_\tau}$$

and

$$\|I - M(\gamma,\eta)\|^*_{\mathcal{B}(s_\tau(\Delta))} < 1.$$

As above we conclude that if (7.25) holds, then the equation (7.26) with $b \in s_\tau(\Delta)$ has the unique solution $\overset{\circ}{x} = [M(\gamma,\eta)]^{-1}b$ in $s_\tau(\Delta)$. If $\tau \in \Gamma$, then $s_\tau(\Delta) = s_\tau$ by Lemma 7.9 and we conclude from the preceding that the equation (7.26) with $b \in s_\tau$ has a unique solution in s_τ.
This completes the proof. □

Finally in this section, we deal with the equation $T_a x = b$, where T_a is a Toeplitz triangle matrix.

For any power series $f(z) = \sum_{n=0}^{\infty} a_n z^n$ where z belongs to an open disc we associate the *Toeplitz triangle*

$$T_a = \varphi(f)^T$$

with φ defined in Sect. 7.3. We denote by $\mathcal{L}' \subset \mathcal{L}$ the set of all infinite Toeplitz triangles. So $T = (t_{nk})_{n,k=1}^{\infty} \in \mathcal{L}'$ if there is a sequence $a = (a_n)_{n=1}^{\infty} \in \omega$ such that $t_{nk} = a_{n-k+1}$ for $k \leq n$ and $t_{nk} = 0$ otherwise. Then we write $T = T_a$; $\mathcal{L}'$ is a subset of the set of Toeplitz matrices, see [1]. So we have $T_a T_{a'} = T_{a'} T_a$ for all $a, a' \in \omega$. We then have the next result.

Proposition 7.6 ([19, Proposition 15, pp. 52–53]) *Let $m \geq 1$ be an integer and $T_a \in \mathcal{L}'$ for $a \in \mathcal{U}$. Then*

(i) $T_a \in \mathcal{B}(s_\tau(\Delta^m))$ *if and only if* $T_a \in S_\tau$;
(ii) $T_a \in \mathcal{B}(s_\tau^0(\Delta^m))$ *if and only if* $T_a \in S_\tau$ *and*

$$\lim_{n\to\infty} \frac{a_{n-k+1}}{\tau_n} = 0 \textit{ for all } k; \tag{7.27}$$

(iii) $T_a \in \mathcal{B}(s_\tau^{(c)}(\Delta^m))$ *if and only if* $T_a \in S_\tau$,

$$\lim_{n\to\infty} \frac{a_{n-k+1}}{\tau_n} = l_k \textit{ for all } k \tag{7.28}$$

and

$$\lim_{n\to\infty} \left(\sum_{k=1}^{n} a_{n-k+1} \frac{\tau_k}{\tau_n} \right) = l; \tag{7.29}$$

(iv) $\|T_a\|^*_{\mathcal{B}(s_\tau^0(\Delta^m))} = \|T_a\|^*_{\mathcal{B}(s_\tau(\Delta^m))} \leq \|T_a\|_{S_\tau}$ *and* $\|T_a\|^*_{\mathcal{B}(s_\tau(\Delta^m))} = \|T_a\|_{S_\tau}$ *for* $\tau \in \Gamma$.
(v) *Let* $1 \leq p < \infty$. *Then* $T_a \in \mathcal{B}((\ell_p)_\tau(\Delta^m))$ *if and only if* $T_a \in ((\ell_p)_\tau, (\ell_p)_\tau)$.

Proof (i) The set $s_\tau(\Delta^m)$ is a BK space by Lemma 7.2 with $T = \Delta^m$. So we have $T_a \in \mathcal{B}(s_\tau(\Delta^m))$ if and only if $T_a \in (s_\tau(\Delta^m), s_\tau(\Delta^m))$. Thus the condition $T_a \in \mathcal{B}(s_\tau(\Delta^m))$ is equivalent to $T_a(\Sigma^m x) \in s_\tau(\Delta^m)$ for all $x \in s_\tau$. So $T_a \in \mathcal{B}(s_\tau(\Delta^m))$ if and only if

$$\Delta^m \left[T_a(\Sigma^m x)\right] \in s_\tau \text{ for all } x \in s_\tau.$$

Since $\Delta^m, \Sigma^m \in \mathcal{L}'$, we have

$$\Delta^m \left[T_a(\Sigma^m x)\right] = \left(\Delta^m T_a \Sigma^m\right) x = \left(T_a (\Delta\Sigma)^m\right) x = T_a x \text{ for all } x \in s_\tau.$$

We conclude that $T_a \in \mathcal{B}(s_\tau(\Delta^m))$ if and only if $T_a \in (s_\tau, s_\tau)$, that is, $T_a \in S_\tau$.
(ii) As above, $s_\tau^0(\Delta^m)$ is a BK space and $T_a \in \mathcal{B}(s_\tau^0(\Delta^k))$ if and only if $T_a \in (s_\tau^0, s_\tau^0)$. Now we have $D_{1/\tau} T_a D_\tau = (\xi_{nk})_{n,k=1}^{\infty}$ with $\xi_{nk} = a_{n-k+1}\tau_k/\tau_n$ for $k \leq n$ and $\xi_{nk} = 0$ otherwise. Thus we obtain $D_{1/\tau} T_a D_\tau \in (c_0, c_0)$ if and only if $T_a \in S_\tau$ and (7.27) holds.

(iii) We obtain Part (iii) by a similar argument. So $T_a \in \mathcal{B}(s_\tau^0(\Delta^m))$ if and only if $T_a \in (s_\tau^0, s_\tau^0)$ and $T_a \in \mathcal{B}(s_\tau^{(c)}(\Delta^m))$ if and only if $T_a \in (s_\tau^{(c)}, s_\tau^{(c)})$.
(iv) We show $\|T_a\|^*_{\mathcal{B}(s_\tau(\Delta^m))} \leq \|T_a\|_{S_\tau}$. We have

$$\|T_a\|^*_{\mathcal{B}(s_\tau(\Delta^m))} = \sup_{x \neq 0} \left(\frac{\|T_a x\|_{s_\tau(\Delta^m)}}{\|x\|_{s_\tau(\Delta^m)}} \right) < \infty.$$

Since $\Delta^m \in \mathcal{L}'$, we have $\Delta^m T_a = T_a \Delta^m$ and, for every $x \in s_\tau(\Delta^m)$

$$\|T_a x\|_{s_\tau(\Delta^k)} = \left\| \Delta^m (T_a x) \right\|_{s_\tau} = \left\| T_a (\Delta^m x) \right\|_{s_\tau},$$

so

$$\|T_a x\|_{s_\tau(\Delta^m)} \leq \|T_a\|_{S_\tau} \|\Delta^m x\|_{s_\tau} = \|T_a\|_{S_\tau} \|x\|_{s_\tau(\Delta^m)}$$

and

$$\|T_a\|^*_{\mathcal{B}(s_\tau(\Delta^m))} \leq \|T_a\|_{S_\tau}.$$

Now it can easily be seen that $\|T_a\|^*_{\mathcal{B}(s_\tau(\Delta^m))} = \|T_a\|^*_{\mathcal{B}(s_\tau^\circ(\Delta^m))}$. In the case when $\tau \in \Gamma$, we have $s_\tau(\Delta^m) = s_\tau$ and we obtain by Part (i) of Proposition 7.1

$$\|T_a\|^*_{\mathcal{B}(s_\tau(\Delta^m))} = \|T_a\|_{S_\tau}.$$

(v) We see by Lemma 7.2 with $T = \Delta^m$ that $(\ell_p)_\tau(\Delta^m)$ is a BK space. We have $T_a \in \mathcal{B}((\ell_p)_\tau(\Delta^m))$ if and only if $T_a \in ((\ell_p)_\tau(\Delta^m), (\ell_p)_\tau(\Delta^m))$. Now $T_a \in ((\ell_p)_\tau(\Delta^m), (\ell_p)_\tau(\Delta^m))$ if and only if $\Delta^m(T_a \Sigma^m x) = T_a x \in (\ell_p)_\tau$ for all $x \in (\ell_p)_\tau$. This means that $T_a \in ((\ell_p)_\tau, (\ell_p)_\tau)$.
This concludes the proof. □

We also have the next result.

Proposition 7.7 ([19, Proposition 16, p. 54]) *Let $T_a \in \mathcal{L}'$ with $a_1 \neq 0$. Then we have*

(i) If $(1/a_1)T_a \in \Gamma_\tau$, then, for any given $b \in s_\tau(\Delta^m)$ the equation $T_a x = b$ has a unique solution in $s_\tau(\Delta^m)$

$$x = T_a^{-1} b. \tag{7.30}$$

(ii) If $(1/a_1)T_a \in \Gamma_\tau$ and (7.27) holds, then, for any given $b \in s_\tau^0(\Delta^m)$, the equation $T_a x = b$ has a unique solution in $s_\tau^0(\Delta^m)$ given by (7.30).
(iii) If $(1/a_1)T_a \in \Gamma_\tau$, (7.28) and (7.29) are satisfied, then, for any given $b \in s_\tau^{(c)}(\Delta^m)$, the equation $T_a x = b$ has a unique solution in $s_\tau^{(c)}(\Delta^m)$ given by (7.30).
(iv) (a) If $(1/a_1)T_a^t \in \Gamma_{1/\tau}$ (here T_a^t denotes the operator represented by the transpose of the matrix T_a that represents T_a), then, for any given $b \in (\ell_1)_\tau(\Delta^m)$, the equation $T_a x = b$ has a unique solution in $(\ell_1)_\tau(\Delta^m)$ given by (7.30).

(b) *Let* $1 < p < \infty$ *and* $(1/a_1)T_a \in \Gamma'_{p,\tau}$. *Then, for any given* $b \in (\ell_p)_\tau(\Delta^m)$, *the equation* $T_a x = b$ *has a unique solution in* $(\ell_p)_\tau(\Delta^m)$ *given by (7.30).*

Proof (i) Since $T_a \in \mathcal{L}$, the equation $T_a x = b$ is equivalent to $x = T_a^{-1}b$. Now the condition $(1/a_1)T_a \in \Gamma_\tau$ implies $T_a^{-1} \in S_\tau$. This means by Part (i) of Proposition 7.6 that $T_a^{-1} \in \mathcal{B}(s_\tau(\Delta^m))$. We conclude $T_a^{-1}b \in s_\tau(\Delta^m)$.
(ii) By Part (ii) of Proposition 7.6, we have $T_a \in \mathcal{B}(s_\tau^0(\Delta^m))$, since $T_a \in (s_\tau^0, s_\tau^0)$. Now $(1/a_1)T_a \in \Gamma_\tau$ and Part (iv) of Proposition 7.6 imply

$$\left\| I - \frac{1}{t_{11}} T_a \right\|^*_{\mathcal{B}(s_\tau(\Delta^m))} = \left\| I - \frac{1}{t_{11}} T_a \right\|^*_{\mathcal{B}(s_\tau^\circ(\Delta^m))} \leq \left\| I - \frac{1}{t_{11}} T_a \right\|_{S_\tau} < 1.$$

Then $T_a^{-1} \in \mathcal{B}(s_\tau^0(\Delta^m))$ and the unique solution of the equation $T_a x = b$ for $b \in s_\tau^0(\Delta^m)$ is $x = T_a^{-1}b \in s_\tau^0(\Delta^m)$.
(iii) We obtain Part (iii) from the characterization of the class $(s_\tau^{(c)}, s_\tau^{(c)})$ using similar arguments as those above.
(iv)

(a) is a direct consequence of Part (ii) (a) of Corollary 7.1 and Part (v) of Proposition 7.6.
(b) Here the condition $(1/a_1)T_a \in \Gamma'_{p,\tau}$ implies $T_a^{-1} \in ((\ell_p)_\tau, (\ell_p)_\tau)$, since $\mathcal{B}((\ell_p)_\tau) = ((\ell_p)_\tau, (\ell_p)_\tau)$. Then we obtain from Part (v) of Proposition 7.6

$$T_a^{-1} \in \big((\ell_p)_\tau(\Delta^m), (\ell_p)_\tau(\Delta^m)\big)$$

and for all $b \in (\ell_p)_\tau(\Delta^m)$, the unique solution $x = T_a^{-1}b$ of the equation $T_a x = b$ belongs to $(\ell_p)_\tau(\Delta^m)$.

□

Remark 7.4 The conditions $(1/a_1)T_a \in \Gamma_\tau$ and $(1/a_1)T_a^t \in \Gamma_{1/\tau}$ (cf. Proposition 7.7 for T_a^t) are equivalent to

$$\sup_{n\geq 2}\left(\sum_{k=1}^{n-1} |a_{n-k+1}|\frac{\tau_k}{\tau_n}\right) < |a_1| \text{ and } \sup_{n\geq 1}\left(\sum_{k=n+1}^{\infty} |a_{k-n+1}|\frac{\tau_n}{\tau_k}\right) < |a_1|,$$

respectively. We note that the condition $(1/a_1)T_a \in \Gamma'_{p,\tau}$ means

$$\sum_{n=2}^{\infty}\left(\sum_{k=1}^{n-1} |a_{n-k+1}|^{p/(p-1)}\right)^{p-1} < |a_1|^{p/(p-1)}.$$

Furthermore, it can be verified from Proposition 7.6 that if $T_a \in \mathcal{L}'$, then

(i) $T_a \in \mathcal{B}(c_0(\Delta^k))$ if and only if $T_a \in S_1$ and $a \in c_0$;
(ii) $T_a \in \mathcal{B}(c(\Delta^k))$ if and only if $T_a \in S_1$ and $a \in cs$.

7.5 The Equation $Ax = b$, Where A Is a Tridiagonal Matrix

In this section, we consider infinite tridiagonal matrices. These matrices are used in many applications, for instance, to continued fractions [13], the finite differences method [26, 32], and the Mathieu equation [21]. We deal with some properties of the matrix map $M(\gamma, a, \eta)$ between certain sequence spaces. Then we explicitly calculate the inverse of $M(\gamma, a, \eta)$.

Let $\gamma = (\gamma_n)_{n=1}^{\infty}$, $\eta = (\eta_n)_{n=1}^{\infty}$ and $a = (a_n)_{n=1}^{\infty}$ be sequences with $a \in \mathcal{U}$. We consider the infinite tridiagonal matrix

$$M(\gamma, a, \eta) = \begin{pmatrix} a_1 & \eta_1 & & & & \\ \gamma_2 & a_2 & \eta_2 & & \mathbf{O} & \\ & & \cdot & \cdot & \cdot & \\ \mathbf{O} & & \gamma_n & a_n & \eta_n & \\ & & & & \cdot & \cdot \end{pmatrix}$$

We then have the next result.

Proposition 7.8 ([19, Proposition 17, pp. 55–56]) *Let $D_{1/a}M(\gamma, a, \eta) \in \Gamma_\tau$, that is,*

$$\sup_n \left[\frac{1}{a_n} \left(|\gamma_n| \frac{\tau_{n-1}}{\tau_n} + |\eta_n| \frac{\tau_{n+1}}{\tau_n} \right) \right] < 1.$$

Then

(i) (a) $M(\gamma, a, \eta) \in (s_\tau, s_{|a|\tau})$;

(b) for any $b \in s_{|a|\tau}$, the equation $M(\gamma, a, \eta)x = b$ has a unique solution in s_τ;

(c) $M(\gamma, a, \eta)$ is invertible and $M(\gamma, a, \eta)^{-1} \in (s_{|a|\tau}, s_\tau)$;

(ii) (a) $M(\gamma, a, \eta) \in (s_\tau^0, s_{|a|\tau}^0)$;

(b) for any $b \in s_{|a|\tau}^0$, the equation $M(\gamma, a, \eta)x = b$ has a unique solution in s_τ^0;

(c) $M(\gamma, a, \eta)$ is invertible and $M(\gamma, a, \eta)^{-1} \in (s_{|a|\tau}^0, s_\tau^0)$.

(iii) If

$$\lim_{n\to\infty} \left[\frac{1}{a_n} \left(\gamma_n \frac{\tau_{n-1}}{\tau_n} + \eta_n \frac{\tau_{n+1}}{\tau_n} \right) \right] = l \neq 0,$$

then

(a) $M(\gamma, a, \eta) \in (s_\tau^{(c)}, s_{|a|\tau}^{(c)})$;

(b) for any $b \in s_{|a|\tau}^{(c)}$, the equation $M(\gamma, a, \eta)x = b$ has a unique solution in $s_\tau^{(c)}$,

(c) $M(\gamma, a, \eta)$ is invertible and $M(\gamma, a, \eta)^{-1} \in (s_{|a|\tau}^{(c)}, s_\tau^{(c)})$.

(iv) Let $p \geq 1$ be a real. If $\widetilde{K}_{p,\tau} = K_1 + K_2 < 1$ with

$$K_1 = \sup_n \left(\left| \frac{\gamma_n}{a_n} \right| \cdot \frac{\tau_{n-1}}{\tau_n} \right) \text{ and } K_2 = \sup_n \left(\left| \frac{\eta_n}{a_n} \right| \cdot \frac{\tau_{n+1}}{\tau_n} \right),$$

then

(a) $M(\gamma, a, \eta) \in ((\ell_p)_\tau, (\ell_p)_{|a|\tau})$;
(b) *for any $b \in (\ell_p)_{|a|\tau}$, the equation $M(\gamma, a, \eta)x = b$ has a unique solution in $(\ell_p)_\tau$;*
(c) *$M(\gamma, a, \eta)$ is invertible and $M(\gamma, a, \eta)^{-1} \in (\ell_p(|a|\tau), (\ell_p)_\tau)$.*

Proof Parts (i), (ii) and (iii) are direct consequences of Corollary 7.1.

(iv) We have

$$\left\| (I - D_{1/a} M(\gamma, a, \eta))x \right\|_{(\ell_p)_\tau} = \left(\sum_{n=1}^\infty \frac{1}{\tau_n^p} \left| \frac{\gamma_n}{a_n} x_{n-1} + \frac{\eta_n}{a_n} x_{n+1} \right|^p \right)^{1/p}$$
$$= \left(\sum_{n=1}^\infty \left| \frac{\gamma_n}{a_n} \cdot \frac{\tau_{n-1}}{\tau_n} \cdot \frac{x_{n-1}}{\tau_{n-1}} + \frac{\eta_n}{a_n} \cdot \frac{\tau_{n+1}}{\tau_n} \cdot \frac{x_{n+1}}{\tau_{n+1}} \right|^p \right)^{1/p}$$
$$\leq \left(\sum_{n=1}^\infty \left(K_1 \left| \frac{x_{n-1}}{\tau_{n-1}} \right| + K_2 \left| \frac{x_{n+1}}{\tau_{n+1}} \right| \right)^p \right)^{1/p}.$$

Applying Minkowski's inequality ((A.4)), we obtain

$$\left(\sum_{n=1}^\infty \left(K_1 \left| \frac{x_{n-1}}{\tau_{n-1}} \right| + K_2 \left| \frac{x_{n+1}}{\tau_{n+1}} \right| \right)^p \right)^{1/p} \leq$$
$$K_1 \left(\sum_{n=1}^\infty \left| \frac{x_{n-1}}{\tau_{n-1}} \right|^p \right)^{1/p} + K_1 \left(\sum_{n=1}^\infty \left| \frac{x_{n+1}}{\tau_{n+1}} \right|^p \right)^{1/p} \leq (K_1 + K_2) \|x|_{(\ell_p)_\tau}.$$

We conclude

$$\left\| I - D_{1/a} M(\gamma, a, \eta) \right\|^*_{\mathcal{B}((\ell_p)_\tau)} \leq (K_1 + K_2) < 1.$$

So $D_{1/a} M(\gamma, a, \eta)$ is invertible in $\mathcal{B}((\ell_p)_\tau)$ and $A = D_a(D_{1/a} M(\gamma, a, \eta))$ is bijective from $(\ell_p)_\tau$ into $(\ell_p)_{|a|\tau}$. Since $\mathcal{B}((\ell_p)_\tau) = ((\ell_p)_\tau, (\ell_p)_\tau)$, we conclude

$$\left[D_{1/a} M(\gamma, a, \eta) \right]^{-1} \in ((\ell_p)_\tau, (\ell_p)_\tau)$$

and $M(\gamma, a, \eta)^{-1} \in ((\ell_p)_{|a|\tau}, (\ell_p)_\tau)$.

This concludes the proof. □

We deduce the next corollary.

Corollary 7.5 *If $\widetilde{K}_{1,\tau} < 1$, then $M(\gamma, a, \eta)$ is bijective from $(\ell_1)_\tau$ to $(\ell_1)_{|a|\tau}$ and bijective from s_τ to $s_{|a|\tau}$.*

Proof First taking $p = 1$ in Part (iv) of Proposition 7.8, we deduce that A is bijective from $(\ell_1)_\tau$ to $(\ell_1)_{|a|\tau}$. Since

$$\left\| I - D_{1/a}M(\gamma, a, \eta) \right\|_{S_\tau} = \widetilde{K}'_\tau \leq \widetilde{K}_{1,\tau} < 1,$$

we conclude that $M(\gamma, a, \eta)$ is bijective from s_τ to $s_{|a|\tau}$.
This completes the proof. □

Remark 7.5 We note that in the case of $p = 1$, the condition

$$\left\| I - \left[D_{1/a}M(\gamma, a, \eta)\right]^t \right\|_{S_\tau} = \sup_n \left(\left| \frac{\gamma_{n+1}}{a_{n+1}} \right| \cdot \frac{\tau_{n+1}}{\tau_n} + \left| \frac{\eta_{n-1}}{a_{n-1}} \right| \cdot \frac{\tau_{n-1}}{\tau_n} \right) < 1$$

also implies that $M(\gamma, a, \eta)$ is bijective from $(\ell_1)_\tau$ to $(\ell_1)_{|a|\tau}$.

Now, among other things, we are interested in the calculation of the inverse of $M(\gamma, \eta)$ defined in Part 7.4.

We obtain the next proposition from the previous results.

Proposition 7.9 ([19, Proposition 20, pp. 57–58]) *Let $\gamma, \eta > 0$ with $\gamma + \eta < 1$. Then we have*

(i) $M(\gamma, \eta) : x \longmapsto M(\gamma, \eta)x$ is bijective from χ into itself for $\chi \in \{s_1, c_0, c\}$.

(ii) (a) Let χ be any of the sets s_1, c_0 or c and

$$u = \frac{1 - \sqrt{1 - 4\gamma\eta}}{2\gamma} \text{ and } v = \frac{1 - \sqrt{1 - 4\gamma\eta}}{2\eta}.$$

Then for any given $b \in \chi$, the equation $M(\gamma, \eta)x = b$ has a unique solution $\overset{\circ}{x} = (\overset{\circ}{x}_n)_{n=1}^\infty$ in χ given by

$$\overset{\circ}{x}_n = \left(\frac{uv+1}{uv-1}\right)(-1)^n v^n \sum_{k=1}^\infty [1 - (uv)^{-l}](-1)^k u^k b_k \text{ for all } n \tag{7.31}$$

with $l = \min\{n, k\}$.

(b) The inverse $[M(\gamma, \eta)]^{-1} = (a'_{nk})_{n,k=1}^\infty$ is determined by

$$a'_{nk} = \left(\frac{uv+1}{uv-1}\right)(-1)^{n+k} v^{n-k} \left[(uv)^l - 1\right]$$

$$\text{for all } n, k \geq 1 \text{ and } l = \min\{n, k\}. \tag{7.32}$$

Proof (i) We have $\|I - M(\gamma, \eta)\|_{S_1} = \gamma + \eta < 1$, so $M(\gamma, \eta) \in \Gamma_1$ and we conclude by Part (i) (a) of Corollary 7.1 with $\tau = e$.
(ii) Let u and v be reals with $0 < u, v < 1$. We consider the matrices

$$\Delta_u^+ = \begin{pmatrix} 1 & u & & \mathbf{0} \\ & 1 & u & \\ \mathbf{0} & & \cdot & \cdot \end{pmatrix} \text{ and } \Delta_v = \begin{pmatrix} 1 & & \mathbf{0} \\ v & 1 & \\ \mathbf{0} & v & 1 \\ & & \cdot \quad \cdot \end{pmatrix}.$$

A short calculation yields

$$\frac{1}{1+uv}\Delta_u^+\Delta_v = \begin{pmatrix} 1 & \frac{u}{1+uv} & & \mathbf{0} \\ \frac{v}{1+uv} & 1 & \frac{u}{1+uv} & \\ \mathbf{0} & \frac{v}{1+uv} & 1 & \frac{u}{1+uv} \\ & & \cdot & \cdot \end{pmatrix}.$$

Thus the identity

$$M(\gamma, \eta) = \frac{1}{1+uv}\Delta_u^+\Delta_v$$

is equivalent to

$$\eta = \frac{u}{1+uv} \text{ and } \gamma = \frac{v}{1+uv}.$$

Putting $\xi = 1 + uv$, we obtain $\eta = u/\xi$ and $\gamma = v/\xi$, so $\xi = 1 + \gamma\eta\xi^2$. Since $0 < \gamma + \eta < 1$, we have $4\gamma\eta < 1$ and $\xi = (1 \pm \sqrt{1 - 4\gamma\eta})/2(\gamma\eta)$ and the conditions $|u|, |v| < 1$ imply

$$u = \frac{1 - \sqrt{1 - 4\gamma\eta}}{2\gamma} \text{ and } v = \frac{1 - \sqrt{1 - 4\gamma\eta}}{2\eta}.$$

Then we have

$$\left\| I - \frac{1}{1+uv}\Delta_u^+\Delta_v \right\|_{S_1} = \gamma + \eta = \frac{u+v}{1+uv} < 1.$$

Furthermore we have $\|I - \Delta_u^+\|_{S_1} = u < 1$ and $\|I - \Delta_v\|_{S_1} = v < 1$. So the matrices Δ_u^+ and Δ_v are invertible in S_1. We have by Lemma 7.6

$$\left(\Delta_u^+\right)^{-1} = \varphi\left(\frac{1}{1+uz}\right) = \varphi\left(\sum_{k=0}^{\infty}(-1)^k u^k z^k\right)$$

and

$$(\Delta_v)^{-1} = \left[\varphi\left(\sum_{k=0}^{\infty}(-1)^k v^k z^k\right)\right]^T,$$

with $|uz|, |vz| < 1$. This means that

$$\left(\Delta_u^+\right)^{-1} = \Sigma_u^+ = \begin{pmatrix} 1 & -u & u^2 & . \\ & 1 & -u & u^2 \\ 0 & & . & . \\ & & & . \end{pmatrix}$$

and

$$(\Delta_v)^{-1} = \Sigma_v = \begin{pmatrix} 1 & & 0 \\ -v & 1 & \\ v^2 & -v & 1 \\ . & . & . & . \end{pmatrix}.$$

Then we have

$$\left(\frac{1}{1+uv}\Delta_u^+\Delta_v\right)^{-1} = (1+uv)\Sigma_v\Sigma_u^+.$$

We successively obtain for any given $b \in s_1$

$$\Sigma_u^+ b = \left(\sum_{k=0}^{\infty}(-1)^k u^k b_{n+k}\right)_{n=1}^{\infty},$$

$$\Sigma_v\left(\Sigma_u^+ b\right) = \left(\sum_{s=1}^{n}(-1)^{n-s}v^{n-s}\left(\sum_{k=0}^{\infty}(-1)^k u^k b_{s+k}\right)\right)_{n=1}^{\infty}$$

and the unique solution is given by

$$\overset{\circ}{x}_n = (1+uv)\sum_{s=1}^{n}\sum_{k=0}^{\infty}(-1)^{n-s+k}v^{n-s}u^k b_{s+k}. \tag{7.33}$$

Hence writing

$$\overset{\circ}{x}_n = (1+uv)(-v)^n\sigma_n \text{ with } \sigma_n = \sum_{s=1}^{n}(uv)^{-s}\sum_{k=0}^{\infty}(-u)^{k+s}b_{s+k}$$

putting $l = \min\{n, k\}$, we obtain

$$\sigma_n = \sum_{s=1}^{n}(uv)^{-s}\sum_{k=1}^{\infty}(-u)^k b_k = \sum_{s=1}^{n}(uv)^{-s}\sum_{k=s}^{n}(-u)^k b_k + \sum_{s=1}^{n}(uv)^{-s}\sum_{k=n+1}^{\infty}(-u)^k b_k$$

$$= \sum_{k=1}^{n} (-u)^k b_k \sum_{s=1}^{k} (uv)^{-s} + \sum_{k=n+1}^{\infty} (-u)^k b_k \sum_{s=1}^{n} (uv)^{-s}$$

$$= \sum_{k=1}^{\infty} (-u)^k b_k \sum_{s=1}^{l} (uv)^{-s} = \sum_{k=1}^{\infty} (-u)^k b_k \frac{(uv)^{-1} - (uv)^{-l-1}}{1 - (uv)^{-1}}$$

$$= \frac{1}{uv - 1} \sum_{k=1}^{\infty} (-u)^k b_k (1 - (uv)^{-l}),$$

that is, (7.31) and Part (a) holds. (b) is a direct consequence of the identity $\overset{\circ}{x}_n = \sum_{k=1}^{\infty} a'_{nk} b_k$ for all n, where $M^{-1} = (a'_{nk})_{n,k=1}^{\infty}$. This concludes the proof.
This concludes the proof. □

7.6 Infinite Linear Systems with Infinitely Many Solutions

In this section, we study systems that are obtained by adding or deleting one or several rows in a given system. Using this method we may determine some properties of the initial system and know if it has a few or infinitely many solutions.

We recall that $A = (a_{nk})_{n,k=1}^{\infty}$ is a *Pòlya matrix* if for any sequence $b \in \omega$, the equation $Ax = b$ has a solution in the space of the sequences $x = (x_n)_{n=1}^{\infty}$ such that the series $\sum_{k=1}^{\infty} a_{nk} x_k$ are absolutely convergent for all n.

The matrix A is called a *Pòlya matrix* if $a_{1k} \neq 0$ for infinitely many values of k and

$$\liminf_{k \to \infty} \sum_{i=1}^{n-1} \left| \frac{a_{ik}}{a_{nk}} \right| = 0 \ (n = 2, 3....).$$

The following result is due to Pòlya.

Theorem 7.2 (Pòlya's theorem) ([9, 37]) *If A is a Pòlya matrix, then for any $b \in \omega$, the equation $Ax = b$ has a solution such that the series $\sum_k a_{nk} x_k$ are absolutely convergent for all n.*

We recall briefly the well-known construction of a solution of such a system (cf. [9]). Here we consider the special case when the finite matrices obtained from A are successively

$$\widetilde{A_1} = (a_{11}), \ \widetilde{A_2} = \begin{pmatrix} a_{12} \ a_{13} \\ a_{22} \ a_{23} \end{pmatrix}, \dots$$

$$\widetilde{A_n} = \begin{pmatrix} a_{1,\alpha_n} & \cdot\ \cdot & a_{1,\alpha_n+n-1} \\ \cdot & \cdot\ \cdot & \cdot \\ \cdot & \cdot\ \cdot & \cdot \\ a_{n,\alpha_n} & \cdot\ \cdot & a_{n,\alpha_n+n-1} \end{pmatrix}, \dots$$

with $\alpha_n = (n^2 - n + 2)/2$ for all $n \geq 2$ and are invertible. Then a solution of $Ax = b$ can be obtained by the following method: The equation

$$\widetilde{A}_1 \widetilde{x}_1 = \widetilde{b}_1$$

with $\widetilde{x}_1 = (x_1)$, $\widetilde{b}_1 = (b_1)$ has $x_1 = b_1/a_{11}$ as a solution. In the same way, with x_1 defined by the preceding equation, we write $\widetilde{b}_2$ where $\widetilde{b}_2^{\,T} = (0, b_2 - a_{21}x_1)$ and consider the equation

$$\widetilde{A}_2 \widetilde{x}_2 = \widetilde{b}_2,$$

with $\widetilde{x}_2^T = (x_2, x_3)$. This equation has a unique solution given by

$$\widetilde{x}_2 = (\widetilde{A}_2)^{-1} \widetilde{b}_2.$$

Step by step, writing

$$\widetilde{b}_i^{\,T} = \left(0, 0, \ldots, b_i - \sum_{k=1}^{\alpha_i - 1} a_{i,k} x_k\right) \text{ for } i \geq 3,$$

where $x_1, \ldots, x_{\alpha_i - 1}$, are determined by $\widetilde{x}_j = (\widetilde{A}_j)^{-1} \widetilde{b}_j$ for $1 \leq j \leq i-1$, it was shown that $z = (\widetilde{x}_n)_{n=1}^{\infty}$ is a solution of $Ax = b$ satisfying Theorem 7.2.

Remark 7.6 In the case when not all of the matrices $\widetilde{A}_n$ are invertible, it is necessary to consider the first integer n_0 for which $\widetilde{A}_{n_0}$ is not invertible. It was shown by Pòlya that there is an integer $k_1 > n_0$ such that the matrix

$$\widetilde{A}'_1 = \begin{pmatrix} a_{1,\alpha_{n_0}} & \cdot\cdot & a_{1,\alpha_{n_0}+k_1-1} \\ \cdot & \cdot\cdot & \cdot \\ \cdot & \cdot\cdot & \cdot \\ a_{k_1,\alpha_{n_0}} & \cdot\cdot & a_{k_1,\alpha_{n_0}+k_1-1} \end{pmatrix}$$

is invertible. So we obtain by induction a strictly increasing sequence of integers $k_1 < k_2 < \cdots < k_n < \ldots$ and a sequence of corresponding matrices. By similar arguments as those above, we can determine a solution satisfying Theorem 7.2.

In fact, a Pòlya system has infinitely many solutions, see [9]. The matrix $A = (k^n)_{n,k=1}^{\infty}$ is an important example of a Pòlya matrix. The matrix

$$A = (a^{(k-n)^2})_{n,k=1}^{\infty},$$

where a is a real with $0 < a < 1$, is also a Pòlya matrix and satisfies $A \in \Gamma_1$. So, for any given bounded sequence b, the equation $Ax = b$ has a unique bounded solution $x = A^{-1}b$.

Now we define the matrix $A\langle t_1, \ldots, t_p \rangle$ with integer $p \geq 1$ obtained from A by adding the following rows

$$t_1 = (t_{11}, t_{12}, \dots),\ t_2 = (t_{21}, t_{22}, \dots), \dots$$
$$t_p = (t_{p1}, t_{p2}, \dots) \text{ with } t_{kk} \neq 0 \text{ for } k = 1, 2, \dots, p,$$

where $t_{nk} \in \mathbb{C}$, that is,

$$A\left\langle t_1, \dots, t_p \right\rangle = \begin{bmatrix} t_{11} & . . & t_{1k} & . . \\ . & . . & . & . . \\ t_{p1} & . . & t_{pk} & . . \\ a_{11} & . . & a_{1k} & . . \\ . & . . & . & . . \\ a_{n1} & . . & a_{nk} & . . \\ . & . . & . & . . \\ & & & \end{bmatrix}.$$

We let

$$b\left\langle u_1, \dots, u_p \right\rangle = (u_1, \dots, u_p, b_1, \dots)^T.$$

If $b = 0$, we write

$$\varpi\left\langle u_1, \dots, u_p \right\rangle = (u_1, ...u_p, 0, \dots)^T.$$

We also write $a'^{-1} = (a'^{-1}_{nn})_{n=1}^{\infty}$ where a'^{-1}_{nn} are the inverses of the diagonal elements of the matrix $A\left\langle t_1, \dots, t_p \right\rangle$. Then we have the following result (cf. [11, Theorem 1, pp. 51–52] and [32, Proposition 3, pp.158–159]).

Proposition 7.10 *Let $\tau \in \mathcal{U}^+$ and*

$$D_{a'^{-1}} A\left\langle t_1, \dots, t_p \right\rangle \in \Gamma_\tau \text{ and } D_{a'^{-1}} b\left\langle u_1, \dots, u_p \right\rangle \in s_\tau. \tag{7.34}$$

Then we have

(i) The solutions of $Ax = b$ in the space s_τ are given by

$$x = \left(D_{a'^{-1}} A\left\langle t_1, \dots, t_p \right\rangle\right)^{-1} D_{a'^{-1}} b\left\langle u_1, \dots, u_p \right\rangle \text{ for } u_1, u_2, \dots, u_p \in \mathbb{C}.$$

(ii) The linear space $Ker(A) \bigcap s_\tau$ of the solutions of $Ax = 0$ in the space s_τ is of dimension p and is given by

$$Ker(A) = \bigcap s_\tau = span\left(x'_1, \dots, x'_p\right),$$

where

$$x'_k = \left[A\left\langle t_1, \dots, t_p \right\rangle\right]^{-1} \varpi \left\langle 0, 0, \dots, 1, 0, \dots, 0 \right\rangle \text{ for } k = 1, 2, \dots, p$$

with 1 being the k-th term of the p-tuple.

Proof Since Part (i) is a direct consequence of (ii), we only show (ii).
For any scalars $u_1, \ldots, u_p$, the system

$$D_{a'^{-1}} A \langle t_1, \ldots, t_p \rangle x = D_{a'^{-1}} \varpi \langle u_1, \ldots, u_p \rangle$$

has one solution in s_τ given by

$$x = \left(D_{a'^{-1}} A \langle t_1, \ldots, t_p \rangle \right)^{-1} D_{a'^{-1}} \varpi \langle u_1, \ldots, u_p \rangle = \sum_{k=1}^{p} u_k x'_k,$$

which is also a solution of $Ax = 0$. It is clear that $(x'_k)_{1 \le k \le p}$ are linearly independent, since x'_k for $k = 1, 2, \ldots, p$) is the k-th column of $(A \langle t_1, \ldots, t_p \rangle)^{-1}$ and by (7.34), $(A \langle t_1, \ldots, t_p \rangle)^{-1}$ is injective in ϕ. Therefore the identity

$$\left(A \langle t_1, \ldots, t_p \rangle \right)^{-1} \varpi \langle u_1, \ldots, u_p \rangle = \sum_{k=1}^{p} u_k x'_k = 0 \text{ for } u_1, \ldots, u_p \in \mathbb{C}$$

implies $u_1 = \cdots = u_p = 0$.
Now we assume

$$\dim \left(Ker(A) \cap s_\tau \right) > p$$

and consider $x'_{p+1}, x'_{p+2}, \ldots, x'_{p+q}$ such that $(x'_1, \ldots, x'_p, x'_{p+1}, x'_{p+2}, \ldots, x'_{p+q})$ are linearly independent in the space $Ker A \cap s_\tau$. Then replacing some linear combination

$$x = \sum_{k=1}^{p+q} \lambda_k x'_k$$

(where λ_k is the unknown) in the system

$$A \langle t_1, \ldots, t_p \rangle x = \varpi \langle u_1, \ldots, u_p \rangle,$$

we obtain the finite system

$$\sum_{k=1}^{p+q} \lambda_k t_n x'_k = u_n \text{ for } n = 1, 2, \ldots, p,$$

where $t_n x'_k = \sum_{j=1}^{\infty} t_{nj} x_{jk}, x'_k = (x_{nk})_{n=1}^{\infty}$. This system cannot have a unique solution since the number of unknowns is strictly larger than the number of equations. So the hypothesis $\dim(Ker A \cap s_\tau) > p$ leads to a contradiction.
We note that this result is independant of the choice of the rows $t_1, t_2, \ldots, t_p$.
This concludes the proof. □

Remark 7.7 The solutions given by (i) can be written as $x = x^* + \sum_{i=1}^{p} u_i x'_i$, where

$$x^* = \left[D_{a'^{-1}} A \left\langle t_1, \ldots, t_p \right\rangle\right]^{-1} D_{a'^{-1}} b \left\langle 0, 0, \ldots, 0 \right\rangle \tag{7.35}$$

is a particular solution of $Ax = b$.

Example 7.4 ([32, Example 5, p. 159]) Let A be the infinite matrix

$$A = \begin{pmatrix} 0 & 1 & a & . & . & . \\ b^2 & 0 & 1 & a & . & . \\ b^3 & b^2 & 0 & 1 & a & . \\ . & . & . & . & . & . \\ & & & & & \end{pmatrix}.$$

We assume $0 < b < 1 < 1/a$. So $A \in S_1$ and it is clear that $A \notin \Gamma_1$, since the diagonal entries are naught. Now we consider the sequence $t_1 = (1, a, a^2, \ldots)$. We have

$$\|I - A\langle t_1 \rangle\|_{S_1} = \frac{b^2}{1-b} + \frac{a}{1-a}$$

and $A\langle t_1 \rangle \in \Gamma_1$ for infinitely many pairs of reals (a, b), (for instance, if $0 < a < 1/3$ and $0 < b < 1/2$). Then the system $Ax = b$ with $b \in s_1$ has infinitely many solutions in s_1 given by

$$x = (A\langle t_1 \rangle)^{-1} b \langle u_1 \rangle + u x_1' \text{ for } u \in \mathbb{C}.$$

Example 7.5 ([32, Example 6, p. 160]) Let A be the following matrix

$$A = \begin{pmatrix} 1 & 0 & \frac{1}{(2!)^2} & 0 & \frac{1}{(4!)^2} & \ldots \\ & 1 & 0 & \frac{1}{(2!)^2} & 0 & \ldots \\ & & 1 & 0 & \frac{1}{(2!)^2} & \ldots \\ & 0 & & 1 & 0 & \ldots \\ & & & & . & \ldots \\ & & & & & . \end{pmatrix}.$$

Since

$$\|I - A\|_{S_r} = \sum_{n=1}^{\infty} \left(\frac{r^n}{(2n!)}\right)^2,$$

we deduce $A \in \Gamma_r$ for $r < \sqrt{2}$. Thus the equation $Ax = b$ for $b \in s_r$ has a unique solution in s_r. Now we let

$$t_1 = \left(\frac{1}{(2!)^2}, 0, \frac{1}{(4!)^2}, 0, \frac{1}{(6!)^2}, 0, \ldots\right),$$

$$t_2 = \left(0, \frac{1}{(2!)^2}, 0, \frac{1}{(4!)^2}, 0, \frac{1}{(6!)^2}, \ldots\right).$$

Then we have $4A\langle t_1, t_2\rangle \in \Gamma_r$ for $r = 3$. So the system $Ax = b$ with $b \in s_3$ has infinitely many solutions in s_3 given by

$$x = (A\langle t_1, t_2\rangle)^{-1} b\langle 0, 0\rangle + u_0 x_0' + u_1 x_1' \text{ for } u_0, u_1 \in \mathbb{C}.$$

We also note that using the map φ we then can locate the zeros of the series

$$f(z) = \sum_{n=0}^{\infty} \frac{z^{2n}}{(2n!)^2}.$$

So we easily see that the equation $f(z) = 0$ has two zeros in the disc $|z| \leq 3$, four zeros in the disc $|z| < 6$, etc.....

Now, we can study the properties of matrices obtained by deleting some rows from A. This problem can be formulated as follows:
Let the matrix A and the sequence b satisfy (7.34). For a given integer q, let $t_n' = (t_{nk}')_{k=1}^{\infty}$ for $n = 1, \dots, q$ and

$$\sum_{k=1}^{\infty} |t_{nk}'| \tau_k < \infty \text{ for } n = 1, 2, \dots, q. \tag{7.36}$$

We note that we have not necessarily $D'A\left\langle t_1', \dots, t_q'\right\rangle \in \Gamma_\tau$, where D' is the diagonal matrix whose entries are the inverses of the diagonal elements of $A\left\langle t_1', \dots, t_q'\right\rangle$. Then we study the new equation for given $u_1', u_2', \dots, u_q' \in \mathbb{C}$

$$A\left\langle t_1', \dots, t_q'\right\rangle x = b\left\langle u_1', \dots, u_q'\right\rangle. \tag{7.37}$$

This equation is equivalent to the finite linear system

$$\sum_{k=1}^{p} \tilde{a}_{nk} u_k = v_n \text{ for } n = 1, 2, \dots, q, \tag{7.38}$$

where $u_1, \dots, u_p$ are the unknowns and $\tilde{a}_{nk} = t_n' x_k'$. Now we state the following result.

Corollary 7.6 *Let ζ denote the rank of the system in (7.38). We assume that hypotheses given (7.34) hold. Then we have*

(i) *If $\zeta = p = q$, then the system in (7.37) has a unique solution in s_τ.*
(ii) *If $\zeta = q < p$, then*

$$\dim Ker(A\left\langle t_1', \dots, t_q'\right\rangle) \cap s_\tau = p - \zeta$$

and the solutions of the system in (7.37) are given by $x = x^* + \sum_{i=1}^{p} u_i x_i'$, *where the scalars* $p - \zeta$ *among* $u_1, \ldots, u_p$ *are arbitrary.*

(iii) *If* $\zeta < q$, *then there is a scalar* χ *and an integer* $i \in \{\zeta + 1, \ldots, q\}$ *such that the system in (7.37) has no solution in* s_τ *for* $u_i' \neq \chi$.

Remark 7.8 Let $d \in \mathcal{U}^+$. If $\zeta < q$, then we see that for a given sequence $b \in \omega$ such that $b \langle u_1', \ldots, u_q' \rangle \in s_d$, the system in (7.37) has no solution in s_τ. This means

$$s_d \not\subset A \langle t_1', \ldots, t_q' \rangle s_\tau.$$

This method is used in the following for the study of the *Hill equation with a second member.*

7.7 The Hill and Mathieu Equations

Another application of the results of Sect. 7.6 is the *Hill equation*, which was studied by Brillouin [8], Ince [29], and more recently by Magnus and Winkler [33], Hochstadt [28] and Rossetto [38]. These authors used an *infinite determinant* and showed that the Hill equation has a nonzero solution if the determinant is zero. B. Rossetto provided an algorithm to calculate the *Floquet exponent* from the generalization of the notion of a characteristic equation and of a *truncated infinite determinant.*

Here we apply our results on infinite linear systems to the Hill equation and obtain a new method to calculate the Floquet exponent. Then we determine the solutions of the equation and give a method to approximate them. Then we consider the Hill equation with a second member.

The *Hill equation* is the well-known differential equation

$$y''(z) + J(z)y(z) = 0, \tag{7.39}$$

where $z \in \Omega$ and $\Omega \subset \mathbb{C}$ is an open set containing the real axis. The coefficient $J(z)$ can be written as an absolutely convergent Fourier series, that is,

$$J(z) = \sum_{k=-\infty}^{+\infty} \theta_k e^{2ikz}, \tag{7.40}$$

with $\theta_n = \theta_{-n}$ for every n. *Floquet's theorem* [36], shows that a solution of (7.39) can be written in the form

$$y(z) = e^{\mu z} \sum_{k=-\infty}^{+\infty} x_k e^{2ikz}, \tag{7.41}$$

where μ is a complex number, called referred to as *Floquet exponent*. To every μ for which the equation in (7.39) has a nonzero solution, we associate the series

$$b(z) = e^{\mu z} \sum_{k=-\infty}^{+\infty} b_k e^{2ikz} \tag{7.42}$$

and study the equation

$$y''(z) + J(z)y(z) = b(z). \tag{7.43}$$

It is natural to assume that the sum of the series $\sum_{k=-\infty}^{+\infty} x_k e^{2ikz}$ has a second derivative which is obtained by a term by termwise differentiation and that the corresponding series are absolutely convergent for all $z \in \Omega$. Substituting (7.41) and (7.42) in (7.43), we obtain the infinite linear system

$$(\mu + 2in)^2 x_n + \sum_{k=-\infty}^{\infty} \theta_{|n-k|} x_k = b_n \text{ for } n \in \mathbb{Z}, \tag{7.44}$$

which can be written in the form $Ax = b$, where the matrix $A = (a_{nk})_{n,k=0}^{\infty}$ is defined by

$$a_{nk} = \begin{cases} \theta_0 + (\mu + 2ni)^2 & \text{for } n = k \in \mathbb{Z} \\ \theta_{|n-k|} & \text{for } k \neq n \end{cases}$$

and

$$x^T = (\dots, x_{-1}, x_0, x_1, \dots) \; b^T = (\dots, b_{-1}, b_0, b_1, \dots).$$

To solve the system in (7.44) we have to determine $\mu \in \mathbb{C}$ for which the equation $Ax = 0$ has a nonzero solution in s_r. We note that Proposition 7.10 can be applied here. In this part, we use natural extensions of the sets s_r and S_r for $r > 0$ and denote the elements of s_r and S_r by $x = (x_n)_{n\in\mathbb{Z}}$ (or $(x_n)_{n=-\infty}^{\infty}$) and $(a_{nk})_{n,k=-\infty}^{\infty}$, respectively.

We recall that when the product Ax exists, it is defined by the one column matrix $(\sum_{k=-\infty}^{+\infty} a_{nk}x_k)_{n\in\mathbb{Z}}$. As it is suggested in the next numerical application, we consider the next hypothesis.

There are reals $\rho, r_0 > 0$ with $0 < \rho < 1$ and an integer s such that $\theta_0 \neq -(\mu + 2is)^2$ *and we have for $k \neq s$*

$$\begin{aligned} \left|\theta_0 + (\mu + 2is)^2\right| &\leq \sum_{n=1}^{\infty} |\theta_n| \left(r_0^n + r_0^{-n}\right) \\ &\leq \rho \left|\theta_0 + (\mu + 2ik)^2\right|. \end{aligned} \tag{7.45}$$

Then $a_{nn} \neq 0$ for every n, and the first inequality of (7.45) means

$$D_{1/a}A \notin \Gamma_{r_0}.$$

Now we consider the second member of the inequalities. The numerical application leads to deal with the case $s = 0$. It is also convenient to consider $s = -1$ or $s = 1$. For instance, we choose $s = 1$. Then it is necessary to consider the matrix A^* obtained by deleting the rows with indices 0 and 1 from A. This implies that the row with index -1 of A becomes the row with index 0 of A^* (and the row with index 2 of A becomes the row with index 1 of A^*). Now let $e'^{(p)} = (\dots, 0, 1, 0, \dots)$, where 1 is in the p-th position. Then $A^*\left\langle e'^{(0)}, e'^{(1)}\right\rangle = (a^*_{nk})^{\infty}_{n,k=-\infty}$ is the matrix obtained by insering the rows $e'^{(0)}$ and $e'^{(1)}$ between the rows with indices 0 and 1 of A^*, that is,

$$A^*\left\langle e'^{(0)}, e'^{(1)}\right\rangle = \begin{pmatrix} \cdot & & & & & \\ \cdot & & & & & \\ .\ \theta_1 & \theta_0 + (\mu - 2i)^2 & \theta_1 & \theta_2\ . & & \cdot \\ .\ . & 0 & 1 & 0\ . & & \cdot \\ & \cdot & 0 & 1 & 0 & \cdot \\ & \theta_3 & \theta_2 & \theta_1 & \theta_0 + (\mu + 4i)^2 & \theta_1 \\ & & & & \cdot & \end{pmatrix}.$$

Now let

$$x'_q = \sum_{n=0}^{\infty}[I - D^*A^*\left\langle e'^{(0)}, e'^{(1)}\right\rangle]^n D^* e'^{(q)^T} \text{ for } q = 0, 1 \qquad (7.46)$$

with $D^* = D_{(1/a^*_{nn})_{n\in\mathbb{Z}}}$. Then it can easily be seen by Proposition 7.10 that $\{x'_0, x'_1\}$ is a basis of $Ker A^* \cap s_{r_0}$. This result follows from Corollary 7.6 and the fact that the second member of the inequalities in (7.45) means $D^*A^*\left\langle e'^{(0)}, e'^{(1)}\right\rangle \in \Gamma_{r_0}$. We are led to state the next remark.

Remark 7.9 It can easily be seen that the identities in (7.45) may be written in the form $x'_q = (D^*A^*\left\langle e'^{(0)}, e^{(1)}\right\rangle)^{-1} e'^{(q)T}$ for $q = 0, 1$. We note that we obtain similar results by deleting the rows with indices -1 and 0 from A instead of those with indices 0 or 1.

Now we put

$$\tilde{a}_{pq} = A_p\left(x'_q\right) = \sum_{k=-\infty}^{\infty} a_{pk}x_{kq}, \qquad (7.47)$$

if $x'_q = (x_{kq})^{\infty}_{k=-\infty}$. By the second inequality in (7.45), the series defined in (7.47) are convergent for $p = 0, 1$ and for all $x \in Ker A^* \cap s_{r_0}$. We will see in the numerical application that there is $\mu = \mu_1$, which satisfies (7.45) and the condition

$$D(\mu) = \begin{vmatrix} \tilde{a}_{00} & \tilde{a}_{01} \\ \tilde{a}_{10} & \tilde{a}_{11} \end{vmatrix} = 0.$$

In the next results we write $A = A_\mu$. We obtain the following result as a direct consequence of Corollary 7.6.

Proposition 7.11 *(i) If there is a pair $(p, q) \in \{0, 1\}^2$ such that $\tilde{a}_{pq} \neq 0$, then*

$$\dim(Ker\, A_{\mu_1} \cap s_{r_0}) = 1.$$

(ii) If for every pair $(p, q) \in \{0, 1\}^2$, we have $\tilde{a}_{pq} = 0$, then

$$\dim(Ker\, A_{\mu_1} \cap s_{r_0}) = 2.$$

Proof (i) The substitution of $x = ux'_0 + vx'_1$ with $u, v \in \mathbb{C}$ in the equation $Ax = 0$ gives a finite linear system, whose terms $\tilde{a}_{pq}$ with $p, q \in \{0, 1\}$ are the coefficients and u and v are the unknowns. Then a special combination of x'_0 and x'_1 yields a basis of $Ker\, A_{\mu_1} \bigcap s_{r_0}$.
(ii) can be obtained similarly.

□

We immediately deduce the following result.

Corollary 7.7 *(i) Given any integer $k \in \mathbb{Z}$,* Proposition *7.11 is still true if we replace μ_1 by $\mu_1 + 2ik$.*
(ii) The equation in (7.39) has infinitely many solutions of the form

$$y(z) = e^{\mu_1 z} \sum_{k=-\infty}^{+\infty} x_k e^{2ikz},$$

where $(x_n)_{n=-\infty}^{\infty} \in s_r$. The space of these solutions is one dimensional if there exist a pair (p, q) for which $\tilde{a}_{pq} \neq 0$, and two dimensional otherwise.

Proof (i) follows from the equivalence of

$$x = (x_n)_{n=-\infty}^{\infty} \in Ker\, A_{\mu_1} \bigcap s_{r_0}$$

and

$$x'_k = (x_{n+k})_{n=-\infty}^{\infty} \in Ker(A_{\mu_1+2ki}) \cap s_{r_0}.$$

(ii) is a direct consequence of Proposition 7.11.
This concludes the proof. □

Remark 7.10 The number μ_1 is the unique value of μ for which (7.45) is satisfied.

Corollary 7.8 Proposition *7.11 is still satisfied when μ_1 is replaced by $\pm\mu_1 + 2ik$ for $k \in \mathbb{Z}$.*

Proof We write

$$A_\mu = \begin{pmatrix} C_\mu & C' \\ C'' & C'_\mu, \end{pmatrix}$$

where

$$C_\mu = (a_{nk})_{\substack{n\geq 0\\ k\leq 0}},\ C'_\mu = (a_{nk})_{\substack{n<0\\ k\geq 1}},\ C' = (a_{nk})_{\substack{n\geq 0\\ k\geq 1}} \text{ and } C'' = (a_{nk})_{\substack{n<0\\ k\leq 0}}.$$

In the same way, we define $\varkappa^T = (\varkappa_1^T, \varkappa_2^T)$, where $\varkappa_1^T = (\dots, x_{-1}, x_0)$ and $\varkappa_2^T = (x_1, x_2, \dots)$. We can easily see that the equation $A_{\mu_1}x = 0$ is equivalent to $C_{\mu_1}\varkappa_1 + C'\varkappa_2 = C''\varkappa_1 + C'_{\mu_1}\varkappa_2 = 0$. Now we put

$$\varkappa_1'^T = (\dots, x_1, x_0),\ \varkappa_2'^T = (x_{-1}, x_{-2}, \dots),$$

and $\varkappa'^T = (\varkappa_1'^T, \varkappa_2'^T)$. We deduce

$$C_{-\mu_1}\varkappa'_1 + C'\varkappa'_2 = C''\varkappa'_1 + C'_{-\mu_1}\varkappa'_2 = 0,$$

which shows $\varkappa' \in Ker A_{-\mu_1} \bigcap s_{r_0}$.
This concludes the proof. □

Now we study the Hill equation with a second member. We will see that there is no real $\rho > 0$ such that $s_\rho \subset As_r$. Let

$$A^*(l) = (a_{nk}(l))_{n,k=-\infty}^\infty \text{ and } b^*(l) = (b_n(l))_{n=-\infty}^\infty$$

be the matrix and the sequence obtained by deleting the rows and terms with indices l and $l+1$ from A and b, respectively. Then we have $A^* = A^*(0)$. We also note that we have for $q = l, l+1$

$$x'_q = (A^*(l)\langle e'^{(l)}, e'^{(l+1)}\rangle)^{-1}(e'^{(q)T}) \in Ker(A^*(l)) \cap s_{r_0}.$$

When $\mu = \mu_1 - 2li$, we obtain the following result, where we write $\kappa_r(l)$ for the set of all sequences $b^*(l)$ such that $D^*(l)b^*(l) \in s_r$ with $D^*(l) = ((\delta_{nk}/a_{nk}))(l))_{n,k=-\infty}^\infty$.

Proposition 7.12 ([15, Proposition 6, p. 24]) *For any given sequence $b^*(l) \in \kappa_r(l)$, we associate a space V_l of dimension $p = 0$ or 1 which satisfies the following property. The dimension of the space of the solutions of (7.43) that belong to s_r is equal to $2 - p$ if and only if $(b_l, b_{l+1}) \in V_l$. Then*

(i) If $p = 0$, then we have

$$x = \left(A^*(l)\langle e'^{(l)}, e'^{(l+1)}\rangle\right)^{-1} b^*(l)\langle 0, 0\rangle + ux_l + vx_{l+1} \text{ for } u, v \in \mathbb{C}.$$

(ii) If $p = 1$, then there are $u_0, v_0 \in \mathbb{C}$ such that

$$x = \left(A^*(l)\left\langle e'^{(l)}, e'^{(l+1)}\right\rangle\right)^{-1} b^*(l)\,\langle 0, 0\rangle + w(u_0 x_l + v_0 x_{l+1})\; for\; w \in \mathbb{C}.$$

Proof First, we deal with the case $l = 0$. By Proposition 7.11, the solutions in s_r of the equation $A^* x = b^*(0)$ are determined by

$$x = \left(A^*\left\langle e'^{(0)}, e'^{(1)}\right\rangle\right)^{-1} b^*(0)\,\langle u, v\rangle = x^* + u x'_0 + v x'_1 \text{ for } u, v \in \mathbb{C},$$

where

$$\left(b^*(0)\,\langle u, v\rangle\right)^T = (\ldots, b_{-1}, u, v, b_2, \ldots),\; x^* = \left(A^*\left\langle e'^{(0)}, e'^{(1)}\right\rangle\right)^{-1} b^*(0)\,\langle 0, 0\rangle .$$

Then the solutions of $Ax = b$ in s_r are given by

$$x = x^* + u x'_0 + v x'_1 \text{ for } u, v \in \mathbb{C}$$

with the additional equations $A_p(x) = b_p$ with $p = 0, 1$. We obtain a finite linear system, where u and v are the unknowns, and V_0 can be determined by similar arguments as those used in Proposition 7.11.
When $l \neq 0$, then a translation of l rows for b and l rows and l columns for A reduces to the preceding case.
This completes the proof. □

Now we give the following numerical application.
The coefficients θ_k of Hill equation are given by

$$\theta_0 = 1.158843,\; \theta_1 = -0.057044,\; \theta_2 = 0.00038323,$$
$$\theta_3 = -0.00000917,\; \theta_4 = -0.000000001049,$$

and so on ... $|\theta_n| \leq 10^{-n}$ for $n \geq 2$. We assume that μ satisfies (7.45). It can easily be seen that the entries of the matrix $(I - D^* A^*\left\langle e'^{(0)}, e'^{(1)}\right\rangle)^n$ for $n \geq 2$ are very " small". Then, for the calculations of x'_i for $i = 0, 1$, it is enough to consider the sum of the two first terms of the series. For a given integer N, the vector defined by

$$x_q'^{(N)} = \sum_{n=0}^{N} \left(I - D^* A^*\left\langle e'^{(0)}, e'^{(1)}\right\rangle\right)^n D^{*t} e'^{(q)^T} \tag{7.48}$$

is an approximation of x'_q for $q = 0, 1$. Now we put

$$\rho = \left\| I - D^* A^*\left\langle e'^{(0)}, e'^{(1)}\right\rangle \right\|_{S_1} .$$

Then from the identity in (7.46), we obtain

$$\left\| x_q'^{(N)} - x'_q \right\|_{s_r} \leq \frac{\rho^{N+1}}{1 - \rho}.$$

In a first approximation, and to illustrate the preceding results, we take $N = 2$. We write $\tau_n = \theta_0 + (\mu + 2in)^2$. Then the computation of $x_q'^{(2)}$ shows that the numbers

$$\tilde{a}_0 = -\frac{\theta_1^2}{\tau_{-1}\tau_0} + 1 - \frac{\theta_2^2}{\tau_0\tau_2}, \quad \tilde{a}_1 = \frac{\theta_1\theta_2}{\tau_{-1}\tau_1} + \frac{\theta_1}{\tau_1} - \frac{\theta_1\theta_2}{\tau_0\tau_2},$$

$$\tilde{a}_2 = \frac{\theta_1\theta_2}{\tau_{-1}\tau_0} + \frac{\theta_1}{\tau_0} - \frac{\theta_1\theta_2}{\tau_0\tau_2}, \quad \tilde{a}_3 = -\frac{\theta_2^2}{\tau_{-1}\tau_1} + 1 - \frac{\theta_1^2}{\tau_1\tau_2}$$

are approximate values of $\tilde{a}_{00}$, $\tilde{a}_{01}$, $\tilde{a}_{10}$ and $\tilde{a}_{11}$. Now we write $\mu = ix$ for $x \in \mathbb{R}$ and put

$$\sigma = 2\sum_{k=1}^{\infty} |\theta_k|.$$

Then the first inequality defined by (7.45) shows that x belongs to one of the intervals

$$\left[-\sqrt{\theta_0 + 2\sigma}, -\sqrt{\theta_0 - 2\sigma}\right], \left[\sqrt{\theta_0 - 2\sigma}, \sqrt{\theta_0 + 2\sigma}\right],$$

$$\left[-2 - \sqrt{\theta_0 + 2\sigma}, -2 - \sqrt{\theta_0 - 2\sigma}\right], \left[-2 + \sqrt{\theta_0 - 2\sigma}, -2 + \sqrt{\theta_0 + 2\sigma}\right].$$

Using the second inequality of (7.45) and the equality $D(\mu) = 0$, we see that if x is chose n in the first interval, then approximate values of $\tilde{a}_{00}$, $\tilde{a}_{01}$, $\tilde{a}_{10}$ and $\tilde{a}_{11}$ are 1, θ_1/τ_1, θ_1/τ_0 and 1, respectively. Then the number $1 - \theta_1^2/\tau_0\tau_1$ is an approximation of $D(\mu)$ and using Corollary 7.8, we obtain $\mu = \pm i \cdot 1.071379 + 2ki$ for $k \in \mathbb{Z}$.

In the case when $k = 1$, the number $\mu = i \cdot 0.9286$ is a good approximation of the Floquet exponent $\mu = i \cdot 0.9284$ obtained by other methods, (see [29, 36, 38]). This result may be improved with $N = 3$. We also have

$$\dim KerA \cap s_{r_0} = 1$$

and information on the non-homogeneous Hill differential equation. Then we obtain a method to approximate the solutions of Hill equation. In fact, we have

$$x^{(N)} = x^{*(N)} + u_N x_0'^{(N)} + v_N x_1'^{(N)},$$

where

$$x^{*(N)} = \sum_{n=0}^{N} \left(I - D^* A^* \langle e'^{(0)}, e'^{(1)} \rangle\right)^n D^* b^*(0) \langle 0, 0 \rangle$$

and u_N, v_N are determined by the system

$$u_N A_0 \left(x_0'^{(N)}\right) + v_N A_0 \left(x_1'^{(N)}\right) = b_0 - a_0 x^{*(N)},$$

$$u_N A_1\left(x_0^{\prime(N)}\right) + v_N A_1\left(x_1^{\prime(N)}\right) = b_1 - a_1 x^{*(N)}.$$

Finally, we consider the Mathieu equation.

Using similar arguments as those above we may deal with the Mathieu equation, (see for instance [21]) defined by

$$y^{(4)}(t) + q y^{(2)}(t) + (p + 2a \cos 2t) y(t) = 0 \ (t \in \mathbb{R}), \tag{7.49}$$

where $a \neq 0$, p and q are reals. The Floquet theorem shows that a solution of (7.49) can be written in the form $y(t) = e^{\mu t} \sum_{m=-\infty}^{+\infty} x_m e^{2imt}$, where μ is the Floquet exponent. As we have seen above, for every μ such that (7.49) has a nontrivial solution, we may consider the equation

$$y^{(4)}(t) + q y^{(2)}(t) + (p + 2a \cos 2t) y(t) = b(t) \text{ for } t \in \mathbb{R},$$

where $b(t) = e^{\mu t} \sum_{m=-\infty}^{+\infty} b_m e^{2imt}$. So we are led to study the infinite linear system

$$a x_{m-1} + \kappa_m(\mu) x_m + a x_{m+1} = b_m \text{ for all } m \in \mathbb{Z}, \tag{7.50}$$

where $\kappa_m(\mu) = p + q(2im + \mu)^2 + (2im + \mu)^4 \neq 0$ for all $m \in \mathbb{Z}$. We see that (7.50) can be written in the form $\mathcal{M}_\mu x = b$, where $x^T = (\ldots, x_{-1}, x_0, x_1, \ldots)$, $b^T = (\ldots, b_{-1}, b_0, b_1, \ldots)$, and $\mathcal{M}_\mu$ is an infinite tridiagonal matrix defined by

$$\mathcal{M}_\mu = \begin{pmatrix} \ddots & & & & \\ a & \kappa_{-1}(\mu) & a & & 0 \\ & a & \kappa_0(\mu) & a & \\ & & a & \kappa_1(\mu) & a \\ 0 & & & & \ddots \\ & & & & \ddots \end{pmatrix}.$$

We may apply the previous method used for the Hill equation to study the equation $\mathcal{M}_\mu x = b$ where $\mathcal{M}_\mu$ maps in ℓ_∞.

References

1. Silbermann, B., Böttcher, A.: Introduction to Large Truncated Toeplitz Matrices. Springer, New York (2000)
2. Akhmedov, A.M., Başar, F.: On the spectra of the difference operator Δ over the sequence space ℓ_p. Demonstr. Math. **39**(3), 585–295 (2006)
3. Akhmedov, A.M., Başar, F.: On the fine spectrum of the operator Δ over the sequence space bv_p $(1 \leq p < \infty)$. Acta Math. Sin. Eng. Ser. **23**(10), 1757–1768 (2007)
4. Akhmedov, A.M., El-Shabrawy, S.R.: On the fine spectrum of the operator $\Delta_{a,b}$ over the sequence space c. Comp. Math. Appl. **61**(10), 2994–3002 (2011)

5. Altay, B., Başar, F.: On the fine spectrum of the generalized difference operator $B(r, s)$ over the sequence spaces c_0 and c. Int. J. Math. Math. Sci. **18**, 3005–3013 (2005)
6. Başar, F., Karaisa, A.: On the fine spectrum of the generalized difference operator defined by a double sequential band matrix over the sequence space ℓ_p $(1<p<\infty)$. Hacet. J. Math. **44**(6), 1315–1332 (2015)
7. Bilgiç, H., Furkan, H.: On the fine spectrum of the generalized difference operator $B(r, s)$ over the sequence spaces ℓ_p and bv_p $(1<p<\infty)$. Nonlinear Anal. **68**(3), 499–506 (2008)
8. Brillouin, L.: A practical method for solving Hill's equation. Q. Appl. Math. **6**(2), 167–179 (1948)
9. Cooke, R.C.: Infinite Matrices. MacMillan and Co., Ltd, London (1950)
10. Das, R.: Spectrum and fine spectrum of the Zweier matrix over the sequence space cs. Bol. Soc. Parana. Mat. **35–2**, 209–221 (2017)
11. de Malafosse, B.: Systèmes linéaires infinis admettant une infinité de solutions. Atti. Accad. Peloritana Pericolanti, Cl. I Fis. Mat. Nat. **65**, 49–59 (1988)
12. de Malafosse, B.: On the spectrum of the Cesàro operator in the space s_r. Comm. Fac. Sci. Univ. Ankara Ser. A1 Math. Stat. **48**, 53–71 (1999)
13. de Malafosse, B.: Some new properties of sequence spaces, and application to the continued fractions. Mat. Vesnik **53**, 91–102 (2001)
14. de Malafosse, B.: Properties of some sets of sequences and application to the spaces of bounded difference sequences of order μ. Hokkaido Math. J. **31**, 283–299 (2002)
15. de Malafosse, B.: Recent results in the infinite matrix theory and application to Hill equation. Demonstr. Math. **35**(1), 11–26 (2002)
16. de Malafosse, B.: On some BK space. Int. J. Math. Math. Sci. **58**, 1783–1801 (2003)
17. de Malafosse, B.: The Banach algebra S_α and applications. Acta Sci. Math. Szeged **70**, 125–145 (2004)
18. de Malafosse, B.: On the Banach algebra $\mathcal{B}(\ell_p(\alpha))$. Int. J. Math. Math. Sci. **60**, 3187–3203 (2004)
19. de Malafosse, B.: The Banach algebra $\mathcal{B}(X)$, where X is a BK space and applications. Mat. Vesnik **57**, 41–60 (2005)
20. de Malafosse, B.: Linear operators mapping in new sequnece spaces. Soochow J. Math. **31**(3), 403–427 (2005)
21. de Malafosse, B.: An application of the infinite matrix theory to Mathieu equation. Comput. Math. Appl. **52**, 1439–1452 (2006)
22. de Malafosse, B.: On some Banach algebras and applications to the spectra of the operator $B(\widetilde{r}, \widetilde{s})$ mapping in new sequence spaces. Appl. Math. Lett. **2**, 7–18 (2014)
23. de Malafosse, B., Malkowsky, E.: Sets of difference sequences of order m. Acta Sci. Math. (Szeged) **70**, 659–682 (2004)
24. de Malafosse, B., Malkowsky, E.: On the Banach algebra $(w_\infty(\lambda), w_\infty(\lambda))$ and applications to the solvability of matrix equations in $w_\infty(\lambda)$. Pub. Math. Debrecen **85**(1–2), 197–217 (2014)
25. de Malafosse, B., Malkowsky, E.: On the spectra of the operator $B(\tilde{r}, \tilde{s})$ mapping in $(w_\infty(\lambda))_a$ and $(w_0(\lambda))_a$ where λ is a nondecreasing exponetially bounded sequence. Mat. Vesn. **70**(1), 30–42 (2020)
26. de Malafosse, B., Medeghri, A.: Numerical scheme for a complete abstract second order differential equation of elliptic type. Commun. Fac. Sci. Ankara, Ser. A1 Math. and Stat. **50**, 43–54 (2001)
27. Furkan, H., Bilgiç, H., Başar, F.: On the fine spectrum of the operator $B(r, s, t)$ over the sequence spaces ℓ_p and bv_p, $(1<p<\infty)$. Comp. Math. Appl. **60**(7), 2141–2152 (2010)
28. Hochstadt, H.: Function theoretic properties of the discriminant of Hill's equation. Math. Zeit. **82**, 237–242 (1963)
29. Ince, E.L.: Periodic solutions of a linear differential equation of the second order. Proc. Cam. Philos. Soc. **23**, 44–46 (1926)
30. Karaisa, A., Başar, F.: Fine spectra of upper triangular triple–band matrices over the sequence space ℓ_p, $(0<p<\infty)$. Abstr. Appl. Anal., Article ID 342682, 10 (2013). https://doi.org/10.1155/2013/342682.

31. Karakaya, V., Altun, M.: Fine spectra of upper triangular double-band matrices. J. Comput. Appl. Math. **234**, 1387–1394 (2010)
32. Labbas, R., de Malafosse, B.: On some Banach algebra of infinite matrices and applications. Demonst. Math. **31**, 153–168 (1998)
33. Magnus, W., Winkler, S.: Hill's Equation. Wiley, New York (1966)
34. Malkowsky, E.: Linear operators in certain BK spaces. Bolyai Soc. Math. Stud. **5**, 259–273 (1996)
35. Malkowsky, E., Rakočević, V.: An introduction into the theory of sequence spaces and measures of noncompactness, vol. 9(17), Zbornik radova, Matematčki institut SANU, pp. 143–234. Mathematical Institute of SANU, Belgrade (2000)
36. Poincaré, H.: Sur les déterminants d'ordre infini. Bull. Soc. Math. Fr. **14**, 87 (1886)
37. Reade, J.B.: On the spectrum of the Cesàro operator. Bull. London Math. Soc. **17**, 263–267 (1985)
38. Rosseto, B.: Détermination des exposants de Floquet de l'équation de Hill d'ordre n . PhD thesis, Doctorat d'Etat Es-Sciences, Univ. Toulon (1983)
39. Srivastava, P.D., Kumar, S.: Fine spectrum of the generalized difference operator Δ_1 on sequence space ℓ_1. Thai J. Math. **8**(2), 7–19 (2010)
40. Srivastava, P.D., Kumar, S.: Fine spectrum of the generalized difference operator Δ_{uv} on sequence space ℓ_1. Appl. Math. Comput. **218**(11), 6407–6414 (2012)
41. Srivastava, P.D., Kumar, S.: Fine spectrum of the generalized difference operator Δ_{uv} on the sequence space c_0. Thai J. Math. **16**(3), 615–633 (2018)
42. Yeşilkayagil, M., Başar, F.: On the fine spectrum of the operator defined by a λ matrix over the sequence spaces of null and convergent sequences. Abstr. Appl. Anal., Article ID 687393, 13 (2013). https://doi.org/10.1155/2013/687393.
43. Yeşilkayagil, M., Başar, F.: On the paranormed Nörlund sequence space of non-absolute type. Abstr. Appl. Anal., Article ID 858704, 9 (2014). https://doi.org/10.1155/2014/858704.
44. Yeşilkayagil, M., Başar, F.: A survey for the spectrum of triangles over sequence spaces. Numer. Funct. Anal. Optim. (2020). (in press)

Appendix
Inequalities

A.1 Inequalities

Let $a, b \in \mathbb{C}$ and $0 < p \leq 1$. Then we have

$$(|a+b|)^p \leq |a|^p + |b|^p. \tag{A.1}$$

Let $a, b \in \mathbb{C}$, $1 < p < \infty$ and $q = p/(p-1)$. Then we have

$$|ab| \leq \frac{|a|^p}{p} + \frac{|b|^q}{q}. \tag{A.2}$$

Hölder Inequality

Let $1 < p < \infty, q = p/(p-1), x \in \ell_p$ and $y \in \ell_q$. Then we have $xy = (x_k y_k)_{k=0}^{\infty} \in \ell_1$ and

$$\sum_{k=0}^{\infty} |x_k y_k| \leq \left(\sum_{k=0}^{\infty} |x_k|^p \right)^{1/p} \left(\sum_{k=0}^{\infty} |y_k|^q \right)^{1/q}. \tag{A.3}$$

Minkowski Inequality

Let $1 \leq p < \infty$ and $x, y \in \ell_p$. Then we have

$$\left(\sum_{k=0}^{\infty} |x_k + y_k|^p \right)^{1/p} \leq \left(\sum_{k=0}^{\infty} |x_k|^p \right)^{1/p} + \left(\sum_{k=0}^{\infty} |y_k|^p \right)^{1/p}. \tag{A.4}$$

Let $\alpha \in \mathbb{R}$ be given. Then there is a constant M such that

$$\left| (k+1)^{\alpha+1} - k^{\alpha} \right| \leq M \cdot (k+1)^{\alpha-1} \text{ for } k = 1, 2 \ldots. \tag{A.5}$$

B. de Malafosse et al., *Operators Between Sequence Spaces and Applications*,
https://doi.org/10.1007/978-981-15-9742-8

Proof The inequality in (A.5) clearly holds for $\alpha = 0$.
So let $\alpha \neq 0$. We define the function $f : [1, \infty) \to \mathbb{R}$ by $f(x) = t^\alpha$ and apply the mean value theorem to the interval $[k, k+1]$ for any fixed $k \in \mathbb{N}$. Then there exists $\xi \in [k, k+1]$ such that

$$|(k+1)^\alpha - k^\alpha| = |f(k+1) - f(k)| = |f'(\xi)| = \alpha \cdot \xi^{\alpha-1}.$$

If $\alpha \geq 1$, then $\xi^{\alpha-1} \leq (k+1)^{\alpha-1}$, and if $\alpha < 1$, then

$$\xi^{\alpha-1} \leq k^{\alpha-1} = \frac{(k+1)^{1-\alpha}}{k^{1-\alpha}} \cdot (k+1)^{\alpha-1} \leq 2^{1-\alpha} \cdot k^{\alpha-1}.$$

Putting

$$M = \begin{cases} \alpha & (\alpha \geq 1) \\ \alpha \cdot 2^{1-\alpha} & (\alpha < 1) \end{cases},$$

we immediately obtain (A.5). □

A.2 Functional Analysis

The following results are well known from functional analysis.

Theorem A.1 (Closed graph lemma) ([1, Theorem 11.1.1]) *Any continuous map into a Hausdorff space has closed graph.*

Theorem A.2 (Closed graph theorem) ([1, Theorem 11.2.2]) *If X and Y are Fréchet spaces and $f : X \to Y$ is a closed linear map, then f is continuous.*

Theorem A.3 (Banach–Steinhaus theorem) ([1, Corollary 11.2.4]) *Let (f_n) be a pointwise convergent sequence of linear functionals on a Fréchet space X. Then f defined by $f(x) = \lim_{n\to\infty} f(x)$ is continuous.*

Theorem A.4 (Hahn–Banach theorem) ([2, 3.0.4]) *Let X be a subspace of a linear topological space Y and f be a linear functional on X which is continuous in the relative topology of Y. Then f can be extended to a continuous linear functional on Y.*

Theorem A.5 (Uniform boundedness principle) ([1, Corollary 11.2.3]) *Let (f_n) be a pointwise convergent sequence of continuous linear functionals on a Fréchet space. Then (f_n) is equicontinuous.*

Theorem A.6 (Convergence lemma) ([2, 7.0.3]) *Let (f_n) be a sequence of equicontinuous linear functionals on a linear topological space X. Then the set $\{x \in X : \lim_{n\to\infty} f_n(x)$ exists $\}$ is a closed linear subspace of X.*

Theorem A.7 (Open mapping theorem) ([1, Theorem 11.2.1]) *Let X and Y be Fréchet spaces, and $f : X \to Y$ a closed linear map onto. Then f is open.*

Corollary A.1 *Let X be a Banach space with respect to the norms $\|\cdot\|_1$ and $\|\cdot\|_2$. If there exists a constant $a > 0$ such that*

$$\|x\|_2 \leq a\|x\|_1 \text{ for all } x \in X,$$

then the norms are equivalent.

References

1. Wilansky, A.: Functional Analysis. Blaisdell Publishing Company, New York (1964)
2. Wilansky, A.: Summability through Functional Analysis. North–Holland Mathematical Studies, vol. 85. North–Holland, Amsterdam (1984)

Index

Symbols
$(X, (p_n))$, 48
(X, Y), 12
(X, p), 3
$(\bar{N}, q)_0$, $(\bar{N}, q)$, $(\bar{N}, q)_\infty$, 58
(ϕ, ψ)-bounded operator, 42
(ϕ, ψ)-measure of noncompactness of an operator, 42
(ϕ, ψ)-operator norm, 42
$A = (a_{nk})_{n,k=0}^{\infty}$, 12
AE, 161
A^*, 27
A_n, A^k, 49
$A_n x$, Ax, 12
$B(\widetilde{r}, \widetilde{s})$, 239
$C(\lambda)$, 166
$C_0^{(1)}$,$C^{(1)}$, $C_\infty^{(1)}$, 57
C_1, 57
D_u, 160
$E * F$, 165
$E^{(r)}$, 64
E_τ, 160
FA, F_A, 161
F_a, 161
G^+, 239
$M(X, Y)$, 19
$\mathcal{M}_X^c$, 36
$N_{p,\tau}(A)$, 318
$N_r(0)$, 15
$S \oplus y$, 21
$S_{\tau,\nu}$, 160
T^-, 66
$W(u, v; X)$, 73
$W_\tau(\Delta(\lambda))$, $W_\tau(C(\lambda))$, $W_\tau(C^+(\lambda))$, 202
W_τ, 202
W_τ^0, 202
X', 12
$X(u\Delta^{(m)})$, 80
X^*, 16
X^α, X^β, X^γ, 19
X^f, 21
X_A, c_A, 48
X_p, 74
$[A_1, A_2]$, 183
$[C, C]$, $[C, \Delta]$, $[\Delta, C]$, $[\Delta, \Delta]$, 184
$[C, \Delta]_{W_\tau}$, $[C, C]_{W_\tau}$, 204
$[C^+, \Delta]_{W_\tau}$, $[C^+, C]_{W_\tau}$, $[C^+, C^+]_{W_\tau}$, 204
$[C^+, \Delta]$, $[C^+, \Delta^+]$, $[C^+, C^+]$, 186
$[\Delta, \Delta^+]$, $[\Delta, C^+]$, $[C, \Delta^+]$, $[\Delta^+ \Delta]$, 186
$[\Delta^+, C]$, $[\Delta^+ \Delta^+]$, $[C^+, C]$, 186
$\Delta(\lambda)$, 166
$\Delta^{(m)}$, 80
$\Delta^{(m)}$, $\Delta = \Delta^{(1)}$, 53
Δ^+, 66, 145
Γ, Γ^+, 167
Γ_1, S_1, s_1, s_1^0, 161
$\Gamma'_{p,\tau}$, 321
Γ_r, S_r, s_r, s_r^0, $s_r^{(c)}$, 161
$\Sigma = (\sigma_{nk})_{n,k=0}^{\infty}$, 51
$\Sigma^{(m)}$, $\Sigma = \Sigma^{(1)}$, 53
α, $\alpha(Q)$, 37
$\bar{N}_q$, 58
χ, $\chi(Q)$, 37
$\ell(p)$, 4
ℓ_∞, c, c_0, ℓ_p, 11
$\ell_\infty(p)$, 4
$\|\cdot\|_\lambda$, 198
$\|\cdot\|^*$, 15
$\|A\|_{(X,\ell_1)}$, 18
$\|A\|_{(X,\ell_\infty)}$, 16

B. de Malafosse et al., *Operators Between Sequence Spaces and Applications*,
https://doi.org/10.1007/978-981-15-9742-8

$\|\cdot\|_M$, 232
$\|\cdot\|_{(\phi,\psi)}$, $\|L\|_{(\phi,\psi)}$, $\|L\|_\phi$, 42
$\|\cdot\|_p$, 11
$\|\cdot\|_\infty$, 11
$\mathbf{n+1}$, 83
$\mathcal{B}(X, Y)$, 16
$\mathcal{I}_a(E, F, F')$, 236
$\mathcal{K}(X, Y)$, $\mathcal{K}(X)$, 35
$\mathcal{L}$, 317
$\mathcal{M}_X$, 35
$\mathcal{T}_Y$, $\mathcal{T}_H|_Y$, 11
$\mathcal{U}$, 56
$\mathcal{U}^+$, 160
$\mathcal{U}_1^+$, 162
ν, $\nu(Q)$, 41
ω, 2
$\|L\|$, 16
$\|\cdot\|_{(w_\infty(\lambda), w_\infty(\lambda))}$, 198
$\|f\|$, 16
ϕ, 12
$\phi(Q)$, 36
$\mathcal{U}_1^+$, 251
$\overline{B}_\delta(x_0)$, $\overline{B}_{X,\delta}(x_0)$, 15
$\overline{\chi}$, 236
$\sup_N$, $\sup_K$, $\sup_{k,N}$, 109
$\widehat{C_1}$, $\widehat{C^+}$, $\widehat{C_1^+}$, 166
$\widehat{C}_\Lambda$, 267
$\widehat{\mathcal{B}}_p(\tau)$, 319
$\sigma(A, X)$, 326
$\tau^\bullet$, 166
a-analytic sequences, 266
a-entire sequences, 266
bs, 19
bv, 55, 262
bv^p, 290
bv_0^+, bv^+, 145
$c(1)$, 162
$c^\bullet$, 171
$c_0(\Delta)$, $c(\Delta)$, $\ell_\infty(\Delta)$, 59
$c_0(\lambda)$, $c_\infty(\lambda)$, $c(\lambda)$, 181
$c_0(p)$, 4
$c_0(p)$, $\ell_\infty(p)$, 269
$c_0(u\Delta)$, $c(u\Delta)$, $\ell_\infty(u\Delta)$, 59
$c_\tau(\lambda, \mu)$, $\overset{\circ}{c}_\tau(\lambda, \mu)$, $c_\tau^\bullet(\lambda, \mu)$, 180
$c_\tau^{+\cdot p}(\lambda, \mu)$, $\widetilde{c_\tau^{+p}}(\lambda, \mu)$, 193
$c_\tau^{+\cdot p}(\lambda, \mu)$, $c_\tau^{+p}(\lambda, \mu)$, 192
$c_\tau^p(\lambda, \mu)$, $c_\tau^{\cdot+p}(\lambda, \mu)$, 192, 193
$cl^{\mathcal{E}}(b)$, 235
cs, 19
d_ω, 4
e_p^r, e_0^r, e_c^r, 64
s_τ, s_τ^0, $s_\tau^{(c)}$, 160
$w_\infty(\lambda)$, $w^0(\lambda)$, 198
$w_\tau(\lambda)$, $\overset{\circ}{w}_\tau(\lambda)$, $w_\tau^\bullet(\lambda)$, 174
$w_\tau^p(\lambda)$, $\overset{\circ}{w}_\tau^p(\lambda)$, $w_\tau^{+p}(\lambda)$, $\overset{\circ}{w}_\tau^{+p}(\lambda)$, 190
$x R_{\mathcal{E}} y$, 235
$x_k \to L(S)$, $st - \lim x = L$, 208
$z^{-1} * Y$, 50
z^α, z^β, z^γ, 51
co(S), 37
cl$_Y(E)$, 13
A-statistical convergence, 208
A transform of a sequence, 12
α-dual, 19
β-dual, 19
γ-dual, 19
m-section of a sequence, 14
$\mathbf{\Gamma}$, 265
$\mathbf{\Lambda}$, 265

A

Absorbing set, 15
AD property, 14
AK property, 14
Analytic sequences, 265
Associated subsequence, 198

B

Ball measure of noncompactness, 37
Ball measure of noncompactness of a bounded set, 37
Banach–Steinhaus theorem, 360
BH space, 10
BK space, 10

C

Cesàro matrix of of oder 1, 57
Cesàro sequence spaces of non-absolute type X_p, 74
Classical sequence spaces, 11
Closed ball in a metric space, 15
Closed graph lemma, 360
Closed graph theorem, 360
Column finite matrix, 49
Compact linear operator, 35
Completely continuous linear operator, 35
Continuity of multiplication by scalars, 3
Continuous dual of a Fréchet space, 12
Convergence domain of A, 48
Convergence lemma, 360
Crone's theorem, 32

D

Dense set, 14
Difference sequence spaces, 59

E

Eigenvalue of a matrix, 27
Eigenvector of a matrix, 27
Entire sequences, 265
Equivalence class $cl^{\mathcal{E}}(b)$, 235
Equivalence relation $R_{\mathcal{E}}$, 235
Equivalent paranoms, 3
Euler matrix of order r E^r, 64
Euler sequence spaces e_p^r, e_0^r, e_c^r, 64
Exponentially bounded sequence, 198
  associated subsequence, 198

F

FH space, 10
Finite sequences ϕ, 12
FK space, 10
  AD space, 14
  AK space, 14
Floquet exponent, 350
Floquet's theorem, 349
Fréchet combination, 3
Fréchet space, 2
F space, 2
Functional dual, 21

G

Generalized Cantor's intersection property, 36
Generalized difference sequence spaces, 59
Goldenštein, Goh'berg, Markus theorem, 39

H

Hahn–Banach theorem, 360
Hardy's Big O Tauberian theorem, 214
Hausdorff measure of noncompactness (HMNC), 37
Hausdorff measure of noncompactness of a bounded set, 37
Hill equation, 349
Hölder inequality, 359

I

Improvement of mapping, 17
Infinite matrix, 12

K

Kuratowski measure of noncompactness (KMNC), 37
Kuratowski measure of noncompactness of a bounded set, 37

L

Lower triangular infinite matrices, 317

M

Mathieu equation, 356
Matrix domain of A in X, 48
Matrix of the arithmetic means, 57
Matrix of the C_1 means, 57
Matrix of the forward differences, 145
Matrix of the forward differences Δ^+, 66
Matrix of the m^-th iterated difference, 53
Matrix of the mth iterated sum, 53
Matrix of the m-th-order difference operator $\Delta^{(m)}$, 80
Matrix of the Riesz means, 58
Matrix of the right shift operator T^-, 66
Matrix of the weighted means, 58
Measure of noncompactness, 36
  ball measure of noncompactness, 37
  ball measure of noncompactness of a bounded set, 37
  generalized Cantor's intersection property, 36
  Hausdorff measure of noncompactness, 37
  Hausdorff measure of noncompactness of a bounded set, 37
  invariance under closure, 36
  Kuratowski measure of noncompactness (KMNC), 37
  Kuratowski measure of noncompactness of a bounded set, 37
  monotonicity, 36
  nonsingularity, 36
  regularity, 36
  semi-additivity, 36
Measure of noncompactness of a bounded set, 36
Minkowski inequality, 359
Monotonous norm, 129
Multiplier space of X in Y, 19

N

Natural metric d_ω for ω, 4
Normal set of sequences, 66

O

Open mapping theorem, 361
Open neighbourhood of 0, 15
Operator norm, 16

P

Paranorm, 2
  continuity of multiplication, 3
  equivalent, 3
  strictly stronger, 3
  strictly weaker, 3
  stronger, 3
  total, 3
  weaker, 3
Paranormed space, 3
  totally paranormed space, 3
Pòlya matrix, 343
Pólya's theorem, 343
Pre-compact set, 36

R

Regular sequence spaces equation, 285
Relatively compact set, 36
Resolvent set of an operator, 326
Right shift operator, 66
Row-finite matrix, 49

S

Schauder basis, 14
Separable metric space, 14
Sequence spaces, 2
  a-analytic sequences, 266
  all complex sequences ω, 2
  a-entire sequences, 266
  analytic sequences, 265
  bounded by the $\bar{N}_q$ method, 58
  bounded sequences ℓ_∞, 11
  bounded series bs, 19
  bounded variation, 55
  bounded variation bv, 261
  Cesàro sequence spaces of non-absolute type X_p, 74
  classical sequence spaces, 11
  convergent sequences c, 11
  convergent series cs, 19
  difference sequence spaces $c_0(\Delta)$, $c(\Delta)$, $\ell_\infty(\Delta)$, 59
  entire sequences, 265
  Euler sequence space, 64
  finite sequences ϕ, 12
  generalized difference sequence spaces $c_0(u\Delta)$, $c(u\Delta)$, $\ell_\infty(u\Delta)$, 59
  generalized weighted means $W(u, v; X)$, 73
  normal, 66
  null sequences c_0, 11
  of generalized m-th-order difference sequences, 80
  p-absolutely summable series ℓ_p, 11
  sequences of p-bounded variation, 290
  summable by the $\bar{N}_q$ method, 58
  summable to 0 by the $\bar{N}_q$ method, 58
Sequences of bounded variation, 55
Sequences of p-bounded variation, 290
Space of all bounded series bs, 19
Space of all convergent series cs, 19
Space of all sequences of bounded variation, 261
Space of bounded linear operators $\mathcal{B}(X, Y)$, 16
Space of the generalized m-th-order difference sequences, 80
Spaces of generalized weighted means, 73
Spectrum of an operator, 326
Statistical convergence, 207
Strictly stronger paranorm, 3
Strictly weaker paranorm, 3
Strong $c(\lambda)$-limit, 181
Stronger paranorm, 3

T

Toeplitz triangle, 335
Totally paranormed space, 3
Total paranorm, 3
Triangle, 49

U

Uniform boundedness principle, 360

W

Weaker paranorm, 3

MIX
Papier aus verantwortungsvollen Quellen
Paper from responsible sources
FSC® C105338

Printed by Books on Demand, Germany